Mass Transfer

Fundamentals and Applications

PRENTICE HALL INTERNATIONAL SERIES
IN THE PHYSICAL AND CHEMICAL ENGINEERING SCIENCES

BALZHISER, SAMUELS, AND ELLIASSEN *Chemical Engineering Thermodynamics*
BEQUETTE *Process Dynamics*
BIEGLER, GROSSMAN, AND WESTERBERG *Systematic Methods of Chemical Process Design*
CROWL AND LOUVER *Chemical Process Safety*
CUTLIP AND SHACHAM *Problem Solving in Chemical Engineering with Numerical Methods*
DENN *Process Fluid Mechanics*
ELLIOTT AND LIRA *Introductory Chemical Engineering Thermodynamics*
FOGLER *Elements of Chemical Reaction Engineering, 3rd Edition*
HANNA AND SANDALL *Computational Methods in Chemical Engineering*
HIMMELBLAU *Basic Principles and Calculations in Chemical Engineering, 6th Edition*
HINES AND MADDOX *Mass Transfer*
KYLE *Chemical and Process Thermodynamics, 3rd Edition*
NEWMAN *Electrochemical Systems, 2nd Edition*
PRAUSNITZ, LICHTENTHALER, AND DE AZEVEDO *Molecular Thermodynamics of Fluid-Phase Equilibria, 3rd Edition*
PRENTICE *Electrochemical Engineering Principles*
SHULER AND KARGI *Bioprocess Engineering*
STEPHANOPOULOS *Chemical Process Control*
TESTER AND MODELL *Thermodynamics and Its Applications, 3rd Edition*
TURTON, BAILIE, WHITING, AND SHAEWITZ *Analysis, Synthesis and Design of Chemical Processes*
WILKES *Fluid Mechanics for Chemical Engineering*

Mass Transfer

Fundamentals and Applications

ANTHONY L. HINES

Oklahoma State University

ROBERT N. MADDOX

Oklahoma State University

PRENTICE HALL PTR
UPPER SADDLE RIVER, NEW JERSEY 07458

Library of Congress Cataloging in Publication Data

HINES, ANTHONY L (date)
Mass transfer.

Includes index.
1. Mass transfer. I. Maddox, Robert N. (Robert Nott), (date). II. title.
TP156.M3H55 1985 660.2'8423 83-19140
ISBN 0-13-559609-2

Editorial/production supervision: *Theresa A. Soler*
Manufacturing buyer: *Anthony Caruso*

Prentice-Hall Inc.
A Pearson Education Company
Upper Saddle River, NJ 07458

Printed in the United States of America

ISBN 0-13-559609-2

Prentice-Hall International (UK) Limited,London
Prentice-Hall of Australia Pty. Limited, Sydney
Prentice-Hall Canada Inc., Toronto
Prentice-Hall Hispanoamericana, S.A., Mexico
Prentice-Hall of India Private Limited, New Delhi
Prentice-Hall of Japan, Inc., Tokyo
Pearson Education Asia Pte. Ltd., Singapore
Editora Prentice-Hall do Brasil, Ltda., Rio de Janeiro

Contents

Preface

Although mass transfer processes are encountered by each of us in our daily lives, we seldom consider the actual transfer process. As an example, both macroscopic and microscopic mass transfer are involved in the simple act of breathing. When we inhale, air is transported into the lungs by macroscopic mass transfer. The air that enters the lungs has a higher concentration of oxygen than does the blood that flows through the lungs. Because of the concentration difference, oxygen diffuses through the walls of the lungs and into the blood, which distributes it throughout the body. Similarly, because the blood has a higher concentration of carbon dioxide than does the lungs, the carbon dioxide diffuses from the blood and into the lungs, where it is exhaled. If we are to properly understand the breathing process, we need information regarding the surface area available for mass transfer, the thickness of the cell walls through which the oxygen is transferred, the concentration difference across the walls of the lungs, and diffusion coefficients for oxygen and carbon dioxide through the walls of the lungs.

In addition to having an interest in understanding diffusional processes such as the one described above, the chemical engineer traditionally has been involved in the design of macroscopic separation processes such as distillation, absorption extraction, and adsorption. The design of these processes requires that we have a thorough understanding of mass transfer fundamentals. Subsequently, we have arranged this textbook to familiarize the student first with concentration units and flux rates, followed by methods to predict diffusion and mass transfer coefficients in a variety of materials.

The recent trend in the teaching of mass transfer has been to introduce the student to mass transfer from the point of view of transport phenomena followed by macroscopic separation processes. To achieve this goal two textbooks are frequently required. Our goal in writing this book is to introduce the student to the fundamentals of both microscopic and macroscopic mass transfer in a single text.

The first part of this book covers the microscopic diffusional processes and the prediction and use of transport coefficients. In Chapter 1, concentration units, flux relationships, and coupled transport processes are discussed. This chapter provides the student with the background necessary to understand the portion of the text that deals with macroscopic separation processes. Chapter 2 introduces the student to diffusion coefficients for gases, liquids, and solids. An attempt is made to expand the discussion of diffusion coefficients as compared to other mass transfer textbooks. In Chapter 3 we present the formulation of macroscopic problems and the formulation of diffusional mass transfer problems by the differential shell balance method (transport method). The differential formulation of diffusion problems is further pursued in Chapter 4 by reducing the equations of change. The major emphasis in Chapter 4, however, is to present methods for solving these problems with several methods of solution discussed. The concept of a convective mass transfer coefficient is introduced in Chapter 5. Theoretical models that describe mass transfer at a fluid–fluid interface are presented. These models include a discussion of the film theory, surface renewal theory, and penetration theory. Interfacial mass transfer and the use of overall mass transfer coefficients are discussed. In Chapter 6, methods for predicting convective mass transfer coefficients for various geometries are presented. The analogies between heat, momentum, and mass transfer are discussed. Boundary layer theory and its application to mass transfer from a solid surface is introduced.

The remainder of the book deals primarily with macroscopic separation processes, including absorption, distillation, and extraction in both tray and packed towers. In Chapter 7 a brief presentation of thermodynamic fundamentals is included to introduce the student to phase equilibrium as applied to mass transfer processes. An interpretation of phase diagrams for liquid–liquid and gas–liquid systems is given in this chapter.

Absorption and stripping calculations for single-component and multicomponent systems are presented in Chapter 8. The student is first introduced to the standard graphical procedures that are applicable to transfer of a single solute. The more complicated multicomponent calculational procedures are then studied. A discussion of distillation of binary systems is given in Chapter 9. The simple graphical McCabe–Thiele and enthalpy–composition methods are presented in this chapter. Distillation of the more complicated multicomponent systems is discussed in Chapter 10. Both the shortcut methods and the more detailed tray-by-tray methods are introduced. Liquid extraction of ternary systems is introduced in Chapter 11. Ternary diagrams are used to determine the number of equilibrium stages for both countercurrent and crosscurrent processes. A discussion of column internals and the sizing of tray-type towers is presented in Chapter 13. Following the discussion of equilibrium-stage calculational methods, absorption and distillation in packed columns is introduced in Chapter 12. The determination of packing heights and column diameters is discussed. A chapter discussing adsorption is presented. Both equilibrium adsorption and adsorption in packed towers are discussed. Methods for designing packed adsorbers are presented in Chapter 14.

The text is intended for both undergraduate and graduate students in chemical engineering, although much of the material can be used as a refresher by practicing engineers. Parts of Chapters 3, 4, and 6 are of an advanced nature and are better suited for a graduate course. The material presented in Chapters 10 and 14 is also typically reserved for the graduate student or the practicing engineer.

The authors wish to express appreciation to their colleagues and former students for their many valuable comments. The suggestions made by Professor James R. Fair of the University of Texas were particularly helpful. The use of the uncompleted manuscript by Professor David O. Cooney of the University of Wyoming and his subsequent comments are also very much appreciated. The authors wish to thank S. L. Kuo and E. O. Pedram for their assistance in the preparation of the solutions manual for the text. We also thank Andrea Heard, Margaret Brecheisen, and Ruth Fisher for their help in typing the manuscript. The careful reading of the final manuscript by Patricia Cordell was valuable and very helpful. Finally we wish to acknowledge the support of our families, who showed a great deal of patience during the preparation of this manuscript.

Anthony L. Hines
Robert N. Maddox

Mass Transfer

Fundamentals and Applications

Mass Transfer Fundamentals 1

1.1 Introduction

Thermal and momentum diffusion are frequently encountered by students throughout their undergraduate careers. These are adequately dealt with by Fourier's law of heat conduction, which relates the flow of heat to a thermal gradient, and Newton's law, which relates the flow of momentum to a velocity gradient.

The molecular transport of matter, often denoted as *ordinary diffusion*, can be described in a manner similar to conductive heat transfer by using Fick's equation. His analogy states that the mass flux of component i per unit cross-sectional area perpendicular to the direction of flow is proportional to its concentration gradient. It can be expressed as

$$j_i = -\rho D \frac{d\omega_i}{dZ} \tag{1-1}$$

where j_i is the mass flux of i in the Z direction with respect to the system moving with the mass average velocity, ρ is the mass concentration, ω_i the mass fraction or driving force, and D a constant of proportionality defined as the diffusion coefficient with the units (length²/time). From this expression we observe that species i diffuses in the direction of decreasing concentration of i. The analogy is thus made with Fourier's law of heat conduction, which relates the flow of heat from a high- to a low-temperature region. The diffusion coefficient in the presence of a concentration gradient as given by Eq. (1-1) is denoted as an *intrinsic* or *interdiffusion coefficient*. Although most intrinsic diffusion coefficients are defined in terms of a concentration gradient, it will be shown later that a gradient of the chemical potential is responsible for the net flux of any species.

The discussion above implies that a gradient must exist before diffusion will take place. On the basis of past studies, this has been shown not to be the case. This is demonstrated by the random motion of a single molecule of species i through gas i. Although this is a diffusion process, in this case a net mass transfer of i will not occur. The diffusion coefficient for i into itself is termed *self-diffusion*. Measurements of this type of coefficient can be made by using radioisotope techniques.

1.2 Concentration and Flux Relationships

Concentrations

The concentration gradient may be expressed in a variety of ways, and before proceeding further, the more popular ones encountered in the literature will be formally defined.

The mass concentration of species i has the same units as density and is expressed as ρ_i with the units (mass i/unit volume). The sum of all mass concentrations within a mixture is equal simply to the overall density and can be expressed as

$$\rho = \sum_{i=1}^{n} \rho_i \tag{1-2}$$

From the mass concentration, the mass fraction of species i can be obtained by dividing the mass concentration of i in the mixture by the total mass density:

$$\omega_i = \frac{\rho_i}{\rho} \tag{1-3}$$

From Eqs. (1-2) and (1-3) it is seen that

$$\sum_{i=1}^{n} \frac{\rho_i}{\rho} = \sum_{i=1}^{n} \omega_i = 1 \tag{1-4}$$

The molar concentration of component i is represented as C_i and has the units (moles i/unit volume). The sum of all molar concentrations for all species can be summed to give the total number of moles or concentration in the system:

$$C = \sum_{i=1}^{n} C_i \tag{1-5}$$

The mole fraction of species i is thus found by dividing the molar concentration of i by C, which gives

$$x_i = \frac{C_i}{C} \tag{1-6}$$

If we sum over all species, then

$$\frac{\sum_{i=1}^{n} C_i}{C} = \sum_{i=1}^{n} x_i = 1 \tag{1-7}$$

It is often necessary to convert from mass to molar concentration. This is accomplished by dividing the mass concentration of i by the molecular weight of that species:

$$C_i = \frac{\rho_i}{M_i} \tag{1-8}$$

Flux relationships

To calculate the flux of a species, it frequently proves convenient to describe the transfer with respect to a fixed set of coordinates. The molar flux of species i can then be expressed as

$$\mathbf{N}_i = C_i \mathbf{U}_i \tag{1-9}$$

where $\mathbf{N}_i$ is the molar flux of i with the units (moles/length^2-time) and $\mathbf{U}_i$ is the velocity of i with respect to a fixed reference frame. Analogously, the mass flux may be expressed as

$$\mathbf{n}_i = \rho_i \mathbf{U}_i \tag{1-10}$$

where $\mathbf{n}_i$ has the units (mass/length^2-time).

In some cases it is more convenient to write the total flux of i with respect to some reference other than a fixed set of coordinates. In general we can reference the flux of i to an arbitrary reference velocity $\mathbf{U}^0$ as

$${}_0\mathbf{j}_i = \rho_i(\mathbf{U}_i - \mathbf{U}^0) \tag{1-11}$$

or

$${}_0\mathbf{J}_i = C_i(\mathbf{U}_i - \mathbf{U}^0) \tag{1-12}$$

where ${}_0\mathbf{j}_i$ and ${}_0\mathbf{J}_i$ represent the mass and molar fluxes of i with respect to the reference velocity $\mathbf{U}^0$. The reference velocity $\mathbf{U}^0$ is chosen such that the following relations are valid:

$$\mathbf{U}^0 = \sum_{i=1}^{n} \chi_i \mathbf{U}_i \tag{1-13}$$

and

$$\sum_{i=1}^{n} \chi_i = 1 \tag{1-14}$$

In the expressions above, values of χ_i are normalized weighting factors which relate the contribution of each species to the reference velocity. Note that

$$\sum_{i=1}^{n} \frac{\chi_i}{C_i} {}_0\mathbf{J}_i = 0 \tag{1-15}$$

Since a system may contain several molecular species, each having a different average velocity, a frame of reference must be chosen before a discussion is undertaken of the transport of a particular component. This may be a fixed axis or the velocity of any other species in the system. The more important and more widely used moving references are the mass average, molar average, and volume average velocities. Relations will be developed using lower case letters for mass fluxes and upper case letters for molar fluxes.

Mass average velocity

The mass average velocity, $\mathbf{U}^m$, can be written in terms of the mass concentration, ρ_i, and the velocity of i with respect to a fixed coordinate axis, $\mathbf{U}_i$, as

$$\mathbf{U}^m \sum_{i=1}^{n} \rho_i = \sum_{i=1}^{n} \rho_i \mathbf{U}_i \tag{1-16}$$

or

$$\mathbf{U}^m = \frac{\sum_{i=1}^{n} \rho_i \mathbf{U}_i}{\sum_{i=1}^{n} \rho_i} \tag{1-17}$$

Substituting Eq. (1-4) into Eq. (1-17) gives

$$\mathbf{U}^m = \sum_{i=1}^{n} \omega_i \mathbf{U}_i \tag{1-18}$$

Thus we see that the weighting factor for the mass average velocity is simply the weight fraction (i.e., $\chi_i = \omega_i$). If we substitute the mass average velocity into Eq. (1-11), the mass flux of i with respect to the mass average velocity results.

$${}_m\mathbf{j}_i = \rho_i(\mathbf{U}_i - \mathbf{U}^m) \tag{1-19}$$

Similarly, by substituting the mass average velocity into Eq. (1-12), we have the molar flux of i with respect to the mass average velocity.

$${}_m\mathbf{J}_i = C_i(\mathbf{U}_i - \mathbf{U}^m) \tag{1-20}$$

Molar average velocity

The molar average velocity is defined by an expression analogous to the mass average velocity.

$$\mathbf{U}^M = \frac{\sum_{i=1}^{n} C_i \mathbf{U}_i}{\sum_{i=1}^{n} C_i} \tag{1-21}$$

or

$$\mathbf{U}^M = \sum_{i=1}^{n} x_i \mathbf{U}_i \tag{1-22}$$

Upon substituting the molar average velocity defined by the equations above for the arbitrary reference velocity of Eq. (1-12), we obtain the molar flux of i with respect to the molar average velocity.

$${}_M\mathbf{J}_i = C_i(\mathbf{U}_i - \mathbf{U}^M) \tag{1-23}$$

An equation for the mass flux of i with respect to the molar average velocity results if the molar average velocity is substituted for the arbitrary reference velocity in Eq. (1-11).

$${}_M\mathbf{j}_i = \rho_i(\mathbf{U}_i - \mathbf{U}^M) \tag{1-24}$$

Volume average velocity

The volume average velocity is probably the most important reference velocity for obtaining experimental data, since for a fixed system of constant volume, this velocity is identically equal to zero. The volume average velocity is defined by

$$\mathbf{U}^v = \sum_{i=1}^{n} \bar{V}_i C_i \mathbf{U}_i \tag{1-25}$$

where $\bar{V}_i$ is defined as the partial molar volume. The molar flux of i with respect to the volume average velocity thus is written as

$${}_v\mathbf{J}_i = C_i(\mathbf{U}_i - \mathbf{U}^v) \tag{1-26}$$

Example 1.1

(a) Based on the definition of $\bar{V}_i$ and $\sum_{i=1}^{n} \bar{V}_i C_i = 1$, prove that $\sum_{i=1}^{n} {}_v\mathbf{J}_i \bar{V}_i = 0$.

(b) Show that $\sum_{i=1}^{n} {}_M\mathbf{J}_i = 0$.

Solution:

(a) Beginning with Eq. (1-26) and multiplying both sides of the equation by $\bar{V}_i$ gives

$${}_v\mathbf{J}_i \bar{V}_i = C_i \bar{V}_i(\mathbf{U}_i - \mathbf{U}^v)$$

Expanding the expression above and summing over all species, we have

$$\begin{aligned}\sum_{i=1}^{n} {}_v\mathbf{J}_i \bar{V}_i &= \sum_{i=1}^{n} (C_i \bar{V}_i \mathbf{U}_i - C_i \bar{V}_i \mathbf{U}^v) \\ &= \mathbf{U}^v - \underbrace{\mathbf{U}^v \sum_{i=1}^{n} C_i \bar{V}_i} \\ &= \mathbf{U}^v\left(1 - \sum_{i=1}^{n} C_i \bar{V}_i\right)\end{aligned}$$

Since $\sum_{i=1}^{n} C_i \bar{V}_i = 1$, we have $\sum_{i=1}^{n} {}_v\mathbf{J}_i \bar{V}_i = 0$.

(b) Starting with Eq. (1-23), we obtain

$$\begin{aligned}\sum_{i=1}^{n} {}_M\mathbf{J}_i &= \sum_{i=1}^{n} C_i(\mathbf{U}_i - \mathbf{U}^M) \\ &= \sum_{i=1}^{n} C_i \mathbf{U}_i - \mathbf{U}^M \sum_{i=1}^{n} C_i\end{aligned}$$

Introducing the definition of the molar average velocity, we have

$$\sum_{i=1}^{n} {}_M\mathbf{J}_i = \sum_{i=1}^{n} C_i \mathbf{U}_i - \frac{\sum_{i=1}^{n} C_i \mathbf{U}_i}{\sum_{i=1}^{n} C_i} \sum_{i=1}^{n} C_i = 0$$

The flux expressions developed previously can be combined in a variety of ways to describe a particular physical system. For example, consider Eq. (1-23), which can be written as

$$_M\mathbf{J}_i = C_i\mathbf{U}_i - C_i\mathbf{U}^M$$

Substituting $\mathbf{N}_i = C_i\mathbf{U}_i$ into the expression above and rearranging gives

$$\mathbf{N}_i = {}_M\mathbf{J}_i + C_i\mathbf{U}^M = \underset{\substack{\text{Diffusion}\\ \text{flux}}}{-CD_i\nabla x_i} + \underset{\substack{\text{Convective}\\ \text{flux}}}{C_i\mathbf{U}^M} \tag{1-27}$$

The total molar flux of i relative to a fixed reference frame is thus equal to the molar flux of i due to diffusion relative to the molar average velocity plus the bulk flow of i with respect to a fixed reference frame. Equation (1-27) can be further modified by replacing the molar average velocity by its definition. Thus we have

$$\mathbf{N}_i = {}_M\mathbf{J}_i + \frac{C_i}{C}\sum_{i=1}^{n} C_i\mathbf{U}_i \tag{1-28}$$

or

$$\mathbf{N}_i = {}_M\mathbf{J}_i + x_i\sum_{i=1}^{n} \mathbf{N}_i \tag{1-29}$$

For mixtures in which x_i is very small, the bulk flow term of Eq. (1-29) can be neglected, and the molar flux with respect to a fixed reference frame becomes equal to the molar flux of i due to diffusion relative to the molar average velocity:

$$\mathbf{N}_i = {}_M\mathbf{J}_i \tag{1-30}$$

Similar developments can be carried out using mass flows and mass average velocities.

Binary systems

Although up to this point general notation has been used, for most of our work we will be interested primarily in binary or pseudobinary systems. First let us consider equivalent but different forms of Fick's first law of diffusion in one dimension for a binary system as expressed by the equations

$$_MJ_A = -CD_{AB}\frac{dx_A}{dZ} \tag{1-31}$$

and

$$_mj_A = -\rho D_{AB}\frac{d\omega_A}{dZ} \tag{1-32}$$

The diffusion coefficients used above are identical and have the units (length^2/time). Now consider Eq. (1-29) written for a binary system.

$$\mathbf{N}_A = {}_M\mathbf{J}_A + x_A(\mathbf{N}_A + \mathbf{N}_B) \tag{1-33}$$

Upon substituting Fick's law written in vector notation, we have

$$\mathbf{N}_A = -CD_{AB}\nabla x_A + x_A(\mathbf{N}_A + \mathbf{N}_B) \tag{1-34}$$

When written in terms of mass units, Eq. (1-34) becomes

$$\mathbf{n}_A = -\rho D_{AB} \nabla \omega_A + \omega_A(\mathbf{n}_A + \mathbf{n}_B) \tag{1-35}$$

A summary of notation for binary systems is given in Tables 1-1 through 1-3.

TABLE 1-1. CONCENTRATIONS IN BINARY SYSTEMS

M_A = molecular weight of A, mass A/mole A

ρ_A = mass concentration of A, mass A/volume solution

$\rho = \rho_A + \rho_B$ = total mass concentration, mass $(A + B)$/volume solution

$\omega_A = \dfrac{\rho_A}{\rho}$ = mass fraction of A

$$\omega_A + \omega_B = 1.0$$

$$\frac{\omega_A}{M_A} + \frac{\omega_B}{M_B} = \frac{1}{M}$$

C_A = molar concentration of A, moles A/volume solution

$C = C_A + C_B$ = total molar concentration, moles $(A + B)$/volume solution

$x_A = \dfrac{C_A}{C}$ = mole fraction of A

$$x_A + x_B = 1.0$$

$$x_A M_A + x_B M_B = M$$

$$x_A = \frac{\omega_A/M_A}{\omega_A/M_A + \omega_B/M_B} = \frac{\omega_A/M_A}{1/M}$$

$$dx_A = \frac{d\omega_A}{M_A M_B (1/M)^2}$$

TABLE 1-2. FLUXES FOR BINARY SYSTEMS (BIRD ET AL., 1960)

Reference System	*Molar Flux*	*Mass Flux*
Fixed coordinates	$\mathbf{N}_A = C_A \mathbf{U}_A$	$\mathbf{n}_A = \rho_A \mathbf{U}_A$
Molar average velocity	${}_M\mathbf{J}_A = C_A(\mathbf{U}_A - \mathbf{U}^M)$	${}_M\mathbf{j}_A = \rho_A(\mathbf{U}_A - \mathbf{U}^M)$
Mass average velocity	${}_m\mathbf{J}_A = C_A(\mathbf{U}_A - \mathbf{U}^m)$	${}_m\mathbf{j}_A = \rho_A(\mathbf{U}_A - \mathbf{U}^m)$
Volume average velocity	${}_v\mathbf{J}_A = C_A(\mathbf{U}_A - \mathbf{U}^v)$	${}_v\mathbf{j}_A = \rho_A(\mathbf{U}_A - \mathbf{U}^v)$

TABLE 1-3. RELATIONSHIP BETWEEN FLUXES (BIRD ET AL., 1960)

Relationship between Mass and Molar Fluxes	*Molar Flux Relative to* $\mathbf{U}^M$	*Mass Flux Relative to* $\mathbf{U}^m$
$\mathbf{n}_A = \mathbf{N}_A M_A$	$\mathbf{N}_A + \mathbf{N}_B = C\mathbf{U}^M$	$\mathbf{n}_A + \mathbf{n}_B = \rho \mathbf{U}^m$
${}_M\mathbf{j}_A = \dfrac{M}{M_B}\,{}_m\mathbf{j}_A$	${}_M\mathbf{J}_A + {}_M\mathbf{J}_B = 0$	${}_m\mathbf{j}_A + {}_m\mathbf{j}_B = 0$
${}_M\mathbf{j}_A = {}_M\mathbf{J}_A M_A$	$\mathbf{N}_A = {}_M\mathbf{J}_A + x_A(\mathbf{N}_A + \mathbf{N}_B)$	$\mathbf{n}_A = {}_m\mathbf{j}_A + \omega_A(\mathbf{n}_A + \mathbf{n}_B)$
${}_m\mathbf{J}_A = \dfrac{M_B}{M}\,{}_M\mathbf{J}_A$	$\mathbf{N}_A = {}_M\mathbf{J}_A + C_A \mathbf{U}^M$	$\mathbf{n}_A = {}_m\mathbf{j}_A + \rho_A \mathbf{U}^m$

In order to change systematically from one reference system to another, Hooyman et al. (1953) derived a general diffusion coefficient expressed as

$$\mathbf{J}_A = -D_{AB}\frac{1-\chi_A}{1-x_A}\frac{1}{V}\nabla x_A \tag{1-36}$$

where V is the molar volume of the mixture, x_A is the mole fraction, and χ_A is the weighting factor relating to the appropriate reference velocity. Using the equation above, the various forms of Fick's first law, as shown in Table 1-4, can be obtained by simply substituting the appropriate term for χ_A, as presented in Table 1-5.

TABLE 1-4. FORMS OF FICK'S FIRST LAW FOR BINARY SYSTEMS

Molar Flux	*Mass Flux*
${}_M\mathbf{J}_A = -CD_{AB}\nabla x_A$	${}_m\mathbf{j}_A = -\rho D_{AB}\nabla\omega_A$
${}_M\mathbf{J}_A = -\frac{\rho^2}{CM_AM_B}D_{AB}\nabla\omega_A$	${}_m\mathbf{j}_A = -\frac{C^2M_AM_B}{\rho}D_{AB}\nabla x_A$
$\mathbf{N}_A = -CD_{AB}\nabla x_A + x_A(\mathbf{N}_A+\mathbf{N}_B)$	$\mathbf{n}_A = -\rho D_{AB}\nabla\omega_A + \omega_A(\mathbf{n}_A+\mathbf{n}_B)$

TABLE 1-5. WEIGHTING FACTORS FOR EQ. (1-36)

Fick's Law	*Reference Velocity*	χ_A
${}_m\mathbf{J}_A$	$\mathbf{U}^m$	ω_A
${}_M\mathbf{J}_A$	$\mathbf{U}^M$	x_A
${}_v\mathbf{J}_A$	$\mathbf{U}^v$	$C_A\bar{V}_A$

Example 1.2

Using Eq. (1-36), obtain (a) the molar flux with respect to the molar average velocity, ${}_M\mathbf{J}_A$; and (b) the mass flux with respect to the mass average velocity, ${}_m\mathbf{j}_A$.

Solution:

Beginning with Eq. (1-36), we have

(a)
$${}_M\mathbf{J}_A = -D_{AB}\frac{1-\chi_A}{1-x_A}\frac{1}{V}\nabla x_A$$

and for

$$\chi_A = x_A$$

$${}_M\mathbf{J}_A = -CD_{AB}\nabla x_A$$

(b)
$${}_m\mathbf{J}_A = -D_{AB}\frac{1-\omega_A}{1-x_A}\frac{1}{V}\nabla x_A$$

From Table 1-1,

$$dx_A = \frac{d\omega_A}{M_AM_B(1/M)^2}$$

$${}_m\mathbf{J}_A = -D_{AB}\frac{\omega_B}{x_B}\frac{CM^2}{M_AM_B}\nabla\omega_A$$

but $$\omega_B = \frac{M_B x_B}{M}$$

and $$\rho = CM$$

Therefore, $${}_m\mathbf{J}_A = -\frac{\rho D_{AB}}{M_A} \nabla \omega_A$$

Since $$M_{A\,m}\mathbf{J}_A = {}_m\mathbf{j}_A$$

we find $${}_m\mathbf{j}_A = -\rho D_{AB} \nabla \omega_A$$

1.3 Coupled Processes

It has been shown that a concentration gradient gives rise to a thermal gradient and analogously a thermal gradient gives rise to a concentration gradient. These coupling phenomena are known respectively as the *Dufour* and *Soret effects.* Furthermore, it has been observed that the diffusion of a particular species is influenced by the presence of pressure gradients and force fields. These effects can be accounted for by the addition of terms to Fick's law. However, these phenomena, which are a part of irreversible thermodynamics, can be systematically described by considering Onsager's theory.

Onsager's theory is based on three basic assumptions:

1. Irreversible processes take place near equilibrium and thermodynamic variables can be used to describe the system.

2. For a system not far displaced from equilibrium, linear relationships between the fluxes and forces are assumed to be valid and are given as

$$\mathbf{J}_i = \sum_{j=1}^{n} L_{ij}\mathbf{X}_j \tag{1-37}$$

where L_{ij} is a phenomenological coefficient and $\mathbf{X}_j$ is the driving force.

3. The phenomenological coefficients are symmetrical; for example,

$$L_{ij} = L_{ji} \tag{1-38}$$

To gain a better understanding of the generality of Eq. (1-37), it will be expanded for the case in which heat, electric current, and a molecular species i are transported. For the heat flux, denoted by $\mathbf{J}_H$, we obtain

$$\mathbf{J}_H = L_{HH}\mathbf{X}_H + L_{HE}\mathbf{X}_E + L_{Hi}\mathbf{X}_i \tag{1-39}$$

In the equation above the product of the phenomenological coefficient, L_{HH}, and the force, $\mathbf{X}_H$, is Fourier's law of heat conduction. The product of the cross-coefficient, L_{HE}, and the force, $\mathbf{X}_E$, represents the heat flow due to the flow of electrical current. This phenomenon is called the *Peltier effect* in honor of its discoverer, who in 1834 found that the flow of current through two isothermal dissimilar metals resulted in heat transfer with the surroundings. The other cross-effect contributing to the flow of heat is described by the product of L_{Hi} and the gradient $\mathbf{X}_i$. It was first encountered in 1872 by Dufour, who noted the presence of a thermal gradient resulting from diffusion in gases.

Let us now consider the flow of a molecular species i described by the equation

$$\mathbf{J}_i = L_{ii}\mathbf{X}_i + L_{iH}\mathbf{X}_H + L_{iE}\mathbf{X}_E \tag{1-40}$$

where $L_{ii}\mathbf{X}_i$ is mass transport due to the chemical potential of i. The term $L_{iH}\mathbf{X}_H$ is the mass transport produced by a thermal gradient and $L_{iE}\mathbf{X}_E$ is the contribution to mass flow produced by an electrical potential. The mass flow that results from a thermal gradient is called the *Soret effect.* It is readily seen that for a system free of thermal, electrical, and gradients other than concentration, Eq. (1-40) may be written as

$$\mathbf{J}_i = L_{ii}\mathbf{X}_i \tag{1-41}$$

The equation above will be used later with Fick's first law to relate the phenomenological coefficient L_{ii} to the diffusion coefficient. In the preceding discussion, only heat, current, and mass flow were considered. A number of other effects, such as pressure and gravity, also contribute to the overall fluxes, and these will be considered when applicable.

To obtain the proper relationship between mass flow by diffusion and conductive heat transport, we take recourse to *irreversible thermodynamics.* Irreversible thermodynamics, usually considered to be the domain of the kineticist, deals with the study of rate processes by considering the rate of entropy production. The production of entropy is then used to relate rate phenomena to classical thermodynamics. The applicability of classical thermodynamics to the rate processes is based on the assumption that the system undergoing a change, such as the transport of heat to or from the system or the transport of mass in a system, is at any time displaced by only a differential amount from equilibrium. This condition is obeyed by a great number of processes under a variety of conditions and will be assumed to be satisfied by the diffusive processes. The reader interested in a detailed study of irreversible thermodynamics is referred to the work of de Groot and Mazur (1963), Prigogine (1967), and Haase (1969).

To find the relationships between fluxes and their driving forces, the production of entropy that results from various irreversible processes is evaluated. Since we are interested primarily in mass flow on a microscopic level and the potential that produces this flow, the following system is used. Consider two reservoirs, 1 and 2, separated by a rigid barrier through which heat and a molecular species i are transported (see Figure 1-1). We will assume that the two reservoirs can be maintained in equilibrium states and that external forces are not

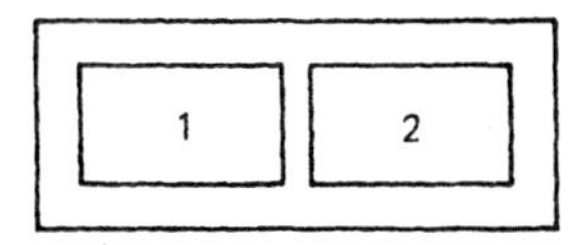

Figure 1-1 Equilibrium System.

present. In addition, the two reservoirs are assumed to be perfectly insulated from the surroundings. If the subsystems are each in equilibrium, the fluxes will produce irreversibilities only in the barrier between the reservoirs. The system is defined such that

$$dV_{sys} = dV_1 = dV_2 = 0 \tag{1-42}$$

$$dE_{sys} = dE_1 + dE_2 = 0 \tag{1-43}$$

$$dm_{sys} = dm_1 + dm_2 = 0 \tag{1-44}$$

$$dS_{sys} = dS_1 + dS_2 = S_p \tag{1-45}$$

The entropy production can be found by applying the fundamental property relationship to each subsystem. The property relationship is

$$dE = T\,dS - P\,dV + \sum_{i=1}^{n} \mu_i\,dm_i \tag{1-46}$$

where P is the pressure, V the volume, m_i the moles, E the energy, S the entropy, and μ_i the chemical potential of component i in the system. Upon rearranging and applying to both subsystems, we have

$$dS_1 = \left(\frac{1}{T}\right)_1 dE_1 + \left(\frac{P}{T}\right)_1 dV_1 - \sum_{i=1}^{n}\left(\frac{\mu_i}{T}\right)_1 dm_{i,1} \tag{1-47}$$

$$dS_2 = \left(\frac{1}{T}\right)_2 dE_2 + \left(\frac{P}{T}\right)_2 dV_2 - \sum_{i=1}^{n}\left(\frac{\mu_i}{T}\right)_2 dm_{i,2} \tag{1-48}$$

By combining the equations above, the entropy production becomes

$$S_p = \left[\left(\frac{1}{T}\right)_1 - \left(\frac{1}{T}\right)_2\right] dE_1 - \sum_{i=1}^{n}\left[\left(\frac{\mu_i}{T}\right)_1 - \left(\frac{\mu_i}{T}\right)_2\right] dm_{i,1} \tag{1-49}$$

Equation (1-49) can be written in rate form by dividing each term by a time interval $d\theta$. Thus

$$\dot{S}_p = \frac{dE_1}{d\theta}\left[\left(\frac{1}{T}\right)_1 - \left(\frac{1}{T}\right)_2\right] - \sum_{i=1}^{n}\frac{dm_{i,1}}{d\theta}\left[\left(\frac{\mu_i}{T}\right)_1 - \left(\frac{\mu_i}{T}\right)_2\right] \tag{1-50}$$

The volumetric rate of entropy production is obtained by dividing by the cross-sectional area, A, and thickness, ΔZ, of the barrier. In the limit as ΔZ approaches zero the equation above becomes

$$\frac{\dot{S}_p}{A\,dZ} = \sigma = J_E\frac{d(1/T)}{dZ} - \sum_{i=1}^{n} J_i\frac{d(\mu_i/T)}{dZ} \tag{1-51}$$

In the equation above, J_E and J_i are the energy and molar flows per unit area per unit time, respectively, and σ is the rate of entropy production. The energy and component transfer potentials corresponding to the energy and molar fluxes are $d(1/T)$ and $d(\mu_i/T)$.

Onsager, however, has chosen the linear driving forces such that when each flow is multiplied by the driving force, the product is equal to the entropy production multiplied by the absolute temperature. Equation (1-51) can be extended to other flows and potentials, and can be written in general in terms of the Onsager-type potentials as

$$T\sigma = \sum_{i=1}^{n} \mathbf{J}_i \mathbf{X}_i \tag{1-52}$$

where $\mathbf{J}_i$ is any flux and $\mathbf{X}_i$ is the conjugate potential or driving force.

Upon expanding Eq. (1-51) and writing the results in the form proposed by Onsager, we have

$$T\sigma = -J_E \frac{1}{T}\frac{dT}{dZ} - \sum_{i=1}^{n} J_i \frac{d\mu_i}{dZ} \tag{1-53}$$

Thus we see that the driving forces for energy and mass transfer are

$$X_{EZ} = \frac{1}{T}\frac{dT}{dZ} \tag{1-54}$$

and

$$X_{iZ} = \frac{d\mu_i}{dZ} \tag{1-55}$$

It was just shown by using irreversible thermodynamics that the driving force for diffusion in a system is the difference in chemical potential and not the difference in composition as one is led to believe by examining Fick's first law. However, it is difficult experimentally to determine the chemical potential gradient, whereas the concentration gradient can usually be readily measured. Thus we define the diffusion coefficient in terms of the concentration gradient. To verify further that the gradient in the chemical potential is indeed the driving force for diffusion, let us consider the following case.

Ethane gas and heptane are placed in a closed container which is then placed in a constant-temperature bath. A portion of the heptane will vaporize into the ethane gas and a portion of the ethane will dissolve into the heptane. After a period of time, the net flux of ethane into the liquid and the net flux of heptane into the gas phase are zero. If the concentrations of ethane in the gas and liquid phases are measured, they will be found to differ by a large amount. Similarly, the concentration of heptane in the two phases will also differ. If concentration is the driving force, a net flux greater than zero would exist. Because the net flux is zero, it can be concluded that concentration is not the driving force for mass transfer.

REFERENCES

BIRD, R. B., W. E. STEWART, and E. N. LIGHTFOOT, *Transport Phenomena*, Wiley, New York, 1960.

DE GROOT, S. R., and P. MAZUR, *Non-equilibrium Thermodynamics*, North-Holland, Amsterdam, 1963.

HAASE, R., *Thermodynamics of Irreversible Processes*, Addison-Wesley, Reading, Mass., 1969.

HOOYMAN, G. J., H. HOLTAN, JR., P. MAZUR, and S. R. DE GROOT. *Physica*, **19**, 1095 (1953).

PRIGOGINE, I., *Thermodynamics of Irreversible Processes*, Wiley-Interscience, New York, 1967.

NOTATIONS*

C = total molar concentration, mol/L^3
C_i = molar concentration of i, mol/L^3
D = diffusion coefficient, L^2/t
D_{AB} = binary diffusivity for system A–B, L^2/t
E_{sys} = internal energy of the system, E
$\mathbf{j}_i$ = mass flux of i, M/tL^2
$_0\mathbf{j}_i$ = mass flux of i with respect to reference velocity U^0, M/tL^2
$_m\mathbf{j}_i$ = mass flux of i with respect to mass average velocity U^m, M/tL^2
$_M\mathbf{j}_i$ = mass flux of i with respect to molar average velocity U^M, M/tL^2
$_v\mathbf{j}_i$ = mass flux of i with respect to volume average velocity U^v, M/tL^2
$\mathbf{J}_i$ = molar flux of i, mol/tL^2
$_0\mathbf{J}_i$ = molar flux of i with respect to reference velocity U^0, mol/tL^2
$_m\mathbf{J}_i$ = molar flux of i with respect to mass average velocity U^m, mol/tL^2
$_M\mathbf{J}_i$ = molar flux of i with respect to molar average velocity U^M, mol/tL^2
$_v\mathbf{J}_i$ = molar flux of i with respect to volume average velocity U^v, mol/tL^2
L_{HE} = cross phenomenological coefficient for heat flow due to the flow of electrical current
L_{HH} = phenomenological coefficient for heat conduction
L_{iE} = phenomenological coefficient for mass transfer due to an electrical potential
L_{iH} = phenomenological coefficient for mass transfer due to a thermal gradient
L_{ii} = phenomenological coefficient for mass transfer due to a chemical potential
M = molecular weight, M/mol
M_i = molecular weight of i, M/mol
$\mathbf{n}_i$ = mass flux of i with respect to a fixed reference frame, M/tL^2
$\mathbf{N}_i$ = molar flux of i with respect to a fixed reference frame, mol/tL^2
P = pressure of the system, F/L^2
S_p = entropy production of the system, E/T
$\dot{S}_p$ = entropy production per unit time, E/Tt
T = absolute temperature, T
$\mathbf{U}_i$ = velocity of i with respect to a fixed reference frame, L/t
$\mathbf{U}^m$ = mass average velocity, L/t
$\mathbf{U}^M$ = molar average velocity, L/t
$\mathbf{U}^v$ = volume average velocity, L/t
$\mathbf{U}^0$ = arbitrary reference velocity, L/t
V = molar volume, L^3/mol
$\bar{V}_i$ = partial molar volume of i, L^3/mol

x_i = mole fraction of i
$\mathbf{X}_E$ = conjugate potential for current flow
$\mathbf{X}_H$ = conjugate potential for heat flow
$\mathbf{X}_i$ = conjugate potential for mass transfer

Greek Letters

θ = time, t
μ_i = chemical potential of i, E/mol
ρ = mass concentration, M/L^3
ρ_i = mass concentration of i, M/L^3
σ = volumetric rate of entropy production, E/TtL^3
χ_i = normalized weighting factor
ω_i = mass fraction of i

---

*Boldface type represents vector quantities.

PROBLEMS

1.1 Show that $D_{AB} = D_{BA}$ for a binary system.

1.2 (a) Prove that

$$\sum_{i=1}^{n} {}_m\mathbf{j}_i = 0$$

(b) Prove that the expressions

$${}_m\mathbf{j}_A = -\rho D_{AB} \nabla \omega_A \quad \text{and} \quad C(\mathbf{U}_A - \mathbf{U}_B) = -\frac{CD_{AB}}{x_A x_B} \nabla x_A$$

are equivalent.

1.3 Determine the relationship between the molar flux with respect to the volume average velocity (${}_v\mathbf{J}_A$) and the molar flux with respect to the molar average velocity (${}_M\mathbf{J}_A$) for a binary system of ideal gases.

1.4 Show that the mass flux with respect to the volume average velocity (${}_v\mathbf{j}_A$) and the molar flux with respect to the mass average velocity (${}_m\mathbf{J}_A$) for a binary system are related by the expression

$${}_v\mathbf{j}_A = \rho \bar{V}_B \frac{M_A}{M_B} {}_m\mathbf{J}_A$$

1.5 From the definitions of concentrations, velocities, and fluxes, show that for a binary system

$${}_v\mathbf{J}_A + {}_v\mathbf{J}_B = \left(\frac{x_B}{x_A}\mathbf{N}_A - \mathbf{N}_B\right)(C\bar{V}_B - 1)$$

1.6 By applying Eq. (1-36), obtain expressions for **(a)** the molar flux with respect to the volume average velocity, ${}_v\mathbf{J}_A = -C^2 D_{AB} \bar{V}_B \nabla x_A$; and **(b)** the molar flux with respect to the mass average velocity, ${}_m\mathbf{J}_A = -CD_{AB}(M_B/M) \nabla x_A$.

1.7 Show that

$$\sum_{i=1}^{n} {}_0\mathbf{J}_i \frac{\chi_i}{C_i} = 0$$

1.8 Obtain the relation between the molar flux with respect to the molar average velocity and the molar flux with respect to the mass average velocity. Compare your result with the expression given in Table 1-3.

1.9 It was shown that the driving force for mass transfer is a gradient in the chemical potential. Thus Fick's equation may be expressed in general terms by using the Onsager coefficient as

$$J_A = L_{AB}\frac{d\mu_A}{dZ}$$

Using the equation above and Eq. (1-1), show how L_{AB} and D_{AB} are related. Assume that the activity coefficient can be described by the expression $\log \gamma_A = Ax_Ax_B$.

1.10 An ideal solution containing 0.1×10^{-3} m^3 of methanol and 0.9×10^{-3} m^3 of benzene moves at a molar average velocity of 0.12 m/s. If the molar flux of benzene relative to the mass average velocity is -1.0 kgmol/m$^2\cdot$s, what is the total molar flux of methanol, N_A, and the mass average velocity?

Methanol (A)	*Benzene (B)*
$M_A = 32.04$ kg/kgmol	$M_B = 78.12$ kg/kgmol
$\rho_A = 792$ kg/m^3	$\rho_B = 879$ kg/m^3

1.11 Methane and hydrogen diffuse through a porous membrane 4.0×10^{-3} m thick at 0°C and 1 atm as shown in Figure 1-2. For each mole of methane diffusing through the membrane into the hydrogen, an equal amount of hydrogen diffuses into the methane (equimolar counter diffusion). The partial pressures of methane on either side of the membrane are maintained at 2.66×10^4 and 1.33×10^4 N/m^2, respectively. Calculate the molar flux of methane into the hydrogen. An effective diffusivity of 1.0×10^{-5} m^2/s may be assumed.

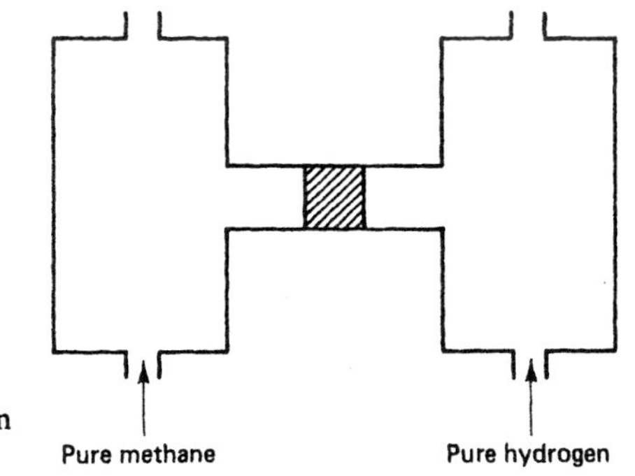

Figure 1-2 Problem 1.11: Diffusion through a Porous Membrane.

1.12 Find the relationship between $\mathbf{U}^v$, the volume average velocity, and $\mathbf{U}^m$, the mass average velocity.

1.13 Determine the relationship between ${}_v\mathbf{J}_A$ and ${}_M\mathbf{j}_A$ for **(a)** a binary mixture of ideal gases, and **(b)** a binary mixture of gases that is described by the equation $PV/nRT = 1 + B'P$.

1.14 Show that

$$dx_A = \frac{d\omega_A}{M_A M_B (1/M)^2}$$

1.15 Find the relationship between $\mathbf{U}^M$ and $\mathbf{U}^v$ for a binary mixture of ideal gases.

1.16 For a binary system show that

$$_v\mathbf{J}_A = \frac{1 - C_A \bar{V}_A}{M_A \omega_B} {_m\mathbf{j}_A}$$

1.17 Prove that for a binary system
(a) $_M\mathbf{j}_A + {_M\mathbf{j}_B} = \rho(\mathbf{U}^m - \mathbf{U}^M)$ and (b) $_m\mathbf{J}_A + {_m\mathbf{J}_B} = C(\mathbf{U}^M - \mathbf{U}^m)$

1.18 Starting with the equation $_MJ_A = -CD_{AB}\, dx_A/dZ$, show that $\mathbf{n}_A = {_m\mathbf{j}_A} + \omega_A(\mathbf{n}_A + \mathbf{n}_B)$.

1.19 When evaluating diffusion coefficients experimentally, consideration must be given to the reference velocity in the experimental system. Starting with Eq. (1-36), obtain an expression relating the diffusivity relative to the molar average velocity to the diffusivity relative to the volume average velocity for an ideal gas.

Diffusion Coefficients 2

2.1 Introduction

Diffusion can be described by a mechanistic approach in which a consideration of atom movement is important or by a continuum approach such as with Fick's first law, where no consideration is given to the actual mechanism by which atom transfer occurs.

Of the three states of matter, gases are the easiest to deal with mathematically because the molecules are far apart and the intermolecular forces can often be disregarded or considered only during collisions. For gases, a molecule is assumed to travel along a straight line until it collides with another molecule, after which its speed and direction are altered. Kinetic theory has provided an accurate means of predicting values for the diffusion coefficient in binary gas mixtures.

As opposed to gases, large intermolecular forces are present between the closely arranged atoms of a solid. Although a great deal of effort has been expended in determining the mechanisms of diffusion, the prediction of diffusion coefficients for solids a priori has met with limited success even for the regular atom arrangement found in crystalline solids.

The intermediate liquid phase has been described as both a dense gas and as an irregular solid because of its close irregular atom arrangement and strong intermolecular forces. For certain types of liquids, diffusion coefficients have been predicted with reasonable accuracy by using continuum models.

2.2 Diffusion Coefficients for Gases

The diffusion coefficient can be derived for an ideal gas by using the simplified discussion presented by Sherwood et al. (1975). From kinetic theory, the diffusion coefficient is assumed to be directly proportional to the mean molecular velocity and the mean free path, λ.

$$D \propto \bar{U}\lambda \tag{2-1}$$

For an ideal gas the motion of the molecule is assumed to be totally random, with the mean free path, λ, being inversely proportional to both the average cross-sectional area of the molecules, A, and the number density, n, of all molecules in a specified volume. Since the number density of an ideal gas varies directly with pressure and inversely with the temperature, this becomes

$$\lambda \propto \frac{1}{nA} \propto \frac{T}{PA} \tag{2-2}$$

But the mean molecular velocity is related to the temperature and molecular weight of the molecule by the expression

$$\bar{U} \propto \left(\frac{T}{M}\right)^{1/2} \tag{2-3}$$

where $\bar{U}$ is the mean velocity and M is the molecular weight. Thus for a binary system the equations above can be combined and extended to yield

$$D_{AB} \propto \left(\frac{T}{M_A} + \frac{T}{M_B}\right)^{1/2} \frac{T}{PA} \tag{2-4}$$

or

$$D_{AB} = \frac{K'T^{3/2}}{PA_{\text{avg}}}\left(\frac{1}{M_A} + \frac{1}{M_B}\right)^{1/2} \tag{2-5}$$

where A_{avg} is the average cross-sectional area of both types of molecules and K' is a constant of proportionality.

Various semiempirical equations similar to the one above have been derived using kinetic theory as a basis. Gilliland (1934) proposed that the diffusivity be calculated from the equation

$$D_{AB} = 4.3 \times 10^{-9} \frac{T^{3/2}}{P(V_A^{1/3} + V_B^{1/3})^2}\left(\frac{1}{M_A} + \frac{1}{M_B}\right)^{1/2} \tag{2-6}$$

where D_{AB} = diffusivity, m^2/s,
T = absolute temperature, K,
M = molecular weight, kg/kgmol,
V = molar volume at the normal boiling point, m^3/kgmol,
P = total pressure, atm (101.3 kN/m^2).

Atomic and molecular volumes are presented in Table 2-1.

TABLE 2-1. ATOMIC AND MOLECULAR VOLUMES AT THE NORMAL BOILING POINT (TREYBAL, 1968)

	Atomic Volume × 10^3 ($m^3/kgatom$)		*Molecular Volume* × 10^3 ($m^3/kgmol$)
Bromine	27.0	Air	29.9
Carbon	14.8	Br_2	53.2
Chlorine	24.6	Cl_2	48.4
Hydrogen	3.7	CO	30.7
Iodine	37.0	CO_2	34.0
Nitrogen	15.6	COS	51.5
Nitrogen in primary amines	10.5	H_2	14.3
Nitrogen in secondary amines	12.0	H_2O	18.9
Oxygen	7.4	H_2S	32.9
Oxygen in methyl esters	9.1	I_2	71.5
Oxygen in higher esters	11.0	N_2	31.2
Oxygen in acids	12.0	NH_3	25.8
Oxygen in methyl ethers	9.9	NO	23.6
Oxygen in higher ethers	11.0	N_2O	36.4
Sulfur	25.6	O_2	25.6
Benzene ring	−15.0	SO_2	44.8
Naphthalene ring	−30.0		

Although Gilliland's equation provides a reasonably accurate means for predicting diffusion coefficients and is satisfactory for engineering work, the semiempirical equation proposed by Fuller et al. (1966) is recommended. Their equation, which results from a curve fit to available experimental data, is

$$D_{AB} = \frac{1.0 \times 10^{-9} T^{1.75}}{P[(\sum v)_A^{1/3} + (\sum v)_B^{1/3}]^2} \left(\frac{1}{M_A} + \frac{1}{M_B} \right)^{1/2} \tag{2-7}$$

where D_{AB} has the units m^2/s; T is the absolute temperature, K; and pressure is given in atmospheres. The diffusion volumes, $\sum v$, are the sum of the atomic volumes of all elements for each molecule. Atomic and diffusion volumes given in Table 2-2 should be used in Eq. (2-7). Fuller et al.'s equation is reported to predict diffusivities to within 7% of experimental values and can be used for both polar and nonpolar gases.

TABLE 2-2. ATOMIC AND MOLECULAR DIFFUSION VOLUMES FOR THE EQUATION OF FULLER ET AL. (1966)[a]

Atomic and Structural Diffusion Volume Increments, $v \times 10^3$ ($m^3/kgatom$)			
C	16.5	(Cl)	19.5
H	1.98	(S)	17.0
O	5.48	Aromatic ring	−20.2
(N)	5.69	Heterocyclic ring	−20.2

TABLE 2-2. (CONTINUED)

Diffusion Volumes for Simple Molecules, $\sum v \times 10^3$ ($m^3/kgmol$)

H_2	7.07	CO	18.9
D_2	6.70	CO_2	26.9
He	2.88	N_2O	35.9
N_2	17.9	NH_3	14.9
O_2	16.6	H_2O	12.7
Air	20.1	(CCl_2F_2)	114.8
Ar	16.1	(SF_6)	69.7
Kr	22.8	(Cl_2)	37.7
(Xe)	37.9	(Br_2)	67.2
Ne	5.59	(SO_2)	41.1

[a]Parentheses indicate that the value listed is based on only a few data points.

Chapman and Enskog (Hirschfelder et al., 1954), working independently, related the properties of gases to the forces acting between the molecules. Using the Lennard-Jones 6-12 potential to relate the attractive and repulsive forces between atoms, Hirschfelder et al. (1949) developed the following equation to predict the diffusivity for nonpolar gas pairs:

$$D_{AB} = \frac{1.858 \times 10^{-27} T^{3/2}}{P \sigma_{AB}^2 \Omega_D} \left(\frac{1}{M_A} + \frac{1}{M_B} \right)^{1/2} \tag{2-8}$$

where D_{AB} = diffusivity, m^2/s,
T = absolute temperature, K,
M = molecular weight, kg/kgmol,
P = absolute pressure, atm,
σ_{AB} = collision diameter, m,
Ω_D = collision integral.

The Lennard-Jones equation is

$$\phi_{AB}(r) = 4\epsilon_{AB}\left[\left(\frac{\sigma_{AB}}{r}\right)^{12} - \left(\frac{\sigma_{AB}}{r}\right)^{6}\right] \tag{2-9}$$

where $\phi_{AB}(r)$ = potential energy,
ϵ_{AB} = maximum energy of attraction,
σ_{AB} = collision diameter,
r = molecular separation distance.

Repulsive and attractive forces are accounted for in the equation above by the r^{-12} and r^{-6} terms, respectively. For a binary system the Lennard-Jones force constants, ϵ_{AB} and σ_{AB}, are evaluated from pure component data using the following relationships:

$$\sigma_{AB} = \frac{\sigma_A + \sigma_B}{2} \tag{2-10}$$

and

$$\frac{\epsilon_{AB}}{k} = \left(\frac{\epsilon_A}{k} \times \frac{\epsilon_B}{k}\right)^{1/2} \tag{2-11}$$

where k is the Boltzmann constant.

Collision diameters and force constants for a number of pure substances are given in Table 2-3. When collision diameters and force constants are not available, they may be estimated from the following relationships:

$$\sigma = 8.33V_c^{1/3} \tag{2-12}$$

or

$$\sigma = 11.8V^{1/3} \tag{2-13}$$

and

$$\frac{\epsilon}{k} = 0.75T_c \tag{2-14}$$

or

$$\frac{\epsilon}{k} = 1.21T_b \tag{2-15}$$

where V_c = critical volume, m^3/kgmol,
T_c = critical temperature, K,
V = molar volume at the normal boiling point, m^3/kgmol,
T_b = normal boiling-point temperature, K,
σ = collision diameter, 10^{-10} m.

TABLE 2-3. COLLISION DIAMETERS AND ENERGY PARAMETERS FOR THE LENNARD-JONES EQUATION (SVEHLA, 1962)

Molecule	*Compound*	$\sigma \times 10^{10}$ (*m*)	ϵ/k (*K*)
Ar	Argon	3.542	93.3
He	Helium	2.551	10.22
Kr	Krypton	3.655	178.9
Ne	Neon	2.820	32.8
Xe	Xenon	4.047	1.0
Air	Air	3.711	78.6
AsH_3	Arsine	4.145	259.8
BCl_3	Boron chloride	5.127	337.7
BF_3	Boron fluoride	4.198	186.3
$B(OCH_3)_3$	Methyl borate	5.503	396.7
Br_2	Bromine	4.296	507.9
CCl_4	Carbon tetrachloride	5.947	322.7
CF_4	Carbon tetrafluoride	4.662	134.0
$CHCl_3$	Chloroform	5.389	340.2
CH_2Cl_2	Methylene chloride	4.898	356.3
CH_3Br	Methyl bromide	4.118	449.2
CH_3Cl	Methyl chloride	4.182	350
CH_3OH	Methanol	3.626	481.8
CH_4	Methane	3.758	148.6
CO	Carbon monoxide	3.690	91.7
COS	Carbonyl sulfide	4.130	336.0
CO_2	Carbon dioxide	3.941	195.2
CS_2	Carbon disulfide	4.483	467
C_2H_2	Acetylene	4.033	231.8
C_2H_4	Ethylene	4.163	224.7
C_2H_6	Ethane	4.443	215.7
C_2H_5Cl	Ethyl chloride	4.898	300

TABLE 2-3. (CONTINUED)

Molecule	*Compound*	$\sigma \times 10^{10}$ (*m*)	ϵ/k (*K*)
C_2H_5OH	Ethanol	4.530	362.6
C_2N_2	Cyanogen	4.361	348.6
CH_3OCH_3	Methyl ether	4.307	395.0
CH_2CHCH_3	Propylene	4.678	298.9
CH_3CCH	Methylacetylene	4.761	251.8
C_3H_6	Cyclopropane	4.807	248.9
C_3H_8	Propane	5.118	237.1
n-C_3H_7OH	*n*-Propyl alcohol	4.549	576.7
CH_3COCH_3	Acetone	4.600	560.2
CH_3COOCH_3	Methyl acetate	4.936	469.8
n-C_4H_{10}	*n*-Butane	4.687	531.4
i-C_4H_{10}	Isobutane	5.278	330.1
$C_2H_5OC_2H_5$	Ethyl ether	5.678	313.8
$CH_3COOC_2H_5$	Ethyl acetate	5.205	521.3
n-C_5H_{12}	*n*-Pentane	5.784	341.1
$C(CH_3)_4$	2,2-Dimethylpropane	6.464	193.4
C_6H_6	Benzene	5.349	412.3
C_6H_{12}	Cyclohexane	6.182	297.1
n-C_6H_{14}	*n*-Hexane	5.949	399.3
Cl_2	Chlorine	4.217	316.0
F_2	Fluorine	3.357	112.6
HBr	Hydrogen bromide	3.353	449
HCN	Hydrogen cyanide	3.630	569.1
HCl	Hydrogen chloride	3.339	344.7
HF	Hydrogen fluoride	3.148	330
HI	Hydrogen iodide	4.211	288.7
H_2	Hydrogen	2.827	59.7
H_2O	Water	2.641	809.1
H_2O_2	Hydrogen peroxide	4.196	289.3
H_2S	Hydrogen sulfide	3.623	301.1
Hg	Mercury	2.969	750
$HgBr_2$	Mercuric bromide	5.080	686.2
$HgCl_2$	Mercuric chloride	4.550	750
HgI_2	Mercuric iodide	5.625	695.6
I_2	Iodine	5.160	474.2
NH_3	Ammonia	2.900	558.3
NO	Nitric oxide	3.492	116.7
NOCl	Nitrosyl chloride	4.112	395.3
N_2	Nitrogen	3.798	71.4
N_2O	Nitrous oxide	3.828	232.4
O_2	Oxygen	3.467	106.7
PH_3	Phosphine	3.981	251.5
SF_6	Sulfur hexafluoride	5.128	222.1
SO_2	Sulfur dioxide	4.112	335.4
SiF_4	Silicon tetrafluoride	4.880	171.9
SiH_4	Silicon hydride	4.084	207.6
$SnBr_4$	Stannic bromide	6.388	563.7
UF_6	Uranium hexafluoride	5.967	236.8

Values of the collision integral, Ω_D, have been evaluated for the Lennard-Jones 6-12 potential by Hirschfelder et al. (1954). These values are presented in Table 2-4. The collision integral can be calculated with good accuracy from the relationship proposed by Neufeld et al. (1972). The Lennard-Jones potential is shown in Figure 2-1.

TABLE 2-4. VALUES OF THE COLLISION INTEGRAL Ω_D BASED ON THE LENNARD-JONES POTENTIAL (HIRSCHFELDER ET AL., 1954[a])

kT/ϵ	Ω_D	kT/ϵ	Ω_D	kT/ϵ	Ω_D
0.30	2.662	1.65	1.153	4.0	0.8836
0.35	2.476	1.70	1.140	4.1	0.8788
0.40	2.318	1.75	1.128	4.2	0.8740
0.45	2.184	1.80	1.116	4.3	0.8694
0.50	2.066	1.85	1.105	4.4	0.8652
0.55	1.966	1.90	1.094	4.5	0.8610
0.60	1.877	1.95	1.084	4.6	0.8568
0.65	1.798	2.00	1.075	4.7	0.8530
0.70	1.729	2.1	1.057	4.8	0.8492
0.75	1.667	2.2	1.041	4.9	0.8456
0.80	1.612	2.3	1.026	5.0	0.8422
0.85	1.562	2.4	1.012	6	0.8124
0.90	1.517	2.5	0.9996	7	0.7896
0.95	1.476	2.6	0.9878	8	0.7712
1.00	1.439	2.7	0.9770	9	0.7556
1.05	1.406	2.8	0.9672	10	0.7424
1.10	1.375	2.9	0.9576	20	0.6640
1.15	1.346	3.0	0.9490	30	0.6232
1.20	1.320	3.1	0.9406	40	0.5960
1.25	1.296	3.2	0.9328	50	0.5756
1.30	1.273	3.3	0.9256	60	0.5596
1.35	1.253	3.4	0.9186	70	0.5464
1.40	1.233	3.5	0.9120	80	0.5352
1.45	1.215	3.6	0.9058	90	0.5256
1.50	1.198	3.7	0.8998	100	0.5130
1.55	1.182	3.8	0.8942	200	0.4644
1.60	1.167	3.9	0.8888	400	0.4170

[a]Hirschfelder et al. use the symbols T^* for kT/ϵ and $\Omega^{(1,1)*}$ for Ω_D.

From the various diffusion equations for gases, it is noted that the diffusivity increases with the 1.5 to 1.8 power of the absolute temperature but is inversely proportional to the system pressure. At low pressures, the diffusion coefficient is not affected by composition. The equation derived by Hirschfelder et al. is recommended for extrapolating experimental data for moderate pressure ranges. Upon simplifying Eq. (2-8), we have

$$D_{AB,T_2,P_2} = D_{AB,T_1,P_1} \frac{P_1}{P_2}\left(\frac{T_2}{T_1}\right)^{3/2} \frac{\Omega_{T_1}}{\Omega_{T_2}} \tag{2-16}$$

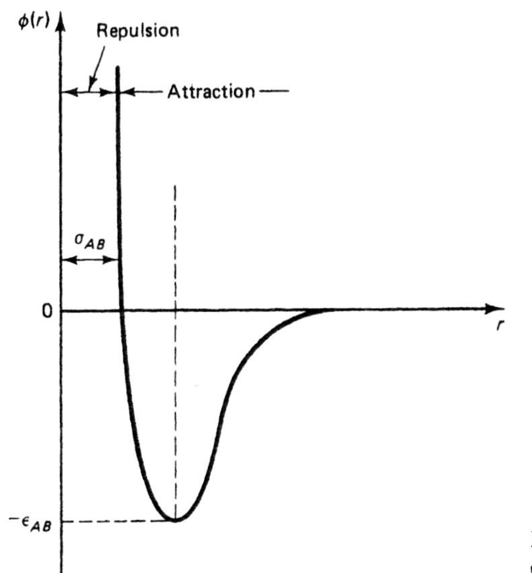

Figure 2-1 Lennard-Jones potential energy function (Lennard-Jones, 1924)

Selected experimental binary gas diffusivities were obtained from the extensive review of Marrero and Mason (1972) and are presented in Table 2-5. Note that for a pressure of 1 atm, the diffusivity has a value on the order of 10^{-5} m^2/s.

TABLE 2-5. INTERDIFFUSION COEFFICIENTS FOR BINARY GASES AT 1 ATM PRESSURE (101.325 KPA) (MARRERO AND MASON, 1972)

System	$T (K)$	$D_{AB} \times 10^4$ (m^2/s)
Air–carbon dioxide	317.2	0.177
Air–ethanol	313	0.145
Air–helium	317.2	0.765
Air–*n*-hexane	328	0.093
Air–*n*-pentane	294	0.071
Air–water	313	0.288
Argon–ammonia	333	0.253
Argon–carbon dioxide	276.2	0.133
Argon–helium	298	0.729
Argon–hydrogen	242.2	0.562
	448	1.76
	806	4.86
	1069	8.10
Argon–methane	298	0.202
Argon–sulfur dioxide	263	0.077
Carbon dioxide–helium	298	0.612

TABLE 2-5. (CONTINUED)

System	$T(K)$	$D_{AB} \times 10^4$ (m^2/s)
Carbon dioxide–nitrogen	298	0.167
Carbon dioxide–nitrous oxide	312.8	0.128
Carbon dioxide–oxygen	293.2	0.153
Carbon dioxide–sulfur dioxide	263	0.064
Carbon dioxide–water	307.2	0.198
	352.3	0.245
Carbon monoxide–nitrogen	373	0.318
Helium–benzene	423	0.610
Helium–ethanol	423	0.821
Helium–methane	298	0.675
Helium–methanol	423	1.032
Helium–nitrogen	298	0.687
Helium–oxygen	298	0.729
Helium–*i*-propanol	423	0.677
Helium–water	307.1	0.902
Hydrogen–acetone	296	0.424
Hydrogen–ammonia	298	0.783
	358	1.093
	473	1.86
	533	2.149
Hydrogen–benzene	311.3	0.404
Hydrogen–cyclohexane	288.6	0.319
Hydrogen–methane	288	0.694
Hydrogen–nitrogen	298	0.784
	573	2.147
Hydrogen–sulfur dioxide	473	1.23
Hydrogen–thiophene	302	0.400
Hydrogen–water	328.5	1.121
Methane–water	352.3	0.356
Nitrogen–ammonia	298	0.230
	358	0.328
Nitrogen–benzene	311.3	0.102
Nitrogen–cyclohexane	288.6	0.0731
Nitrogen–sulfur dioxide	263	0.104
Nitrogen–water	307.5	0.256
	352.1	0.359
Oxygen–benzene	311.3	0.101
Oxygen–carbon tetrachloride	296	0.0749
Oxygen–cyclohexane	288.6	0.0746
Oxygen–water	352.3	0.352

Marrero and Mason present a thorough review of the literature, with tables of values for gas pairs not included here and references to reported experimental values of D_{AB} for several hundred systems.

Example 2.1

Estimate the diffusion coefficient for the hydrogen–ammonia system at 25°C and 1 atm pressure. Use the collision integrals to find the diffusivity at 85°C. Compare the results with experimental values.

Solution:

$$P = 1 \text{ atm}$$

$$T = 298 \text{ K}$$

$$M_A = 2 \qquad M_B = 17$$

From Table 2-3,

$$\sigma_A = 2.827 \times 10^{-10} \text{ m} \qquad \frac{\epsilon_A}{k} = 59.7 \text{ K}$$

$$\sigma_B = 2.900 \times 10^{-10} \text{ m} \qquad \frac{\epsilon_B}{k} = 558.3 \text{ K}$$

$$\sigma_{AB} = \frac{\sigma_A + \sigma_B}{2} = \frac{2.827 \times 10^{-10} + 2.900 \times 10^{-10}}{2} = 2.8635 \times 10^{-10} \text{ m}$$

$$\frac{\epsilon_{AB}}{k} = \left(\frac{\epsilon_A}{k} \times \frac{\epsilon_B}{k}\right)^{1/2} = [(59.7)(558.3)]^{1/2} = 182.57 \text{ K}$$

$$\frac{kT}{\epsilon_{AB}} = \frac{298}{182.57} = 1.632$$

From Table 2-4,

$$\Omega_D = 1.158$$

$$\frac{1}{M_A} + \frac{1}{M_B} = \frac{1}{2} + \frac{1}{17} = 0.5588$$

From Eq. (2-8),

$$\begin{aligned} D_{AB} &= \frac{1.858 \times 10^{-27} T^{3/2}}{P\sigma_{AB}^2 \Omega_D}\left(\frac{1}{M_A} + \frac{1}{M_B}\right)^{1/2} \\ &= \frac{1.85 \times 10^{-27}(298)^{3/2}}{1(2.8635 \times 10^{-10})^2(1.158)}(0.5588)^{1/2} \\ &= 7.52 \times 10^{-5} \text{ m}^2/\text{s} \end{aligned}$$

Temperature correction:

$$D_{AB,T_2} = D_{AB,T_1}\left(\frac{T_2}{T_1}\right)^{3/2} \frac{\Omega_{T_1}}{\Omega_{T_2}}$$

$$\frac{kT_2}{\epsilon_{AB}} = \frac{358}{182.57} = 1.961$$

Table 2-4 gives

$$\Omega_D = 1.082$$

$$\begin{aligned} D_{AB,T_2} &= 7.52 \times 10^{-5}\left(\frac{358}{298}\right)^{3/2} \frac{1.158}{1.082} \\ &= 1.06 \times 10^{-4} \text{ m}^2/\text{s} \end{aligned}$$

From Table 2-5 the experimental values at 25 and 85°C are 7.83×10^{-5} m²/s and 1.093×10^{-4} m²/s, respectively.

2.3 Diffusion Coefficients for Liquids

Liquids are characterized by a combination of strong intermolecular forces in addition to the irregular chaotic arrangement of atoms present in the gas phase. This combination has compounded the problems of accurately describing the liquid state quantitatively and has made it difficult to predict transport properties. Several attempts to treat liquid data have generated various structural pictures for the liquid state varying from a dense gas to a solid. The use of gaseous models to describe the liquid state was due in part to the fact that atom movement in a liquid is a random process. By contrast, the solid-type models have been used for liquids such as the liquid metals because of atom clusters which appear as nearly regular atom arrangements that exist just above the melting point and because of the large intermolecular forces. The difficulty in treating the liquid state becomes readily apparent since liquids in reality consist of clusters of irregularly arranged atoms or molecules which exist in an activated state and exhibit random motion. Although relatively accurate values of diffusion coefficients can be obtained from solutions to Fick's law for gases, liquids, and solids, very little is known about the actual mechanism of liquid diffusion as opposed to gases and solids. Because of the lack of an accurate theoretical structural model for liquids, considerable work has been done in this area and a wide variety of experimental techniques for measuring diffusion coefficients exists in the literature.

Numerous attempts to predict molecular diffusivities in the liquid state have been made using one of three general approaches. These are the hydrodynamical, quasi-crystalline, and fluctuation theories. A variety of equations are based on the hydrodynamical theories, which relate the diffusion coefficient to the viscosity or to a friction constant, which can then be related to viscosity. A physical interpretation of the diffusivity for spherical solute particles that are large compared to the solvent in which they move is given by the Stokes–Einstein equation (see Frenkel, 1946). The Stokes–Einstein equation is based on the assumption that the particles are hard spheres that move at a uniform velocity in a continuum under the action of a unit force F. Stokes' law describing the force acting on an atom is

$$F = 6\pi\mu r U \tag{2-17}$$

Einstein proposed an equation relating the diffusion coefficient to the mobility, M, as

$$D = kTM \tag{2-18}$$

Since the mobility has units of velocity per unit force, Eqs. (2-17) and (2-18) can be solved to obtain

$$D = \frac{kT}{6\pi r\mu} \tag{2-19}$$

where D = self-diffusion coefficient,
k = Boltzmann's constant,
T = absolute temperature,
r = radius of the diffusing particle,
μ = viscosity.

Although there are many arguments against this model, its vindication lies in its use as a model that predicts, in a number of cases, diffusion coefficients with the correct order of magnitude.

An improvement to the Stokes–Einstein equation was introduced by Sutherland (1905) in the form of an empirical correction to Stokes' law. Sutherland's modification is expressed in terms of the drag force on the diffusing atom as

$$F = 6\pi r \mu U\left(\frac{\beta r + 2\mu}{\beta r + 3\mu}\right) \tag{2-20}$$

where β is the coefficient of sliding friction between the atom and the medium. The substitution of Eq. (2-20) into Eq. (2-18) gives the resulting Sutherland–Einstein equation:

$$D = \frac{kT}{6\pi r \mu}\left(\frac{\beta r + 3\mu}{\beta r + 2\mu}\right) \tag{2-21}$$

Some points regarding Eq. (2-21) are to be noted. If the no-slip condition exists between a diffusing atom and the medium, β becomes infinitely large and Eq. (2-21) reverts to the original Stokes–Einstein equation. If, on the other hand, the diffusing atoms are assumed to move into voids that exist in the liquid, then β becomes equal to zero and Eq. (2-21) is reduced to

$$D = \frac{kT}{4\pi r \mu} \tag{2-22}$$

It has been noted that Eq. (2-22) fits some liquid metal diffusion data quite well if the Pauling (1960) univalent ionic radius is used for r.

Another diffusion model which has received a good deal of notoriety is described by Glasstone et al. (1941) and Eyring et al. (1960). It is based on their theory of an activation state in conjunction with what may be described as the "hole theory" for liquids, as first proposed by Frenkel (1946). This theory, which is reviewed in depth by Walls and Upthegrove (1964) and Edwards et al. (1968), may be expressed for self-diffusion as

$$D = \frac{kT}{2r\mu} \tag{2-23}$$

Diffusion coefficients calculated by Eq. (2-23) are in very poor agreement with experimental data. A modified form of Eyring's theory was developed on the basis of a consideration of nearest neighbors lying in the same plane as the diffusing atom. His second model has been shown to predict experimental data more accurately than did his first.

Dilute nonelectrolyte solutions

Because of a lack of success in using the previous models to predict liquid diffusion coefficients accurately, several empirical relations have been proposed using the Stokes–Einstein equation as the basic model. For example, assuming spherical molecules with molar volume V_A, Eq. (2-19) becomes

$$D_{AB} = K' \frac{T}{\mu V_A^{1/3}} \tag{2-24}$$

where μ is the viscosity of the solution and K' is a constant of proportionality.

Using Eq. (2-24) as the starting point, Wilke and Chang (1955) obtained an equation for predicting diffusivities in dilute solutions of nonelectrolytes.

$$D_{AB}^{\circ} = \frac{1.17 \times 10^{-13}(\xi_B M_B)^{1/2} T}{V_A^{0.6} \mu} \tag{2-25}$$

where D_{AB}° = interdiffusion coefficient in dilute solutions, m^2/s,
μ = viscosity of solution, cP (10^{-3} kg/m·s or mPa·s),
V_A = molar volume of solute at the normal boiling point, m^3/kgmol,
M_B = molecular weight of solvent, kg/kgmol,
T = absolute temperature, K,
ξ_B = association factor of solvent B.

Recommended values for the association factor are: 1.0 for nonpolar solvents such as benzene, ether, and the aliphatic hydrocarbons; 1.5 for ethanol; 1.9 for methanol; and 2.6 for water.

Example 2.2

Using the equation of Wilke and Chang, determine the diffusivity at low concentration for the following diffusing pairs at 25°C and compare the results with experimental values: (a) methanol–water, and (b) ethanol–water.

Solution:

$$T = 298 \qquad \xi_B = 2.6 \qquad M_B = 18$$

From Appendix A,

$$\mu \simeq \mu_B(25°\text{C}) = 0.89 \text{ cP (mPa·s)}$$

From Table 2-1,

$$V_A(\text{methanol}) = 0.0148 + 4(0.0037) + 0.0074 = 0.037$$

$$V_A(\text{ethanol}) = 2(0.0148) + 6(0.0037) + 0.0074 = 0.0592$$

Substituting into the Wilke–Chang equation gives

Methanol–water:

$$D_{AB}^{\circ} = \frac{(1.17 \times 10^{-13})[(2.6)(18)]^{1/2}(298)}{(0.037)^{0.6}(0.89)} = 1.94 \times 10^{-9} \text{ m}^2/\text{s}$$

Ethanol–water:

$$D^{\circ}_{AB} = \frac{(1.17 \times 10^{-13})[(2.6)(18)]^{1/2}(298)}{(0.0592)^{0.6}(0.89)} = 1.46 \times 10^{-9} \text{ m}^2/\text{s}$$

Experimental diffusivities obtained from Perry and Chilton's *Chemical Engineers' Handbook* (1963) are 1.6×10^{-9} m²/s for methanol–water and 1.28×10^{-9} m²/s for ethanol–water.

On the basis of comparisons made by Reid et al. (1977), several semiempirical equations can be used to obtain a reasonable fit to experimental data. Although the Wilke–Chang equation is not recommended when water is the solute, diffusion coefficients can be predicted to within 11% using water as the solvent and to within 27% for organic solvents. Errors greater than 200% are possible when water is used as the solute.

A more general type of equation was presented by Sitaraman et al. (1963). Their equation, which is less restrictive than that of Wilke and Chang, is recommended when water is the solute. Skelland (1974) indicated that when using water as the solute, the error should be no greater than 12%. Sitaraman et al.'s equation is

$$D^{\circ}_{AB} = 16.79 \times 10^{-14}\left(\frac{M_B^{1/2}\,\Delta H_B^{1/3} T}{\mu_B V_A^{1/2}\,\Delta H_A^{0.3}}\right)^{0.93} \tag{2-26}$$

where ΔH_A and ΔH_B are the latent heats of vaporization of the solute and solvent at their normal boiling points, with the units J/kg, and μ_B is the viscosity of the solvent in centipoise (mPa·s). The remaining terms have the same units as Eq. (2-25).

Example 2.3

Determine the diffusivity of ethanol in benzene at 30°C using the equation of Wilke and Chang and the equation of Sitaraman et al. Convert the diffusivity to 15°C and compare the result with the experimental value given in Table 2-6.

Solution:

$$T = 303 \text{ K} \qquad \xi_B = 1 \qquad M_B = 78$$

From the *Chemical Engineers' Handbook*,

$$\Delta H_A = 85.52 \times 10^4 \text{ J/kg} \qquad \Delta H_B = 43.33 \times 10^4 \text{ J/kg}$$

From Table 2-1,

$$V_A = 2(0.0148) + 6(0.0037) + 0.0074 = 0.0592$$

From Appendix A,

$$\mu_B(30°\text{C}) = 0.56 \text{ cP} \qquad \mu_B(15°\text{C}) = 0.70 \text{ cP}$$

Substituting into the Wilke–Chang equation yields

$$D^\circ_{AB} = \frac{(1.17 \times 10^{-13})(78)^{1/2}(303)}{(0.0592)^{0.6}(0.56)} = 3.06 \times 10^{-9}\ \text{m}^2/\text{s}$$

Substituting into the equation of Sitaraman et al. gives us

$$D^\circ_{AB} = 16.79 \times 10^{-14}\left[\frac{(78)^{1/2}(43.33 \times 10^4)^{1/3}(303)}{(0.56)(0.0592)^{1/2}(85.52 \times 10^4)^{0.3}}\right]^{0.93}$$

$$= 2.04 \times 10^{-9}\ \text{m}^2/\text{s}$$

Convert to 15°C using the equation of Wilke and Chang:

$$\frac{D^\circ_{AB_2}\mu_2}{T_2} = \frac{D^\circ_{AB_1}\mu_1}{T_1}$$

$$D_{AB} = \frac{288}{303}\left(\frac{0.56}{0.7}\right)(3.06 \times 10^{-9}) = 2.33 \times 10^{-9}\ \text{m}^2/\text{s}$$

The experimental value at 15°C is 2.25×10^{-9} m²/s. If the experimental value given in Table 2-6 is corrected to 30°C by using the Wilke–Chang equation, then at 30°C, $D^\circ_{AB} = 2.96 \times 10^{-9}$ m²/s. This compares favorably with the value predicted by the Wilke–Chang equation.

Skelland (1974) presents a summary of semiempirical equations that may be used for predicting diffusion coefficients in dilute solutions of nonelectrolytes. Experimental diffusion coefficients for hydrocarbon liquid solutions at infinite dilution are given in Table 2-6. Liquid diffusivities are usually of the order of 10^{-9} m²/s.

TABLE 2-6. EXPERIMENTAL DIFFUSION COEFFICIENTS AT INFINITE DILUTION (SHERWOOD ET AL., 1975)

Solute A	*Solvent B*	$T(K)$	$D^\circ_{AB} \times 10^9\ (m^2/s)$
Acetic acid	Acetone	298	3.31
Benzoic acid	Acetone	298	2.62
Carbon dioxide	Amyl alcohol	298	1.91
Water	Aniline	293	0.70
Acetic acid	Benzene	298	2.09
Carbon tetrachloride	Benzene	298	1.92
Cinnamic acid	Benzene	298	1.12
Ethanol	Benzene	280.6	1.77
Ethylene chloride	Benzene	288	2.25
Methanol	Benzene	298	3.82
Napthalene	Benzene	280.6	1.19
Carbon dioxide	*i*-Butanol	298	2.20
Acetone	Carbon tetrachloride	293	1.86
Benzene	Chlorobenzene	293	1.25
Acetone	Chloroform	288	2.36
Benzene	Chloroform	288	2.51
Ethanol	Chloroform	288	2.20

TABLE 2-6. (CONTINUED)

Solute A	*Solvent B*	*T (K)*	$D^{\circ}_{AB} \times 10^9$ *(m²/s)*
Carbon tetrachloride	Cyclohexane	298	1.49
Azobenzene	Ethanol	293	0.74
Camphor	Ethanol	293	0.70
Carbon dioxide	Ethanol	290	3.20
Carbon dioxide	Ethanol	298	3.42
Glycerol	Ethanol	293	0.51
Pyridine	Ethanol	293	1.10
Urea	Ethanol	285	0.54
Water	Ethanol	298	1.132
Water	Ethylene glycol	293	0.18
Water	Glycerol	293	0.0083
Carbon dioxide	Heptane	298	6.03
Carbon tetrachloride	*n*-Hexane	298	3.70
Toluene	*n*-Hexane	298	4.21
Carbon dioxide	Kerosene	298	2.50
Tin	Mercury	303	1.60
Water	*n*-Propanol	288	0.87
Water	1,2-Propylene glycol	293	0.0075
Acetic acid	Toluene	298	2.26
Acetone	Toluene	293	2.93
Benzoic acid	Toluene	293	1.74
Chlorobenzene	Toluene	293	2.06
Ethanol	Toluene	288	3.00
Carbon dioxide	White spirit	298	2.11

Concentrated nonelectrolyte solutions

Diffusion coefficients for liquids have been found to be strongly dependent on concentration. Since concentrated solutions are often nonideal, we might expect that the nonideal behavior is partially responsible for the strong concentration effects on the diffusivity. For binary systems, the concentration dependent diffusivity can be expressed as

$$D_{AB} = D^{\circ}_{AB}\frac{\partial \log a_A}{\partial \log c_A} = D^{\circ}_{AB}\left(1 + \frac{\partial \log \gamma_A}{\partial \log c_A}\right) \tag{2-27}$$

where a_A is the activity, D°_{AB} is the concentration independent diffusion coefficient, and γ_A is the activity coefficient. Using the equation above as a basis, Vignes (1966) proposed an empirical equation that provided a good fit to experimental data for both ideal and nonideal solutions.

$$D_{AB} = (D^{\circ}_{AB})^{x_B}(D^{\circ}_{BA})^{x_A}\left(1 + \frac{\partial \log \gamma_A}{\partial \log x_A}\right) \tag{2-28}$$

From the development of the Stokes–Einstein equation it was shown that the mobility is related to viscosity. Thus it might be concluded that the concentration dependence of the mobility also contributes to how the diffusivity changes with concentration. Leffler and Cullinan (1970a) modified the equation of Vignes by using the viscosities of the mixture to obtain a better correlation with experimental data. Their equation is

$$D_{AB}\mu_m = (D^\circ_{AB}\mu_B)^{x_B}(D^\circ_{BA}\mu_A)^{x_A}\left(1 + \frac{\partial \ln \gamma_A}{\partial \ln x_A}\right) \tag{2-29}$$

When viscosity data are available, the equation of Leffler and Cullinan is recommended for predicting diffusivities in concentrated solutions.

Example 2.4

Using the relationship proposed by Vignes to account for the effects of concentration on the diffusivity, predict the diffusivity of ethanol–water for concentrations up to 50% ethanol at 25°C.

Solution:

$$D^\circ_{EW} = 1.28 \times 10^{-9}\text{m}^2/\text{s} \qquad D^\circ_{WE} = 1.132 \times 10^{-9}\text{ m}^2/\text{s}$$

Activity coefficients for ethanol must be known as a function of composition in order to use the equation of Vignes. If experimental values are not available, they can be calculated from either the van Laar or the two- or three-suffix Margules equation. Using the van Laar equation and constants obtained from the *Chemical Engineers' Handbook*, we get

$$\ln \gamma_1 = \frac{A_{12}}{[1 + (A_{12}x_1/A_{21}x_2)]^2} = \frac{0.67}{[1 + (0.67)(x_1)/(0.42)(x_2)]^2}$$

x_1	$\ln \gamma_1$	γ_1	$\partial \ln \gamma_1/\partial \ln x_1$ (Figure 2-2)
0.1	0.483	1.621	−0.246
0.2	0.342	1.408	—
0.3	0.236	1.266	−0.249
0.4	0.157	1.170	—
0.5	0.099	1.104	−0.244
0.6	0.058	1.060	—

Using the equation of Vignes:

At $x_1 = 0.1$ $\quad D_{EW} = (1.28 \times 10^{-9})^{0.9}(1.132 \times 10^{-9})^{0.1}(1 - 0.246)$

$= 9.53 \times 10^{-10}\text{ m}^2/\text{s}$

At $x_1 = 0.3$ $\quad D_{EW} = (1.28 \times 10^{-9})^{0.7}(1.132 \times 10^{-9})^{0.3}(1 - 0.249)$

$= 9.26 \times 10^{-10}\text{ m}^2/\text{s}$

At $x_1 = 0.5$ $\quad D_{EW} = (1.28 \times 10^{-9})^{0.5}(1.132 \times 10^{-9})^{0.5}(1 - 0.244)$

$= 9.10 \times 10^{-10}\text{ m}^2/\text{s}$

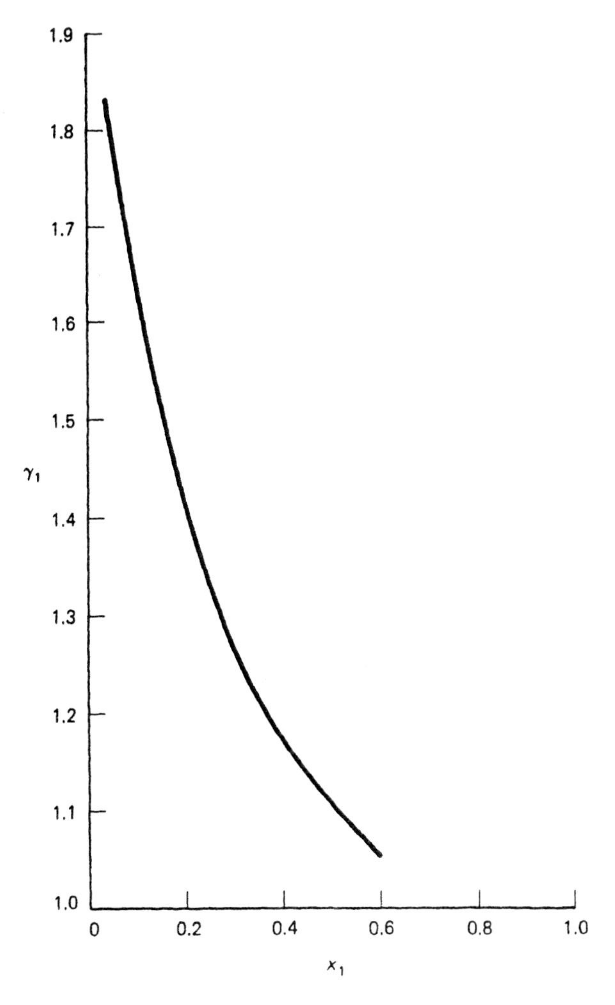

Figure 2-2 Activity coefficient for ethanol in water

Dilute electrolyte solutions

In an electrolyte solution the solute dissociates into cations and anions. Because the size of the ions are different than the original molecule, their mobility through the solvent will also be different. On the basis of previous discussions, the smaller ion might be expected to diffuse faster than the larger ion. However, so that a separation of electric charge does not occur, both ionic species must diffuse at the same rate.

Nernst (1888) developed the first equation for predicting diffusion coefficients in electrolyte solutions by relating the diffusivity to electrical conductivities. His equation, which is valid at infinite dilution, is

$$D^\circ_{AB} = \frac{RT}{F^2} \frac{\lambda^\circ_+ \lambda^\circ_-}{\lambda^\circ_+ + \lambda^\circ_-} \frac{|Z_-| + |Z_+|}{|Z_+ Z_-|} \tag{2-30}$$

$$= 8.931 \times 10^{-14} T \frac{\lambda_+ \lambda^\circ_-}{\lambda^\circ_+ + \lambda^\circ_-} \frac{|Z_-| + |Z_+|}{|Z_+ Z_-|} \tag{2-31}$$

where
F = Faraday constant, A·s/g-equiv.,
D°_{AB} = diffusion coefficients at infinite dilution, m^2/s,
λ°_+ = cationic conductance at infinite dilution, $(A/cm^2)(cm/V)(cm^3/g\text{-equiv.})$,
λ°_- = anionic conductance at infinite dilution, $(A/cm^2)(cm/V)(cm^3/g\text{-equiv.})$,
$\lambda^\circ_+ + \lambda^\circ_-$ = electrolyte conductance at infinite dilution, $(A/cm^2)(cm/V)(cm^3/g\text{-equiv.})$,
Z_+ = cation valence,
Z_- = anion valence,
T = absolute temperature, K.

The equation of Nernst has been verified experimentally for dilute solutions. Selected ionic conductances at infinite dilution are given in Table 2-7 for use with Eq. (2-31). An extensive list of values is given by Robinson and Stokes (1959).

TABLE 2-7. IONIC CONDUCTANCES AT INFINITE DILUTION IN WATER AT 25°C (ROBINSON AND STOKES, 1959)

Cation	λ°_+	Anion	λ°_-
Ag^+	61.9	Br^-	78.4
H^+	349.8	Cl^-	76.35
Li^+	38.7	ClO_3^-	64.6
Na^+	50.1	ClO_4^-	67.4
K^+	73.5	F^-	55.4
NH_4^+	73.6	I^-	76.8
Ca^{2+}	59.5	NO_3^-	71.46
Cu^{2+}	56.6	OH^-	198.6
Mg^{2+}	53.0	CO_3^{2-}	69.3
Zn^{2+}	52.8	SO_4^{2-}	80.0

Liquid metals

Liquid metals may hold the key to describing the liquid state due to their simple structural nature. Self-diffusion coefficients for liquid metals have been calculated using hydrodynamical, quasi-crystalline, and the more recent fluctuation theories. The most accurate phenomenological equation was developed by Walls and Upthegrove (1964), who modified the Stokes–Einstein equation to incorporate a geometric parameter, b, which they defined as the ratio of the atomic radius of a diffusing atom to its interatomic spacing. Their equation for predicting self-diffusion coefficients is

$$D_{AA} = \frac{kT}{2\pi r(2b + 1)\mu} \tag{2-32}$$

From a curve fit of experimental data, Walls and Upthegrove found b to equal 0.419. Their equation provides an excellent correlation of experimental data provided that Pauling's univalent ionic radius is used for r.

In 1959, Swalin (1959) derived a theory for liquid diffusion on the basis of a fluctuation model in which he postulated the atom movement to be the results of local density fluctuations. Using the Morse intermolecular potential to determine the fluctuation energy, Swalin presented an equation which can be reduced to

$$D_{AA} = \frac{k^2 Z T^2}{8hK} \tag{2-33}$$

where K = Waser and Pauling (1950) force constant,
h = Planck's constant,
k = Boltzmann's constant,
Z = number of nearest neighbors.

Swalin set the number of nearest neighbors equal to the number of possible diffusion paths and used a value of 8 for liquid metals. The equation above, with few exceptions, has not generally been successful in predicting diffusion coefficients with acceptable accuracy despite its formulation within the context of small fluctuations as the basic diffusion mechanism.

Since the force constant used above was determined from solid-state data, this sometimes has been thought to be the source of the generally poor agreement between predictions from this equation and experimental data. Hines et al. (1975) have recently shown that by modifying the definition of the number of diffusion paths to be the number of unoccupied sites surrounding a central atom, improved prediction of diffusion data is obtained. Their equation is

$$D_{AA} = \frac{k^2 T^2}{hK} \frac{12 - Z}{Z} \tag{2-34}$$

The most recent fluctuation theory for liquid metals was derived by Hines and Walls (1979) using the basic Einstein equation and fluctuation distances obtained from radial distribution curves. From their model, the diffusion coefficient is expressed as

$$D_{AA} = \frac{1}{6} \bar{j}^2 \nu_s \left(\frac{\sigma_l}{\sigma_s}\right)^{1/2} \frac{T}{T_m} \frac{\mu_m}{\mu} \frac{12 - Z}{2} \tag{2-35}$$

where T = absolute temperature, K,
T_m = melting-point temperature, K,
$\bar{j}$ = mean fluctuation distance, m,
μ = absolute viscosity, kg/m·s,
μ_m = melting-point viscosity, kg/m·s,

σ_l, σ_s = conductivities of the liquid and solid at the melting point, $\Omega^{-1} \cdot m^{-1}$,

ν_s = vibrational frequency of the solid calculated from the Debye temperature, s^{-1}.

Parameters for use in Eqs. (2-33) through (2-35) are given in Table 2-8. Equation (2-35) provides excellent agreement with experimental diffusivities over a wide temperature range with approximately the same accuracy as the Walls and Upthegrove equation.

TABLE 2-8. DATA FOR CALCULATING LIQUID METAL SELF-DIFFUSION COEFFICIENTS

Element	$\bar{j} \times 10^{11}$ (*m*)	*Z*	*Debye Temperature* $\theta_D = h\nu_s/k$	ΔH_F (*kcal/mol*)	$\frac{\sigma_s}{\sigma_l}$	*K* (J/m^2)
Na	6.49	7.6	158	0.622	1.45	4.05
K	7.89	7.7	91	0.562	1.56	2.201
Ag	5.15	8.7	225	2.855	2.09	30.4
Pb	7.32	9.1	105	1.14	1.94	15.31
Sn	3.91	6.0	200	1.67	2.10	33.13
In	5.74	7.8	108	0.78	2.18	12.526
Zn	4.06	6.8	327	1.765	2.24	7.975

Hines and Walls developed an alternative form of their equation for cases in which the electrical conductivities of the solid and liquid at the melting point were not available. The alternative form of their equation is

$$D_{AA} = \frac{1}{6}\bar{j}^2 \nu_s \left[\exp\left(\frac{-\Delta H_F}{RT_m}\right)\right]^{1/3} \frac{T}{T_m} \frac{\mu_m}{\mu} \frac{12 - Z}{2} \tag{2-36}$$

where ΔH_F is the heat of fusion.

Example 2.5

Calculate self-diffusion coefficients for sodium and potassium at their melting-point temperatures using (a) the equation of Swalin, and (b) the equation of Hines and Walls.

Solution:

(a) Swalin equation:

$$D_{AA} = \frac{k^2 Z T^2}{8hK}$$

Sodium:

$$T = 371 \text{ K}$$

$$K = 4.05 \text{ J/m}^2$$

Potassium:

$$T = 336.4 \text{ K}$$

$$K = 2.201 \text{ J/m}^2$$

$$D_{Na} = \frac{(1.38 \times 10^{-23})^2(8)(371)^2}{8(6.62 \times 10^{-34})(4.05)}$$

$$= 9.78 \times 10^{-9} \text{ m}^2/\text{s}$$

$$D_K = \frac{(1.38 \times 10^{-23})^2(8)(336.4)^2}{8(6.62 \times 10^{-34})(2.201)}$$

$$= 14.8 \times 10^{-9} \text{ m}^2/\text{s}$$

(b) Hines–Walls equation:
Sodium: From Table 2-8,

$$\bar{j} = 6.49 \times 10^{-11} \text{ m}$$

$$Z = 7.6$$

$$\nu_s = \frac{\theta_D k}{h} = \frac{158(1.38 \times 10^{-23})}{6.62 \times 10^{-34}} = 3.294 \times 10^{12} \text{ s}^{-1}$$

$$D_{Na} = \frac{1}{6}(6.49 \times 10^{-11})^2(3.294 \times 10^{12})\left(\frac{1}{1.45}\right)^{0.5}\left(\frac{12 - 7.6}{2}\right)$$

$$= 4.22 \times 10^{-9} \text{ m}^2/\text{s}$$

Potassium:

$$\bar{j} = 7.89 \times 10^{-11} \text{ m}$$

$$Z = 7.7$$

$$\nu_s = \frac{91(1.38 \times 10^{-23})}{6.62 \times 10^{-34}} = 1.897 \times 10^{12} \text{ s}^{-1}$$

$$D_K = \frac{1}{6}(7.89 \times 10^{-11})^2(1.897 \times 10^{12})\left(\frac{1}{1.56}\right)^{0.5}\left(\frac{12 - 7.7}{2}\right)$$

$$= 3.39 \times 10^{-9} \text{ m}^2/\text{s}$$

In Example 2.3, the diffusion coefficient was changed from one temperature to another by using the relationship

$$\left(\frac{D^\circ_{AB}\mu}{T}\right)_{T_2} = \left(\frac{D^\circ_{AB}\mu}{T}\right)_{T_1} \tag{2-37}$$

Because of the presence of viscosity, the expression above provides little insight into the actual effects of temperature on the diffusivity. However, over moderate temperature ranges, diffusion coefficients can be related to temperature with the conventional Arrhenius expression

$$D_{AB} = D_0 \exp\left(\frac{-Q}{RT}\right) \tag{2-38}$$

where the preexponential or frequency factor, D_0, typically has the units m²/s and the activation energy, Q, has the units energy/mol. The activation energy may be visualized simply as the energy necessary to raise an atom over a barrier. For liquids this may be interpreted as the energy required to break one-half of the bonds between a diffusing molecule and its nearest neighbors. Self-diffusion data for elemental liquid metals are presented in Table 2-9 in terms of their

TABLE 2-9. SELF-DIFFUSION DATA FOR LIQUID METAL ELEMENTS[a]

Element	$D_0 \times 10^8$ (m^2/s)	$Q \times 10^{-6}$ ($J/kgmol$)	Fusion Temperature (K)
Na	11.0	10.25	371.0
K	16.7	10.67	336.4
Ag	5.8	32.47	1234.0
Pb	2.369	13.04	600.6
Sn	3.24	11.55	505.0
In	2.89	10.17	429.3
Zn	8.2	21.30	692.7

[a] Data taken from Walls (1970).

preexponential factors and activation energies. Experimentally, D_0 and Q can be found by plotting ln D versus $1/T$. The slope of the graph gives the activation energy as

$$\frac{d \ln D}{d(1/T)} = -\frac{Q}{R} \tag{2-39}$$

The frequency factor is the intercept at $1/T = 0$.

2.4 Diffusion Coefficients for Multicomponent Systems

An effective diffusion coefficient in multicomponent gas systems can be accurately predicted by extending the Stefan–Maxwell equation as shown. For a binary system in which component A diffuses through component B, Maxwell assumed that the difference in the partial pressure of A was proportional to:

1. The molar concentration of A and B
2. The length of the diffusion path, dZ
3. The relative velocities, U_{AZ} and U_{BZ}

Therefore at constant system temperature and pressure,

$$-\frac{d\bar{P}_A}{dZ} = K'C_AC_B(U_{AZ} - U_{BZ}) \tag{2-40}$$

or

$$-\frac{d\bar{P}_A}{dZ} = K'C_AC_BU_{AZ} - K'C_AC_BU_{BZ} \tag{2-41}$$

Using Eq. (1-9) for both components, we have

$$-\frac{d\bar{P}_A}{dZ} = K'C_BN_{AZ} - K'C_AN_{BZ} \tag{2-42}$$

For a binary ideal gas, $C = C_A + C_B$ and $\bar{P}_A = C_A RT$; thus

$$-RT \frac{dC_A}{dZ} = K'CN_{AZ} - K'C_A N_{AZ} - K'C_A N_{BZ} \tag{2-43}$$

or

$$-RT \frac{dC_A}{dZ} = -K'C_A(N_{AZ} + N_{BZ}) + K'CN_{AZ} \tag{2-44}$$

Rearranging and writing the equation above in terms of the molar flux, N_{AZ}, gives

$$N_{AZ} = -\frac{RT}{K'C}\frac{dC_A}{dZ} + \frac{C_A}{C}(N_{AZ} + N_{BZ}) \tag{2-45}$$

Since $C = P/RT$, Eq. (2-45) becomes

$$N_{AZ} = -\frac{R^2T^2}{K'P}\frac{dC_A}{dZ} + x_A(N_{AZ} + N_{BZ}) \tag{2-46}$$

Comparison of Eqs. (1-34) and (2-46) shows that the diffusion coefficient is

$$D_{AB} = \frac{R^2T^2}{K'P} = \frac{RT}{CK'} \tag{2-47}$$

The derivation above for a binary system can be generalized for multicomponent mixtures by writing Eq. (2-42) for all components i and j. This becomes

$$-RT \frac{dC_i}{dZ} = \sum_{j=1}^{n} K'(C_j N_{iZ} - C_i N_{jZ}) \tag{2-48}$$

which gives

$$-\frac{dx_i}{dZ} = \sum_{j=1}^{n} \frac{K'}{RT}(x_j N_{iZ} - x_i N_{jZ}) \tag{2-49}$$

Upon defining K' in terms of the diffusion coefficient as shown in Eq. (2-47),

$$-\frac{dx_i}{dZ} = \sum_{j=1}^{n} \frac{1}{CD_{ij}}(x_j N_{iZ} - x_i N_{jZ}) \tag{2-50}$$

When written in general notation, Eq. (2-50) is

$$\nabla x_i = \sum_{j=1}^{n} \frac{1}{CD_{ij}}(x_i \mathbf{N}_j - x_j \mathbf{N}_i) \tag{2-51}$$

The binary flux equation, Eq. (1-34), can be written for the transfer of component i through a multicomponent mixture as

$$\mathbf{N}_i = -CD_{im}\nabla x_i + x_i \sum_{j=1}^{n} \mathbf{N}_j \tag{2-52}$$

Combining Eqs. (2-51) and (2-52) gives the definition of the diffusion coefficient for i through the mixture.

$$D_{im} = -\frac{\mathbf{N}_i - x_i \sum_{j=1}^{n} \mathbf{N}_j}{\sum_{j=1}^{n} \frac{x_i \mathbf{N}_j - x_j \mathbf{N}_i}{D_{ij}}} \tag{2-53}$$

From the equation above, D_{im} can be seen to vary with composition and as a result is difficult to use. An approximate equation that can be used to obtain multicomponent diffusion coefficients results if we assume that i is diffusing through a stagnant mixture. Writing component A for i, we get

$$D_{Am} = -\frac{\mathbf{N}_A - x_A\mathbf{N}_A - x_A\sum_{j=2}^{n}\mathbf{N}_j}{\sum_{j=2}^{n}\frac{x_A\mathbf{N}_j - x_j\mathbf{N}_A}{D_{Aj}}} \quad \text{where } j \neq A \tag{2-54}$$

For a stagnant mixture $\mathbf{N}_j = 0$. Thus we have

$$D_{Am} = \frac{1 - x_A}{\sum_{j=2}^{n}\frac{x_j}{D_{Aj}}} \tag{2-55}$$

Expansion of Eq. (2-55) for the diffusion of component A through a stagnant mixture of gases A, B, C, and D becomes

$$D_{Am} = \frac{1 - x_A}{x_B/D_{AB} + x_C/D_{AC} + x_D/D_{AD}} \tag{2-56}$$

Example 2.6

Calculate the effective diffusivity for carbon dioxide at 25°C and 1 atm in a mixture of nondiffusing gases in which $x_{CO_2} = 0.2$, $x_{N_2} = 0.5$, and $x_{He} = 0.3$.

Solution:

From Table 2-5,

$$D_{CO_2-He} = 6.12 \times 10^{-5} \quad \text{and} \quad D_{CO_2-N_2} = 1.67 \times 10^{-5}$$

Therefore,

$$D_{CO_2-m} = \frac{1 - 0.2}{\frac{0.5}{1.67 \times 10^{-5}} + \frac{0.3}{6.12 \times 10^{-5}}} = 2.3 \times 10^{-5}\ \text{m}^2/\text{s}$$

Although effective diffusivities can be predicted for multicomponent gas mixtures with reasonable accuracy, the prediction of liquid diffusivities in multicomponent systems has been less successful, particularly for concentrated solutions that depart significantly from ideality. Upon considering the difficulty of predicting binary diffusivities in concentrated systems, this is not surprising. Diffusion coefficients in a ternary liquid system in which a dilute solute A diffuses through a concentrated mixture of solvents B and C can be estimated by an expression suggested by Leffler and Cullinan (1970b):

$$D^\circ_{Am}\mu_{ABC} = (D^\circ_{AB}\mu_B)^{x_B}(D^\circ_{AC}\mu_C)^{x_C} \tag{2-57}$$

where μ_{ABC} is the viscosity of the solution. For a detailed discussion of multicomponent diffusion, the reader is referred to the book by Cussler (1976).

2.5 Diffusion in Solids

Transport in crystals

In an attempt to describe the diffusional process in solids, investigators have proposed a variety of mechanisms that depended on the structure of the solid and the nature of the process.

For crystalline solids, the simplest mechanism is the exchange mechanism in which two atoms exchange positions in the lattice. In crystals with a closely packed structure, this mechanism, as shown in Figure 2-3a, is unlikely due to the nearness of neighboring atoms. Diffusion by this mechanism would distort the crystalline structure and require a large activation energy. The ring mechanism shown in Figure 2-3b is a variation of the direct exchange mechanism. Although less distortion in the lattice is necessary, diffusion by this method is also very unlikely.

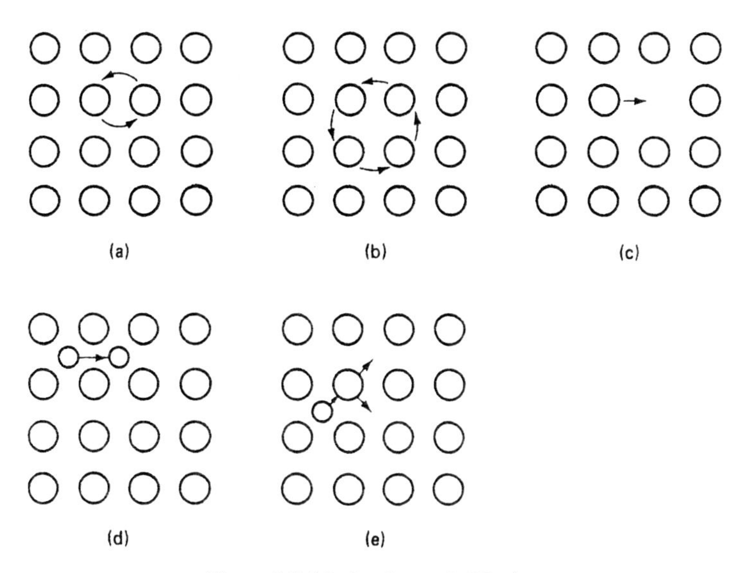

Figure 2-3 Mechanisms of diffusion

One of the more favorable diffusion mechanisms involves the presence of vacancies on regular lattice sites as seen in Figure 2-3c. Transfer by this mechanism occurs by an atom moving from its original position to the vacant site. The vacated position thus is available as a site for another diffusing atom. As might be expected, a large activation energy is not required by this mechanism. Other methods by which diffusion is likely to occur are the interstitial and interstitialcy mechanisms presented in Figures 2-3d and e. The *interstitial mechanism* is usually associated with the diffusion of small solute atoms through the lattice of the larger solvent atoms and is characterized by a low activation energy. The *interstitialcy mechanism* occurs when the solute and solvent atoms are approximately equal in size. This mechanism takes place by one atom moving from an interstitial position to a lattice site and thereby forcing the atom originally on that site to another interstitial position.

The mechanisms above describe atom transport in perfect crystals. However, for real solids diffusion may also occur along a high diffusivity path such as a grain boundary or along a surface. Diffusion coefficients for these mechanisms are significantly larger than the values found for diffusion by the other mechanisms. Diffusion coefficients in crystalline solids range from about 10^{-10} m^2/s to as low as 10^{-37} m^2/s. However, it is not uncommon for diffusion along a path of high diffusivity to be four or five orders of magnitude greater than diffusion in a perfect crystal.

When an atom moves from one site to another, an activation energy is required to overcome the energy barrier associated with breaking the interatomic bonds between the diffusing atom and those surrounding it. As opposed to liquids, however, diffusion coefficients for solids have been shown empirically to obey an Arrhenius relationship over a wide temperature range. On this basis, Sherby and Simnad (1961) developed an empirical equation for predicting self-diffusion coefficients in solids. From the plot of log D versus T_m/T shown in Figure 2-4, they found that most solid diffusion data fell into three distinct

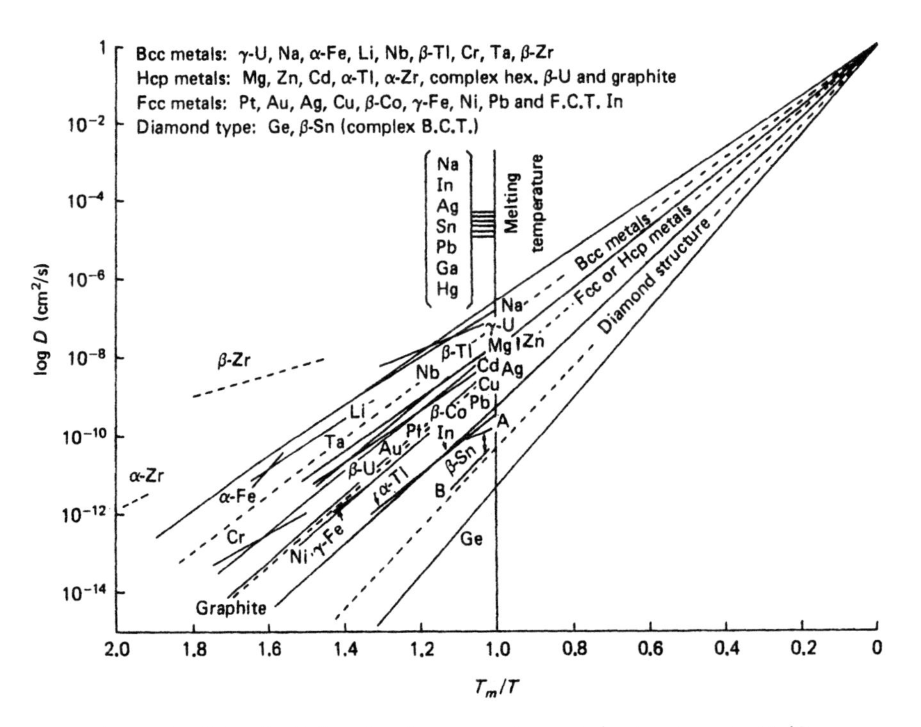

Figure 2-4 Rate of self-diffusion for pure melts plotted as log D versus T_m/T (from Sherby and Simnad, 1961)

groups: (1) body-centered cubic (bcc), (2) face-centered cubic (fcc) and hexagonal closed-packed (hcp), and (3) diamond structure. The resulting equation proposed as a result of their work is given as

$$D_{AA} = D_0 \exp\left[-(K_0 + V)\frac{T_m}{T}\right] \tag{2-58}$$

where V is the valence; T_m is the melting-point temperature, K; and K_0 is 14 for bcc, 17 for fcc and hcp metals, and 21 for the diamond structure.

In the equation above, the activation energy is equal to $(K_0 + V)RT_m$. From the correlation of activation energy for self-diffusion data versus absolute temperature shown in Figure 2-5, the activation energy can be related to tem-

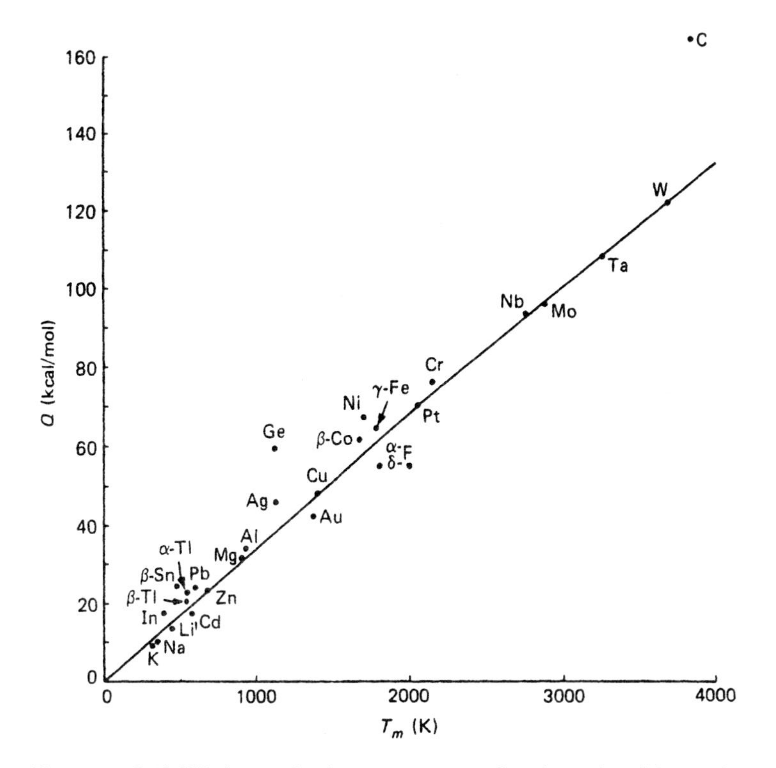

Figure 2-5 Self-diffusion activation energy as a function of melting point temperature (from Askill, 1970)

perature by the expression $Q = AT_m$. From a study of data for all types of crystal structures, the coefficient is $A = 33.7$ cal/mol·K. For fcc and bcc structures the activation energy is better correlated with A values of 38.0 cal/mol·K and 32.5 cal/mol·K, respectively. An extrapolation of the diffusion data in Figure 2-4 to $T_m/T = 0$ gives an approximate value for the frequency factor of

$D_0 = 1.0\ cm^2/s$ ($1.0 \times 10^{-4}\ m^2/s$). In the absence of any data, the frequency factor above may be used in Eq. (2-58) to obtain a reasonable estimate of the self-diffusion coefficient.

Selected diffusion data are presented in Table 2-10. An extensive tabulation of diffusion data for metals, alloys, and simple oxides is given by Askill (1970). Comprehensive discussions of diffusion in solids are given by Barrer (1941), Manning (1968), and Shewmon (1963).

TABLE 2-10. DIFFUSION DATA FOR SOLIDS[a]

Solute	*Solvent*	$D_0 \times 10^4$ (m^2/s)	$Q \times 10^{-8}$ $(J/kgmol)$	T_m *(solvent)* (°C)
Cu	Cu (fcc)	0.33	2.02	1083
Au	Au (fcc)	0.15	1.91	1063
Mo	Mo (bcc)	0.5	4.05	2625
Ni	Ni (fcc)	1.9	2.79	1455
Ag	Ag (fcc)	0.34	1.82	960.5
Na	Na (bcc)	0.145	0.42	97.5
Zn	Zn (hcp)	0.031	0.86	419.46
Ag	Al (fcc)	0.08	1.17	660.2
Au	Ag (fcc)	0.41	1.94	960.5
Cu	Ag (fcc)	1.23	1.93	960.5
Hg	Ag (fcc)	0.079	1.60	960.5
C	Co (hcp)	0.21	1.42	1495
Ni	Co (hcp)	3.35	2.97	1495

[a]Data taken from Askill (1970).

Porous solids

This section is included to provide an introduction to the subject of diffusion in porous materials because of its importance in the study of catalysis and reaction in solids. Detailed discussions are presented by Satterfield (1970) and by Smith (1970).

Diffusion through porous materials is typically described as either ordinary, Knudsen, or surface diffusion and has been found to play an important role in catalyzed reactions. *Ordinary diffusion* occurs when the pore diameter of the material is large in comparison to the mean free path of the molecules of the gas. Molecular transport through pores which are small in comparison to the mean free path of the gas is described as *Knudsen-type diffusion*. Surface diffusion is the third type of mechanism for molecular transport in porous materials and has been found to be the most difficult to characterize. In *surface diffusion*, molecules are adsorbed on the surface of the material and are subsequently transported from one site to another in the direction of decreasing concentration. Since for many cases surface diffusion has been assumed to contribute little to the overall transport, it will not be discussed here.

For Knudsen diffusion, molecules collide more often with the pore walls than with other molecules. Upon collision, the atoms are instantly adsorbed on the surface and then are desorbed in a diffuse manner. As a result of the frequent collisions with the wall of the pore, the transport of the molecule is impeded. The Knudsen diffusion coefficient can be predicted from kinetic theory by relating the diameter of the pore and the mean free path of the gas by the expression

$$D_{A,K} = \frac{\bar{U}d}{3} \tag{2-59}$$

where $\bar{U}$ is the velocity of the gas molecule and d is the pore diameter. For straight, round pores the diffusivity becomes

$$D_{A,K} = 97.0r\left(\frac{T}{M_A}\right)^{1/2} \tag{2-60}$$

where r = pore radius, m,
T = temperature, K,
$D_{A,K}$ = Knudsen diffusion coefficient, m^2/s,
M_A = molecular weight of component A.

The mean pore radius can be evaluated from the bulk density, the surface area of the porous solid, and the porosity by the expression

$$r = \frac{2\epsilon}{S\rho_B} = \frac{2V_p}{S} \tag{2-61}$$

where S = surface area of the porous solid, m^2/kg,
ρ_B = bulk density of the solid particle, kg/m^3,
V_p = specific pore volume of a solid particle, m^3/kg,
ϵ = porosity of the solid.

In order to account for the tortuous path of the molecule and the porosity of the material, an effective Knudsen diffusivity may be expressed as

$$D_{A,K,e} = D_{A,K}\frac{\epsilon}{\tau} \tag{2-62}$$

where τ is the tortuosity factor related to the path of the molecule. Bulk diffusion also contributes to the total molecular transport through the pores, but it also is decreased because of the porosity of the particle and the length of the path along which the molecule travels. The effective diffusivity can be described in terms of the ordinary diffusion coefficient by the equation

$$D_{AB_e} = D_{AB}\frac{\epsilon}{\tau} \tag{2-63}$$

Obvious cases exist where bulk and Knudsen diffusion both contribute to the effective diffusion coefficient. For self-diffusion or equimolar counter transfer, the effective diffusivity in a porous material can be written as

$$\frac{1}{D_{A,e}} = \frac{1}{D_{A,K,e}} + \frac{1}{D_{AB,e}} \tag{2-64}$$

In porous solids, in which transfer takes place primarily by Knudsen diffusion, the effective diffusivity can be used to estimate the tortuosity. A number of these systems have been studied experimentally and are shown in Table 2-11. An extensive list of values is presented by Satterfield (1970). Although ordinary diffusion coefficients for gases are influenced significantly by pressure, Knudsen diffusion coefficients are independent of pressure as shown by Eq. (2-60). The extent to which the effective diffusivity depends on pressure is a function of the relative magnitudes of the bulk and Knudsen diffusivities. Typical Knudsen diffusion coefficients may be an order of magnitude smaller than bulk coefficients for gases, particularly at low pressures.

TABLE 2-11. TRANSFER IN POROUS SOLIDS

Material	*Gases*	$T(K)$	$r \times 10^{10}$ (m)	τ	ϵ
Alumina pellets: Henry et al. (1961)	N_2, He, CO_2	303	96	0.85	0.812
Silica gel: Schneider and Smith (1968)	C_2H_6	323–473	11	3.35	0.486
Silica-alumina cracking catalyst: Barrer and Gabor (1959)	He, Ne, Ar, N_2	273–323	16	0.725	0.40
Vycor glass: Gilliland et al. (1958)	H_2, He, Ar, N_2	298	30.6	5.9	0.31

Example 2.7

Calculate the effective Knudsen diffusivity of sulfur dioxide in activated carbon at 20°C. The activated carbon has a surface area of 700 × 10³ m²/kg, a porosity of 0.5, a particle density of 1.2 × 10³ kg/m³, and a tortuosity of 2.

Solution:

The Knudsen diffusivity is obtained from Eq. (2-60):

$$D_{A,K} = 97.0r\left(\frac{T}{M_A}\right)^{1/2}$$

Thus the effective diffusivity is

$$D_{A,K,e} = 97.0\frac{r\epsilon}{\tau}\left(\frac{T}{M_A}\right)^{1/2}$$

The radius of the pore is found from Eq. (2-61). Thus the effective diffusivity becomes

$$D_{A,K,e} = 194.0\frac{\epsilon^2}{\tau S \rho_B}\left(\frac{T}{M_A}\right)^{1/2}$$

$$= 194.0\frac{(0.5)^2}{2(700 \times 10^3)(1.2 \times 10^3)}\left(\frac{293}{64.06}\right)^{1/2}$$

$$= 6.18 \times 10^{-8}\ \mathrm{m^2/s}$$

2.6 Phenomenological and Coupled Coefficients

For a system at uniform pressure with no external forces, Haase (1969) presented a general expression similar to Eq. (1-36) that relates the molar flux to gradients in concentration and temperature as

$$\mathbf{J}_i = -\frac{1-\chi_i}{1-x_i}\frac{1}{V}\left(D_{ij}\,\nabla x_i \pm D_T\frac{\nabla T}{T}\right) \tag{2-65}$$

where D_T is defined as the thermal diffusion coefficient. The upper sign is used when $i = 1$ and the lower sign is for $i = 2$. The thermal diffusion coefficient is frequently written as the thermal diffusion ratio given by

$$k_T = \frac{D_T}{D_{12}} \tag{2-66}$$

or the thermal diffusion factor $\alpha_T = k_T/x_1x_2$. The Soret coefficient is defined in terms of the thermal diffusion factor by the equation

$$\gamma_s = \frac{\alpha_T}{T} \tag{2-67}$$

The coupled coefficients can be related to the phenomenological coefficients by expanding Eq. (1-53) for a two-component system. Considering heat transfer by conduction only, we have

$$T\sigma = -\frac{J_H}{T}\frac{dT}{dZ} - J_1\frac{d\mu_1}{dZ} - J_2\frac{d\mu_2}{dZ} \tag{2-68}$$

Since $\sum_{i=1}^{n} J_i = 0$, we have $J_1 = -J_2$. The equation above can be written as

$$T\sigma = -\frac{J_H}{T}\frac{dT}{dZ} - J_1\frac{d(\mu_1 - \mu_2)}{dZ} \tag{2-69}$$

Introducing the Gibbs–Duhem equation for a binary system, the chemical potential of component 2 can be eliminated. Thus

$$T\sigma = -\frac{J_H}{T}\frac{dT}{dZ} - J_1\frac{d\mu_1}{dZ}\left(1 + \frac{x_1}{x_2}\right) \tag{2-70}$$

or

$$T\sigma = -\frac{J_H}{T}\frac{dT}{dZ} - J_1\frac{1}{x_2}\frac{d\mu_1}{dx_1}\frac{dx_1}{dZ} \tag{2-71}$$

By expanding Eq. (1-37) for heat and mass transfer, we obtain

$$\mathbf{J}_H = L_{HH}\mathbf{X}_H + L_{H1}\mathbf{X}_1 \tag{2-72}$$

and

$$\mathbf{J}_1 = L_{1H}\mathbf{X}_H + L_{11}\mathbf{X}_1 \tag{2-73}$$

Upon introducing Eqs. (1-54) and (1-55) into the equations above, the flux expressions become

$$J_H = -\frac{L_{HH}}{T}\frac{dT}{dZ} - \frac{L_{H1}}{x_2}\frac{d\mu_1}{dx_1}\frac{dx_1}{dZ} \tag{2-74}$$

and

$$J_1 = -\frac{L_{1H}}{T}\frac{dT}{dZ} - \frac{L_{11}}{x_2}\frac{d\mu_1}{dx_1}\frac{dx_1}{dZ} \tag{2-75}$$

To obtain the relationships between the phenomenological and coupled coefficients, we expand the general equation proposed by Haase for the molar flux as

$$J_1 = -CD_{12}\frac{dx_1}{dZ} - \frac{CD_T}{T}\frac{dT}{dZ} \tag{2-76}$$

$$= -CD_{12}\left(\frac{dx_1}{dZ} - \frac{k_T}{T}\frac{dT}{dZ}\right) \tag{2-77}$$

A comparison of Eqs. (2-75) and (2-76) gives

$$\frac{L_{1H}}{C} = D_T \quad \text{(thermal diffusion coefficient)} \tag{2-78}$$

and

$$\frac{L_{11}}{Cx_2}\frac{d\mu_1}{dx_1} = D_{12} \quad \text{(mass diffusivity)} \tag{2-79}$$

The Onsager coefficients given in Eq. (2-74) can be identified by writing an extended form of Fourier's law to include heat transfer due to the presence of a concentration gradient.

$$J_H = -k\frac{dT}{dZ} - \beta_T\frac{dx_1}{dZ} \tag{2-80}$$

From a comparison of Eqs. (2-74) and (2-80), we see that

$$\frac{L_{HH}}{T} = k \quad \text{(thermal conductivity)} \tag{2-81}$$

and

$$\frac{L_{H1}}{x_2}\frac{d\mu_1}{dx_1} = \beta_T \quad \text{(Dufour coefficient)} \tag{2-82}$$

The thermal diffusion factor, α_T, is usually employed to characterize thermal diffusion in binary gas mixtures. This quantity is dimensionless and less concentration dependent than the thermal diffusion coefficient. It has been shown experimentally that even for the same binary system, α_T is a function of temperature, pressure, and concentration. Although it can be either positive or negative, a positive value for α_T indicates enrichment of the lighter component in the warmer region.

For liquids, a thermal diffusion factor greater than zero again means enrichment of the lighter component in the warmer region. When one liquid is associated, α_T usually changes sign at some concentration. Typical values of the thermal diffusion ratio are shown in Table 2-12.

TABLE 2-12. THERMAL DIFFUSION RATIOS FOR GASES AND LIQUIDS[a]

Gases	$\bar{T}\,(K)$	x_1	α_T[b]
CH_4–N_2	360	0.5	0.10
N_2–CO_2	360	0.5	0.04
H_2–CO_2	360	0.5	0.31
H_2–N_2	360	0.5	0.36
C_6H_6–CCl_4	313	0.2	1.37
C_6H_6–CCl_4	313	0.5	1.43
C_6H_6–CCl_4	313	0.8	1.48
Sn–Bi	568	0.5	0.10
Sn–Zn	648	0.5	4.10

[a]Data are taken from Haase (1969), Tichacek et al. (1956), and Winter and Drickamer (1955).
[b]α_T is positive when the lighter component accumulates in the warmer region.

2.7 Transport Coefficients in Membranes

Membranes have been used for many years in the biomedical area and have gained wide acceptance. Their use in the purification of blood in artificial kidney machines and for blood oxygenation in heart–lung machines are but two typical examples. In recent years, membrane processes have gained increased acceptance as a means for carrying out a number of separation processes that are of commercial importance. Specific examples are: the demineralization of water (e.g., reverse osmosis), purification of industrial effluent streams, the recovery of heavy metals, and gas and liquid separations.

For our purposes, a membrane will be defined as a thin barrier that separates two fluids. In contrast to the porous solids previously discussed, convective effects are unimportant and mass transfer through the membrane will take place by diffusion only. Although most membranes are typically made of long-chain polymers, a detailed description is beyond the scope of this treatment. Comprehensive discussions of membranes and membrane separation processes are presented by Tuwiner (1962), Crank and Park (1968), Kotyk and Janáček (1975), and Hwang and Kammermeyer (1975). Since membranes may be either solid or liquid, the actual transport mechanism is frequently described by the same models that were presented earlier in this chapter. Consequently, these models will not be reintroduced here.

Let us consider the transport of solute A across the membrane shown in Figure 2-6. By using Fick's law, we can write the diffusional flux across the

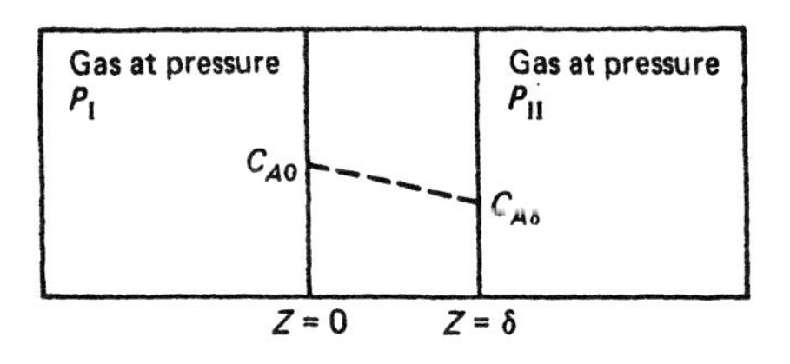

Figure 2-6 Transport across a membrane

membrane as

$$N_A = -D_A \frac{dC_A}{dZ} \tag{2-83}$$

Upon integrating and introducing the concentrations of A on the surfaces of the membrane, we obtain

$$N_A = \frac{D_A(C_{A0} - C_{A\delta})}{\delta} \tag{2-84}$$

The diffusion coefficient can be readily determined if values for N_A, C_{A0}, $C_{A\delta}$, and δ are available for a given process. Although frequently we will not know the concentrations of A on the surface of the membrane, we can use the partial pressures of A in the gas on either side of the membrane and rewrite Eq. (2-84) as

$$N_A = \frac{P_M(\bar{P}_{A\text{I}} - \bar{P}_{A\text{II}})}{\delta} \tag{2-85}$$

where $\bar{P}_{A\text{I}}$ and $\bar{P}_{A\text{II}}$ are the partial pressures of A in the gas and P_M is defined as the permeability constant. If the pressure of A on either side of the membrane can be related to the concentration of A inside the membrane by a linear relationship, we can write the permeability as

$$P_M = D_A S \tag{2-86}$$

where S is the solubility. In the equation above the permeability is defined as the cm^3 of gas at 0°C and 1 atm pressure that diffuses per second through a membrane 1 cm^2 in area and 1 cm thick due to a pressure gradient of 1 atm. This gives dimensions of (cm^3 solute)/[(cm^2 surface·s) (atm/cm thickness)]. Another method of defining the permeability is in terms of the diffusion across a 1-mm-thick membrane due to a pressure gradient of 1 cmHg($\frac{1}{76}$ atm). If the latter definition is used, Eq. (2-86) must be divided by 7.6. For this case the units for permeability are (cm^3 solute)/[(cm^2 surface·s)(cmHg/mm thickness)]. Permeabilities and diffusivities were obtained by Heilman et al. (1956) for the transfer of hydrogen sulfide through several membranes; these data are shown in Table 2-13.

The diffusion coefficient can be related to temperature by the Arrhenius expression given in Eq. (2-38):

$$D_A = D_0 \exp\left(\frac{-Q}{RT}\right) \tag{2-38}$$

TABLE 2-13. PERMEABILITY, DIFFUSIVITY, AND SOLUBILITY OF HYDROGEN SULFIDE IN POLYMER MEMBRANES

Film	Temperature (°C)	Pressure (mmHg)	P_M $\left[\frac{cm^3(NTP)}{cm^2 \cdot s(atm/cm)}\right]$	D $\left(\frac{cm^2}{s}\right)$	S $\left(\frac{cm^3 A}{cm^3 \cdot atm}\right)$
Nylon	30	110	2.37×10^{-9}	3.0×10^{-10}	7.9
		153	2.34×10^{-9}	6.0×10^{-10}	3.9
		226	2.50×10^{-9}	3.9×10^{-10}	6.4
		270	2.70×10^{-9}	5.4×10^{-10}	5.0
		404	2.61×10^{-9}	4.5×10^{-10}	5.8
		621	2.60×10^{-9}	4.9×10^{-10}	5.3
		707	2.60×10^{-9}	5.0×10^{-10}	5.2
	45	741	7.52×10^{-9}	1.6×10^{-9}	4.7
	60	670	1.80×10^{-8}	6.0×10^{-9}	3.0
		720	1.86×10^{-8}	5.3×10^{-9}	3.5
	75	603	3.30×10^{-8}	1.5×10^{-8}	2.2
		651	4.59×10^{-8}	1.7×10^{-8}	2.7
	80	688	6.00×10^{-8}	2.0×10^{-8}	3.0
Plasticized cellulose acetate (15% dibutyl phthalate)	0	711	1.65×10^{-8}	7.5×10^{-10}	22.0
	30	68	4.62×10^{-8}	2.2×10^{-9}	21.0
		151	4.20×10^{-8}	2.4×10^{-9}	17.5
		355	4.06×10^{-8}	2.8×10^{-9}	14.5
		728	4.42×10^{-8}	3.4×10^{-9}	13.0
	45	742	5.83×10^{-8}	7.2×10^{-9}	8.1
	60	728	9.07×10^{-8}	14.4×10^{-9}	6.3
Unplasticized cellulose acetate	0	716	9.99×10^{-9}	3.7×10^{-10}	27.0
	15	704	1.55×10^{-8}	7.4×10^{-10}	21.0
	30	176	2.70×10^{-9}	1.0×10^{-10}	27.0
		462	2.38×10^{-9}	1.4×10^{-10}	17.0
		703	2.56×10^{-9}	1.6×10^{-10}	16.0
	45	735	3.90×10^{-9}	3.0×10^{-10}	13.0
	60	695	4.64×10^{-9}	5.1×10^{-10}	9.1
Poly(vinyl butyral)	0	758	2.25×10^{-8}	1.5×10^{-9}	15.0
	15	716	3.40×10^{-8}	3.4×10^{-9}	10.0
	30	30	7.92×10^{-8}	4.4×10^{-9}	18.0
		156	4.84×10^{-8}	4.4×10^{-9}	11.0
		332	4.95×10^{-9}	5.5×10^{-9}	9.0
		641	5.04×10^{-8}	6.3×10^{-9}	8.0
Poly(vinyl trifluoroacetate)	30	244	2.20×10^{-9}	5.5×10^{-9}	0.4
		453	1.96×10^{-9}	4.9×10^{-9}	0.4
		752	2.04×10^{-9}	6.8×10^{-9}	0.3
	45	643	3.00×10^{-9}	1.5×10^{-8}	0.2
	60	653	4.40×10^{-9}	2.2×10^{-8}	0.2
		668	5.80×10^{-9}	2.9×10^{-8}	0.2
Mylar A (DuPont polyester film)	0	715	1.43×10^{-10}	1.3×10^{-11}	11.0
	30	394	5.67×10^{-10}	8.1×10^{-11}	7.0
		694	5.00×10^{-10}	1.0×10^{-10}	5.0
	45	734	1.12×10^{-9}	2.8×10^{-10}	4.0
	60	747	1.74×10^{-9}	6.2×10^{-10}	2.8
Saran	30	397	2.48×10^{-10}	8.0×10^{-11}	3.1
		695	2.69×10^{-10}	9.6×10^{-11}	2.8
	45	734	1.13×10^{-9}	2.9×10^{-10}	3.9
	60	743	4.08×10^{-9}	2.4×10^{-9}	1.7
	75	672	7.80×10^{-9}	6.0×10^{-9}	1.3

Diffusion coefficients for gases in several membranes can be calculated by using Eq. (2-38) and the data presented in Table 2-14.

TABLE 2-14. DIFFUSION DATA FOR MEMBRANES

Polymer	*Gas*	D_0 (cm^2/s)	Q (kcal/gmol)
Polypropylene[a] (isotactic)	He	4.1	7.3
	H_2	2.4	8.3
Polypropylene[a] (atactic)	He	24	7.9
	H_2	15	8.8
Isoprene[b] acrylonitrile 74/26	CO_2	1.15×10^3	14.4
	He	3.1×10^{-2}	4.9
	H_2	0.67	7.4
	N_2	1.88×10^3	14.5
	O_2	70	12.7
Butyl rubber[b]	CO_2	36	12.0
	He	1.5×10^{-2}	5.8
	H_2	1.36	8.1
	N_2	34	12.1
	O_2	43	11.9
Poly(dimethyl[b] butadiene)	CO_2	1.6×10^2	12.8
	H_2	1.3	7.5
	N_2	1.05×10^2	12.4
	O_2	20	11.1
Butadiene[b] acrylonitrile (Perbunan)	CO_2	13.5	10.7
	He	7.7×10^{-2}	5.2
	H_2	0.52	6.9
	N_2	10.7	10.4
	O_2	2.4	9.2
Poly(methyl[b] pentadiene)[b]	N_2	42	11.1
	O_2	8.5	9.8
Polybutadiene[b]	CO_2	0.24	7.3
	H_2	5.3×10^{-2}	5.1
	N_2	0.22	7.2
	O_2	0.15	6.8

[a]From Jechke and Stuart (1961).
[b]From Van Amerongen (1950).

We can also relate permeability to temperature by an Arrhenius expression:

$$P_M = P_0 \exp\left(\frac{-Q_p}{RT}\right) \tag{2-87}$$

In the equation above, P_0 is a preexponential factor and Q_p is the activation energy associated with the permeability. The difference in the activation energies for diffusion and permeability may be described as a heat of solution.

REFERENCES

ASKILL, J., *Tracer Diffusion Data for Metals, Alloys, and Simple Oxides*, Plenum, New York, 1970.

BARRER, R. M., *Diffusion in and through Solids*, Cambridge University Press, New York, 1941.

BARRER, R. M., and T. GABOR, *Proc. R. Soc. (Lond.)*, *A251*, 353 (1959).

CRANK, J., and G. S. PARK, Ed., *Diffusion in Polymers*, Academic Press, New York, 1968.

CUSSLER, E. L., *Multicomponent Diffusion*, Elsevier, Amsterdam, 1976.

EDWARDS, J. B., E. E. HUCKE, and J. J. MARTIN, *Met. Rev.*, *13* (1968).

EYRING, H., T. REE, D. M. GRANT, and R. C. HIRST, *Z. Electrochem.*, *64*, 146 (1960).

FRENKEL, J., *Kinetic Theory of Liquids*, Clarendon Press, Oxford, 1946.

FULLER, E. N., P. D. SCHETTLER, and J. C. GIDDINGS, *Ind. Eng. Chem.*, *58*, 19 (1966).

GILLILAND, E. R., *Ind. Eng. Chem.*, *26*, 681(1934).

GILLILAND, E. R., R. F. BADDOUR, and J. L. RUSSELL, *AIChE J.*, *4*, 90 (1958).

GLASSTONE, S. K., J. LAIDLER, and H. EYRING, *The Theory of Rate Processes*, MCGRAW-Hill, New York, 1941.

HAASE, R., *Thermodynamics of Irreversible Processes*, Addison-Wesley, Reading, Mass., 1969.

HEILMAN, W., V. TAMMELA, J. A. MEYER, V. STANNETT, and M. SZWARC, *Ind. Eng. Chem.*, *48*, 821 (1956).

HENRY, J. P., B. CHENNAKESAVAN, and J. M. SMITH, *AIChE J.*, *7*, 10 (1961).

HINES, A. L., and H. A. WALLS, *Metall. Trans. A*, *10*, 1365 (1979).

HINES, A. L., H. A. WALLS, and D. W. ARNOLD, *Metall. Trans. B*, *6*, 484 (1975).

HIRSCHFELDER, J. O., R. B. BIRD, and E. L. SPOTZ, *Chem. Rev.*, *44*, 205 (1949).

HIRSCHFELDER, J. O., C. F. CURTIS, and R. B. BIRD, *Molecular Theory of Gases and Liquids*, Wiley, New York, 1954.

HWANG, S., and K. KAMMERMEYER, *Membranes in Separations*, Wiley, New York, 1975.

JECHKE, D., and H. A. STUART, *Z. Naturforsch., Teil A*, *16*, 37 (1961).

JOST, W., *Diffusion in Solids, Liquids, Gases*, p. 245, Academic Press, New York, 1960.

KOTYK, A., and K. JANÁČEK, *Cell Membrane Transport*, 2nd ed., Plenum Press, New York, 1975.

LEFFLER, J., and H. T. CULLINAN, *Ind. Eng. Chem. Fundam.*, *9*, 84 (1970a).

LEFFLER, J., and H. T. CULLINAN, *Ind. Eng. Chem. Fundam.*, *9*, 88 (1970b).

LENNARD-JONES, J. E., *Proc. Roy. Soc.* (London), *A106*, 463 (1924).

MANNING, J. R., *Diffusion Kinetics for Atoms in Crystals*, Van Nostrand, Princeton, N.J., 1968.

MARRERO, T. R., and E. A. MASON, *J. Phys. Chem. Ref. Data*, *1*, 3 (1972).

NERNST, W., *Z. Phys. Chem.*, *2*, 613 (1888).

NEUFELD, P. D., A. R. JANZEN, and R. A. AZIZ, *J. Chem. Phys.*, *57*, 1100 (1972).

PAULING, L., *Nature of the Chemical Bond*, Cornell University Press, Ithaca, N.Y., 1960.

PERRY, R. H., and C. H. CHILTON, Eds., *Chemical Engineers' Handbook*, 5th ed., McGraw-Hill, New York, 1973.

REID, R. C., J. M. PRAUSNITZ, and T. K. SHERWOOD, *The Properties of Gases and Liquids*, McGraw-Hill, New York, 1977.

ROBINSON, R. A., and R. H. STOKES, *Electrolyte Solutions*, 2nd ed., Academic Press, New York, 1959.

SATTERFIELD, C. N., *Mass Transfer in Heterogeneous Catalysis*, MIT Press, Cambridge, Mass., 1970.

SCHNEIDER, P., and J. M. SMITH, *AIChE J.*, *14*, 886 (1968).

SHERBY, O. D., and M. T. SIMNAD, *Trans. ASM*, *54*, 227 (1961).

SHERWOOD, T. K., R. L. PIGFORD, and C. R. WILKE, *Mass Transfer*, McGraw-Hill, New York, 1975.

SHEWMON, P. G., *Diffusion in Solids*, McGraw-Hill, New York, 1963.

SITARAMAN, R., S. H. IBRAHIM, and N. R. KULOOR, *J. Chem. Eng. Data*, *8*, 198 (1963).

SKELLAND, A. H. P., *Diffusional Mass Transfer*, Wiley, New York, 1974.

SMITH, J. M., *Chemical Engineering Kinetics*, 2nd ed., McGraw-Hill, New York, 1970.

SUTHERLAND, W., *Phil. Mag.*, *2*, 781 (1905).

SVEHLA, R. A., *NASA Tech. Rep. R-132*, Lewis Research Center, Cleveland, Ohio, 1962.

SWALIN, R. A., *Acta Met.*, *7*, 736 (1959).

TICHACEK, L. J., W. S. KMAK, and H. G. DRICKAMER, *J. Phys. Chem.*, *60*, 660 (1956).

TREYBAL, R. E., *Mass Transfer Operations*, 2nd ed., McGraw-Hill, New York, 1968.

TUWINER, S. B., *Diffusion and Membrane Technology*, Reinhold, New York, 1962.

VAN AMERONGEN, G. J., *J. Polym. Sci.*, *5*, 307 (1950).

VIGNES, A., *Ind. Eng. Chem. Fundam.*, *5*, 189 (1966).

WALLS, H. A., in R. A. RAPP, Ed., *Physiochemical Measurements in Metals Research*, Vol. 4, pp. 459–492, Wiley, New York, 1970.

WALLS, H. A., and W. R. UPTHEGROVE, *Acta Met.*, *12*, 461 (1964).

WASER, J., and L. PAULING, *J. Chem. Phys.*, *18*, 747 (1950).

WILKE, C. R., and P. CHANG, *AIChE J.*, *1*, 264 (1955).

WINTER, F. R., and H. G. DRICKAMER, *J. Phys. Chem.*, *59*, 1229 (1955).

NOTATIONS

a_A = activity of component A

A = average cross-sectional area, L^2

C = total molar concentration, mol/L^3

C_A, C_B = molar concentration of A and B, mol/L^3

d = pore diameter, L

D = diffusion coefficient, L^2/t
D_{AA} = self-diffusion coefficient of A, L^2/t
D_{AB} = binary diffusivity for system A–B, L^2/t
D_{AB}° = interdiffusion coefficient in dilute solution, L^2/t
$D_{AB,e}$ = effective binary diffusivity for system A–B, L^2/t
$D_{A,e}$ = effective diffusivity for self-diffusion or equimolar counter diffusion, L^2/t
$D_{A,K}$ = Knudsen diffusion coefficient, L^2/t
$D_{A,K,e}$ = effective Knudsen diffusivity, L^2/t
D_{BA} = binary diffusivity for system B–A, L^2/t
D_{im} = effective binary diffusivity of i in a multicomponent mixture, L^2/t
D_T = thermal diffusion coefficient, L^2/t
D_0 = frequency factor, L^2/t
F = Faraday's constant, C/M
h = Planck's constant, M/tL^2
ΔH_A = latent heat of vaporization, ML^2/t^2
ΔH_F = heat of fusion, ML^2/t^2
$\bar{j}$ = mean fluctuation distance, L
k = Boltzmann constant, ML^2/t^2T; thermal conductivity, ML/t^3T
k_T = thermal diffusion ratio
K = Waser and Pauling force constant, EL^2
K_0 = constant relating to solid structure
M = molecular weight, M/mol; mobility, L/tF
M_A, M_B = molecular weight of A and B, M/mol
n = number density, molecules/L^3
$\mathbf{N}_i$ = molar flux of i with respect to a fixed reference frame, mol/tL^2
P = pressure of the system, F/L^2
P_M = permeability, L^4/tF
P_0 = preexponential factor, L^4/tF
$\bar{P}_A$ = vapor pressure of A, F/L^2
Q = activation energy, ML^2/t^2
R = universal gas constant
S = surface area of porous solid, L^2; solubility, L^2/F
T = temperature of the system
T_c = critical temperature
T_m = melting-point temperature
$\bar{U}$ = mean molecular velocity, L/t
$\mathbf{U}_A$, $\mathbf{U}_B$ = velocity of A and B with respect to a fixed reference frame, L/t
V = molar volume at the normal boiling point, L^3/mol; valence
V_A, V_B = molecular volume of A and B
V_c = critical volume, L^3
V_p = specific pore volume of a solid particle, L^3/M
Z = coordination number
Z_+, Z_- = cation and anion valence

Greek Letters

α_T = thermal diffusion factor
β = coefficient of sliding friction between an atom and the medium
β_T = Dufour coefficient
γ_A = activity coefficient of component A
γ_S = Soret effect, T^{-1}
δ = membrane thickness, L
ϵ = porosity of the solid
ϵ_{AB} = maximum energy of attraction between A and B, ML^2/t^2
θ_D = Debye temperature, T
λ = mean free path, L
$\lambda_+^\circ, \lambda_-^\circ$ = cationic and anionic conductance at infinite dilution, AL^2/MV
μ_i = viscosity of i, M/Lt
μ_m = viscosity of mixture, M/Lt
$\nu = \mu/\rho$ = kinematic viscosity, L^2/t
ν_s = vibrational frequency of the solid calculated from the Debye temperature, t^{-1}
ξ_B = association factor of solvent B
ρ_B = bulk density, M/L^3
σ = volumetric rate of entropy production
σ_{AB} = collison diameter, L
σ_l, σ_s = conductivity of liquid and solid at the melting point
τ = tortuosity factor
$\phi_{AB}(r)$ = potential energy between A and B at a separation distance, r, ML^2/t^2
χ_i = normalized weighting factor
Ω_D = collision integral

PROBLEMS

2.1 Calculate the diffusivity for the air–carbon dioxide system at 317.2 K and 1 atm using (a) Eq. (2-6), (b) Eq. (2-7), and (c) Eq. (2-8). Compare the results with the experimental value presented in Table 2-5.

2.2 The diffusion coefficient for carbon dioxide into nitrogen at 25°C and 1 atm is 1.67×10^{-5} m²/s. Determine the diffusivity for the system at 125°C and 10 atm using (a) the equation of Fuller et al., and (b) the Chapman–Enskog equation.

2.3 Calculate the self-diffusion coefficient for zinc at its melting-point temperature by using the Hines and Walls equation. Compare your result with the value predicted from the Swalin equation.

2.4 Determine the diffusion coefficients for chloroform in ethanol and for ethanol in chloroform in very dilute solutions at 25°C using the Wilke–Chang equation and the equation of Sitaraman et al.

2.5 Calculate the diffusion coefficient for carbon dioxide in nitrogen at 25°C. The molar volume of nitrogen at 25°C is 27.17 liters/gmol.

2.6 Calculate the diffusivity of methanol in water at 25°C and 1 atm for a methanol concentration $x_A = 0.4$. The equation of Leffler and Cullinan is recommended.

The activity coefficient for methanol can be calculated by using the van Laar equation with the constants $A_{AB} = 0.3861$ and $A_{BA} = 0.2349$.

2.7 Diffusion coefficients for nitrogen in steel at several temperatures are shown below. Assuming that the diffusion coefficients can be related to temperature by an Arrhenius expression, determine the frequency factor and activation energy.

Temperature (°C)	800	850	900	950	1000	1050	1100
$D \times 10^{12}$ (m^2/s)	1.2	3.0	6.0	10.8	13.5	25.0	40.0

2.8 Calculate the Knudsen diffusivity of hydrogen sulfide through activated carbon at 20°C. The mean pore radius is 15×10^{-10} m.

2.9 Predict the diffusion coefficient for nitrogen through a mixture containing 50% hydrogen, 30% ammonia, 15% nitrogen, and 5% water at 25°C and 1 atm.

2.10 The diffusion flux can be written in general terms as

$$J_A = -\frac{C_A M}{N}\frac{\partial \mu_A}{\partial Z}$$

where N is Avogadro's number. Using the expression above, derive Eq. (2-27).

2.11 Diffusion coefficients for zinc in brass were measured at 750°C for several different concentrations of zinc (Jost, 1960). Using the data below, calculate D_{AB}°. Discuss your results.

Wt%	γ_{Zn}	$D_{AB} \times 10^{14}$ (m^2/s)
2	0.056	2.5
10	0.070	5.5
20	0.115	19.5
26	0.158	61.0

2.12 Activity coefficients for several solutions can be predicted with reasonable accuracy with Margules equation given by $\ln \gamma_A = x_B^2[A_{AB} + 2(A_{BA} - A_{AB})x_A]$. **(a)** Combine the equations of Margules and Vignes to obtain an expression for the diffusion coefficient as a function of concentration, and **(b)** calculate the diffusion coefficient for a 50 mol% mixture of ethanol–water at 25°C. Compare your answer with the value given in Example 2.4. $A_{AB} = 0.6848$, and $A_{BA} = 0.3781$.

2.13 Liquid diffusion coefficients for chlorobenzene in bromobenzene were obtained at 10°C as a function of concentration.

x_A	0.0332	0.2642	0.5122	0.7617	0.9652
$D_{AB} \times 10^9$ (m^2/s)	1.007	1.069	1.146	1.226	1.291

(a) Assuming that chlorobenzene and bromobenzene form an ideal solution, determine if the data above can be described by the equation of Vignes. **(b)** Obtain the diffusivity of chlorobenzene in bromobenzene at infinite dilution, D_{AB}°, and compare your result with the value calculated from the equation of Wilke and Chang. **(c)** Calculate D_{BA}° from the data above.

2.14 Assuming that $D_{AB} = D_{BA}$, verify that the Onsager coefficients L_{AB} and L_{BA} are equal.

2.15 Calculate the effective diffusivity of nitrogen at 300°C and 1 atm through a catalyst which has a mean pore radius of 45×10^{-10} m and a porosity of 0.4. Assume that the tortuosity is equal to 2 and the pores are filled with nitrogen.

2.16 Determine the effective diffusivity of carbon dioxide in nitrogen at 303 K and 3 atm through an alumina pellet. Assume equimolar counter diffusion of the two diffusing species.

2.17 Estimate the diffusion coefficient of sulfuric acid at infinite dilution in water at 25°C.

2.18 Consider the following reaction, in which species A_2 and B diffuse through a gas film to the surface of a solid catalyst and react to form A and C.

$$2A_2 + B \longrightarrow A + 3C$$

Species A and C then desorb from the catalyst surface and diffuse back through the gas film. Using Eq. (2-53), derive an equation to calculate $D_{A,m}$ in terms of the binary diffusion coefficients.

2.19 Estimate the diffusion coefficient for nickel in a cobalt rod at 1000°C. Compare your answer to the diffusivity of carbon through cobalt at the same temperature.

2.20 Calculate the diffusion coefficient of ethanol in a mixture containing 50 mol% benzene and 50 mol% chloroform at 288 K.

2.21 Predict the effective Knudsen diffusivity and the effective interdiffusivity for nitrogen diffusing through carbon dioxide in alumina catalysts which have pore radii of (1000, 500, 100, 50, and 10) $\times 10^{-10}$ m. Diffusion takes place at 298 K and 1 atm. For a pore diameter of 100×10^{-10} m, the effective Knudsen diffusivity comprises what fraction of the total effective diffusivity?

2.22 Estimate the effective diffusivity of nitrogen at 200°C and 1.2 atm through a catalyst which has pores filled with 30% helium, 45% hydrogen, 15% carbon dioxide, and 10% nitrogen. Data: $r = 44 \times 10^{-10}$ m, $\tau = 5.9$, and $\epsilon = 0.31$.

2.23 Simultaneous ordinary and pressure diffusion in an isothermal field can be expressed in a manner similar to that used for simultaneous ordinary and thermal diffusion by the equation

$$J_i = -\frac{1 - \chi_i}{1 - x_i}\frac{1}{\bar{V}}\left(D_{ij}\nabla x_i \mp D_p \frac{\nabla P}{P}\right)$$

where the upper sign is used for $i = 1$ and the lower sign is for $i = 2$. The relationship between ordinary and pressure diffusion for a binary system is

$$\frac{D_p}{D_{AB}} = \frac{M_A M_B}{M} P x_A \left(\frac{\bar{V}_B}{M_B} - \frac{\bar{V}_A}{M_A}\right)\frac{1}{(\partial \mu_B/\partial x_B)_{T,P}}$$

For a binary system of ideal gases, obtain the following equation:

$$_m\mathbf{j}_A = -\rho D_{AB}\left[\nabla\omega_A - \frac{M_A M_B}{M^3}\frac{x_A x_B (M_A - M_B)\nabla P}{P}\right]$$

Formulation of Mass Transfer Models 3

3.1 Introduction

Modeling of mass transfer problems depends to a great extent on the type of information desired from the model. For example, the macroscopic models, although widely used in describing industrial processes, do not provide a detailed description of the process since the properties of the system are averaged over position. Differential equations derived for this type of model are usually easy to solve because the only independent variable is time. By contrast, more detailed information about a process can be gained by using a microscopic model since the spacial coordinates are also included. However, the extra information is gained usually at the price of more extensive mathematical manipulation. Although the model and resulting differential equations describing the process may be easily obtained, the equation may be the type that requires a computer solution. In this chapter both macroscopic and microscopic models are discussed.

3.2 Macroscopic Material Balance

The material balance is based on the laws of conservation. These laws, which make use of the principle of accountability, are assumed to be valid since so far as we know, they have never been refuted. In the derivation of the material balance presented here, the interconversion of mass and energy by nuclear transformations will not be considered.

By applying the law of conservation of matter to a stream flowing into a volume element fixed in space, as shown in Figure 3-1, a general material balance may be written in the rate form:

rate of accumulation within the element	=	rate of transport into the element	−	rate of transport leaving the element	+	rate of generation within the element	−	rate of consumption within the element	(3-1)

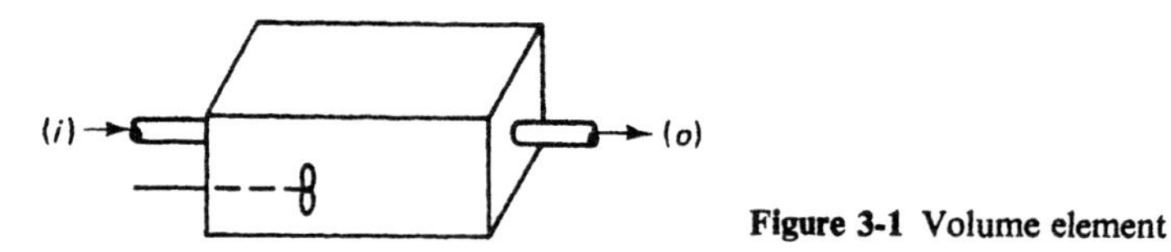

Figure 3-1 Volume element

Applying the general material balance for the total flow of mass, we obtain

$$\rho V|_{t+\Delta t} - \rho V|_t = (\rho \bar{U} S|_i - \rho \bar{U} S|_o)\Delta t \tag{3-2}$$

where $\rho V|_{t+\Delta t}$ = mass of material within the system at time $t + \Delta t$,
$\rho V|_t$ = mass of material within the system at time t,
V = volume of the system,
ρ = mass concentration,
S = cross-sectional area for the inlet and outlet of the volume element,
$\bar{U}$ = average velocity of the fluid.

Upon dividing both sides by Δt and shrinking the time element Δt to zero, Eq. (3-2) becomes

$$\frac{d}{dt}(\rho V) = (\rho \bar{U} S)_i - (\rho \bar{U} S)_o = -\Delta(\rho \bar{U} S) \tag{3-3}$$

Since the balance above considered the total mass, which can be neither created nor destroyed, the generation and consumption terms shown in the general expression, given by Eq. (3-1), are not present. Further, if the assumption is made of perfect mixing within the volume element, the concentration leaving the system will be the same as that inside. Since the exiting stream is a variable, the subscript on that term will be dropped. Thus for the case in which the cross-sectional areas of the inlet and outlet are the same, the mass balance becomes

$$\frac{1}{S}\frac{d}{dt}(\rho V) = (\rho \bar{U})_i - \rho \bar{U} \tag{3-4}$$

In the equations above, the absence of spacial coordinates should be noted.

Example 3.1

After leaving the refinery, number 2 fuel oil is piped to a tank farm, where it is pumped into a large cylindrical storage tank. The tank is 15 m in diameter and has a height of 12.20 m. If the tank is initially empty, determine the time required to fill it to a height of 11.6 m using a pumping rate of 90 m^3/h. The density of the fuel oil at the pumping conditions is 881 kg/m^3.

Solution:

Since no fuel oil leaves the tank, the material balance can be written as

$$\frac{d}{dt}\rho V = (\rho \bar{U} S)_i$$

Integrating the equation above gives

$$\rho V = (\rho \bar{U} S)_i t + K$$

where K is a constant of integration. Since the tank is initially empty, the boundary condition $V = 0$ at $t = 0$ applies. Thus the constant of integration is zero and the equation describing the rate of filling is

$$\rho V = (\rho \bar{U} S)_i t$$

Solving for time gives

$$t = \frac{\rho V}{(\rho \bar{U} S)_i}$$

The volume of fuel oil contained in the tank when the filling level reaches 11.6 m is

$$V = \frac{\pi D^2 Z}{4} = \frac{\pi (15)^2 (11.6)}{4} = 2049.9 \text{ m}^3$$

Thus

$$t = \frac{881(2049.9)}{881(90)}$$

$$= 22.8 \text{ h}$$

The problem above demonstrates the case in which there is a volume change but no concentration change within the system. Now let us consider the case in which volume is constant and the concentration changes.

Example 3.2

A tank initially contains 2.8 m^3 of brine solution with a concentration of 32 kg of salt per cubic meter of solution. If water enters the tank at a rate of 1.0 m^3/h and the brine solution leaves at the same rate, determine the salt concentration in the tank after 20 min. Assume perfect mixing.

Solution:

For the case in which no salt enters the tank and the total volume entering and leaving are the same, the material balance for the salt concentration is written as

$$\frac{d}{dt}\rho V = -\rho \bar{U} S$$

where the subscript for the exit stream has been dropped since perfect mixing was assumed. At constant volume, the variables can be separated and the equation above can be expressed as

$$\frac{d\rho}{\rho} = -\frac{\bar{U}S}{V}dt$$

Following integration, we have

$$\ln \rho = -\frac{\bar{U}S}{V}t + \ln K$$

The constant of integration, K, can be determined from the initial salt concentration in the tank, which is

$$t = 0 \qquad \rho = 32 \text{ kg/m}^3$$

Thus

$$\ln \rho = -\frac{\bar{U}S}{V}t + \ln 32$$

or

$$\rho = 32 \exp\left(-\frac{\bar{U}S}{V}t\right)$$

The salt concentration in the tank after 20 min is

$$\rho = 32 \exp\left[\left(-\frac{1}{2.8}\right)\frac{20}{60}\right]$$

$$= 28.4 \text{ kg salt/m}^3 \text{ solution}$$

Consider now the application of Eq. (3-1) to a chemical species j. Although the total mass transferred through the volume element is conserved, the mass or number of moles of a particular species is not conserved if a chemical reaction occurs. A mole balance for component j thus becomes

$$C_jV|_{t+\Delta t} - C_jV|_t = (C_j\bar{U}S|_i - C_j\bar{U}S|_o)\Delta t + (\dot{R}_j^v V)_{\text{gen}}\Delta t - (\dot{R}_j^v V)_{\text{con}}\Delta t \qquad (3\text{-}5)$$

where $C_j\bar{U}S$ = molar flow rate either entering or leaving the system,
C_jV = moles of species j within the system at t or $t + \Delta t$,
$\dot{R}_j^v$ = molar rate of generation or consumption per unit volume.

Again, dividing both sides by Δt and shrinking the time element Δt to zero gives

$$\frac{d}{dt}(C_jV) = (C_j\bar{U}S)_i - (C_j\bar{U}S)_o + (\dot{R}_j^v V)_{\text{gen}} - (\dot{R}_j^v V)_{\text{con}} \qquad (3\text{-}6)$$

Assuming that perfect mixing exists within the volume element, the concentration inside the element and that leaving will be the same. Thus the subscript o at the exit may be dropped. The use of Eq. (3-6) can be demonstrated by considering the following example.

Example 3.3

The irreversible reaction $2A \xrightarrow{k_1} B$ is carried out at high temperature in a reactor with constant volume V. If the reactor is initially charged with a concentration C_{Ai}, obtain an expression for the concentration of A as a function of time. If we assume an elementary reaction, the rate of consumption can be expressed by $\dot{R}_A^v = k_1C_A^2$.

Solution:

For a reactor in which material does not enter or leave, Eq. (3-6) is written for component A as

$$\frac{d}{dt}(C_A V) = -(\dot{R}_A^v V)_{con}$$

Since the volume of the reactor is constant, we have

$$\frac{dC_A}{dt} = -\dot{R}_A^v$$

Upon substituting $k_1 C_A^2$ for the rate of consumption, the equation above becomes

$$\frac{dC_A}{dt} = -k_1 C_A^2$$

Separating the variables and integrating gives

$$-\frac{1}{C_A} = -k_1 t + K$$

where K is a constant of integration. Using the initial condition in the reactor, $C_A = C_{Ai}$ at $t = 0$, gives

$$C_A = \frac{C_{Ai}}{1 + C_{Ai} k_1 t}$$

The equation above gives the concentration of species A in the reactor as a function of time.

3.3 General Microscopic Material Balance

Derivation of the continuity equation will be made by considering the mass of component A moving at a velocity, U_A, through a fixed volume element. This method, often called the *Eulerian approach*, can easily be extended to cases in which the volume element moves with some reference velocity. We begin by considering a differential volume element of fixed shape and account for the flow entering and leaving the element. Upon substituting the mass flux terms for species A, shown in Figure 3-2, into the general balance given by Eq. (3-1), we obtain

$$\begin{aligned}(n_{AX}|_X - n_{AX}|_{X+\Delta X})\Delta Y\,\Delta Z + (n_{AY}|_Y - n_{AY}|_{Y+\Delta Y})\Delta X\,\Delta Z \\ + (n_{AZ}|_Z - n_{AZ}|_{Z+\Delta Z})\Delta X\,\Delta Y + \dot{r}_A^v\,\Delta X\,\Delta Y\,\Delta Z = \Delta X\,\Delta Y\,\Delta Z \frac{\partial \rho_A}{\partial t}\end{aligned} \tag{3-7}$$

where the reaction term $\dot{r}_A^v$ represents generation minus consumption. If we divide the equation above by the volume of the element, $\Delta X\,\Delta Y\,\Delta Z$, and shrink the volume element to zero, we have the general statement of the conservation of species A:

$$\frac{\partial \rho_A}{\partial t} + \frac{\partial n_{AX}}{\partial X} + \frac{\partial n_{AY}}{\partial Y} + \frac{\partial n_{AZ}}{\partial Z} - \dot{r}_A^v = 0 \tag{3-8}$$

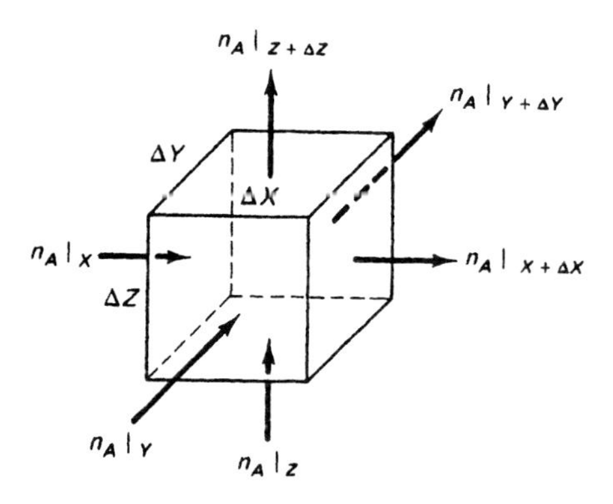

Figure 3-2 Differential volume element

In vector notation

$$\frac{\partial \rho_A}{\partial t} + \nabla \cdot \mathbf{n}_A - \dot{r}_A^v = 0 \tag{3-9}$$

or

$$\frac{\partial \rho_A}{\partial t} + \nabla \cdot \rho_A \mathbf{U}_A - \dot{r}_A^v = 0 \tag{3-10}$$

In Eq. (3-7), n_{AX}, n_{AY}, and n_{AZ} are mass flux terms with the units (mass/area-time), $\dot{r}_A^v$ is the net rate of volumetric production of A (mass/volume-time), and ρ_A is the mass concentration (mass/volume). An examination of the terms in Eqs. (3-3) and (3-7) shows that each term has the units (mass/time).

Similarly, the conservation equation for species B can be written as

$$\frac{\partial \rho_B}{\partial t} + \nabla \cdot \mathbf{n}_B - \dot{r}_B^v = 0 \tag{3-11}$$

Adding Eqs. (3-9) and (3-11) gives

$$\frac{\partial(\rho_A + \rho_B)}{\partial t} + \nabla \cdot (\mathbf{n}_A + \mathbf{n}_B) - (\dot{r}_A^v + \dot{r}_B^v) = 0 \tag{3-12}$$

In order to obey the conservation of mass for a binary system, the increase in the mass of species A will result in the depletion of the mass of species B. Thus

$$\dot{r}_A^v = -\dot{r}_B^v \tag{3-13}$$

Since $\rho_A + \rho_B = \rho$ and $\mathbf{n}_A + \mathbf{n}_B = \rho \mathbf{U}^m$, Eq. (3-12) reduces to

$$\frac{\partial \rho}{\partial t} + \nabla \cdot \rho \mathbf{U}^m = 0 \tag{3-14}$$

Equation (3-14) is the continuity equation for a multicomponent system. Continuity equations for species A are given in Table 3-1 for the various coordinate systems in terms of both mass and molar flux rates.

The conservation of mass for species A, shown in Eq. (3-10), can be written in terms of the arbitrary reference velocity $\mathbf{U}^0$ by expanding the expression given in Eq. (1-11) and writing it in terms of $\rho_i \mathbf{U}_i$. Thus upon combining these two

TABLE 3-1

A. Continuity Equations of Species A

Rectangular coordinates:

$$\frac{\partial \rho_A}{\partial t} + \left(\frac{\partial n_{AX}}{\partial X} + \frac{\partial n_{AY}}{\partial Y} + \frac{\partial n_{AZ}}{\partial Z}\right) = \dot{r}_A^v \tag{A}$$

Cylindrical coordinates:

$$\frac{\partial \rho_A}{\partial t} + \left[\frac{1}{r}\frac{\partial (r n_{Ar})}{\partial r} + \frac{1}{r}\frac{\partial n_{A\theta}}{\partial \theta} + \frac{\partial n_{AZ}}{\partial Z}\right] = \dot{r}_A^v \tag{B}$$

Spherical coordinates:

$$\frac{\partial \rho_A}{\partial t} + \left[\frac{1}{r^2}\frac{\partial (r^2 n_{Ar})}{\partial r} + \frac{1}{r \sin\theta}\frac{\partial (n_{A\theta}\sin\theta)}{\partial \theta} + \frac{1}{r\sin\theta}\frac{\partial n_{A\phi}}{\partial \phi}\right] = \dot{r}_A^v \tag{C}$$

*B. Continuity Equations of Species A for Constant ρ and D_{AB}**

Rectangular coordinates:

$$\frac{\partial \rho_A}{\partial t} + \left(U_X^m \frac{\partial \rho_A}{\partial X} + U_Y^m \frac{\partial \rho_A}{\partial Y} + U_Z^m \frac{\partial \rho_A}{\partial Z}\right) = D_{AB}\left(\frac{\partial^2 \rho_A}{\partial X^2} + \frac{\partial^2 \rho_A}{\partial Y^2} + \frac{\partial^2 \rho_A}{\partial Z^2}\right) + \dot{r}_A^v \tag{D}$$

Cylindrical coordinates:

$$\frac{\partial \rho_A}{\partial t} + \left(U_r^m \frac{\partial \rho_A}{\partial r} + U_\theta^m \frac{1}{r}\frac{\partial \rho_A}{\partial \theta} + U_Z^m \frac{\partial \rho_A}{\partial Z}\right)$$
$$= D_{AB}\left[\frac{1}{r}\frac{\partial}{\partial r}\left(r\frac{\partial \rho_A}{\partial r}\right) + \frac{1}{r^2}\frac{\partial^2 \rho_A}{\partial \theta^2} + \frac{\partial^2 \rho_A}{\partial Z^2}\right] + \dot{r}_A^v \tag{E}$$

Spherical coordinates:

$$\frac{\partial \rho_A}{\partial t} + \left(U_r^m \frac{\partial \rho_A}{\partial r} + U_\theta^m \frac{1}{r}\frac{\partial \rho_A}{\partial \theta} + U_\phi^m \frac{1}{r\sin\theta}\frac{\partial \rho_A}{\partial \phi}\right)$$
$$= D_{AB}\left[\frac{1}{r^2}\frac{\partial}{\partial r}\left(r^2\frac{\partial \rho_A}{\partial r}\right) + \frac{1}{r^2 \sin\theta}\frac{\partial}{\partial \theta}\left(\sin\theta\frac{\partial \rho_A}{\partial \theta}\right) + \frac{1}{r^2\sin^2\theta}\frac{\partial^2 \rho_A}{\partial \phi^2}\right] + \dot{r}_A^v \tag{F}$$

*The mass concentrations can be changed to molar concentrations by dividing each term by the molecular weight of A.

equations, we obtain

$$\frac{\partial \rho_A}{\partial t} + \underset{\text{Diffusion flux}}{\nabla \cdot {}_0\mathbf{j}_A} + \underset{\text{Convective flux}}{\nabla \cdot (\rho_A \mathbf{U}^0)} - \dot{r}_A^v = 0 \tag{3-15}$$

The continuity equations for species A are written in terms of the mass, molar, and volume average velocities by simply substituting the appropriate reference velocity into Eq. (3-15).

$$\frac{\partial \rho_A}{\partial t} + \nabla \cdot {}_m\mathbf{j}_A + \nabla \cdot (\rho_A \mathbf{U}^m) - \dot{r}_A^v = 0 \tag{3-16}$$

$$\frac{\partial \rho_A}{\partial t} + \nabla \cdot {}_M\mathbf{j}_A + \nabla \cdot (\rho_A \mathbf{U}^M) - \dot{r}_A^v = 0 \tag{3-17}$$

$$\frac{\partial \rho_A}{\partial t} + \nabla \cdot {}_v\mathbf{j}_A + \nabla \cdot (\rho_A \mathbf{U}^v) - \dot{r}_A^v = 0 \tag{3-18}$$

Inserting Fick's first law into Eq. (3-16) gives

$$\frac{\partial \rho_A}{\partial t} - \nabla \cdot \rho D_A \nabla \omega_A + \nabla \cdot (\rho_A \mathbf{U}^m) - \dot{r}_A^v = 0 \tag{3-19}$$

For many experimental observations, the term containing the mass average velocity can be neglected. Thus in the absence of a chemical reaction Eq. (3-19) reduces to

$$\frac{\partial \rho_A}{\partial t} = \nabla \cdot \rho D_A \nabla \omega_A \tag{3-20}$$

Equations similar to those previously developed can also be derived in terms of molar concentrations. An expression analogous to Eq. (3-8) written in terms of molar fluxes is

$$\frac{\partial C_A}{\partial t} + \frac{\partial N_{AX}}{\partial X} + \frac{\partial N_{AY}}{\partial Y} + \frac{\partial N_{AZ}}{\partial Z} - \dot{R}_A^v = 0 \tag{3-21}$$

or

$$\frac{\partial C_A}{\partial t} + \nabla \cdot \mathbf{N}_A - \dot{R}_A^v = 0 \tag{3-22}$$

where $\mathbf{N}_A$ is the molar flux (moles/area-time) and $\dot{R}_A^v$ is the net molar rate of production of A (moles/volume-time). For multicomponent systems, the general continuity equation is

$$\frac{\partial C}{\partial t} + \nabla \cdot C\mathbf{U}^M - \sum_{i=1}^{n} \dot{R}_i^v = 0 \tag{3-23}$$

A generalized conservation equation in molar units is obtained from Eq. (3-22) by introducing the arbitrary reference velocity from Eq. (1-12).

$$\frac{\partial C_A}{\partial t} + \underbrace{\nabla \cdot {}_0\mathbf{J}_A}_{\text{Diffusion flux}} + \underbrace{\nabla \cdot (C_A \mathbf{U}^0)}_{\text{Convective flux}} - \dot{R}_A^v = 0 \tag{3-24}$$

The continuity equations for species A written in terms of the mass, molar, and volume average velocities are

$$\frac{\partial C_A}{\partial t} + \nabla \cdot {}_m\mathbf{J}_A + \nabla \cdot (C_A \mathbf{U}^m) - \dot{R}_A^v = 0 \tag{3-25}$$

$$\frac{\partial C_A}{\partial t} + \nabla \cdot {}_M\mathbf{J}_A + \nabla \cdot (C_A \mathbf{U}^M) - \dot{R}_A^v = 0 \tag{3-26}$$

$$\frac{\partial C_A}{\partial t} + \nabla \cdot {}_v\mathbf{J}_A + \nabla \cdot (C_A \mathbf{U}^v) - \dot{R}_A^v = 0 \tag{3-27}$$

The equations above are quite difficult to solve as presented. However, when modeling diffusional processes, several simplifying assumptions may be appropriate.

1. For the case of no chemical reaction, substitution of Fick's first law into Eq. (3-27) gives

$$\frac{\partial C_A}{\partial t} + \nabla \cdot \left[\left(\frac{1 - C_A \bar{V}_A}{1 - x_A}\right) C D_A \nabla x_A\right] + \nabla \cdot (C_A \mathbf{U}^v) = 0 \tag{3-28}$$

2. In the absence of convective flow and for an ideal gas, Eq. (3-28) further reduces to

$$\frac{\partial C_A}{\partial t} = \nabla \cdot C D_A \nabla x_A \tag{3-29}$$

3. For the case of constant overall concentration and diffusivity, Eq. (3-29) becomes

$$\frac{\partial C_A}{\partial t} = D_A \nabla^2 C_A \tag{3-30}$$

Equation (3-30) is called *Fick's second law* and has found wide application in the experimental determination of diffusion coefficients. A number of diffusion systems described by the equation above will be solved in the next chapter. It is important to note that Fick's second law can be obtained from Eq. (3-26) by simply neglecting chemical reaction and convective flow.

4. Finally, for a steady-state process (no accumulation of species A), Eq. (3-30) becomes Laplace's equation in terms of the molar concentration.

$$\nabla^2 C_A = 0 \tag{3-31}$$

Laplace's equation is restricted to cases in which

$$C = \text{constant}$$
$$\mathbf{U}^0 = 0$$
$$\frac{\partial C_A}{\partial t} = 0$$
$$D_A = \text{constant}$$
$$\dot{R}_A^v = 0$$

Various forms of Eq. (3-26) are presented in Table 3-2 for a binary system.

After deriving the differential equation that describes the diffusion process, appropriate boundary conditions must be selected to complete the mathematical model of the physical process. For steady-state diffusion problems the boundary conditions most often encountered are as follows:

1. The concentration at a surface may be specified.

$$C_A = C_{A1} \quad \text{at } Z = Z_1 \tag{3-32}$$

2. The mass flux at a surface can be related to a difference in the surface concentration and the concentration of the surrounding media as

$$N_{AZ}\big|_{Z=Z_1} = k_c(C_{A1} - C_{A\infty}) \tag{3-33}$$

where k_c is a convective mass transfer coefficient with the units (length/time).

3. The flux at a surface may be related to the reaction rate constant, which for a first-order reaction is

$$N_{AZ}\big|_{Z=Z_1} = \dot{R}^s_{A,Z_1} = k_1^s C_{A1} \tag{3-34}$$

where k_1^s is a rate constant with the units (length/time).

TABLE 3-2

A. Molar Flux of Species A in Various Coordinate Systems

Rectangular coordinates:

$$\frac{\partial C_A}{\partial t} + \left(\frac{\partial N_{AX}}{\partial X} + \frac{\partial N_{AY}}{\partial Y} + \frac{\partial N_{AZ}}{\partial Z}\right) = \dot{R}_A^v \tag{A}$$

Cylindrical coordinates:

$$\frac{\partial C_A}{\partial t} + \left[\frac{1}{r}\frac{\partial(rN_{Ar})}{\partial r} + \frac{1}{r}\frac{\partial N_{A\theta}}{\partial \theta} + \frac{\partial N_{AZ}}{\partial Z}\right] = \dot{R}_A^v \tag{B}$$

Spherical coordinates:

$$\frac{\partial C_A}{\partial t} + \left[\frac{1}{r^2}\frac{\partial(r^2 N_{Ar})}{\partial r} + \frac{1}{r\sin\theta}\frac{\partial(N_{A\theta}\sin\theta)}{\partial \theta} + \frac{1}{r\sin\theta}\frac{\partial N_{A\phi}}{\partial \phi}\right] = \dot{R}_A^v \tag{C}$$

*B. Continuity Equations of Species A for Constant ρ and D_{AB}**

Rectangular coordinates:

$$\frac{\partial C_A}{\partial t} + \left(U_X^m\frac{\partial C_A}{\partial X} + U_Y^m\frac{\partial C_A}{\partial Y} + U_Z^m\frac{\partial C_A}{\partial Z}\right) = D_{AB}\left(\frac{\partial^2 C_A}{\partial X^2} + \frac{\partial^2 C_A}{\partial Y^2} + \frac{\partial^2 C_A}{\partial Z^2}\right) + \dot{R}_A^v \tag{D}$$

Cylindrical coordinates:

$$\frac{\partial C_A}{\partial t} + \left(U_r^m\frac{\partial C_A}{\partial r} + U_\theta^m\frac{1}{r}\frac{\partial C_A}{\partial \theta} + U_Z^m\frac{\partial C_A}{\partial Z}\right)$$
$$= D_{AB}\left[\frac{1}{r}\frac{\partial}{\partial r}\left(r\frac{\partial C_A}{\partial r}\right) + \frac{1}{r^2}\frac{\partial^2 C_A}{\partial \theta^2} + \frac{\partial^2 C_A}{\partial Z^2}\right] + \dot{R}_A^v \tag{E}$$

Spherical coordinates:

$$\frac{\partial C_A}{\partial t} + \left(U_r^m\frac{\partial C_A}{\partial r} + U_\theta^m\frac{1}{r}\frac{\partial C_A}{\partial \theta} + U_\phi^m\frac{1}{r\sin\theta}\frac{\partial C_A}{\partial \phi}\right)$$
$$= D_{AB}\left[\frac{1}{r^2}\frac{\partial}{\partial r}\left(r^2\frac{\partial C_A}{\partial r}\right) + \frac{1}{r^2\sin\theta}\frac{\partial}{\partial \theta}\left(\sin\theta\frac{\partial C_A}{\partial \theta}\right) + \frac{1}{r^2\sin^2\theta}\frac{\partial^2 C_A}{\partial \phi^2}\right] + \dot{R}_A^v \tag{F}$$

*For low concentrations of A in B the mass average velocity can be replaced by the local velocity, which is the velocity for a pure fluid.

The use of each of the boundary conditions above and the formulation of simple mass transfer models will be demonstrated in the remainder of this chapter.

3.4 Simple Diffusion Models

Systems in which diffusion takes place can be modeled by starting with one of the material balance equations appearing in either Table 3-1 or 3-2 and eliminating the terms that do not apply to the system to be modeled. The other method for modeling diffusing systems is the *shell balance method* made popular by Bird et al. (1960). Since this method requires that a material balance be made over a differential element in the direction of mass transfer, it is usually employed to model one-dimensional systems. For systems of complicated geometry, the elimination of terms from the general material balance is probably the better approach. In this section both methods will be used to formulate models for one-dimensional diffusion systems in rectangular, cylindrical, and spherical coordinates. The reduction of the material balance equations and the shell balance method will be demonstrated for the case in which a single component is transferred.

Reducing the general material balance equations

Let us consider the steady-state evaporation of a liquid from a small-diameter tube through a stationary gas film as shown in Figure 3-3. If we assume

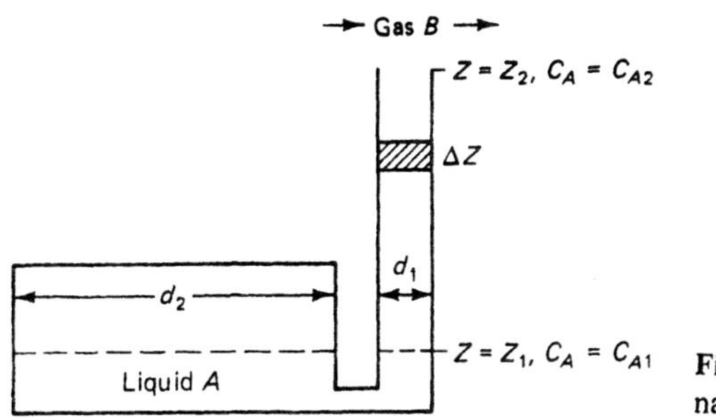

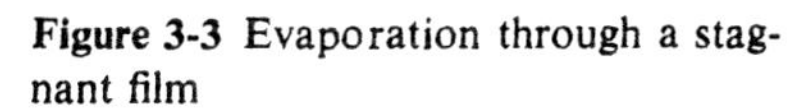
Figure 3-3 Evaporation through a stagnant film

that $d_1 \ll d_2$, then the level of the fluid in the tube will remain constant. For the gas film to remain stationary, gas B must be insoluble in liquid A or component B must be at its solubility limit in A. Further, we will assume that there is no chemical reaction, the system remains at constant temperature and pressure (constant total concentration), and the gas mixture is ideal. Starting with Eq. (A) of Table 3-2 and noting that mass is transferred only in the Z direction, $\partial N_{AX}/\partial X$ and $\partial N_{AY}/\partial Y$ are equal to zero. At steady-state conditions $\partial C_A/\partial t = 0$. Thus we have

$$\frac{dN_{AZ}}{dZ} = 0 \tag{3-35}$$

But Eq. (1-34) for the gas phase in the Z direction is

$$N_{AZ} = -CD_{AB}\frac{dy_A}{dZ} + y_A(N_{AZ} + N_{BZ})$$

If component B is stagnant, then substituting $N_{BZ} = 0$ into the equation above and rearranging gives

$$N_{AZ} = -\frac{CD_{AB}}{1 - y_A}\frac{dy_A}{dZ} \tag{3-36}$$

Upon substituting Eq. (3-36) into Eq. (3-35) and assuming that the total molar concentration and diffusivity are constant, we obtain

$$\frac{d}{dZ}\left(\frac{1}{1 - y_A}\frac{dy_A}{dZ}\right) = 0 \tag{3-37}$$

In order to complete the model, boundary conditions giving the concentrations of component A at the limits of the diffusion path (the top of the tube and the gas–liquid interface) must be established. As shown in Figure 3-3, the boundary conditions are

$$y_A = y_{A1} \quad \text{at } Z = Z_1 \tag{3-38}$$

and

$$y_A = y_{A2} \quad \text{at } Z = Z_2 \tag{3-39}$$

It is important to note that the concentration of A at the gas–liquid interface is the concentration of A in the vapor that is in equilibrium with the liquid. Thus the equilibrium distribution of A in gas B must be known if numerical calculations are to be carried out.

Integrating Eq. (3-37) once gives

$$\frac{1}{1 - y_A}\frac{dy_A}{dZ} = K_1 \tag{3-40}$$

Upon separating the variables and integrating a second time, we obtain

$$-\ln(1 - y_A) = K_1 Z + K_2 \tag{3-41}$$

After substituting the boundary conditions given by Eqs. (3-38) and (3-39) into Eq. (3-41), the constants of integration, K_1 and K_2, can be evaluated. The concentration profile for species A over the length of the tube is given as

$$\ln\frac{1 - y_A}{1 - y_{A1}} = \frac{Z - Z_1}{Z_2 - Z_1}\ln\frac{1 - y_{A2}}{1 - y_{A1}} \tag{3-42}$$

The molar flux, N_{AZ}, at the liquid surface can be obtained by evaluating Eq. (3-36) at $Z = Z_1$.

$$N_{AZ}\big|_{Z=Z_1} = -\frac{CD_{AB}}{1 - y_A}\frac{dy_A}{dZ}\bigg|_{Z=Z_1} \tag{3-43}$$

Since the gradient in mole fraction must be known, we differentiate Eq. (3-42) and evaluate the result at $Z = Z_1$. The molar flux at the gas–liquid interface is represented by

$$N_{AZ}\big|_{Z=Z_1} = \frac{CD_{AB}}{Z_2 - Z_1}\ln\frac{1 - y_{A2}}{1 - y_{A1}} \tag{3-44}$$

or

$$N_{AZ}\big|_{Z=Z_1} = \frac{CD_{AB}}{Z_2 - Z_1}\ln\frac{y_{B2}}{y_{B1}} \tag{3-45}$$

The flux can be equated to a mass transfer coefficient and a linear driving force by defining the logarithmic mean mole fraction as

$$(y_B)_M = \frac{y_{B2} - y_{B1}}{\ln(y_{B2}/y_{B1})} \tag{3-46}$$

Thus

$$N_{AZ}\big|_{Z=Z_1} = \frac{CD_{AB}}{(Z_2 - Z_1)(y_B)_M}(y_{B2} - y_{B1}) \tag{3-47}$$

Therefore, a mass transfer coefficient can be defined in terms of the diffusion coefficient and the length of the diffusion path by the expression

$$k_y = \frac{CD_{AB}}{(Z_2 - Z_1)(y_B)_M} \tag{3-48}$$

Subsequently, the flux is given in terms of a mass transfer coefficient and a linear driving force as

$$N_{AZ}|_{Z=Z_1} = k_y(y_{B2} - y_{B1}) \tag{3-49}$$

The equation above is similar to the boundary condition given in Eq. (3-33). Detailed discussions of convective mass transfer coefficients are presented in Chapters 5 and 6.

If the gas mixture in the tube is ideal, the flux can be written in terms of partial pressures at Z_1 and Z_2. Since the total pressure is constant,

$$P = \bar{P}_{A1} + \bar{P}_{B1} = \bar{P}_{A2} + \bar{P}_{B2} \tag{3-50}$$

and

$$\bar{P}_A = y_A P \qquad \bar{P}_B = y_B P \tag{3-51}$$

Thus for a binary system the molar flux can be expressed as

$$N_{AZ}|_{Z=Z_1} = \frac{PD_{AB}}{RT(Z_2 - Z_1)(P_B)_M}(\bar{P}_{B2} - \bar{P}_{B1}) \tag{3-52}$$

Shell balance method

Let us reconsider the preceding problem but with the condition that gas B is soluble in liquid A, and for each mole of A vaporized 1 mole of B will condense into liquid A. We will also assume that the apparatus can be modified so that the process can be carried out at steady-state conditions. Rather than starting with the equations presented in Table 3-2, the differential equation describing the diffusion process will be derived by using the shell balance technique and the definition of the first derivative from elementary calculus. The shell balance involves a mass or mole balance across a differential element of the medium such that consideration is given to homogeneous chemical reaction and accumulation in the element. Carrying out the balance for species A over the differential element ΔZ shown in Figure 3-3 gives

molar rate of A in at Z: $\quad SN_{AZ}|_Z \quad$ (3-53)

molar rate of A out at $Z + \Delta Z$: $\quad SN_{AZ}|_{Z+\Delta Z} \quad$ (3-54)

where S is the cross-sectional area of the tube. In the absence of a homogeneous chemical reaction and for a steady-state process, the general molar rate balance can be written as

$$SN_{AZ}|_Z - SN_{AZ}|_{Z+\Delta Z} = 0 \tag{3-55}$$

When formulating the differential equation, the shell balance is divided by constants and differential values. Thus dividing Eq. (3-55) by the cross-sectional area S and the differential thickness ΔZ, we obtain

$$\frac{N_{AZ}|_Z - N_{AZ}|_{Z+\Delta Z}}{\Delta Z} = 0 \tag{3-56}$$

In the limit as ΔZ approaches zero, we have from the definition of the first derivative

$$-\frac{dN_{AZ}}{dZ} = 0 \tag{3-57}$$

Equation (3-57) is identical to Eq. (3-35). Thus the difference between the preceding model and the one just developed must be due to the difference in the relative fluxes. For equimolar counter transfer $N_{AZ} = -N_{BZ}$. Thus upon substituting the fluxes for A and B into Eq. (1-34), we get

$$N_{AZ} = -CD_{AB}\frac{dy_A}{dZ} \tag{3-58}$$

Substitution of the molar flux term shown above into the differential flux expression given by Eq. (3-57) for the case of constant diffusivity and constant total concentration gives

$$\frac{d}{dZ}\left(\frac{dy_A}{dZ}\right) = \frac{d^2y_A}{dZ^2} = 0 \tag{3-59}$$

Equation (3-59) can be integrated twice to give

$$y_A = K_1Z + K_2 \tag{3-60}$$

Application of the boundary conditions given by Eqs. (3-38) and (3-39) gives the concentration profile for species A over the length of the diffusion path.

$$\frac{y_A - y_{A1}}{Z - Z_1} = \frac{y_{A1} - y_{A2}}{Z_1 - Z_2} \tag{3-61}$$

The molar flux can be found as shown in the preceding example or by simply separating the variables in Eq. (3-58) and integrating.

$$N_{AZ}\int_{Z_1}^{Z_2} dZ = -CD_{AB}\int_{y_{A1}}^{y_{A2}} dy_A \tag{3-62}$$

Thus

$$N_{AZ} = \frac{CD_{AB}}{Z_2 - Z_1}(y_{A1} - y_{A2}) \tag{3-63}$$

If the gas mixture is assumed to be ideal, the flux can be written in terms of partial pressure as

$$N_{AZ} = \frac{D_{AB}}{RT(Z_2 - Z_1)}(\bar{P}_{A1} - \bar{P}_{A2}) \tag{3-64}$$

where $\bar{P}_{A1}$ and $\bar{P}_{A2}$ are the partial pressures of species A at the gas–liquid interface and at the top of the tube, respectively.

3.5 Steady-State Diffusion with Heterogeneous Chemical Reaction

A wide variety of processes involves mass transfer accompanied by chemical reaction. Heterogeneous reactions are described as occurring on the surface or on a boundary. An example of a heterogeneous reaction is the hydrogenation of ethylene on a porous nickel catalyst. The purpose of the catalyst is to provide a large surface area for the reacting components, in addition to providing an energetically more favorable reaction path. This process involves the diffusion of the reactants to internal sites of the catalyst, where they are adsorbed and then chemically react. Following the reaction, the products desorb from the surface and diffuse away from the catalyst. The reader is referred to the work of Satterfield (1970) for a detailed discussion of diffusion and heterogeneous reaction.

In this treatment of diffusion accompanied by heterogeneous chemical reactions, only first-order kinetics will be considered in order to simplify our discussion. When formulating differential equations to model a process in which heterogeneous chemical reaction is occurring, the molar flux at a surface is related to the product of the rate constant and the concentration as a boundary condition. Two limiting types of boundary conditions must be considered: one is the case of slow reaction in which a finite concentration will exist on the surface and the second is for the case of very rapid reaction. For the second case the concentration of the reacting species on the surface will be essentially zero. Application of the boundary conditions for a slow heterogeneous reaction, given in Eq. (3-34), is very similar to the application of the type 2 boundary condition given by Eq. (3-33) in which the flux is specified on the surface. For a very fast reaction, the boundary condition is the same as a type 1, given by Eq. (3-32), except that the concentration on the surface is equal to zero.

Slow reaction on a surface

Consider the system shown in Figure 3-4 in which gas A diffuses through a stagnant gas film to the surface of a nonporous spherical catalyst particle where it reacts irreversibly according to the expression

$$nA \xrightarrow{k_1} B \tag{3-65}$$

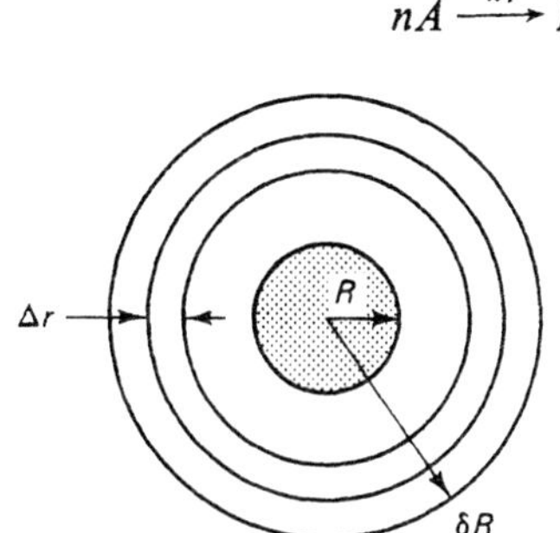

Figure 3-4 Nonporous spherical catalyst particle

The rate at which A reacts is proportional to the rate at which it diffuses through the stagnant film, and to the concentration of A on the surface. For first-order kinetics the flux at the surface is given by

$$N_A = k_1^s C_A = C k_1^s y_A \tag{3-66}$$

Since the reaction takes place on the surface, the reaction term will not appear in the shell balance. The balance, which is made over a differential element, Δr, in the stagnant gas film surrounding the spherical particle, is

$$4\pi(r^2 N_{Ar})|_r - 4\pi[(r + \Delta r)^2 N_{Ar}]|_{r+\Delta r} = 0 \tag{3-67}$$

where N_{Ar} is the molar flux of A in the r direction. In the expression above we do not divide by r^2 since this is the independent variable. Dividing by $4\pi\, \Delta r$ and shrinking the differential element to zero gives

$$-\frac{d}{dr}(r^2 N_{Ar}) = 0 \tag{3-68}$$

The flux of A in the r direction is related to the flux of B by writing Eq. (1-34) in the form

$$N_{Ar} = -CD_{AB}\frac{dy_A}{dr} + y_A(N_{Ar} + N_{Br}) \tag{3-69}$$

From the kinetics of the problem, we see that for n moles of A diffusing to the catalyst surface, $1/n$ moles of component B diffuse away. Therefore, the relationship between the flux terms is

$$N_{Br} = -\frac{1}{n}N_{Ar} \tag{3-70}$$

The negative sign in the equation above is the result of species A and B diffusing in opposite directions. Combining Eqs. (3-69) and (3-70) gives the flux term in the radial direction.

$$N_{Ar} = -\frac{CD_{AB}}{1 - y_A(1 - 1/n)}\frac{dy_A}{dr} \tag{3-71}$$

Substitution of the flux expression into Eq. (3-68) and assuming constant diffusivity and constant total concentration gives

$$\frac{d}{dr}\left(\frac{r^2}{1 - y_A(1 - 1/n)}\frac{dy_A}{dr}\right) = 0 \tag{3-72}$$

Integration of the equation above, after separating the variables, gives

$$-\frac{1}{1 - 1/n}\ln\left[1 - y_A\left(1 - \frac{1}{n}\right)\right] = -\frac{K_1}{r} + K_2 \tag{3-73}$$

The boundary conditions for a slow heterogeneous reaction are obtained from Figure 3-4. At the surface of the catalyst, the concentration of A is related to the flux of A and a rate constant as shown in Eq. (3-66). At the outer edge of the stagnant film, which is assumed to have a thickness of δ, the concentration of A is equal to the bulk gas concentration. The resulting boundary conditions are

$$\text{BC1:}\quad y_A = \frac{N_{Ar}}{Ck_1^s} \quad \text{at } r = R \tag{3-74}$$

and

$$\text{BC2:}\quad y_A = y_{A\delta} \quad \text{at } r = \delta R \tag{3-75}$$

Evaluation of the constants gives the concentration profile in the stagnant film.

$$\frac{1 - y_A(1 - 1/n)}{1 - y_{A\delta}(1 - 1/n)} = \left[\frac{1 - (N_{Ar}/Ck_1^s)(1 - 1/n)}{1 - y_{A\delta}(1 - 1/n)}\right]^{(R/r)(\delta R - r)/(\delta R - R)} \tag{3-76}$$

The molar flux of A at any r can be found by first integrating Eq. (3-68) as follows:

$$r^2 N_{Ar} = K_1 \tag{3-77}$$

For transfer in a sphere we note that the molar flow rate in the r direction is the product of the surface area and molar flux. At the catalyst surface this becomes

$$r^2 N_{Ar} = R^2 N_{Ar}|_R \tag{3-78}$$

The molar flux can thus be determined at the surface of the sphere by integrating $r^2 N_{Ar}$ over the thickness of the stagnant film. Upon combining Eqs. (3-71) and (3-78), we obtain

$$R^2 N_{Ar}|_R = r^2 N_{Ar} = -\frac{r^2 C D_{AB}}{1 - y_A(1 - 1/n)} \frac{dy_A}{dr} \tag{3-79}$$

Separating the variables and introducing the limits of integration gives

$$R^2 N_{Ar}|_R \int_R^{\delta R} \frac{dr}{r^2} = -C D_{AB} \int_{N_A/Ck_1^s}^{y_{A\delta}} \frac{dy_A}{1 - y_A(1 - 1/n)} \tag{3-80}$$

After integration, the molar flux may be expressed as

$$N_{Ar}|_R = \frac{\delta R}{\delta R - R} \frac{C D_{AB}}{R} \frac{n}{n-1} \ln\left[\frac{1 - y_{A\delta}(1 - 1/n)}{1 - (N_A/Ck_1^s)(1 - 1/n)}\right] \tag{3-81}$$

Instantaneous reaction on a surface

In a number of experimental cases, it has been shown that the rate of catalyzed chemical reaction can be predicted if the assumption is made that the controlling mechanism is the diffusion of the species A through a stagnant film surrounding the catalyst particle. Therefore, it may be assumed that A will react instantly upon reaching the surface of the catalyst. For a nonporous spherical catalyst, the concentration profile through the stagnant film and the molar flux at the surface of the particle can be determined in a manner similar to that previously shown.

The differential equation obtained in the previous example also applies for the case of instantaneous reaction. Thus the only difference between this and the previous model is the boundary condition at the surface. If the assumption is made that the reaction is very fast, then species A will be consumed the instant it contacts the catalyst. Since A will not accumulate on the surface, the new boundary conditions are

$$\text{BC1:} \quad y_A = 0 \quad \text{at } r = R \tag{3-82}$$

and

$$\text{BC2:} \quad y_A = y_{A\delta} \quad \text{at } r = \delta R \tag{3-83}$$

Therefore, to obtain the molar flux we integrate as in Eq. (3-80) but using the new boundary conditions. After integration we have

$$N_{Ar}|_R = \frac{\delta R}{\delta R - R}\frac{CD_{AB}}{R}\frac{n}{n-1}\ln\left[1 - y_{A\delta}\left(1 - \frac{1}{n}\right)\right] \tag{3-84}$$

A simple comparison of Eqs. (3-81) and (3-84) shows that if the rate constant is very large, Eq. (3-81) reduces to the limiting case given by Eq. (3-84).

Example 3.4

Gas A diffuses through a stagnant gas film to the surface of a nonporous cylindrical catalyst, as shown in Figure 3-5, where it undergoes the reaction $2A \xrightarrow{k_1} B$. Gas B then diffuses from the catalyst surface and is swept away. Neglecting diffusion and reaction on the ends of the particle, derive an equation for the molar flux of A if the reaction is very fast.

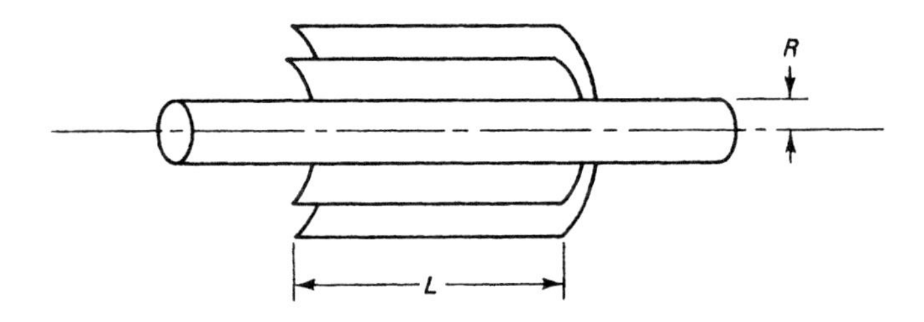

Figure 3-5 Diffusion through a stagnant film

Solution:

At steady-state conditions in which there is no accumulation, the shell balance in the radial direction for a differential element in the stagnant film is

$$2\pi(rLN_{Ar})|_r - 2\pi[(r + \Delta r)LN_{Ar}]|_{r+\Delta r} = 0 \tag{3-85}$$

Dividing by $2\pi L\,\Delta r$ and taking the limit as Δr approaches zero gives

$$-\frac{d}{dr}(rN_{Ar}) = 0 \tag{3-86}$$

From the kinetics of the problem, it is seen that for every 2 mol of A diffusing to the surface only 1 mol of B diffuses away. Thus

$$N_{Br} = -\tfrac{1}{2}N_{Ar} \tag{3-87}$$

From Eq. (1-34) we thus have

$$N_{Ar} = -\frac{CD_{AB}}{1 - \frac{1}{2}y_A}\frac{dy_A}{dr} \tag{3-88}$$

Combining Eqs. (3-86) and (3-88) gives

$$\frac{d}{dr}\left(\frac{CD_{AB}r}{1 - \frac{1}{2}y_A}\frac{dy_A}{dr}\right) = 0 \tag{3-89}$$

The boundary conditions for this model are the same as those for the spherical catalyst.

$$\text{BC1:} \quad y_A = 0 \quad \text{at } r = R \tag{3-90}$$

and

$$\text{BC2:} \quad y_A = y_{A\delta} \quad \text{at } r = \delta R \tag{3-91}$$

The molar flux is obtained by integrating Eq. (3-86) to give $rN_{Ar} = K_1$. At the surface of the catalyst this becomes

$$rN_{Ar} = RN_{Ar}|_R \tag{3-92}$$

Multiplying both sides of Eq. (3-88) by r and combining with Eq. (3-92) gives

$$RN_{Ar}|_R \int_R^{\delta R} \frac{dr}{r} = -\int_0^{y_{A\delta}} CD_{AB} \frac{dy_A}{1 - \frac{1}{2}y_A} \tag{3-93}$$

For constant diffusivity and constant total molar concentration, the flux at the surface is found by integrating the equation above. Thus

$$N_{Ar}|_R = \frac{2CD_{AB}}{R \ln (\delta R/R)} \ln \left(1 - \tfrac{1}{2}y_{A\delta}\right) \tag{3-94}$$

3.6 Steady-State Diffusion Accompanied by Homogeneous Chemical Reaction

Diffusion accompanied by homogeneous reaction occurs in a variety of chemical processes involving both gas–liquid and liquid–liquid systems. This process consists of the diffusion of a reacting species from one phase to the interface of the other phase followed by transfer from the interface into the bulk of the second phase, where reaction takes place. The process continues until both phases are in physical and chemical equilibrium.

An alternative process can be used in some cases for modeling systems in which diffusion and heterogeneous reaction take place. In the two preceding examples, chemical reaction occurred on the surface of the catalyst particle and was described by the use of an appropriate boundary condition. Since a catalyst particle typically has a large internal surface area, the chemical reaction can be described as being distributed homogeneously throughout the particle. Under these conditions, the shell balance is made over a differential element in the solid and the reaction terms thus appear in the balance equation.

Reaction in a porous catalyst

Consider the spherical catalyst shown in Figure 3-6 in which species A diffuses into the catalyst and undergoes an irreversible reaction $A \xrightarrow{k_1} B$ (Satterfield, 1970). Assuming that the reaction is distributed homogeneously throughout the particle and that the conversion of A to B can be expressed by first-order kinetics, the shell balance for a differential volume of the spherical catalyst including the volumetric reaction term is

$$4\pi r^2 N_{Ar}|_r - 4\pi(r + \Delta r)^2 N_{Ar}|_{r+\Delta r} + 4\pi r^2 \, \Delta r \dot{R}_A^v = 0 \tag{3-95}$$

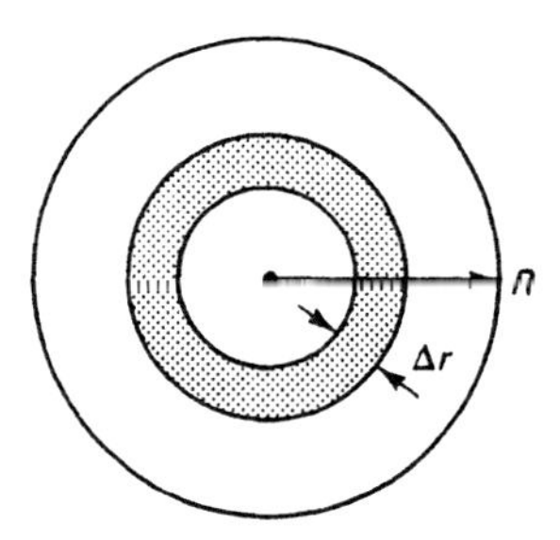

Figure 3-6 Homogeneous chemical reaction in a porous catalyst

where $4\pi r^2 \, \Delta r$ is the differential volume of the sphere. Dividing by constants and differential values, and shrinking the differential element to zero gives

$$-\frac{d}{dr}(r^2 N_{Ar}) + r^2 \dot{R}_A^v = 0 \tag{3-96}$$

In the absence of information regarding the mechanism of transport of A within the pores of the catalyst, the molar flux is related by Fick's first law with an effective diffusivity. The effective diffusivity accounts for the path of the molecules through the pores of the catalyst. For this case the flux is expressed as

$$N_{Ar} = -D_{Ae}\frac{dC_A}{dr} \tag{3-97}$$

For first-order kinetics the rate of reaction is $\dot{R}_A^v = -k_1^v C_A$. Substitution of the rate expression given by Eq. (3-97) into Eq. (3-96) gives the differential equation that describes diffusion in a sphere accompanied by homogeneous chemical reaction.

$$D_{Ae}\frac{1}{r^2}\frac{d}{dr}\left(r^2\frac{dC_A}{dr}\right) = k_1^v C_A \tag{3-98}$$

The boundary conditions that complete the description of this model are obtained by inspection of Figure 3-6. Because of symmetry in the r direction, a concentration gradient will not exist at the center of the particle. Thus we have

$$\text{BC1:} \quad \frac{dC_A}{dr} = 0 \quad \text{or} \quad C_A \text{ is finite} \quad \text{at } r = 0 \tag{3-99}$$

From the figure we see that the concentration at the surface can be specified.

$$\text{BC2:} \quad C_A = C_{AS} \quad \text{at } r = R \tag{3-100}$$

A solution to Eq. (3-98) as shown in its present form is not straightforward. Since a change of variable makes it more amenable to solution, the transformation $f(r) = rC_A$ is introduced. Thus Eq. (3-98) becomes

$$\frac{d^2 f}{dr^2} - \frac{k_1^v}{D_{Ae}} f = 0 \tag{3-101}$$

Equation (3-101) is a second-order, linear, homogeneous differential equation. The solution can be written in terms of concentration as

$$C_A = \frac{K_1}{r}\cosh\left(\sqrt{\frac{k_1^v}{D_{Ae}}}\, r\right) + \frac{K_2}{r}\sinh\left(\sqrt{\frac{k_1^v}{D_{Ae}}}\, r\right) \tag{3-102}$$

where K_1 and K_2 are constants of integration. Upon applying the boundary conditions to evaluate these constants, Eq. (3-102) becomes

$$\frac{C_A}{C_{AS}} = \frac{R}{r} \frac{\sinh \sqrt{k_1^v/D_{Ae}}\, r}{\sinh \sqrt{k_1^v/D_{Ae}}\, R} \tag{3-103}$$

The equation above relates the concentration C_A to distance in the r direction.

In the model above, we assumed that the reaction was distributed throughout the particle, and subsequently the reaction term was included in the material balance. However, the volumetric reaction rate may be expressed in terms of the reaction occurring at a surface. This can be done by relating the rate constants in terms of the surface area per unit volume of catalyst as

$$k_1^v = k_1^s a \tag{3-104}$$

where a is the surface area per unit volume.

The extent to which diffusion effects are important in a reacting system can be expressed by the magnitude of a modified Thiele modulus defined as

$$\phi = \frac{V_p}{S_x}\sqrt{\frac{k_1^v}{D_{Ae}}} = \frac{V_p}{S_x}\sqrt{\frac{ak_1^s}{D_{Ae}}} \tag{3-105}$$

where V_p is the volume of the particle and S_x is the external surface area. For our case (i.e., first-order reaction), the concentration profile is expressed in terms of the Thiele modulus by the equation

$$\frac{C_A}{C_{AS}} = \frac{R \sinh (3\phi r/R)}{r \sinh (3\phi)} \tag{3-106}$$

We see from the equation above that as the Thiele modulus increases, the concentration gradient of A inside the sphere also increases.

An important parameter used to describe mass transfer and chemical reaction in a porous catalyst is the *effectiveness factor*. The effectiveness factor is defined as the ratio of the actual consumption of species A within the particle to the consumption of A if the entire catalyst surface was exposed to the exterior concentration C_{AS}. The total transfer to a catalyst is

$$-k^v \int_0^L SC_A(x)\, dx = -S_x D_{Ae} \frac{dC_A}{dx}\bigg|_{x=L} \tag{3-107}$$

where S is the surface area. For a sphere this becomes

$$-k^v \int_0^R 4\pi r^2 C_A(r)\, dr = -4\pi R^2 D_{Ae} \frac{dC_A}{dr}\bigg|_{r=R} \tag{3-108}$$

If the entire catalyst surface area is exposed to the exterior concentration, the rate is equal to

$$-k^v C_{AS} \int_0^L S\, dx = -k^v C_{AS} \int_0^R 4\pi r^2\, dr \tag{3-109}$$

Therefore, the effectiveness factor for a spherical particle is

$$\eta = \frac{k^v \int_0^R 4\pi r^2 C_A(r)\, dr}{k^v C_{AS} \int_0^R 4\pi r^2\, dr} = \frac{4\pi R^2 D_{Ae}(dC_A/dr)|_{r=R}}{\frac{4}{3}\pi R^3 k^v C_{AS}} \tag{3-110}$$

After evaluating either the integral or the derivative in the equation above, we obtain

$$\eta = \frac{1}{3\phi^2}(3\phi \coth 3\phi - 1) \tag{3-111}$$

The effectiveness factor can be simply described as a term that corrects the maximum possible reaction rate for diffusional resistance inside a catalyst particle. Effectiveness factors for spheres, long cylinders, and flat slabs sealed along the edges are shown in Figure 3-7. When using the modified Thiele modulus given by Eq. (3-105), the curves for η are approximately the same for various values of ϕ for all geometries.

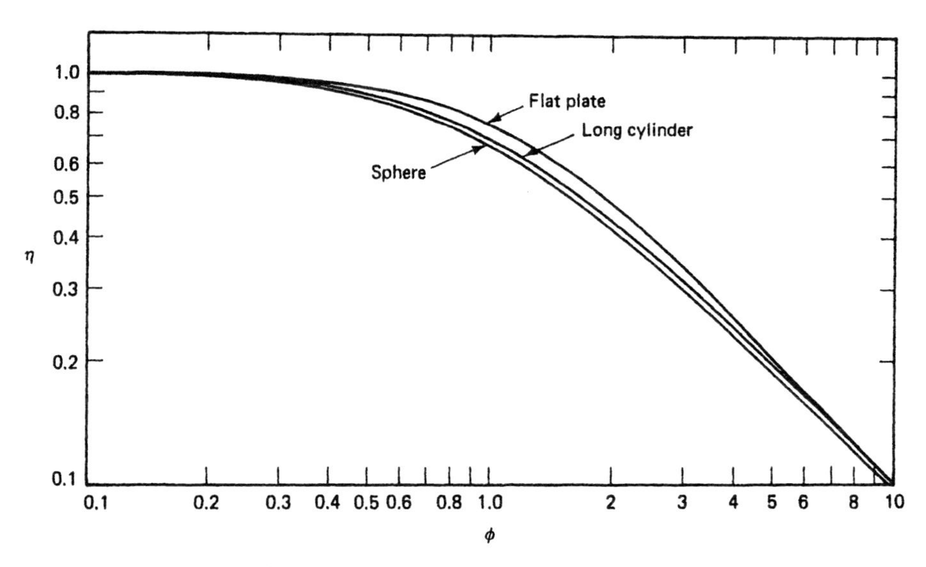

Figure 3-7 Effectiveness factors for porous catalysts (from Aris, 1957)

Diffusion and chemical reaction in a liquid film

Another important example of mass transfer with chemical reaction is the absorption of a solute into a liquid in which it reacts. The absorption of either hydrogen sulfide or carbon dioxide into monoethanolamine followed by chemical reaction are two such examples.

Consider the case discussed by Sherwood and Pigford (1952) in which solute A diffuses from a gas mixture and into a liquid where the following reaction takes place.

$$A + B \xrightarrow{k_1} P \tag{3-112}$$

For the system under consideration, we will assume that A is only slightly soluble in the liquid and its diffusivity is constant. Further, if the concentration of B is much greater than that of A, the reaction can be assumed to be first order with respect to A. If B is nonvolatile, diffusion and the subsequent reaction can be assumed to occur in a thin liquid film.

The concept of a stagnant film of thickness δ on either side of the gas–liquid interface was first proposed by Whitman (1923) and by Lewis and Whitman (1924), who assumed that the transfer of a solute from the gas phase to the liquid phase could be described totally by molecular diffusion. The convective effects associated with the transfer of the solute are thus accounted for by varying the film thickness δ. Although the film model does not provide a realistic description of the physical process for many cases, it is simple to use and generally provides an acceptable description of the overall transport process when the gas–liquid contact time is short. A physically more realistic model will be presented in Chapter 5. The film model is shown in Figure 3-8 for the case in which solute A is transferred from a gas to a liquid phase, where it reacts in the stagnant film.

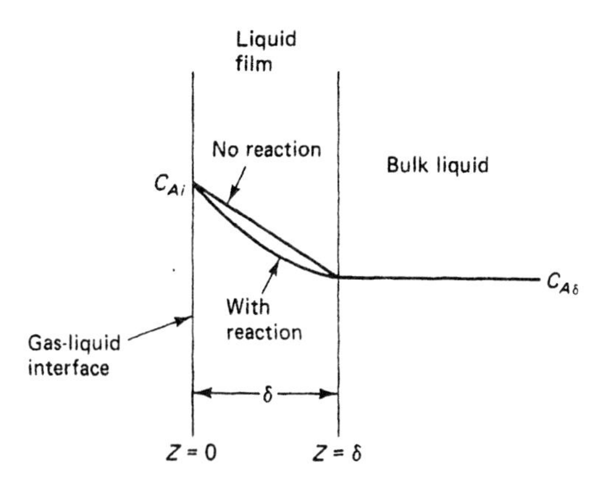

Figure 3-8 Diffusion and chemical reaction in a liquid film

The model that describes diffusion through the film can be obtained by using Eq. (D) from Table 3-2. For the Z direction,

$$\frac{\partial C_A}{\partial t} + U_Z^m \frac{\partial C_A}{\partial Z} - D_{AB} \frac{\partial^2 C_A}{\partial Z^2} = \dot{R}_A^v \tag{3-113}$$

Since the diffusion and reaction of A are assumed to be at steady-state conditions in this model, the first term can be eliminated. Further, the assumption that A is only slightly soluble in the liquid permits us to neglect the bulk flow term,

$U_Z^m(\partial C_A/\partial Z)$. The resulting equation for the film model in which diffusion is accompanied by chemical reaction is

$$D_{AB}\frac{d^2C_A}{dZ^2} + \dot{R}_A^v = 0 \tag{3-114}$$

For a first-order reaction

$$\dot{R}_A^v = \text{generation} - \text{consumption} = -k_1^v C_A \tag{3-115}$$

From the figure the boundary conditions are

$$\text{BC1:} \quad C_A = C_{Ai} \quad \text{at } Z = 0 \tag{3-116}$$

and

$$\text{BC2:} \quad C_A = C_{A\delta} \quad \text{at } Z = \delta \tag{3-117}$$

The concentration profile can be found by solving the differential equation above with the appropriate boundary conditions. The concentration profile is

$$\frac{C_A}{C_{Ai}} = \frac{(C_{A\delta}/C_{Ai})\sinh \delta[\sqrt{k_1^v/D_{AB}}\,(Z/\delta)] + \sinh[\delta\sqrt{k_1^v/D_{AB}}\,(1 - Z/\delta)]}{\sinh(\delta\sqrt{k_1^v/D_{AB}})} \tag{3-118}$$

The effects of the chemical reaction on the concentration profile can be seen by solving Eq. (3-114) for the case in which reaction does not occur, that is,

$$\frac{d^2C_A}{dZ^2} = 0 \tag{3-119}$$

Using the same boundary conditions given above, we obtain the concentration profile

$$\frac{C_A}{C_{Ai}} = \frac{C_{A\delta}}{C_{Ai}}\frac{Z}{\delta} + \left(1 - \frac{Z}{\delta}\right) \tag{3-120}$$

The upward-concave concentration profile shown in Figure 3-8 is the result of chemical reaction. For very slow reaction, in which the rate constant approaches zero, the concentration profile given by Eq. (3-118) becomes linear and can be represented by Eq. (3-120). By contrast, as the term $\delta\sqrt{k_1^v/D_{AB}}$ increases, the curve becomes more concave.

The effects of chemical reaction taking place in a system can be better observed by comparing the flux rates for a reacting and a nonreacting system. In the absence of a chemical reaction, the flux is found by using the concentration profile given in Eq. (3-120).

$$N_{AZ}|_{Z=0} = -D_{AB}\frac{dC_A}{dZ}\bigg|_{Z=0} = \frac{D_{AB}}{\delta}(C_{Ai} - C_{A\delta}) \tag{3-121}$$

For chemical reaction, Eq. (3-118) is used to give

$$N_{AZ,R}|_{Z=0} = -D_{AB}\frac{dC_A}{dZ}\bigg|_{Z=0} = \frac{D_{AB}}{\delta}\frac{\delta\sqrt{k_1^v/D_{AB}}}{\tanh(\delta\sqrt{k_1^v/D_{AB}})} \times \left[C_{Ai} - \frac{C_{A\delta}}{\cosh(\delta\sqrt{k_1^v/D_{AB}})}\right] \tag{3-122}$$

Since $(\delta\sqrt{k_1^v/D_{AB}})/\tanh(\delta\sqrt{k_1^v/D_{AB}})$ and $\cosh(\delta\sqrt{k_1^v/D_{AB}})$ are always equal to or greater than 1, the flux for a first-order irreversible chemical reaction will be greater than the flux for the case of no reaction.

Example 3.5

Component A diffuses from an inert gas into a liquid film, where it reacts reversibly as shown in Figure 3-8. For the reaction

$$A \underset{k_2}{\overset{k_1}{\rightleftharpoons}} P \tag{3-123}$$

obtain the differential equations that describe the diffusion and reaction of component A. Assume that the solution is sufficiently dilute, such that the diffusion coefficients are constant and can be considered to be those of the binary system consisting of the particular species and the solvent.

Solution:

Neglecting bulk flow and assuming steady-state conditions, Eq. (D) from Table 3-2 can be written for each component in the Z direction as

$$D_A \frac{d^2 C_A}{dZ^2} + \dot{R}_A^v = 0 \tag{3-124}$$

and

$$D_P \frac{d^2 C_P}{dZ^2} + \dot{R}_P^v = 0 \tag{3-125}$$

For first-order reversible reactions we have

$$\dot{R}_A^v = k_2^v C_P - k_1^v C_A \tag{3-126}$$

and

$$\dot{R}_P^v = k_1^v C_A - k_2^v C_P \tag{3-127}$$

Differential equations (3-124) and (3-125) thus become

$$D_A \frac{d^2 C_A}{dZ^2} - k_1^v C_A + k_2^v C_P = 0 \tag{3-128}$$

and

$$D_P \frac{d^2 C_P}{dZ^2} + k_1^v C_A - k_2^v C_P = 0 \tag{3-129}$$

Using the appropriate boundary conditions, Eqs. (3-128) and (3-129) can be solved simultaneously to give the concentration profile for each species. Typical boundary conditions for species A are

$$\text{BC1:} \quad C_A = C_{Ai} \quad \text{at } Z = 0 \tag{3-130}$$

and

$$\text{BC2:} \quad C_A = C_{A\delta} \quad \text{at } Z = \delta \tag{3-131}$$

Boundary conditions for P must also be introduced before a solution to Eqs. (3-128) and (3-129) can be obtained.

Olander (1960) provided a solution to the set of equations above for the case in which the reaction rates were fast enough so that equilibrium was maintained. At equilibrium the reaction rate equals zero; thus the concentrations can be related by the mass action constant:

$$K = \frac{k_1^v}{k_2^v} = \frac{C_P}{C_A} \tag{3-132}$$

Equations (3-128) and (3-129) can be combined to give

$$D_A \frac{d^2 C_A}{dZ^2} + D_P \frac{d^2 C_P}{dZ^2} = 0 \tag{3-133}$$

The equation above can be solved by using the boundary conditions given in Eqs. (3-130) and (3-131) and Eq. (3-132). The concentration profile is

$$D_A C_A + D_P C_P = \frac{Z}{\delta}(C_{A\delta} - C_{Ai})(D_A + KD_P) + C_{Ai}(D_A + KD_P) \tag{3-134}$$

The transfer rate of species A from the inert gas phase is equal to the flux of the total A component within the reacting phase. Thus

$$N_{AZ,R} = -D_A \frac{dC_A}{dZ} - D_P \frac{dC_P}{dZ} \tag{3-135}$$

$$= \frac{D_A}{\delta}(C_{Ai} - C_{A\delta})\left(1 + K\frac{D_P}{D_A}\right) \tag{3-136}$$

A comparison of Eqs. (3-121) and (3-136) shows that the increase in the flux for a reacting system is related to the magnitude of the mass action constant.

Frequently, the increase in the flux for reacting systems is related to a mass transfer coefficient. For cases in which bulk flow can be neglected and in the absence of chemical reaction, the mass transfer coefficient is defined as the ratio of the flux to the concentration gradient. From Eq. (3-121) we obtain

$$\frac{N_{AZ}}{C_{Ai} - C_{A\delta}} = \frac{D_A}{\delta} = k_c \tag{3-137}$$

Thus from Eq. (3-136) the mass transfer coefficient is

$$\frac{N_{AZ,R}}{C_{Ai} - C_{A\delta}} = \frac{D_A}{\delta}\left(1 + K\frac{D_P}{D_A}\right) = k_{c,R} \tag{3-138}$$

Therefore, the mass transfer coefficients are related as shown:

$$k_{c,R} = k_c\left(1 + K\frac{D_P}{D_A}\right) \tag{3-139}$$

The ratio $k_{c,R}/k_c$ is defined as the *enhancement factor*. It is generally expressed graphically as a function of the reaction rate, as shown by Sherwood et al. (1975). Extensive discussions of mass transfer with homogeneous chemical reaction are given by Astarita (1966) and Danckwerts (1970).

REFERENCES

ARIS, R., *Chem. Eng. Sci.*, *6*, 265 (1957).

ASTARITA, G., *Mass Transfer with Chemical Reaction*, Elsevier, Amsterdam, 1966.

BIRD, R. B., W. E. STEWART, and E. N. LIGHTFOOT, *Transport Phenomena*, Wiley, New York, 1960.

DANCKWERTS, P. V., *Gas–Liquid Reactions*, McGraw-Hill, New York, 1970.

LEWIS, W. K., and W. G. WHITMAN, *Ind. Eng. Chem.*, *16*, 1215 (1924).

OLANDER, D. R., *AIChE J.*, *6*, 233 (1960).

SATTERFIELD, C. N., *Mass Transfer in Heterogeneous Catalysis*, MIT Press, Cambridge, Mass., 1970.

SHERWOOD, T. K., and R. L. PIGFORD, *Absorption and Extraction*, 2nd ed., McGraw-Hill, New York, 1952.

SHERWOOD, T. K., R. L. PIGFORD, and C. R. WILKE, *Mass Transfer*, McGraw-Hill, New York, 1975.

WHITMAN, W. G., *Chem. Met. Eng.*, *29*, 146 (1923).

NOTATIONS

a = surface area per unit volume, L^2/L^3

C = total molar concentration, mol/L^3

C_A = molar concentration of A, mol/L^3

C_{Ai} = initial molar concentration of A, mol/L^3

C_{As} = molar concentration of A at interface, mol/L^3

$C_{A\delta}$ = molar concentration of A in bulk fluid, mol/L^3

D_{AB} = binary diffusivity for system A–B, L^2/t

D_{Ae} = effective diffusivity of A, L^2/t

k_c, k_y = convective mass transfer coefficients, mol/tL^2 (concentration difference)

k^s, k^v = reaction rate constants, L/t, t^{-1}

K = integration constant

n = number of moles, mol

N_{AX}, N_{AY}, N_{AZ} = molar flux of A with respect to a fixed reference frame in the X, Y, and Z directions, mol/tL^2

$\mathbf{N}_A$, $\mathbf{N}_B$ = molar flux of A and B with respect to a fixed reference frame, mol/tL^2

P = pressure of the system, F/L^2

$(P_B)_M$ = logarithmic mean pressure of B, F/L^2

$\bar{P}_A$, $\bar{P}_B$ = partial pressures of species A and B, F/L^2

$\dot{r}_A^v$, $\dot{r}_B^v$ = net mass rate of production of A and B, M/tL^3

R = gas constant; radius of a tube or cylinder, L

$\dot{R}_A^s$ = net molar rate of production of A at a surface, mol/tL^2

$\dot{R}_j^v$ = net molar rate of production of j, mol/tL^3

S = cross-sectional area, L^2

S_x = external surface area, L^2

t = time, t

T = temperature, T

U_Z^m = velocity in the Z direction, L/t

$\bar{U}$ = average velocity, L/t

$\mathbf{U}_A$ = velocity of A, L/t

U^m = mass average velocity, L/t
U^M = molar average velocity, L/t
U^v = volume average velocity, L/t
U^0 = general reference velocity, L/t
V = volume of the system, L^3
V_p = particle volume, L^3
$\bar{V}_A$ = partial molar volume of A, L^3
$(y_B)_M$ = logarithmic mean mole fraction
Z_l = length of reactor, L

Greek Letters

δ = film thickness, L
η = effectiveness factor
ϕ = modified Thiele modulus
ρ_A, ρ_B = mass concentration of A and B, M/L^3
ω_A = mass fraction of A

PROBLEMS

3.1 A tank initially contains 1.4 m³ of brine solution with a concentration of 16.0 kg of salt per cubic meter of solution. A brine solution containing 48.0 kg of salt per cubic meter of solution enters the tank at a rate of 8.5 m³/h and is mixed perfectly with the existing solution. If the resulting brine solution leaves the tank at a rate of 8.5 m³/h, determine the time required for the salt in the tank to reach a concentration of 32.0 kg per cubic meter of solution.

3.2 A tank containing 3.8 m³ of 20% NaOH solution is to be purged by adding pure water at a rate of 4.5 m³/h. If the solution leaves the tank at a rate of 4.5 m³/h, determine the time necessary to purge 90% of the NaOH by mass from the tank. Assume perfect mixing. The specific gravity of NaOH is 1.22.

3.3 In Problem 3.1 the flow rates entering and leaving the tank were equal. **(a)** If the exiting stream is reduced to 5 m³/h, determine the time required for the salt in the tank to reach a concentration of 24.0 kg per cubic meter of solution. **(b)** What is the volume of solution in the tank when this concentration is reached?

3.4 Pure water enters the first of two tanks connected in series at a flow rate of 4.5 m³/h. The effluent from the first tank then enters the second tank also at a rate of 4.5 m³/h. Initially, the first tank contains 1.9 m³ of 20% NaOH and the second tank contains 1.15 m³ of 40% NaOH. Determine the time required for the stream leaving the second tank to reach a concentration of 5% NaOH if the exiting flow rate is 4.5 m³/h. Assume perfect mixing. $\rho(NaOH) = 1220$ kg/m³.

3.5 The irreversible reaction $A \xrightarrow{k_1} P$ is carried out in a reactor that initially contains 1.5 m³ of solution with a concentration of 1 kgmol of A per cubic meter of solution. A solution containing 5×10^{-3} kgmol of A/m³ enters the tank at a rate of 10 m³/h and is mixed perfectly with the existing solution. If the resulting product solution leaves the reactor at 5 m³/h, obtain an expression for concentration of A as a function of time. Assume an elementary reaction.

3.6 The irreversible reaction $A \xrightarrow{k_1} B$ is carried out in a continuous flow reactor. The inlet and outlet flow rates are 5.0 m³/h, with an inlet concentration of $C_{Ai} = 0.18$ kgmol/m³. The reaction is first order, with $k_1 = 8.5$ m³/h. Initially, the tank contains 5.6 m³ of solvent B and no A. Calculate the outlet concentration **(a)** solute 1 h, and **(b)** after 3 h. Assume perfect mixing.

3.7 Consider the series of three tanks shown in Figure 3-9. Initially each tank contains V_0 m³ of solution with a concentration of C_{A0}. If a solution with a concentration of C_{Ai} enters the first tank at a rate of l m³/h and the solution leaves each tank at the same rate, determine an equation that allows us to calculate the concentration of solute A in the solution leaving the last tank as a function of time. Assume perfect mixing.

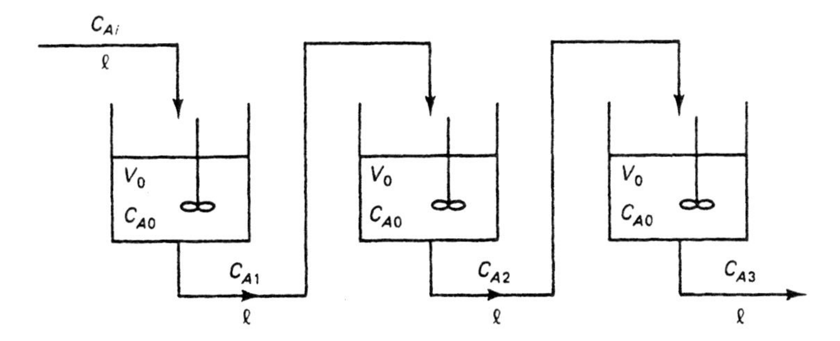

Figure 3-9 Problem 3.7: Tanks in series

3.8 A 0.15-m-long test tube containing ethanol is left open in the laboratory. The level of ethanol is initially 0.1 m below the top. The temperature in the laboratory is 26°C and the atmospheric pressure is 0.987 atm. The vapor pressure of ethanol is 0.08 atm. Determine **(a)** an expression for the concentration profile of ethanol in air in the test tube if the liquid level can be held constant, **(b)** an expression for the instantaneous molar flux of ethanol, and **(c)** the time required for the level of ethanol to decrease by 0.005 m if the evaporation rate does not change with time. The test tube diameter is 0.015 m. ρ(ethanol) = 784 kg/m³.

3.9 A 0.20-m-long test tube was used to study the diffusion process in which liquid A diffuses into gas B. In one study the level of liquid A was initially 0.1 m below the top of the tube. The temperature of the diffusion process was 25°C and the total pressure was maintained at 1 atm. The molar flux of component A at the top of the test tube was found to be 1.6×10^{-3} kgmol/m²·h. Find the diffusion coefficient for A into gas B. Assume that gas B is insoluble in liquid A. The partial pressure of A at the surface of the liquid was 0.06 atm.

3.10 Consider the steady-state evaporation of liquid A through a small-diameter tube into a stagnant mixture of gases B, C, and D as shown in Figure 3-10. Assume that the gas mixture is insoluble in liquid A and there is no chemical reaction. The system temperature and pressure are constant. Obtain the concentration profile for species A. (*Hint:* Use the Stefan–Maxwell equation presented in Chapter 2.)

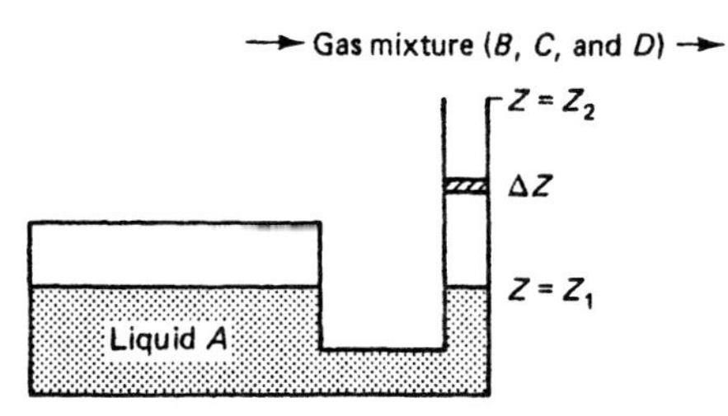

Figure 3-10 Problem 3.10: Evaporation into a multicomponent gas mixture

3.11 Gas A diffuses through two immiscible liquids contained in a capillary tube as shown in Figure 3-11. The concentration of A at the bottom of the capillary is maintained at a constant value. The partial pressure of A in the gas is 0.05 atm. The equilibrium relationships for A in the two liquids are

$$C_A^{\mathrm{I}} = 200P_A$$

and

$$C_A^{\mathrm{I}} = 2C_A^{\mathrm{II}}$$

where P_A has units of atm and C_A has units of kgmol/m³. The concentration of A at Z_2 is $C_{AI,2}^{\mathrm{II}} = 3.0$ kgmol/m³. $D_{A\mathrm{I}} = 1.5 \times 10^{-9}$ m²/s. $D_{A\mathrm{II}} = 7.5 \times 10^{-9}$ m²/s. $Z_2 - Z_1 = 0.05$ m. $Z_1 - Z_0 = 0.02$ m. Assume that convective effects at the bottom of the capillary are negligible and liquid I has a very low vapor pressure. Calculate the molar flux of A dissolving into liquid II.

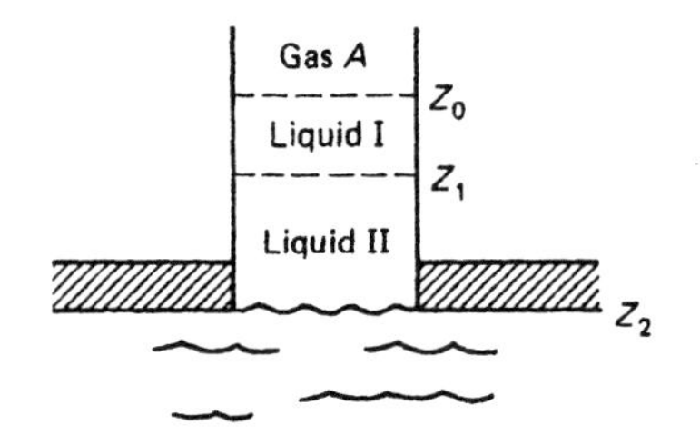

Figure 3-11 Problem 3.11: Mass transfer through immiscible liquids

3.12 As shown in Figure 3-12, a binary mixture of gases A and B are circulated through the center of a hollow sphere. Although the sphere is porous, only species A can diffuse through the walls. The concentration of A at the inner surface is maintained constant at C_{A1}. The flux of A at the outer surface can be described by the expression $N_A = k_c^*(C_{A2} - C_{A\infty})$, where C_{A2} is the concentration of A at R_2 and $C_{A\infty}$ is the concentration of A in the fluid that flows past the sphere. Obtain an expression for the molar flow rate of A through the porous sphere. Your solution should not be expressed in terms of the concentration of A at R_2.

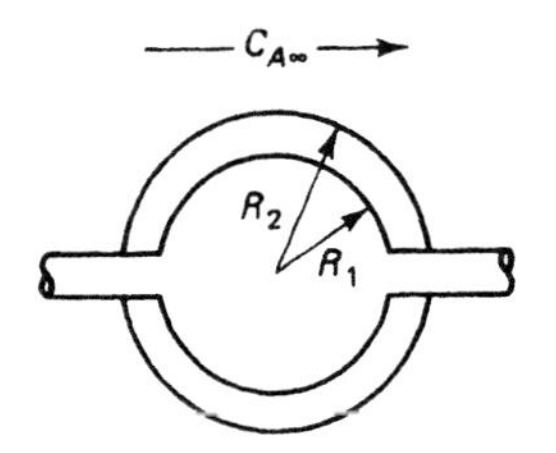

Figure 3-12 Problem 3.12: Diffusion through a porous sphere

3.13 In vacuum systems it is common practice to attach the vacuum pump to the apparatus to be evacuated with a heavy-walled rubber hose. Using the shell balance method, derive **(a)** an expression for the concentration profile of air in the rubber hose, and **(b)** an equation to predict the instantaneous rate at which air diffuses through the tube wall into the vacuum system. The concentration of air at the outer and inner tube surfaces, R_o and R_i, are C_{Ao} and C_{Ai}, respectively.

3.14 Gas A diffuses through a gas film to the surface of a cylindrical catalyst particle where it undergoes the reaction $2A \xrightarrow{k_1} B$. Gas B then diffuses from the catalyst surface and is swept away. **(a)** Neglecting diffusion and reaction on the ends of the particle, derive an expression for the concentration profile of A in the stagnant film surrounding the cylinder for a slow reaction. **(b)** Obtain an expression for the molar flux of B. Assume that the thickness of the gas film is δR.

3.15 The mass flux that results from an external force F can be expressed as

$$J' = C_A MF$$

where C_A is the molar concentration, M is the mobility, and F is the force. Obtain an expression for the steady-state concentration profile of species A if a cylindrical test tube filled with a solution containing A and B is centrifuged. Assume that the centrifuge rotates with an angular velocity ω.

3.16 The reaction $3A \xrightarrow{k_1} B$ takes place in the interior of a cylindrical catalyst. Neglecting diffusion through the ends of the porous cylinder, obtain an expression for the concentration profile and molar flux of A in the radial direction. The concentration of A at the outer surface of the catalyst is C_{AS}.

3.17 Derive an expression for the effectiveness factor for a cylindrical porous catalyst in which the reaction $A \xrightarrow{k_1} B$ occurs. Assume that the reaction is first order. Neglect diffusion in the axial direction. At $r = R$, $C_A = C_{AS}$.

3.18 Obtain an equation for the effectiveness factor for the first-order chemical reaction, $A \xrightarrow{k_1} B$, taking place in the thin rectangular slab shown in Figure 3-13. Assume that mass is transferred in only the X direction and neglect diffusion through the edges of the catalyst. The concentration at $X = \pm L$ is C_{AS}.

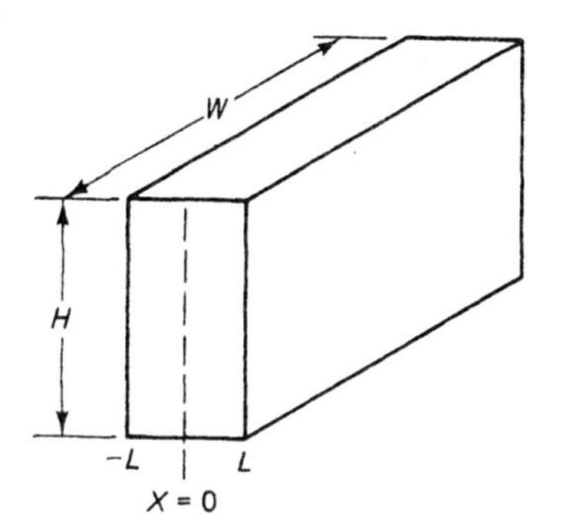

Figure 3-13 Problem 3.18: Effectiveness factor for a thin rectangular slab

3.19 The liquid-phase isomerization $A \xrightarrow{k_1} B$ takes place in a metal sponge catalyst as shown in Figure 3-14. The process can be modeled as the diffusion of A in a long, thin tube. Upon reaching the end of the tube it reacts and B diffuses back. Assume that A and B form a solution which is nearly ideal and can be described by the

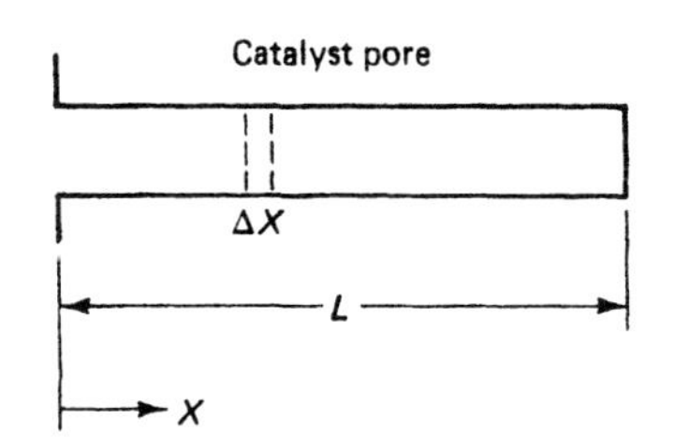

Figure 3-14 Problem 3.19: Diffusion and chemical reaction in a long, thin tube

expression

$$\ln \gamma_A = A x_B^2$$

where γ_A is the activity coefficient of A. Obtain the concentration profile for A in the pore as a function of position. Assume that the pores are large and ordinary diffusion occurs. At $Z = 0$, $x_A = 1.0$ and at $Z = L$, $x_A = x_{AL}$.

3.20 The gas phase reaction $A \xrightarrow{k_1} B$ was studied at 3 atm and 100°C over an MOS on alumina catalyst. Two experiments were performed in a batch reactor that contained 0.02 kg of catalyst, and the following data were obtained.

Particle Size	*Conversion, x*	*Feed Rate (m^3/h)*
Powder	0.9	0.085
9.5×10^{-3} m (sphere)	0.6	0.085

The reaction is known to be first order in A. Find the effective diffusivity of A into the catalyst particle. (*Note:* For the case of the powder, the effectivenes factor η is 1.) The density of the catalyst is 399 kg/m³.

3.21 A first-order reaction $A \xrightarrow{k_1} P$ takes place inside a spherical catalyst. The catalyst is 6×10^{-3} m in diameter and has a surface area of 2×10^8 m²/m³ catalyst. **(a)** Obtain an expression for the concentration profile of A inside the catalyst, and **(b)** determine the value of the effectiveness factor. Data:

Effective diffusivity $D_{Ae} = 1.26 \times 10^{-4}$ m²/h

Reaction rate constant $= 1.728 \times 10^{-7} \dfrac{1}{(\text{h})(\text{m}^2/\text{m}^3 \text{ cat.})}$

Concentration of A at the catalyst surface is 28 kgmol/m³

3.22 The slow reaction $A \xrightarrow{k_1} B + 2C$ takes place on the surface of a flat catalyst. The reaction is first order in A. Obtain **(a)** the concentration profile of A in the film surrounding the surface and **(b)** an expression for the flux of A at $Z = \delta$.

3.23 In Example 3.5 a set of equations was derived for the reversible reaction

$$A \underset{k_2}{\overset{k_1}{\rightleftharpoons}} P$$

It was further shown that for a very fast reaction, in which equilibrium was maintained, a solution could be readily found. For the case in which equilibrium is not maintained, obtain an expression for the flux of A. Assume that the diffusivities for the reactant and product are constant and equal.

3.24 Component A reacts reversibly in a liquid film as shown by the expression

$$A \underset{k_2}{\overset{k_1}{\rightleftharpoons}} 2P$$

Assuming that the reaction is very fast and equilibrium is maintained in the film, **(a)** obtain an expression for the concentration profile of A. **(b)** Also obtain an expression for the transfer rate of A. Assume that the diffusivities are constant and equal for both species. At the gas–liquid interface the concentration of A is C_{Ai}; at the edge of the film the concentration is $C_{A\delta}$.

3.25 Solute A diffuses from a liquid phase I, where its molar concentration is C_{Ai}, into liquid II where it reacts. The reaction is first order with respect to the solute A. The molar flux at a depth L in liquid II falls to one-fourth of its value at the liquid–liquid interface:

$$N_{AZ}|_{Z=0} = 4N_{AZ}|_{Z=L}$$

Obtain an expression for **(a)** the concentration of A in liquid II as a function of the depth from the surface, and **(b)** the molar flux at the liquid–liquid interface. Assume that the process is carried out at steady-state. The process is shown in Figure 3-15.

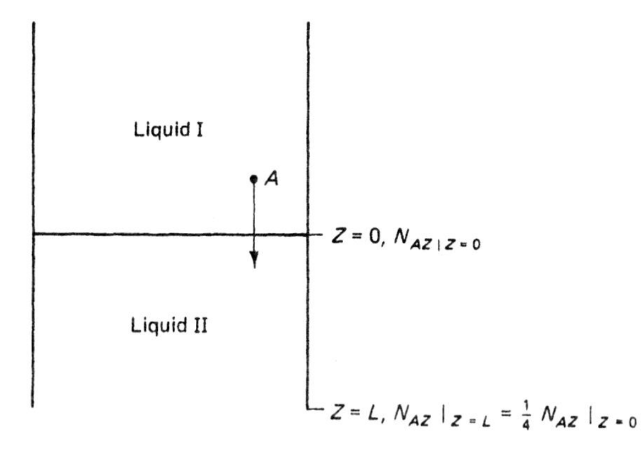

Figure 3-15 Problem 3.25: Diffusion and chemical reaction in immiscible liquids

4 Partial Differential Equations of Diffusion

4.1 Introduction

In the preceding chapter two basic types of problems were considered. These were unsteady-state problems in which the concentration was independent of spacial coordinates, and simple diffusion problems in which mass was transferred in one dimension under steady-state conditions. Although these simple models provide certain types of information, they quite often fail to give a complete picture of the transport process. In this chapter, complex models will be developed and several different mathematical techniques will be employed to obtain solutions to the partial differential equations that describe the mass transfer process. These techniques will include combination of variables, separation of variables, Laplace transforms, and the method of weighted residuals. Both steady- and unsteady-state problems in rectangular, cylindrical, and spherical coordinates will be discussed. Because of the level of mathematics, this chapter should be reserved for seniors and first-year graduate students. For those interested in a detailed treatment of a wide variety of diffusion problems, the treatise of Crank (1956) is recommended.

4.2 Combination of Variables

The success of the combination of variables method depends on the selection of a similarity transformation that can be used to reduce the partial differential equation to an ordinary one, with the appropriate reduction in the number of boundary conditions. The similarity transformation to be used here can be found by a dimensional analysis of the partial differential equation. It has been used successfully in solving unsteady-state problems when the spacial dimension extends to infinity.

Two-dimensional mass transfer in wetted-wall columns

To demonstrate the combination of variables technique, we will consider the absorption of solute A from an inert gas into a liquid film B flowing in laminar flow inside a wetted-wall column. Because the thickness of the liquid film is very small when compared to the radius of the column, rectangular coordinates may be used, as shown by Figure 4-1. Further, we will assume that

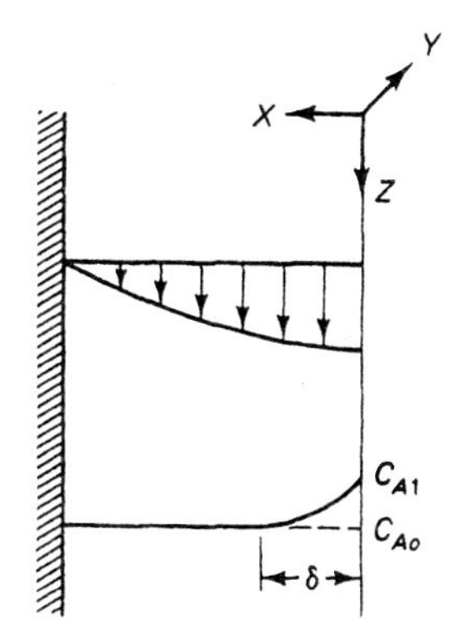

Figure 4-1 Diffusion into a falling film

the velocity profile in the Z direction is fully developed and the process is carried out under steady-state conditions. From the figure we see that species A is transferred in both the X and Z directions. Thus we can obtain the differential equation for mass transfer by making a shell balance on a differential element of the falling film shown in Figure 4-2. At steady-state and in the absence of chemical reaction, the molar rate balance is

$$W\,\Delta Z\,N_{AX}|_X - W\,\Delta Z\,N_{AX}|_{X+\Delta X} + W\,\Delta X\,N_{AZ}|_Z - W\,\Delta X\,N_{AZ}|_{Z+\Delta Z} = 0 \tag{4-1}$$

where W is the thickness of the film in the Y dimension. Dividing by $W\,\Delta X\,\Delta Z$ and letting ΔX and ΔZ approach zero, we obtain the partial differential equation

$$\frac{\partial N_{AX}}{\partial X} + \frac{\partial N_{AZ}}{\partial Z} = 0 \tag{4-2}$$

The equation above could also have been obtained by deleting terms in Eq. (A) of Table 3-2. The concentration of A in the liquid at the gas–liquid interface is C_{A1} and the concentration of the entering liquid can be set at some arbitrary

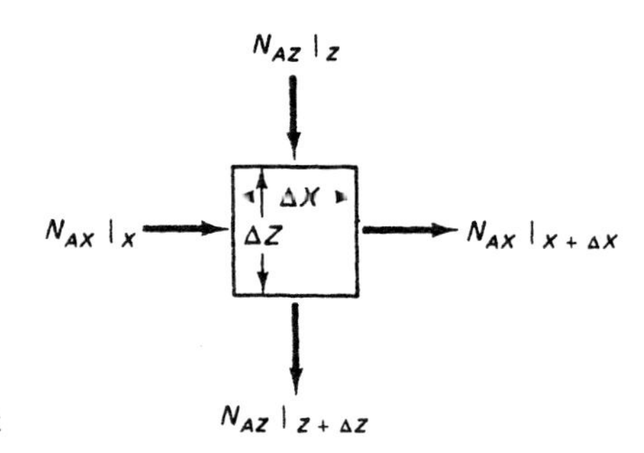

Figure 4-2 Differential element

value, C_{A0}. If we assume that A is only slightly soluble in liquid B and that the rate of diffusion in the X direction is slow, then A will penetrate a very small distance δ during the time the gas and liquid are in contact. The concentration of A in the film will be equal to C_{A0} for distances greater than δ. Thus the total film thickness can be considered to be infinite with respect to the penetration distance δ. The boundary conditions for the described process are

$$\text{BC1:} \quad C_A = C_{A0} \quad \text{at } Z = 0 \quad \text{for all } X \tag{4-3}$$

$$\text{BC2:} \quad C_A = C_{A1} \quad \text{at } X = 0 \quad \text{for all } Z \tag{4-4}$$

$$\text{BC3:} \quad C_A = C_{A0} \quad \text{at } X = \infty \quad \text{for all } Z \tag{4-5}$$

Consideration must now be given to the flux terms in order to complete the model. From Eq. (1-27) for species A, the flux term in the X direction is

$$N_{AX} = -D_{AB}\frac{\partial C_A}{\partial X} + C_A U_X^M \tag{4-6}$$

or

$$N_{AX} = -D_{AB}\frac{\partial C_A}{\partial X} + x_A(N_{AX} + N_{BX}) \tag{4-7}$$

as shown in Table 1-2. Since A is only slightly soluble in B, the mole fraction x_A will be very small. As a result, the second term of Eq. (4-6) can be neglected, thus permitting us to write the flux in the X direction as

$$N_{AX} = -D_{AB}\frac{\partial C_A}{\partial X} \tag{4-8}$$

The flux term in the Z direction is

$$N_{AZ} = -D_{AB}\frac{\partial C_A}{\partial Z} + C_A U_Z^M \tag{4-9}$$

Since we have convective flow in the Z direction, the contribution of the diffusional flux in the Z dimension compared to the overall flux in that dimension will be very small and subsequently can be neglected. Therefore, Eq. (4-9) becomes

$$N_{AZ} \cong C_A U_Z^M \tag{4-10}$$

Due to the small penetration of A into the liquid, we will be concerned only with the velocity at the surface which can be replaced by $U_{Z,\max}$, which is a constant. Substituting Eqs. (4-8) and (4-10) into Eq. (4-2) and assuming that the diffusion coefficient is constant gives

$$D_{AB}\frac{\partial^2 C_A}{\partial X^2} = U_{Z,\max}\frac{\partial C_A}{\partial Z} \tag{4-11}$$

or

$$D_{AB}\frac{\partial^2 C_A}{\partial X^2} = \frac{\partial C_A}{\partial(Z/U_{Z,\max})} \tag{4-12}$$

It is interesting to note that if $Z/U_{Z,\max}$ is replaced by t, an unsteady-state partial differential equation results. The unsteady-state form of the equation could have been derived by letting the volume element shown in Figure 4-1 move with the velocity of the film near the gas–liquid interface.

A solution to Eq. (4-12) will be obtained by using the combination of variables technique by assuming that the dependent variable, concentration, can be related to only one independent variable with a similarity transformation. Before proceeding with the solution, the concentration will be written in dimensionless form by defining a new variable:

$$f = \frac{C_A - C_{A0}}{C_{A1} - C_{A0}} \tag{4-13}$$

If we write the concentration as shown above, the differential equation is nondimensional and two of the boundary conditions are homogeneous. Upon introducing the dimensionless variable given by Eq. (4-13), Eq. (4-12) can be written as

$$D_{AB}\frac{\partial^2 f}{\partial X^2} = \frac{\partial f}{\partial(Z/U_{Z,\max})} \tag{4-14}$$

with the new boundary conditions

$$\text{BC1:}\quad f = 0 \quad \text{at } Z = 0 \quad \text{for all } X \tag{4-15}$$

$$\text{BC2:}\quad f = 1 \quad \text{at } X = 0 \quad \text{for all } Z \tag{4-16}$$

$$\text{BC3:}\quad f = 0 \quad \text{at } X = \infty \quad \text{for all } Z \tag{4-17}$$

Equation (4-14) can be transformed to an ordinary differential equation by combining the X and Z variables in terms of the following similarity variable:

$$\eta = \frac{X}{(4D_{AB}Z/U_{Z,\max})^{1/2}} \tag{4-18}$$

Thus

$$\frac{\partial f}{\partial X} = \frac{df}{d\eta}\frac{\partial \eta}{\partial X} \tag{4-19}$$

$$\frac{\partial^2 f}{\partial X^2} = \frac{\partial}{\partial X}\left(\frac{\partial f}{\partial X}\right) = \frac{d^2 f}{d\eta^2}\left(\frac{\partial \eta}{\partial X}\right)^2 \tag{4-20}$$

and

$$\frac{\partial f}{\partial(Z/U_{Z,\max})} = \frac{df}{d\eta}\left[\frac{\partial \eta}{\partial(Z/U_{Z,\max})}\right] \tag{4-21}$$

After taking the derivative of η with respect to the appropriate variable and substituting these values into Eqs. (4-20) and (4-21), we obtain

$$\frac{d^2f}{d\eta^2} + 2\eta\frac{df}{d\eta} = 0 \tag{4-22}$$

The boundary conditions, which must also be transformed with the similarity transformation, are

$$\text{BC1:} \quad f = 0 \quad \text{at } \eta = \infty \tag{4-23}$$

$$\text{BC2:} \quad f = 1 \quad \text{at } \eta = 0 \tag{4-24}$$

It should be noted that following the transformation, only two independent boundary conditions remain.

To solve Eq. (4-22) let $\psi = df/d\eta$. Thus

$$\frac{d\psi}{d\eta} + 2\eta\psi = 0 \tag{4-25}$$

Integration gives

$$\psi = K_1 e^{-\eta^2} \tag{4-26}$$

After substituting $df/d\eta$ for ψ and integrating a second time, we have

$$f = K_1 \int_0^\eta e^{-\eta^2}\, d\eta + K_2 \tag{4-27}$$

The second boundary condition gives $K_2 = 1$, and from the first boundary condition

$$K_1 = -\frac{1}{\int_0^\infty e^{-\eta^2}\, d\eta} = -\frac{2}{\sqrt{\pi}} \tag{4-28}$$

Therefore, the solution to Eq. (4-22) can be written as

$$f = 1 - \frac{2}{\sqrt{\pi}} \int_0^\eta e^{-\eta^2}\, d\eta \tag{4-29}$$

or

$$\frac{C_A - C_{A0}}{C_{A1} - C_{A0}} = 1 - \operatorname{erf}(\eta) = 1 - \operatorname{erf}\left(\frac{X}{\sqrt{4D_{AB}Z/U_{Z,\max}}}\right) \tag{4-30}$$

where the error function is defined as

$$\operatorname{erf}(\eta) = \frac{2}{\sqrt{\pi}} \int_0^\eta e^{-\eta^2}\, d\eta \tag{4-31}$$

The error function shown in Figure 4-3 appears frequently in unsteady-state diffusion problems. A list of values is presented in Table 4-1.

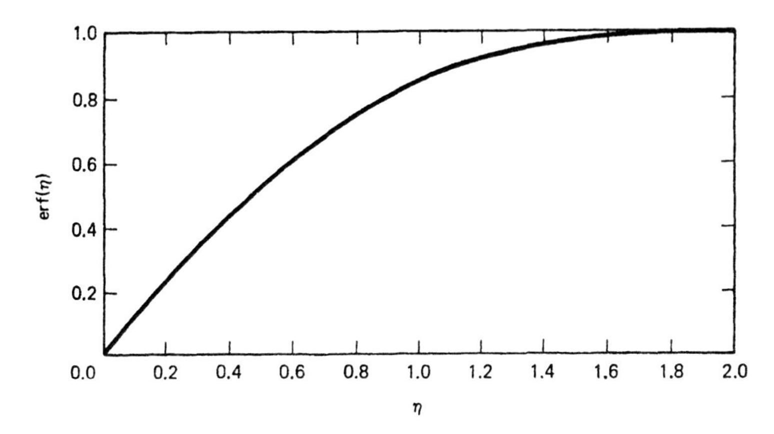

Figure 4-3 Error function

TABLE 4-1. VALUES OF THE GAUSS ERROR FUNCTION

η	$erf(\eta)$	η	$erf(\eta)$	η	$erf(\eta)$
0.00	0.0000	0.4	0.4284	1.3	0.9340
0.01	0.0113	0.5	0.5205	1.4	0.9523
0.02	0.0226	0.6	0.6039	1.5	0.9661
0.04	0.0451	0.7	0.6778	1.6	0.9763
0.06	0.0676	0.8	0.7421	1.8	0.9891
0.08	0.0901	0.9	0.7969	2.0	0.9953
0.10	0.1125	1.0	0.8427	2.2	0.9981
0.20	0.2227	1.1	0.8802	2.5	0.9996
0.30	0.3286	1.2	0.9103	3.0	1.0000

Example 4.1

The winter-kill of fish in mountain lakes has been attributed in part to the depletion of oxygen from the water due to the frozen surface. At the end of the winter, following the thaw, the oxygen concentration in the water in one case was found to be 3.0×10^{-5} kgmol/m^3. However, in the spring the water is again oxygenated as a result of its contact with air. If the lake is at an elevation of 2133 m and is "very" deep, determine the concentration of oxygen in the water at a depth of 0.06 m due to diffusion after (a) 1 day, and (b) 3 days. (c) Determine the penetration distance of oxygen after 30 days. The temperature of the lake is uniform at 5°C. Assume that the concentration of oxygen in the water at the surface is in equilibrium with air.

Solution:

An unsteady-state material balance for O_2 gives

$$D_{AB} \frac{\partial^2 C_A}{\partial X^2} = \frac{\partial C_A}{\partial t}$$

with boundary conditions

$$C_A = C_{A1} \quad \text{at } X = 0 \quad \text{for } t > 0$$
$$C_A = C_{A0} \quad \text{at } X = \infty \quad \text{for all } t$$
$$C_A = C_{A0} \quad \text{at } t = 0 \quad \text{for } X > 0$$

A comparison of the formulation above with the model for absorption into a falling liquid film shows that if we substitute $t = Z/U_{Z,\max}$, we can write the solution to the problem above as

$$\frac{C_A - C_{A0}}{C_{A1} - C_{A0}} = 1 - \text{erf}\left(\frac{X}{\sqrt{4D_{AB}t}}\right)$$

$$D_{AB}(5°\text{C}) = 1.58 \times 10^{-9} \text{ m}^2/\text{s}$$

$$C_{A0} = 3.0 \times 10^{-5} \text{ kgmol/m}^3$$

The pressure at 2133 m is calculated as shown.

$$P = P_0 \exp\left(\frac{-Mhg}{g_c RT}\right)$$
$$= 1 \exp\left[\frac{-(29)(2133)(9.807)}{(1.01 \times 10^5)(0.08205)(278)}\right]$$
$$= 0.769 \text{ atm}$$
$$\bar{P}_{O_2} = 0.21(0.769 \text{ atm}) = 0.16 \text{ atm}$$

Using Henry's law to obtain the interfacial concentration gives

$$\bar{P}_{O_2} = mx_{O_2}$$

where Henry's constant, m, is

$$m = 2.91 \times 10^4 \frac{\text{atm}}{\text{kgmol O}_2/\text{kgmol solution}}$$

and $$x_{O_2} = \frac{0.16}{2.91 \times 10^4} = 5.55 \times 10^{-6} \text{ kgmol O}_2/\text{kgmol solution}$$

$$C_{A1} = 3.08 \times 10^{-4} \text{ kgmol O}_2/\text{m}^3 \text{ solution}$$

Thus $$\frac{C_A - 3.0 \times 10^{-5}}{3.08 \times 10^{-4} - 3.0 \times 10^{-5}} = 1 - \text{erf}\left(\frac{X}{\sqrt{4D_{AB}t}}\right)$$

(a) $t = 1$ day $= 86{,}400$ s, $X = 0.06$ m. From the equation above we find that

$$C_A = 3.01 \times 10^{-5} \text{ kgmol/m}^3$$

(b) $t = 3$ days $= 259{,}200$ s, $X = 0.06$ m.

$$C_A = 4.0 \times 10^{-5} \text{ kgmol/m}^3$$

(c) At the penetration distance, $C_A = 3.0 \times 10^{-5}$ kgmol/m³. Therefore,

$$\text{erf}\left(\frac{X}{\sqrt{4D_{AB}t}}\right) = 1$$

and from the table containing the error function,

$$\frac{X}{\sqrt{4D_{AB}t}} = 3.0$$

Thus after 30 days,

$$X = (3.0)[\sqrt{4(1.58 \times 10^{-9})(86{,}400)(30)}] = 0.384 \text{ m}$$

4.3 Separation of Variables

The separation-of-variables method can be used to solve a wide variety of homogeneous and nonhomogeneous linear second-order partial differential equations found in diffusional mass transfer. The general class of equations solvable by this method can be represented as

$$a_1(\zeta)\frac{\partial^2 C_A}{\partial \zeta^2} + a_2(\zeta)\frac{\partial C_A}{\partial \zeta} + a_3(\zeta)C_A + b_1(\gamma)\frac{\partial^2 C_A}{\partial \gamma^2} + b_2(\gamma)\frac{\partial C_A}{\partial \gamma} + b_3(\gamma)C_A = 0 \tag{4-32}$$

where the independent variables ζ and γ represent either time or a spacial coordinate. Separation of variables can readily be used to solve steady two-dimensional problems if after separating the variables, the two differential equations are homogeneous and three of the four boundary conditions are also homogeneous. Unsteady-state problems can usually be solved by separation of variables if the differential equation and all but one of the boundary conditions are homogeneous. For certain types of nonhomogeneous boundary conditions and partial differential equations, the principle of superposition can be employed to make the variables separable. Several problems will be considered to demonstrate the separation-of-variables technique.

Unsteady diffusion with constant surface concentration

Diffusion in a Slab: The large rectangular sheet of wood shown in Figure 4-4 is to be treated prior to use by immersing it into a large container of a chemical species A. If the surface area on the two faces is much greater than the area along

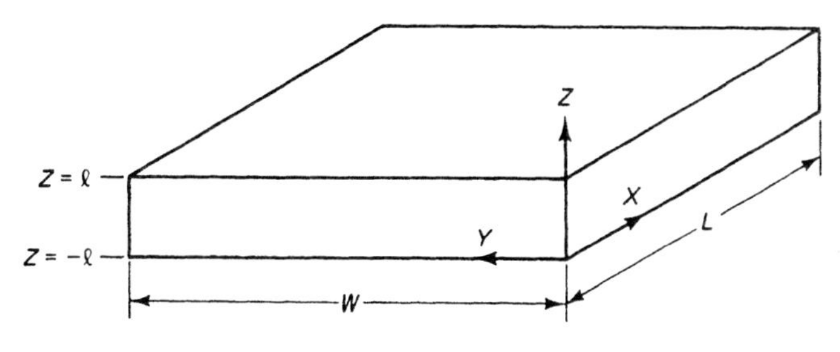

Figure 4-4 Diffusion in a slab

the edge, it is reasonable to assume that mass transfer takes place primarily in the direction perpendicular to the faces. The initial concentration of the chemical in the wood is zero. During the time the sheet is in the chemical the concentration on the surface is constant.

Transfer in the Z direction can be described by the unsteady diffusion equation

$$D\frac{\partial^2 C_A}{\partial Z^2} = \frac{\partial C_A}{\partial t} \tag{4-33}$$

where D is the effective diffusivity of the chemical through the wood. The boundary conditions are

$$\text{BC1:} \quad \frac{\partial C_A}{\partial Z} = 0 \quad \text{at } Z = 0 \quad \text{for all } t \tag{4-34}$$

$$\text{BC2:} \quad C_A = C_{A1} \quad \text{at } Z = l \quad \text{for } t > 0 \tag{4-35}$$

$$\text{BC3:} \quad C_A = C_{A0} = 0 \quad \text{at } t = 0 \quad \text{for } -l < Z < l \tag{4-36}$$

As a matter of convenience, the concentrations will be written in dimensionless form by using the expression

$$\theta = \frac{C_{A1} - C_A}{C_{A1} - C_{A0}} \tag{4-37}$$

Equation (4-33) thus becomes

$$D\frac{\partial^2 \theta}{\partial Z^2} = \frac{\partial \theta}{\partial t} \tag{4-38}$$

The new boundary conditions are

$$\text{BC1:} \quad \frac{\partial \theta}{\partial Z} = 0 \quad \text{at } Z = 0 \quad \text{for all } t \tag{4-39}$$

$$\text{BC2:} \quad \theta = 0 \quad \text{at } Z = l \quad \text{for } t > 0 \tag{4-40}$$

$$\text{BC3:} \quad \theta = 1 \quad \text{at } t = 0 \quad \text{for } -l < Z < l \tag{4-41}$$

Equation (4-38) and the boundary conditions given by Eqs. (4-39) through (4-41) conform to the criteria for using the separation of variables method. Thus a solution will be sought by applying this technique.

Since the concentration $\theta(Z, t)$ is a function of time and distance, a product solution will be assumed in which one function depends only on distance and the other function depends on time. The product solution thus is assumed to take the form

$$\theta(Z, t) = \zeta(Z)\tau(t) \tag{4-42}$$

The forms of the functions ζ and τ are not known but can be determined from the boundary conditions. A product solution thus exists only if it is possible to fit the boundary conditions to the function.

Taking derivatives of Eq. (4-42), we obtain

$$\frac{\partial \theta}{\partial t} = \zeta\frac{d\tau}{dt} \tag{4-43}$$

and

$$\frac{\partial^2 \theta}{\partial Z^2} = \tau\frac{d^2\zeta}{dZ^2} \tag{4-44}$$

Substituting the derivatives above into Eq. (4-38) yields

$$\tau\frac{d^2\zeta}{dZ^2} = \frac{\zeta}{D}\frac{d\tau}{dt} \tag{4-45}$$

By dividing both sides of Eq. (4-45) by ζ and τ we can separate the variables such that one side of the equation is a function of Z and the other side of the equation is a function of t only. Thus we have

$$\frac{1}{\zeta}\frac{d^2\zeta}{dZ^2} = \frac{1}{\tau D}\frac{d\tau}{dt} \tag{4-46}$$

The left side of the equation depends only on Z and the right side is a function of t only. Because neither side of the equation can change as Z and t vary independently, both sides must be equal to some constant a. Therefore,

$$\frac{1}{\zeta}\frac{d^2\zeta}{dZ^2} = \frac{1}{\tau D}\frac{d\tau}{dt} = a \tag{4-47}$$

which gives

$$\frac{d^2\zeta}{dZ^2} = a\zeta \tag{4-48}$$

and

$$\frac{d\tau}{dt} = aD\tau \tag{4-49}$$

The sign of the separation constant must be consistent with the physical conditions of the problem. The sign of the separation constant can be determined by considering the solution to Eq. (4-49).

$$\tau = K\exp(aDt) \tag{4-50}$$

If a were equal to zero, the concentration inside the sheet would not change with time. We must therefore reject this possibility because it is not consistent with the physical description of the problem. The second case would be for a being equal to some positive value. However, if a were positive, then τ and the concentration of A inside the slab would become infinitely large after a very long period of time. Since this is not physically possible, it too must be rejected. We can thus conclude that a must be a negative value. It can be readily seen that if a is negative, τ will no longer change at large times. Therefore, if we replace a by $-\lambda^2$ for convenience, Eqs. (4-48) and (4-49) can be written as

$$\frac{d^2\zeta}{dZ^2} + \lambda^2\zeta = 0 \tag{4-51}$$

and

$$\frac{d\tau}{dt} = -\lambda^2 D\tau \tag{4-52}$$

Equations (4-51) and (4-52) are both ordinary linear differential equations with solutions

$$\zeta = K_1\sin\lambda Z + K_2\cos\lambda Z \tag{4-53}$$

and

$$\tau = K_3\exp(-\lambda^2 Dt) \tag{4-54}$$

After substituting the solutions above into the original product solution of Eq. (4-42), the concentration is given as

$$\theta(Z, t) = (K_1\sin\lambda Z + K_2\cos\lambda Z)K_3\exp(-\lambda^2 Dt) \tag{4-55}$$

or

$$\theta(Z, t) = (A\sin\lambda Z + B\cos\lambda Z)\exp(-\lambda^2 Dt) \tag{4-56}$$

where K_1, K_2, and K_3 are arbitrary constants, and $A = K_1K_3$ and $B = K_2K_3$. Values for A, B, and λ must be chosen to satisfy the boundary conditions.

Upon applying boundary condition 1 to Eq. (4-56), we see that for all time $\sin 0 = 0$ and $\cos 0 = 1$. Thus the first boundary condition is satisfied only if $A = 0$. Therefore,

$$\theta(Z, t) = B(\cos \lambda Z) \exp(-\lambda^2 Dt) \tag{4-57}$$

In order to satisfy boundary condition 2 we have

$$B \cos \lambda l = 0 \tag{4-58}$$

If B is equal to zero, then a trivial solution exists. Consequently, if $B \neq 0$,

$$\cos \lambda l = 0 \tag{4-59}$$

The cosine function is equal to zero for multiple values. Thus

$$\lambda l = \frac{(2n + 1)\pi}{2} \tag{4-60}$$

or

$$\lambda = \frac{(2n + 1)\pi}{2l} \tag{4-61}$$

Since a particular solution of Eq. (4-57) exists for each value of λ, the constant B must be evaluated for each solution. The general solution for n values of λ is thus written as

$$\theta(Z, t) = \sum_{n=0}^{\infty} B_n\left[\cos \frac{(2n + 1)\pi Z}{2l}\right] \exp\left[-\frac{(2n + 1)^2\pi^2 Dt}{4l^2}\right] \tag{4-62}$$

The constants represented by B_n can be evaluated by using the third boundary condition, which gives the initial concentration of the chemical A in the sheet. Thus we have

$$1 = \sum_{n=0}^{\infty} B_n\left[\cos \frac{(2n + 1)\pi Z}{2l}\right] \tag{4-63}$$

The characteristic function, $\cos [(2n + 1)\pi Z/2l]$, can be shown to be orthogonal with respect to a weighting factor equal to 1 in the region $Z = 0$ to $Z = +l$:

$$\int_0^{+l} \left[\cos \frac{(2n + 1)\pi Z}{2l} \cos \frac{(2m + 1)\pi Z}{2l}\right] dZ = 0 \quad \text{if } m \neq n \tag{4-64}$$

and

$$\int_0^{+l} \left[\cos \frac{(2n + 1)\pi Z}{2l} \cos \frac{(2m + 1)\pi Z}{2l}\right] dZ \neq 0 \quad \text{if } m = n \tag{4-65}$$

Therefore, upon multiplying both sides of Eq. (4-63) by $\cos [(2n + 1)\pi Z/2l]$, we have

$$\int_0^{+l} \cos \frac{(2n + 1)\pi Z}{2l}\, dZ = \int_0^{+l} B_n \cos^2 \frac{(2n + 1)\pi Z}{2l}\, dZ \tag{4-66}$$

After integrating Eq. (4-66) and simplifying, we obtain

$$B_n = \frac{4(-1)^n}{\pi(2n + 1)} \tag{4-67}$$

The general solution to Eq. (4-38) is given by

$$\theta(Z, t) = \frac{4}{\pi} \sum_{n=0}^{\infty} \frac{(-1)^n}{(2n+1)} \cos \frac{(2n+1)\pi Z}{2l} \exp\left[-\frac{(2n+1)^2 \pi^2 D t}{4l^2}\right] \tag{4-68}$$

or

$$\frac{C_A - C_{A0}}{C_{A1} - C_{A0}} = 1 - \frac{4}{\pi} \sum_{n=0}^{\infty} \frac{(-1)^n}{(2n+1)} \cos \frac{(2n+1)\pi Z}{2l} \exp\left[-\frac{(2n+1)^2 \pi^2 D t}{4l^2}\right] \tag{4-69}$$

Graphical solutions of Eq. (4-68) are given by Crank (1956) and by Carslaw and Jaeger (1959) for the analogous heat transfer problem. These curves are presented in Figure 4-5 using the nomenclature from this chapter.

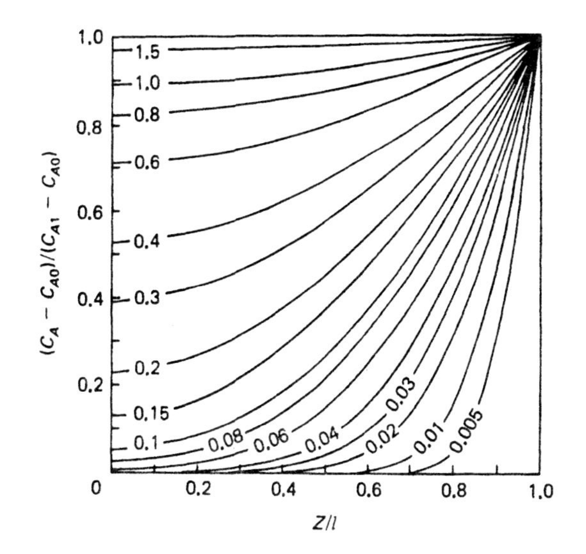

Figure 4-5 Concentration profiles in a slab. The numbers on the curves are for various values of Dt/l^2. (from Crank, 1956)

The amount of chemical transferred into the sheet at any given time can be determined by integrating the concentration over the thickness. Therefore, the mass transferred at any time is

$$M_t = 2(C_{A1} - C_{A0}) \int_0^l \left\{1 - \frac{4}{\pi} \sum_{n=0}^{\infty} \frac{(-1)^n}{(2n+1)} \cos \frac{(2n+1)\pi Z}{2l} \times \exp\left[-\frac{(2n+1)^2 \pi^2 D t}{4l^2}\right]\right\} dZ \tag{4-70}$$

Following integration, Eq. (4-70) becomes

$$M_t = 2(C_{A1} - C_{A0}) l \left\{1 - \frac{8}{\pi^2} \sum_{n=0}^{\infty} \frac{1}{(2n+1)^2} \exp\left[-\frac{(2n+1)^2 \pi^2 D t}{4l^2}\right]\right\} \tag{4-71}$$

The maximum uptake by the sheet is found by writing Eq. (4-71) as t becomes infinitely large:

$$M_\infty = M_t|_{t\to\infty} = 2l(C_{A1} - C_{A0}) \tag{4-72}$$

The uptake relative to the maximum is given as

$$\frac{M_t}{M_\infty} = 1 - \frac{8}{\pi^2}\sum_{n=0}^{\infty}\frac{1}{(2n+1)^2}\exp\left[-\frac{(2n+1)^2\pi^2 Dt}{4l^2}\right] \tag{4-73}$$

The plot of M_t/M_∞ versus $(Dt/l^2)^{1/2}$ shown in Figure 4-6 demonstrates how the mass of solute in the slab changes with respect to time.

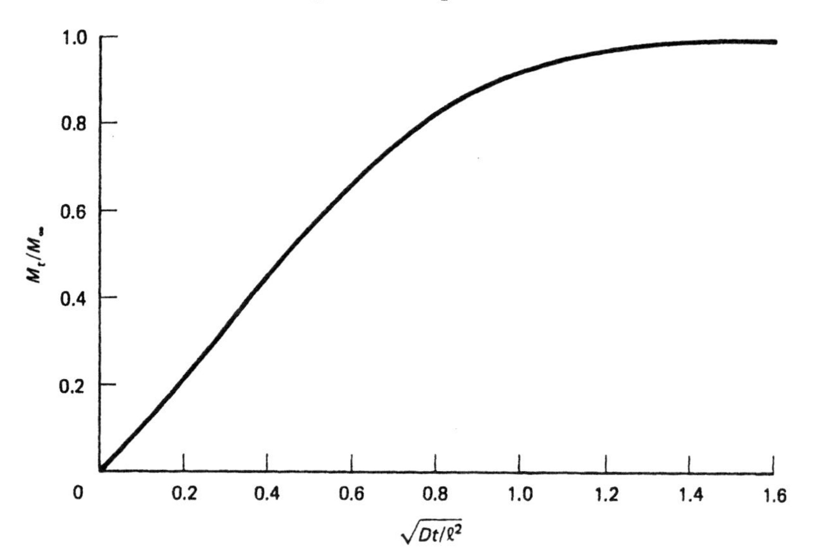

Figure 4-6 Mass transfer into a slab with constant surface concentration

Diffusion in Systems with Spherical and Cylindrical Geometry: An adsorption mechanism was assumed by Wicke (1939) in which the adsorbent consisted of numerous spheres each with large macropores beginning at the surface and passing radially to the center of the sphere. He assumed that transport was controlled by diffusion in the macropores and described the overall adsorption process in terms of unsteady diffusion into a sphere. If we assume that convective fluxes are absent and that the effective diffusivity of a solute A is constant through the sphere, we can write the differential equation that describes the diffusion process for solute A as

$$D\frac{\partial}{\partial r}\left(r^2\frac{\partial C_A}{\partial r}\right) = r^2\frac{\partial C_A}{\partial t} \tag{4-74}$$

For the case in which the convective mass transfer coefficient is very large, the concentration of solute on the surface of the sphere is constant. Since this case is similar to diffusion in a slab, the boundary conditions are

$$\text{BC1:} \quad \frac{\partial C_A}{\partial r} = 0 \quad \text{at } r = 0 \quad \text{for all } t \tag{4-75}$$

$$\text{BC2:} \quad C_A = C_{A1} \quad \text{at } r = R \quad \text{for } t > 0 \tag{4-76}$$

$$\text{BC3:} \quad C_A = C_{A0} \quad \text{at } t = 0 \quad \text{for } r < R \tag{4-77}$$

If we introduce the dimensionless expression given by Eq. (4-37), Eq. (4-74) can be written as

$$D \frac{\partial}{\partial r}\left(r^2 \frac{\partial \theta}{\partial r}\right) = r^2 \frac{\partial \theta}{\partial t} \tag{4-78}$$

Difficulties arise when attempting a solution by separation of variables. However, problems with spherical geometry can be transformed to cartesian coordinates, whose solution can be expressed in terms of circular functions, by introducing the transformation

$$\psi(r, t) = r\theta(r, t) \tag{4-79}$$

The new equation is thus

$$D \frac{\partial^2 \psi}{\partial r^2} = \frac{\partial \psi}{\partial t} \tag{4-80}$$

with transformed boundary conditions

$$\text{BC1:} \quad \psi = 0 \quad \text{at } r = 0 \quad \text{for all } t \tag{4-81}$$

$$\text{BC2:} \quad \psi = 0 \quad \text{at } r = R \quad \text{for } t > 0 \tag{4-82}$$

$$\text{BC3:} \quad \psi = r \quad \text{at } t = 0 \quad \text{for } r < R \tag{4-83}$$

The separation-of-variables method, when applied to Eq. (4-80) with the boundary conditions above, gives

$$\psi(r, t) = -\frac{2R}{\pi} \sum_{n=1}^{\infty} \frac{(-1)^n}{n} \sin \frac{n\pi r}{R} \exp\left(-\frac{n^2\pi^2 Dt}{R^2}\right) \tag{4-84}$$

The final form of the solution is expressed in terms of concentration as

$$\frac{C_A - C_{A0}}{C_{A1} - C_{A0}} = 1 + \frac{2R}{\pi r} \sum_{n=1}^{\infty} \frac{(-1)^n}{n} \sin \frac{n\pi r}{R} \exp\left(-\frac{n^2\pi^2 Dt}{R^2}\right) \tag{4-85}$$

The mass of adsorbate A that has entered the sphere, M_t, relative to the amount that would be adsorbed after infinite time, M_∞, is expressed by the relationship

$$\frac{M_t}{M_\infty} = 1 - \frac{6}{\pi^2} \sum_{n=1}^{\infty} \frac{1}{n^2} \exp\left(-\frac{n^2\pi^2 Dt}{R^2}\right) \tag{4-86}$$

The concentration profile given by Eq. (4-85) is presented in Figure 4-7.

Consider the case of diffusion of solute A in the radial direction of an infinitely long cylinder. The initial concentration of solute in the cylinder is C_{A0}. If the cylinder is suddenly immersed in pure solute A and the concentration on the surface is constant, then the boundary conditions are similar to those obtained for the slab and sphere. A solution to the equations that describes

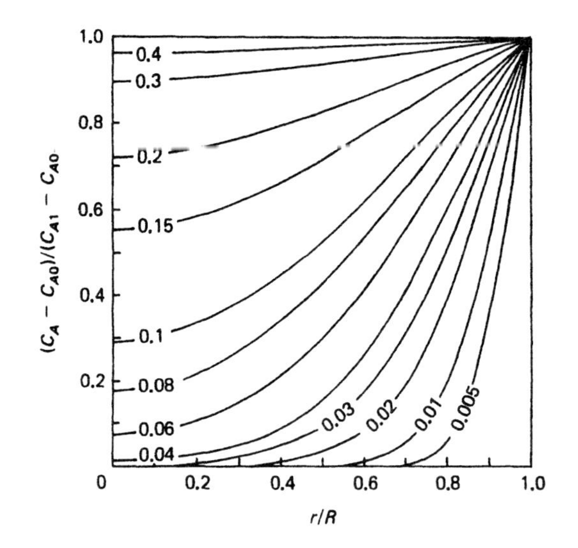

Figure 4-7 Concentration profiles in a sphere. The numbers on the curves are for various values of Dt/R^2. (from Crank, 1956)

diffusion in a cylinder is also readily obtained by separation of variables. After separating the variables, the solution to the ordinary differential equation in the radial direction gives a zero-order Bessel's function of the first kind which is orthogonal with respect to a weighting function r. The solution is shown graphically in Figure 4-8.

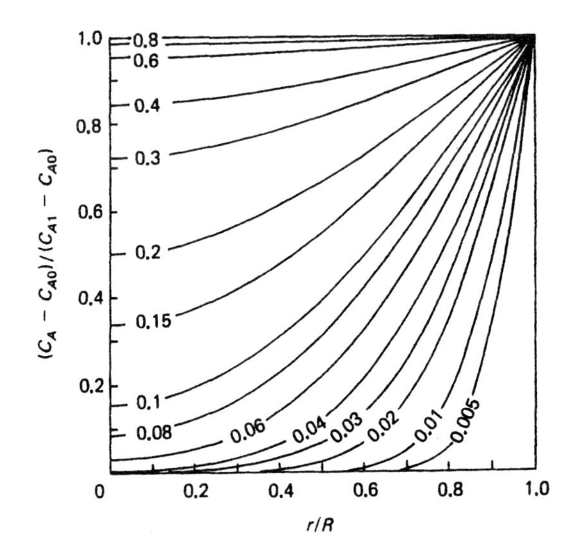

Figure 4-8 Concentration profiles in a cylinder. The numbers on the curves are for various values of Dt/R^2. (from Crank, 1956)

Unsteady diffusion with convection at the surface

The use of the boundary condition that describes convection at a surface can be demonstrated by considering the drying of a porous solid where the solid surface is exposed to a flowing gas. Let us consider the case shown in Figure 4-9, in which species A diffuses only in the Z direction to the surface, where it is removed by convection into the gas stream. We will assume that the concentration of A in the gas is zero and the diffusion coefficient through the solid is constant.

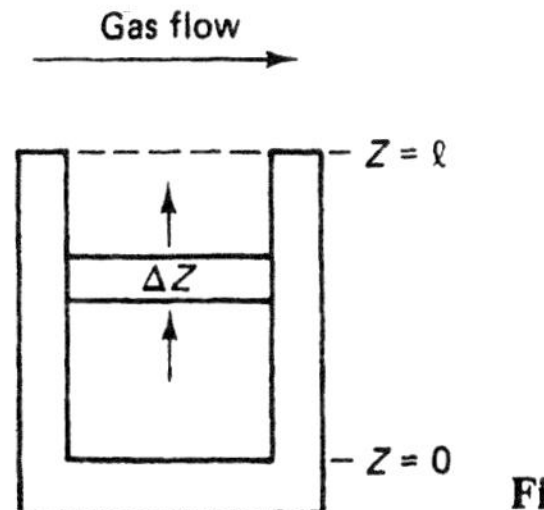

Figure 4-9 Convection at a surface

Since the concentration of A in the solid will continuously change, the unsteady-state drying process is described by

$$D \frac{\partial^2 C_A}{\partial Z^2} = \frac{\partial C_A}{\partial t} \tag{4-87}$$

where D is the effective diffusion coefficient. If the bottom of the solid is sealed, the concentration gradient of A is zero at $Z = 0$. At the top of the solid, the mass flux evaluated at the gas–solid interface due to diffusion through the solid is equated to a mass transfer coefficient and a linear driving force. Thus the boundary conditions for the drying process are

$$\text{BC1:} \quad \frac{\partial C_A}{\partial Z} = 0 \quad \text{at } Z = 0 \quad \text{for all } t \tag{4-88}$$

$$\text{BC2:} \quad -D\frac{\partial C_A}{\partial Z} = k_c(C_{A,g} - C_{A\infty}) \quad \text{at } Z = l \quad \text{for } t > 0 \tag{4-89}$$

$$\text{BC3:} \quad C_A = C_{A0} \quad \text{at } t = 0 \quad \text{for } 0 < Z < l \tag{4-90}$$

where $C_{A,g}$ is the concentration of A in the gas on the surface and $C_{A\infty}$ is equal to zero. The concentration of A in the gas at the interface must be related to A in the solid at the interface before the second boundary condition can be used. The relationship between these values can be expressed as

$$C_A = KC_{A,g} \tag{4-91}$$

where K is the distribution coefficient. A linear relationship between the concentrations in the two phases is usually found when the concentrations are very low. By combining Eqs. (4-89) and (4-91), the second boundary condition becomes

$$\text{BC2:}\quad \frac{\partial C_A}{\partial Z} + \frac{k_c}{DK} C_A = 0 \quad \text{at } Z = l \quad \text{for } t > 0 \tag{4-92}$$

If we apply the separation-of-variables method, in which the product solution $C_A(Z, t) = \zeta(Z)\tau(t)$ is introduced, we obtain

$$\tau(t) = C_1 \exp(-\lambda^2 Dt) \tag{4-93}$$

and

$$\zeta(Z) = C_2 \cos \lambda Z + C_3 \sin \lambda Z \tag{4-94}$$

Upon introducing the first boundary condition, we find that $C_3 = 0$. Hence the product solution for $C_A(Z, t)$ is

$$C_A(Z, t) = A \cos \lambda Z \exp(-\lambda^2 Dt) \tag{4-95}$$

In order to satisfy the second boundary condition, we equate the mass of A that diffuses to the surface to the mass carried away by convection in the gas stream. Thus at $Z = l$,

$$-\lambda A \sin \lambda l \exp(-\lambda^2 Dt) + \frac{k_c}{DK} A \cos \lambda l \exp(-\lambda^2 Dt) = 0 \tag{4-96}$$

Since the relation above must hold for all values of t, it can be rearranged to give

$$\lambda l \tan \lambda l = \frac{lk_c}{DK} \tag{4-97}$$

The formula above is a transcendental equation which can be satisfied by an infinite number of values for λ. Roots to Eq. (4-97) are presented by Crank (1956). The product solution for all values of λ_n is given as

$$C_A(Z, t) = \sum_{n=1}^{\infty} A_n \cos \lambda_n Z \exp(-\lambda_n^2 Dt) \tag{4-98}$$

Introducing the third boundary condition and using the orthogonality property over the thickness 0 to l, we obtain a relationship for A_n in terms of the characteristic value λ_n. Substitution of this value into Eq. (4-98) gives

$$\frac{C_A(Z, t)}{C_{A0}} = \sum_{n=1}^{\infty} \frac{2 \sin(\lambda_n l) \cos(\lambda_n Z)}{\lambda_n l + \sin(\lambda_n l) \cos(\lambda_n l)} \exp(-\lambda_n^2 Dt) \tag{4-99}$$

Graphical solutions to the equation above will be presented in a later section for the general case in which $C_{A\infty}$ is not equal to zero. Solutions to problems with spherical and cylindrical geometry will also be shown.

Example 4.2

Prior to treating pine logs to prevent them from warping, the logs are kiln dried to remove the moisture. The moisture content of the solid surface is in equilibrium with the air and can be assumed to be constant at 5%. The initial moisture content is 30% and the maximum permissible concentration is 15%.

If the logs are 12.2 m long with a diameter of 0.3 m, determine the time necessary for them to dry. The effective diffusivity of water in pine may be assumed to be constant with a value of 1.0×10^{-7} m²/s.

Solution:

The time required to dry the logs can be found by using Figure 4-8. Since the logs are very long with respect to their diameter, transfer only in the radial direction will be considered. For drying, we customarily write the concentration on a solute-free basis. Thus

$$C_{A1} = \frac{0.05}{1 - 0.05} = 0.0525 \qquad C_{A0} = \frac{0.3}{1 - 0.3} = 0.429$$

The maximum concentration, which will be at the center, is

$$C_A = \frac{0.15}{1 - 0.15} = 0.1764$$

The unaccomplished concentration change is

$$\frac{C_A - C_{A0}}{C_{A1} - C_{A0}} = \frac{0.1764 - 0.429}{0.0525 - 0.429} = 0.67$$

From Figure 4-8, we see that at the center of the log

$$\frac{Dt}{R^2} = 0.28$$

Therefore,

$$t = \frac{(0.28)(0.15)^2}{1.0 \times 10^{-7}}$$

$$= 6.3 \times 10^4 \text{ s}$$

$$= 17.5 \text{ h}$$

4.4 The Laplace Transform Method

A large number of diffusion problems lead to linear partial differential equations with constant coefficients. For this type of problem, the Laplace transform provides a useful and important method for finding the solution. Although many problems can be solved by separation of variables, the boundaries must be finite in the orthogonal directions. The conditions of finite geometry are not required when the Laplace transform method is employed. In this section the solution to partial differential equations of diffusion will be demonstrated by using this method.

Definition of the Laplace transform

The Laplace transform of a function $F(t)$ is defined by the expression

$$\mathcal{L}[F(t)] = f(s) = \int_0^\infty e^{-st} F(t)\, dt \qquad (4\text{-}100)$$

where s can be any complex number and $F(t)$ is continuous. As an example, if $F(t) = a$,

$$\mathcal{L}[a] = \int_0^\infty ae^{-st}\,dt = -\frac{a}{s}e^{-st}\Big|_0^\infty = \frac{a}{s} \tag{4-101}$$

Before considering a particular problem, the Laplace transform of a function containing two independent variables such as $F(X, t)$ must be defined. Since there are two independent variables, the variable X will be treated as a parameter and the Laplace transform will be taken as though $F(X, t)$ is a function of t only. Thus we have

$$\mathcal{L}[F(X, t)] = f(X, s) = \int_0^\infty e^{-st}F(X, t)\,dt \tag{4-102}$$

The transform of the time derivative is given by

$$\mathcal{L}\left[\frac{\partial F(X, t)}{\partial t}\right] = s\mathcal{L}[F(X, t)] - F(X, 0) \tag{4-103}$$

The transform of the partial derivative with respect to X is somewhat different, however. Using the definition of the transform given by Eq. (4-100) and assuming that it is permissible to interchange the order of integration and differentiation, we have

$$\begin{aligned}\mathcal{L}\left[\frac{\partial F(X, t)}{\partial X}\right] &= \int_0^\infty e^{-st}\frac{\partial F(X, t)}{\partial X}\,dt \\ &= \frac{\partial}{\partial X}\int_0^\infty e^{-st}F(X, t)\,dt \\ &= \frac{\partial f(X, s)}{\partial X}\end{aligned} \tag{4-104}$$

In general, the equation above can be written as

$$\mathcal{L}\left[\frac{\partial^n F(X, t)}{\partial X^n}\right] = \frac{\partial^n f(X, s)}{\partial X^n} \tag{4-105}$$

The transformation of a partial differential equation gives an ordinary differential equation that can be solved in the s domain. However, to obtain a solution to the problem, we must transform from the s domain back into the time domain. Although this step is frequently very difficult, extensive tables of Laplace transforms have been developed and are useful in obtaining solutions (see Roberts and Kaufman, 1966). The use of the inversion integral provides us with another method of inverting the transform. However, this technique is beyond the scope of this book and will not be discussed here. A detailed development of Laplace transforms is given by Churchill (1958).

Application of Laplace transforms to mass transfer in a wetted-wall column

The application of the Laplace transform method can be demonstrated by solving the partial differential equation that describes the absorption of a solute by a liquid film in a wetted-wall column. If $Z/U_{Z,\max}$ is replaced by t, then Eq. (4-14) becomes

$$D\frac{\partial^2 f}{\partial X^2} = \frac{\partial f}{\partial t} \tag{4-106}$$

with the following boundary conditions:

$$\text{BC1:}\quad f = 0 \quad \text{at } t = 0 \quad \text{for all } X \tag{4-107}$$

$$\text{BC2:}\quad f = 1 \quad \text{at } X = 0 \quad \text{for } t > 0 \tag{4-108}$$

$$\text{BC3:}\quad f = 0 \quad \text{at } X \longrightarrow \infty \quad \text{for all } t \tag{4-109}$$

If we regard X as a parameter, the transform of the left-hand side of Eq. (4-106) is found by applying Eq. (4-105). Thus

$$\mathcal{L}\left[D\frac{\partial^2 f(X, t)}{\partial X^2}\right] = D\frac{d^2\bar{f}(X, s)}{dX^2} \tag{4-110}$$

where $\bar{f}(X, s)$ is the transform of $f(X, t)$. Applying Eq. (4-103), we can write the transform of the right-hand side as

$$\mathcal{L}\left[\frac{\partial f(X, t)}{\partial t}\right] = s\bar{f}(X, s) - f(X, 0) \tag{4-111}$$

$$= s\bar{f}(X, s) \tag{4-112}$$

where the first boundary condition has been used to eliminate $f(X, 0)$ and give Eq. (4-112). The partial differential equation is thus transformed into the homogeneous ordinary differential equation

$$D\frac{d^2\bar{f}(X, s)}{dX^2} = s\bar{f}(X, s) \tag{4-113}$$

The general solution of Eq. (4-113) is

$$\bar{f}(X, s) = K_1 e^{-\sqrt{s/D}X} + K_2 e^{\sqrt{s/D}X} \tag{4-114}$$

It is interesting to note that if Eq. (4-12) and the nonhomogeneous boundary conditions given by Eqs. (4-3) through (4-5) had been transformed, a nonhomogeneous ordinary differential equation would have been obtained.

Before the constants of integration, K_1 and K_2, can be evaluated, the second and third boundary conditions must be transformed. Using Eq. (4-101), we have

$$\text{BC2:}\quad \mathcal{L}[1] = \int_0^\infty 1e^{-st}\,dt = \frac{1}{s} = \bar{f}(0, t) \tag{4-115}$$

and

$$\text{BC3:} \quad \mathcal{L}[0] = 0 = \bar{f}(\infty, t) \tag{4-116}$$

Introducing BC3 into Eq. (4-114), we obtain

$$\bar{f}(\infty, t) = 0 = K_1 e^{-\infty} + K_2 e^{\infty} \tag{4-117}$$

Thus K_2 must equal zero. Therefore,

$$\bar{f}(X, s) = K_1 e^{-\sqrt{s/D}\,X} \tag{4-118}$$

The constant K_1 is found by applying BC2. Thus

$$\frac{1}{s} = K_1 e^0$$

or

$$K_1 = \frac{1}{s} \tag{4-119}$$

The final form of Eq. (4-114) is

$$\bar{f}(X, s) = \frac{e^{-\sqrt{s/D}\,X}}{s} \tag{4-120}$$

The dimensionless concentration profile is found by inverting the solution above into the time domain with the help of Table 4-2. After inverting we have

TABLE 4-2. LAPLACE TRANSFORMS

Number	*Transform*	*Function*
1	$\frac{1}{s}$	1
2	$\frac{1}{s^2}$	t
3	$\frac{1}{s^n}$	$\frac{t^{n-1}}{(n-1)!}$
4	$\frac{1}{\sqrt{s}}$	$\frac{1}{\sqrt{\pi t}}$
5	$\frac{1}{s-a}$	e^{at}
6	$\frac{1}{(s-a)^2}$	te^{at}
7	$\frac{a}{s^2+a^2}$	$\sin at$
8	$\frac{s}{s^2+a^2}$	$\cos at$
9	$\frac{a}{s^2-a^2}$	$\sinh at$
10	$\frac{s}{s^2-a^2}$	$\cosh at$
11	$\frac{1}{s}e^{-(k/s)}$	$J_0(2\sqrt{kt})$
12	$\frac{1}{s}e^{-k\sqrt{s}}\,(k > 0)$	$\operatorname{erfc}\left(\frac{k}{2\sqrt{t}}\right)$

$$\frac{C_A - C_{A0}}{C_{A1} - C_{A0}} = \text{erfc}\left(\frac{X}{\sqrt{4Dt}}\right) \tag{4-121}$$

where erfc is defined as the complementary error function and is equal to $1 - \text{erf}(X/\sqrt{4Dt})$.

4.5 Application of Graphical Solutions

Concentration profiles and mass flows have been calculated for several simple shapes and are presented in the form of charts and tables by Crank (1956) for one-dimensional problems. Solutions to multidimensional unsteady-state problems can be found by expressing two- or three-dimensional problems in terms of two or three one-dimensional problems. For example, a two-dimensional unsteady-state problem will be expressed as the product of two one-dimensional problems as shown.

$$\theta(X, Y, t) = \bar{X}(X, t)\bar{Y}(Y, t) \tag{4-122}$$

In this section, graphical solutions will be used to obtain concentration profiles and mass transfer rates for both one-dimensional and multidimensional problems with a convective boundary condition.

Flat plate

Let us first consider the porous flat plate shown in Figure 4-10 that extends to infinity in the Y and X directions. Initially the plate is saturated with component A, but at $t = 0$ the plate is exposed to an air stream. The solute diffuses through the solid to the surface and is carried away by convection. If convective effects are neglected inside the porous plate, the transient diffusion problem can be described by Fick's second law. From Figure 4-10 we have

$$D\frac{\partial^2 C_A}{\partial Z^2} = \frac{\partial C_A}{\partial t} \tag{4-123}$$

with boundary conditions

BC1: $$\frac{\partial C_A}{\partial Z} = 0 \quad \text{at } Z = 0 \quad \text{for all } t \tag{4-124}$$

BC2: $$C_A = C_{A0} \quad \text{at } t = 0 \quad \text{for } -l < Z < l \tag{4-125}$$

BC3: $$-D\frac{\partial C_A}{\partial Z} = k_c(C_{A,g} - C_{A\infty}) \quad \text{at } Z = l \quad \text{for } t > 0 \tag{4-126}$$

As shown previously, the concentration of A in the gas must be related to the concentration within the solid. Assuming the linear equilibrium relationship

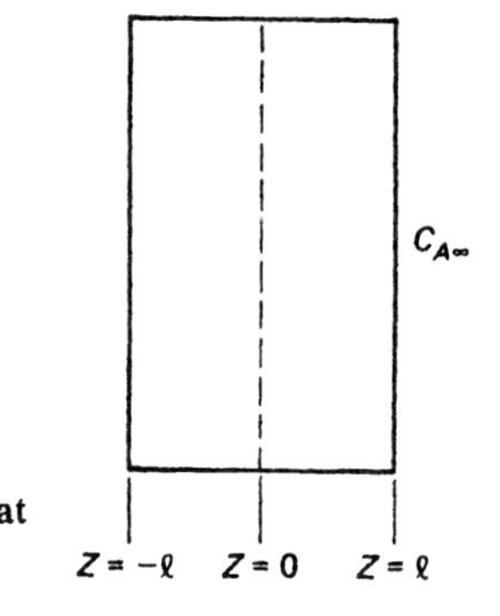

Figure 4-10 Unsteady diffusion in a flat plate

between the concentration of A in the gas and in the solid, $C_A = KC_{A,g}$, BC3 becomes

$$-\frac{\partial C_A}{\partial Z} = \frac{k_c}{DK}(C_A - KC_{A\infty}) \tag{4-127}$$

where K is the equilibrium distribution constant. The term $KC_{A\infty}$ can be replaced by the concentration in the solid that is in equilibrium with the bulk concentration $C_{A\infty}$ by using the Henry's law relationship $C_A^* = KC_{A\infty}$. The third boundary condition is thus written as

$$\text{BC3:}\quad -\frac{\partial C_A}{\partial Z} = \frac{k_c}{DK}(C_A - C_A^*) \quad \text{at } Z = l \quad \text{for } t > 0 \tag{4-128}$$

The concentration profile for the system above is

$$\frac{C_A - C_A^*}{C_{A0} - C_A^*} = \sum_{n=1}^{\infty} \frac{2\alpha \cos [\beta_n(Z/l)]}{(\beta_n^2 + \alpha^2 + \alpha) \cos \beta_n} \exp\left(-\frac{\beta_n^2 Dt}{l^2}\right) \tag{4-129}$$

where the β_n values are the roots of

$$\beta_n \tan \beta_n = \alpha = \frac{k_c l}{KD} \tag{4-130}$$

The roots to the equation above are tabulated by Crank (1956).

The mass transferred from the plate relative to the total amount transferred after infinite time is expressed as

$$\frac{M_t}{M_\infty} = 1 - \sum_{n=1}^{\infty} \frac{2\alpha^2}{\beta_n^2(\beta_n^2 + \alpha^2 + \alpha)} \exp\left(-\frac{\beta_n^2 Dt}{l^2}\right) \tag{4-131}$$

Concentration profiles calculated from Eq. (4-129) are shown in Figure 4-11 as a function of α and position within the slab. A graphical representation of Eq. (4-131) is shown in Figure 4-12. For the case of convective mass transfer from both surfaces, l is defined as the distance from the center of the slab to the surface. If one surface is covered and is impermeable to the diffusing solute, l is defined as the total thickness of the slab.

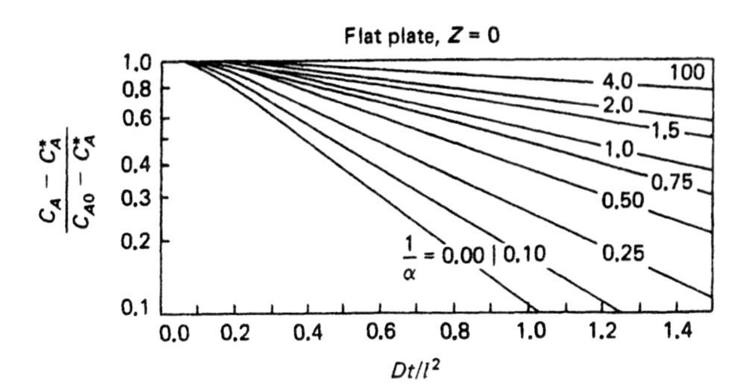

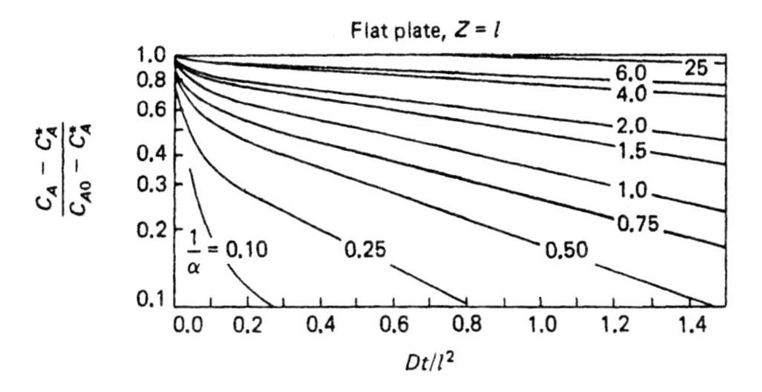

Figure 4-11 Unsteady concentration profiles for a flat plate (from Boelter et al., 1965)

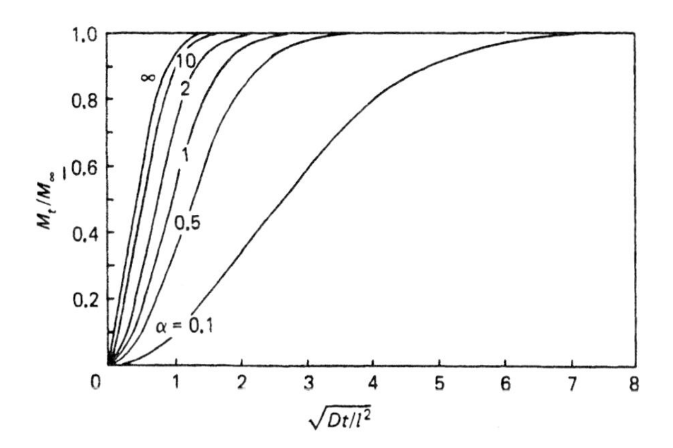

Figure 4-12 Unsteady mass transfer from a slab for different values of α (from Crank, 1956)

Cylinder

Instead of a flat plate, let us consider the infinitely long cylinder shown in Figure 4-13. For an infinitely long cylinder, end effects can be neglected and we need only consider diffusion in the radial direction. Fick's second law for this

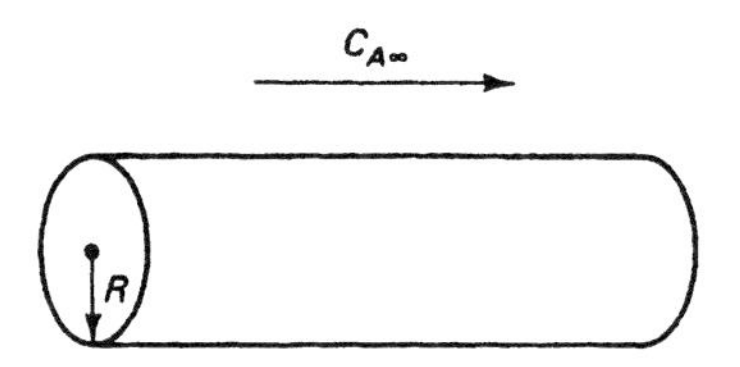

Figure 4-13 Unsteady diffusion in a cylinder

case becomes

$$\frac{1}{r}\frac{\partial}{\partial r}\left(rD\frac{\partial C_A}{\partial r}\right) = \frac{\partial C_A}{\partial t} \tag{4-132}$$

The boundary conditions for the cylinder with convection at the surface are similar to those for the flat plate. These are:

BC1: $$\frac{\partial C_A}{\partial r} = 0 \quad \text{at } r = 0 \quad \text{for all } t \tag{4-133}$$

BC2: $$C_A = C_{A0} \quad \text{at } t = 0 \quad \text{for } r < R \tag{4-134}$$

BC3: $$-\frac{\partial C_A}{\partial r} = \frac{k_c}{DK}(C_A - C_A^*) \quad \text{at } r = R \quad \text{for } t > 0 \tag{4-135}$$

The dimensionless concentration profile is

$$\frac{C_A - C_A^*}{C_{A0} - C_A^*} = \sum_{n=1}^{\infty} \frac{2\alpha J_0[\beta_n(r/R)]}{(\beta_n^2 + \alpha^2)J_0(\beta_n)} \exp\left(-\frac{\beta_n^2 Dt}{R^2}\right) \tag{4-136}$$

Values for β_n are the roots of the equation

$$\beta_n J_1(\beta_n) - \alpha J_0(\beta_n) = 0 \tag{4-137}$$

where $J_0(\beta_n)$ and $J_1(\beta_n)$ are Bessel functions of the first kind of zero order and first order, respectively. Roots of Eq. (4-137) are given by Crank (1956) for several values of α.

The total quantity of solute leaving the cylinder up to a given time is given as

$$\frac{M_t}{M_\infty} = 1 - \sum_{n=1}^{\infty} \frac{4\alpha^2}{\beta_n^2(\beta_n^2 + \alpha^2)} \exp\left(-\frac{\beta_n^2 Dt}{R^2}\right) \tag{4-138}$$

Graphical representations of Eqs. (4-136) and (4-138) are shown in Figures 4-14 and 4-15.

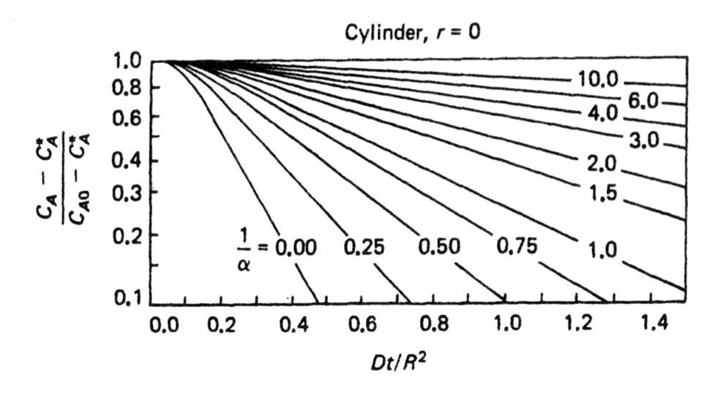

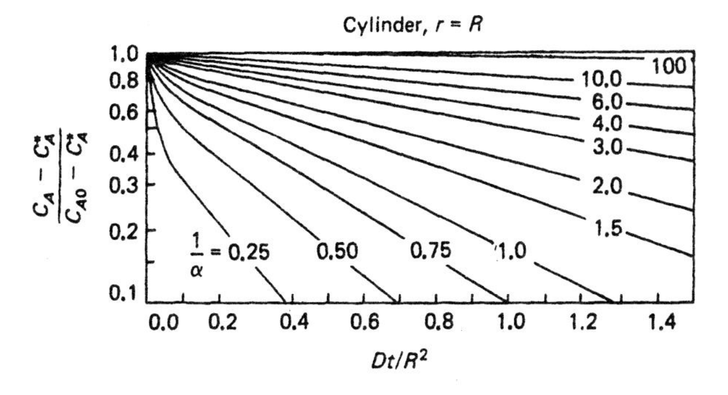

Figure 4-14 Unsteady concentration profiles for an infinitely long cylinder (from Boelter et al., 1965)

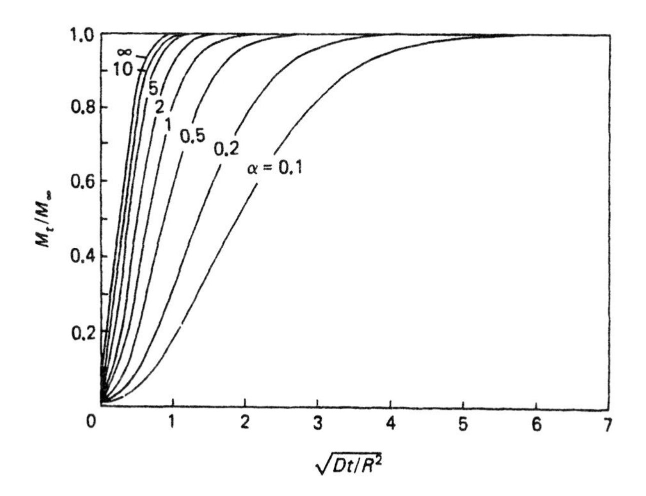

Figure 4-15 Unsteady mass transfer from an infinitely long cylinder for different values of α (from Crank, 1956)

Sphere

For the same boundary conditions as previously shown for the cylinder, the concentration profile inside a sphere is

$$\frac{C_A - C_A^*}{C_{A0} - C_A^*} = \frac{2\alpha R}{r} \sum_{n=1}^{\infty} \frac{\sin [\beta_n(r/R)]}{(\beta_n^2 + \alpha^2 - \alpha) \sin (\beta_n)} \exp\left(-\frac{\beta_n^2 Dt}{R^2}\right) \tag{4-139}$$

where β_n's are the roots of

$$\beta_n \cot \beta_n + \alpha - 1 = 0 \tag{4-140}$$

The total mass transferred from the sphere up to time t relative to the total mass transferred is given by the expression

$$\frac{M_t}{M_\infty} = 1 - \sum_{n=1}^{\infty} \frac{6\alpha^2}{\beta_n^2(\beta_n^2 + \alpha^2 - \alpha)} \exp\left(-\frac{\beta_n^2 Dt}{R^2}\right) \tag{4-141}$$

Concentration and mass transfer profiles are shown in Figures 4-16 and 4-17.

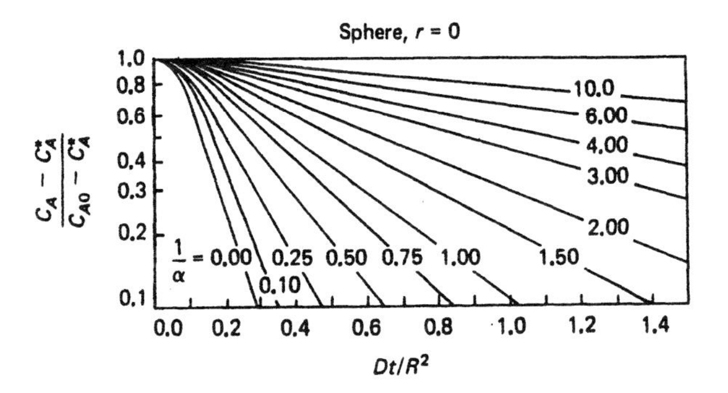

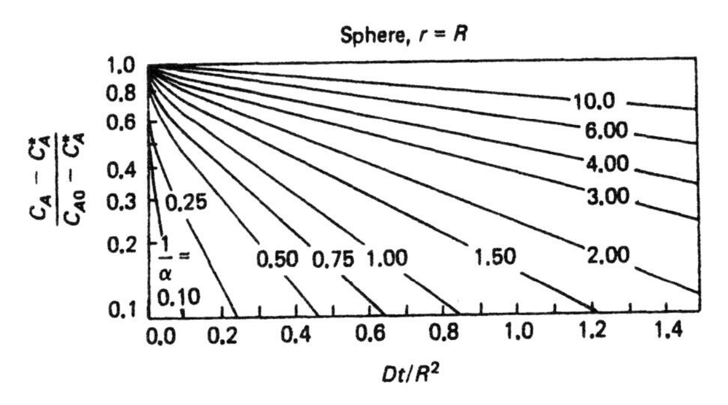

Figure 4-16 Unsteady concentration profiles in a sphere (from Boelter et al., 1965)

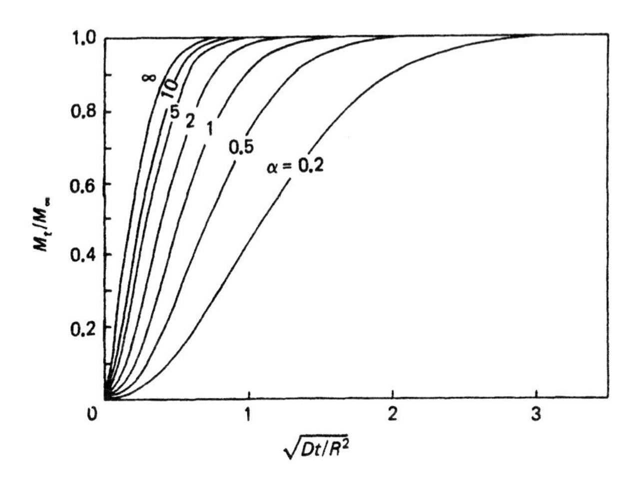

Figure 4-17 Unsteady mass transfer in a sphere as a function of α (from Crank, 1956)

Example 4.3

Determine the quantity of moisture removed from the pine log in Example 4.2 when the maximum concentration of the log is reduced to 15%. The density of dry pine is 550 kg/m³. Neglect mass transfer from the ends of the log.

Solution:

From Example 4.2

$$\frac{Dt}{R^2} = 0.28$$

Thus from Figure 4-15 for $\alpha = \infty$, which corresponds to a constant surface concentration,

$$\frac{M_t}{M_\infty} = 0.87$$

But

$$M_\infty = \pi R^2 L(C_{A0} - 0)\rho_s$$

$$= \pi(0.15)^2(12.2)\left(\frac{0.3}{1 - 0.3}\right)(550)$$

$$= 203.3 \text{ kg of water}$$

Thus

$$M_t = 0.87 M_\infty$$

$$= 176.85 \text{ kg of water}$$

On the basis of Eq. (4-122), solutions obtained from the unsteady one-dimensional charts can readily be extended to multidimensional problems. For example, if we wish to predict the concentration profile in a short cylinder, consideration must be given to mass transfer in both the radial and axial directions. Thus the solution for a short cylinder is given as

$$\left[\frac{C_A - C_A^*}{C_{A0} - C_A^*}\right]_{\substack{\text{short}\\ \text{cylinder}}} = \left[\frac{C_A - C_A^*}{C_{A0} - C_A^*}\right]_{\substack{\text{infinite}\\ \text{cylinder}}} \times \left[\frac{C_A - C_A^*}{C_{A0} - C_A^*}\right]_{\substack{\text{flat plate}\\ 2l \text{ thickness}}} \quad (4\text{-}142)$$

For a three-dimensional object such as a parallelepiped, the concentration profile becomes

$$\left[\frac{C_A - C_A^*}{C_{A0} - C_A^*}\right]_{\substack{\text{paralle-}\\ \text{lepiped}}} = \left[\frac{C_A - C_A^*}{C_{A0} - C_A^*}\right]_{\substack{\text{flat plate}\\ 2l_1}} \times \left[\frac{C_A - C_A^*}{C_{A0} - C_A^*}\right]_{\substack{\text{flat plate}\\ 2l_2}} \times \left[\frac{C_A - C_A^*}{C_{A0} - C_A^*}\right]_{\substack{\text{flat plate}\\ 2l_3}} \quad (4\text{-}143)$$

The application of Eqs. (4-142) and (4-143) greatly enhances the usefulness of the unsteady graphical solutions.

4.6 Method of Weighted Residuals

Approximation techniques

Thus far in this chapter we have considered methods that yielded exact solutions to partial differential equations. In this section approximate solutions will be obtained by applying the method of weighted residuals. This method includes several approximation techniques that utilize a trial solution in terms of undetermined parameters or undetermined functions. Since the trial function actually represents a family of solutions, the undetermined parameters are chosen to provide the best solution to the partial differential equation. This is done by defining the difference between the actual solution and approximate solution as an error or residual and evaluating the undetermined parameter such that the weighted residual is zero:

$$\int_0^1 R_N(a, x_j)W_j(X)\,dX = 0 \quad j = 1, 2, \ldots, N \quad (4\text{-}144)$$

where R_N is the residual, a is the undetermined parameter, $W_j(X)$ is the weighting function, and X is the independent variable. The choice of the weighting function prescribes the method of weighted residual that is to be used. The most important of these are (1) collocation, (2) subdomain, (3) method of moments, and (4) Galerkin's method. Since the purpose here is only to apply these techniques to solving partial differential equations of diffusion, the reader is referred to the texts of Finlayson (1972) and Villadsen and Michelsen (1978) for a detailed description of these methods.

Collocation: For this case, the weighting factor is defined as the delta function:

$$W_j(X) = \delta(X - X_j) \quad j = 1, 2, \ldots, N \quad (4\text{-}145)$$

where X_j are N points in the interval 0 to 1. Equation (4-145) implies that the residual vanishes at selected points $X_1, \ldots, X_N$. Thus

$$R_N(a, X_j) = 0 \quad j = 1, 2, \ldots, N \quad (4\text{-}146)$$

When applying collocation, we select the number of points in the interval from 0 to 1 to equal the number of undetermined parameters.

Subdomain: In the subdomain method $W_j(X) = 1$. To apply this technique, the interval 0 to 1 is divided such that the number of subdomains equals the number of undetermined parameters. For example, if the trial solution contains two undetermined parameters, then for $W_j(X) = 1$, Eq. (4-144) can be expressed by two integrals each equal to zero.

$$\int_0^{X_1} R(a, X)\, dX = 0 \tag{4-147}$$

and

$$\int_{X_1}^{1} R(a, X)\, dX = 0 \tag{4-148}$$

The integrals above provide two equations containing the two undetermined parameters.

Method of Moments: The weighting factor for this method can be expressed as X^{j-1}, where $j = 1, 2, \ldots, N$. Thus we have

$$\int_0^1 R(a, X)X^{j-1}\, dX = 0 \quad j = 1, 2, \ldots, N \tag{4-149}$$

If $j = 1$, the method of moments becomes equal to the subdomain method for the case of one undetermined parameter. For a second-order approximation, two undetermined parameters must be evaluated. This is done by evaluating Eq. (4-144) for $j = 1$ and then for $j = 2$. Thus we have

$$\int_0^1 R(a, X)\, dX = 0 \quad \text{and} \quad \int_0^1 R(a, X)X\, dX = 0 \tag{4-150}$$

Galerkin's Method: In Galerkin's method the weighting functions are taken to be equal to the trial solutions used to obtain the residuals. Thus we have

$$\int_0^1 R(a_1, a_2, \ldots, a_N, X)\phi_j(X)\, dX = 0 \quad j = 1, 2, \ldots, N \tag{4-151}$$

where $\phi_j(X)$ is the trial solution in terms of the independent variable X. For two undetermined parameters, Eq. (4-151) becomes

$$\int_0^1 R(a_1, a_2, X)\phi_1\, dX = 0 \tag{4-152}$$

and

$$\int_0^1 R(a_1, a_2, X)\phi_2\, dX = 0 \tag{4-153}$$

Selection of trial solutions

The most important step in applying the method of residuals is the selection of a family of trial solutions. This selection is even more important for low-order approximations since the trial solution may be valid only at the boundaries.

For higher-order approximations, the choice of a trial solution is less critical because the residual is required to vanish at each point in the interval of interest. Past experience and a good understanding of the physical problem is very important when selecting a low-order trial solution.

For one-dimensional problems, we seek a solution in the form of a polynomial expressed in terms of the independent variable X such as

$$\theta(X) = \sum_{i=0}^{N+1} a_i X^i \tag{4-154}$$

Expanding the expression above and fitting the boundary conditions to the equation provides a trial function. This approach can be extended to partial differential equations by requiring that a_i be a function of the other independent variable as proposed by Kantorovich and Krylov (1958). Thus for diffusion in both the X and Z directions, we have

$$\theta(X, Z) = \sum_{i=0}^{N+1} a_i(Z) X^i \tag{4-155}$$

A more general representation of the equation above is

$$\theta(X, Z) = \sum_{i=0}^{N+1} a_i(Z)\phi_i(X) \tag{4-156}$$

The direction in which the polynomial is expanded depends on the length of the object in each dimension. For first-order approximations, the polynomial should be expanded in the dimension corresponding to the shortest interval between boundaries. The application of the integral approximation techniques can be readily demonstrated by considering the following example and using a simple first-order approximation.

Example 4.4

A very long rectangular section of wood, shown in Figure 4-18, is placed in a reservoir. Because of its low density, we will assume that it floats on top of the water. After a period of time some water will be adsorbed by the wood and

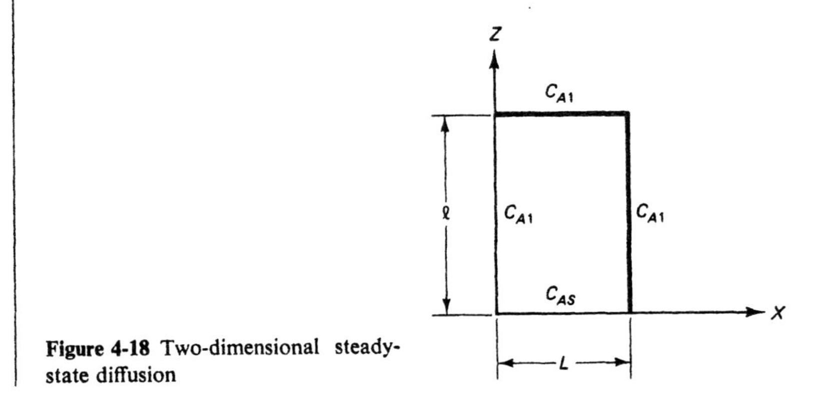

Figure 4-18 Two-dimensional steady-state diffusion

ultimately a steady-state concentration profile will exist. The concentration on the top and two sides of the wood will be in equilibrium with the air and will be assumed to have a constant value of C_{A1}. The concentration on the bottom is C_{AS}. Since the slab is assumed to be very long in the Y direction, mass transfer from the ends will be neglected. Determine concentration profiles of water by using the following methods: collocation, subdomain, method of moments, and Galerkin.

Solution:

The equation describing steady-state mass transfer for the problem above is

$$\frac{\partial^2 C_A}{\partial X^2} + \frac{\partial^2 C_A}{\partial Z^2} = 0$$

The boundary conditions are

$$\text{BC1:} \quad C_A = C_{A1} \quad \text{at } X = 0 \quad \text{for all } Z$$

$$\text{BC2:} \quad C_A = C_{A1} \quad \text{at } X = L \quad \text{for all } Z$$

$$\text{BC3:} \quad C_A = C_{A1} \quad \text{at } Z = l \quad \text{for all } X$$

$$\text{BC4:} \quad C_A = C_{AS} \quad \text{at } Z = 0 \quad \text{for all } X$$

The equations above are made dimensionless by introducing the new variables given below:

$$\bar{\theta} = \frac{C_A - C_{A1}}{C_{AS} - C_{A1}} \qquad \xi = \frac{X}{L} \qquad \delta = \frac{Z}{L}$$

After introducing the groups above, the new nondimensional form of the equation is

$$\frac{\partial^2 \bar{\theta}}{\partial \xi^2} + \frac{\partial^2 \bar{\theta}}{\partial \delta^2} = 0$$

with

$$\bar{\theta} = 0 \quad \text{at } \xi = 0 \qquad \bar{\theta} = 0 \quad \text{at } \xi = 1$$

$$\bar{\theta} = 0 \quad \text{at } \delta = l/L \qquad \bar{\theta} = 1 \quad \text{at } \delta = 0$$

Since the problem is symmetrical in the X direction, we select a second-order polynomial in that direction such that the coefficients are a function of δ. Using Eq. (4-155), we obtain

$$\bar{\theta}(\xi, \delta) = \sum_{i=0}^{k} a_i(\delta)\xi^i = a_0 + a_1\xi + a_2\xi^2$$

Applying the boundary conditions, we find that $a_0 = 0$ and $a_1 = -a_2$. The final form of our trial solution can be expressed as

$$\bar{\theta} = a_1(\delta)[\xi(1 - \xi)]$$

Higher-order approximations can be introduced by simply writing the above equation in general terms as

$$\bar{\theta}_N = \sum_{j=1}^{N} a_j(\delta)[\xi^j(1 - \xi)^j] = \sum_{j=1}^{N} a_j(\delta)\phi_j(X) \tag{4-157}$$

In our problem only a first-order approximation will be considered. The residual for $N = 1$ is found by substituting the trial solution $\theta = a_1(\delta)[\xi(1 - \xi)$

into the partial differential equation. Thus

$$R = \frac{\partial^2\bar{\theta}}{\partial\xi^2} + \frac{\partial^2\bar{\theta}}{\partial\delta^2}$$

gives

$$R = -2a_1 + \xi(1-\xi)\frac{d^2a_1}{d\delta^2}$$

After obtaining the residual, we may use any of the previously described approximation methods to find a solution for the problem. Each method will be demonstrated for this example.

Collocation:

Applying collocation about $\xi = \frac{1}{2}$, the residual becomes

$$R = \frac{d^2a_1}{d\delta^2} - 8a_1 = 0$$

The solution to the ordinary differential equation above is

$$a_1 = C_1e^{-\sqrt{8}\delta} + C_2e^{\sqrt{8}\delta}$$

After introducing the boundary conditions we have

$$a_1 = -\frac{e^{2\sqrt{8}\,l/L}}{1-e^{2\sqrt{8}\,l/L}}e^{-\sqrt{8}\delta} + \frac{1}{1-e^{2\sqrt{8}\,l/L}}e^{\sqrt{8}\delta}$$

The final form of the solution is found by substituting a_1 back into the trial solution. Thus we have

$$\bar{\theta}(\xi,\delta) = \xi(1-\xi)\left(\frac{1}{1-e^{2\sqrt{8}\,l/L}}e^{\sqrt{8}\delta} - \frac{e^{2\sqrt{8}\,l/L}}{1-e^{2\sqrt{8}\,l/L}}e^{-\sqrt{8}\delta}\right)$$

Subdomain:

Substituting the trial solution into Eq. (4-147) for the case of only one undetermined parameter, we obtain

$$\int_0^1 \left[\xi(1-\xi)\frac{d^2a_1}{d\delta^2} - 2a_1\right]d\xi = 0$$

Integrating the equation above and applying the limits gives

$$\frac{d^2a_1}{d\delta^2} - 12a_1 = 0$$

After solving the differential equation above and introducing the boundary conditions, we can write the solution as

$$a_1 = \frac{1}{1-e^{2\sqrt{12}\,l/L}}e^{\sqrt{12}\delta} - \frac{e^{2\sqrt{12}\,l/L}}{1-e^{2\sqrt{12}\,l/L}}e^{-\sqrt{12}\delta}$$

The approximate solution for the subdomain method becomes

$$\bar{\theta}(\xi,\delta) = \xi(1-\xi)\left(\frac{1}{1-e^{2\sqrt{12}\,l/L}}e^{\sqrt{12}\delta} - \frac{e^{2\sqrt{12}\,l/L}}{1-e^{2\sqrt{12}\,l/L}}e^{-\sqrt{12}\delta}\right)$$

Method of moments:

According to Eq. (4-150), the weighting factor, X^{j-1}, for the first approximation, $j = 1$, is equal to 1. Thus for the first approximation, the subdomain method and method of moments are the same. Therefore, the approximate solution for this method is given by the equation above.

Galerkin's method:

For a first-order approximation, Galerkin's method is

$$\int_0^1 \left[\xi(1-\xi)\frac{d^2a_1}{d\delta^2} - 2a_1\right][\xi(1-\xi)]\,d\xi = 0$$

Integration and subsequent solution of the ordinary differential equation results in the final approximate solution. An examination reveals that each solution is exact on the boundaries.

$$\bar{\theta}(\xi,\delta) = \xi(1-\xi)\left(\frac{1}{1-e^{2\sqrt{10}\,l/L}}e^{\sqrt{10}\,\delta} - \frac{e^{2\sqrt{10}\,l/L}}{1-e^{2\sqrt{10}\,l/L}}e^{-\sqrt{10}\,\delta}\right)$$

When applying Eq. (4-155) to transient problems, the polynomial should be expanded in the spacial coordinate. The constant a then becomes a function of time. If we are concerned about changes over very small time intervals, an approximate solution utilizing the penetration distance should be used. Each of these methods will be demonstrated here. However, we will limit our discussion to a first-order approximation in the following examples.

Example 4.5

The application of the method of weighted residuals to transient problems can be demonstrated by considering the diffusion and subsequent adsorption of a gas into a thin slab of activated carbon as discussed by Finlayson (1972). The initial gas concentration inside the adsorbent is C_{A0}. After the adsorption process is initiated, the concentration on the solid surface is constant at C_{A1}. Determine an approximate concentration profile.

Solution:

The partial differential equation describing the unsteady diffusion of the gas into the solid in one dimension is

$$D\frac{\partial^2 C_A}{\partial Z^2} = \frac{\partial C_A}{\partial t}$$

where D is an effective diffusivity. The boundary conditions for this case are obtained from Figure 4-4. Introducing the dimensionless variables $\tau = tD/l^2$, $\bar{\theta} = (C_A - C_{A0})/(C_{A1} - C_{A0})$, and $\xi = (l - Z)/l$ into the equation above gives

$$\frac{\partial^2\bar{\theta}}{\partial\xi^2} = \frac{\partial\bar{\theta}}{\partial\tau}$$

with boundary conditions $\delta\bar{\theta}/\partial\xi = 0$ at $\xi = 1$, $\bar{\theta} = 1$ at $\xi = 0$, and $\bar{\theta} = 0$ at $\tau = 0$. A trial solution can be obtained by substituting the boundary conditions into the following equation:

$$\bar{\theta}(\tau,\xi) = \sum_{i=0}^{k} a_i(\tau)\xi^i = a_0 + a_1\xi + a_2\xi^2 \tag{4-158}$$

The trial solution, which is found by fitting the boundary conditions to the equation above, is

$$\bar{\theta} = 1 + a(\xi^2 - 2\xi)$$

where the parameter a is a function of τ.

The residual is obtained by substituting the trial solution into the differential equation as shown.

$$R = \frac{\partial^2 \bar{\theta}}{\partial \xi^2} - \frac{\partial \bar{\theta}}{\partial \tau}$$

or

$$R = 2a - (\xi^2 - 2\xi)\frac{da}{d\tau}$$

Applying the integral method gives

$$\int_0^1 \left[2a - (\xi^2 - 2\xi)\frac{da}{d\tau}\right] d\xi = 0$$

After integrating and applying the limits, we obtain the ordinary differential equation

$$\frac{da}{d\tau} + 3a = 0$$

The solution to the equation above is

$$a = K \exp(-3\tau)$$

and the trial profile thus becomes

$$\bar{\theta} = 1 + K(\xi^2 - 2\xi)\exp(-3\tau)$$

The constant of integration K can be found by introducing the initial condition. To do this, however, the equation above is integrated over the region of interest, $0 \leq \xi \leq 1$, for $\tau = 0$. Thus

$$0 = \int_0^1 [1 + K(\xi^2 - 2\xi)\exp(0)]\, d\xi$$

After integrating we find that $K = \frac{3}{2}$. Thus the first approximation is

$$\bar{\theta} = 1 + \tfrac{3}{2}(\xi^2 - 2\xi)\exp(-3\tau)$$

The first-order approximation above was used only to demonstrate the integral method for solving transient problems. An improvement in the concentration profile can be obtained by using higher-order trial functions as well as other types of approximations. An improvement in the predicted concentration profile for short times can be achieved by solving the problem for the time domain prior to when the gas reaches the other boundary. The first time domain can be described in terms of the *penetration depth* to which the gas has diffused at any time. As shown by Finlayson (1972), the concentration, which is a function of position and time, can be expressed in terms of a penetration depth as

$$\bar{\theta} = \phi(\eta) \quad \text{where } \eta = \frac{Z}{q(t)} \tag{4-159}$$

and $q(t)$ is the penetration depth. The use of the penetration depth is readily demonstrated by solving the preceding problem for short times.

Example 4.6

Obtain the concentration profile inside the adsorbent discussed in Example 4.5 for the time domain prior to when the gas diffuses to the center of the slab.

Solution:

As in Example 4.5, we must first obtain a trial solution that satisfies the boundary conditions. The new boundary conditions can be found by considering Figure 4-19. At the surface of the slab, $\xi = 0$, $\bar{\theta} = 1$. At the edge of the

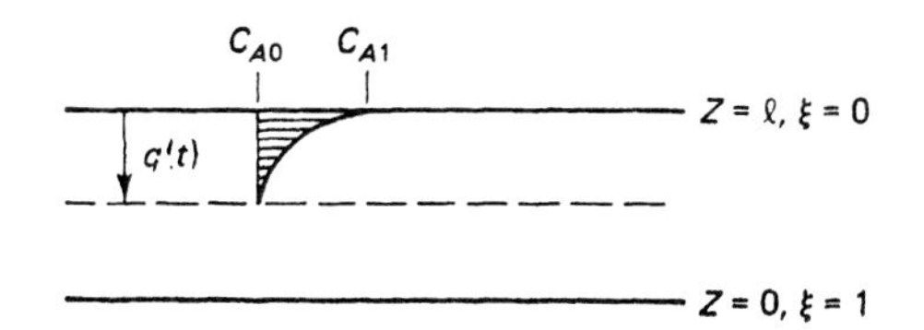

Figure 4-19 Penetration model for diffusion into a slab

penetration depth, the concentration of the gas in the solid is C_{A0} and the concentration gradient is zero. The initial condition is expressed in terms of the penetration depth, which is zero at the beginning of the diffusion process. Introducing Eq. (4-159), but replacing Z with ξ, gives the boundary conditions

$$\bar{\theta} = 1 \quad \text{at } \xi = 0 \quad \text{or} \quad \eta = 0$$

$$\bar{\theta} = 0, \qquad \frac{\partial \bar{\theta}}{\partial \xi} = 0 \quad \text{at } q(t) = \xi \quad \text{or} \quad \eta = 1$$

$$q = 0 \quad \text{at } \tau = 0$$

A first-order trial function for this case is

$$\bar{\theta} = a_0 + a_1\eta + a_2\eta^2$$

After applying the boundary conditions, we obtain the trial function

$$\bar{\theta} = (1 - \eta)^2$$

The residual is expressed by

$$R = \frac{\partial^2 \bar{\theta}}{\partial \xi^2} - \frac{\partial \bar{\theta}}{\partial \tau}$$

$$= \frac{d^2\bar{\theta}}{d\eta^2}\left(\frac{\partial \eta}{\partial \xi}\right)^2 - \frac{d\bar{\theta}}{d\eta}\frac{\partial \eta}{\partial \tau}$$

Since $\eta = \xi/q(t)$, $\partial\eta/\partial\xi = 1/q$, and $\partial\eta/\partial\tau = -(\xi/q^2)(dq/d\tau)$. Therefore,

$$R = \frac{1}{q^2}\frac{d^2\bar{\theta}}{d\eta^2} + \frac{\xi}{q^2}\frac{dq}{d\tau}\frac{d\bar{\theta}}{d\eta}$$

$$= \frac{2}{q^2} - \frac{2\xi(1-\eta)}{q^2}\frac{dq}{d\tau}$$

and

$$q^2 R = 2 - 2(1 - \eta)\eta q \frac{dq}{d\tau} = 0$$

Applying the integral method from 0 to 1 gives

$$q\frac{dq}{d\tau} = 6$$

Integrating the equation above and introducing the initial condition gives the change in penetration depth with time.

$$q = \sqrt{12\tau}$$

Therefore, the approximate solution for short times is

$$\bar{\theta} = \left(1 - \frac{\xi}{\sqrt{12\tau}}\right)^2$$

When the gas reaches the center of the slab, the penetration depth is $q = 1$. This corresponds to a time of $\tau = \frac{1}{12}$.

A solution for all times can be found by combining the two approximate solutions at $\tau = \frac{1}{12}$ and reevaluating the constant of integration in the first approximate solution.

$$(1 - \xi)^2 = 1 + K(\xi^2 - 2\xi)\exp(-\tfrac{3}{12})$$

Integrating both sides of the equation above from $\xi = 0$ to 1 gives $K = 1.284$. Thus the approximate solution is

$$\bar{\theta} = 1 + 1.284(\xi^2 - 2\xi)\exp(-3\tau)$$

Nonplanar geometry

Although approximate solutions can also be obtained for problems involving nonplanar geometry, the simple parabolic trial functions used in the previous examples yield poor results. This is to be expected since the volume into which the mass diffuses changes with the radius. For diffusion into a cylinder, a suitable trial function can be obtained by multiplying the parabolic profile by the logarithm of the radius. The logarithmic term is the result of $(1/r)$ appearing in the steady-state solution for diffusion into a cylinder. Concentration profiles for problems involving spherical geometry can be found by dividing the parabolic profile by r. Since a problem can be readily transformed from spherical to Cartesian coordinates with Eq. (4-79), this modification should be expected. If a problem is first transformed into Cartesian coordinates, a parabolic trial function should yield a reasonable concentration profile.

4.7 Similarity Analysis

If a mathematical model can be written to describe a mass transfer process, the minimum number of dimensionless groupings necessary to decribe the process can be determined by writing the model in nondimensional form. This method of analyzing physical processes is frequently referred to as a similarity analysis. In many cases this type of analysis leads directly to similarity transformations.

Various methods of determining similarity transforms have been considered. Schlichting (1968) assumed the transformations of the variables to be of a general form which were defined by introduction into the differential equation that described the process. Birkhoff (1961) used a different approach based on group theory. His method, called the *method of search for symmetric solutions*, obtains groups of transformations under which the differential equation for the process and the associated boundary conditions are invariant. The general method of similarity analysis outlined by Hellums and Churchill (1964), however, will be presented here.

The method consists of the following steps:

1. System variables, parameters, boundary conditions, and initial conditions are written in dimensionless form by introducing arbitrary reference variables. For each independent variable this is $X = \bar{X}X_a$, where X is the independent variable, $\bar{X}$ is the dimensionless variable, and X_a is the reference quantity. The dependent variable is conveniently defined as $C = \bar{\theta}C_a + K_a$, where C is the original dependent variable, $\bar{\theta}$ is the new dimensionless variable, and C_a and K_a are arbitrary reference quantities.
2. After all the equations are written in nondimensional form, the properties and reference quantities are factored out of one term in each equation. Each dimensionless parameter is then equated to a constant. The constant usually selected is unity. However, for the dependent variable which may exhibit an additive reference quantity, the constant is equal to zero. This gives a system of algebraic equations that can be solved to yield expressions for the reference quantities in terms of the parameters present in the original problem.
3. If the system is over determined, all the parameters cannot be eliminated by choice of the reference quantities; one parameter will appear in the problem for each algebraic equation that cannot be satisfied.
4. If the system is under determined (i.e., if all the independent algebraic equations can be satisfied without specifying all the reference quantities), the number of independent variables can be reduced. The reference variables can then be combined to eliminate the remaining arbitrary reference quantities. This procedure will thus give any possible similarity transformation.
5. The remaining dimensionless dependent variable can then be written as a function of the remaining independent variables and parameters.

The application of this procedure is easily demonstrated by considering the following example.

Example 4.7

Consider a fluid flowing over a flat surface as shown in Figure 4-20. At $Z = 0$ the surface is coated with a solid that dissolves into the fluid. Assuming that the solid is only slightly soluble and the flow is fully developed, the dissolution process can be described by the following differential equation and boundary conditions.

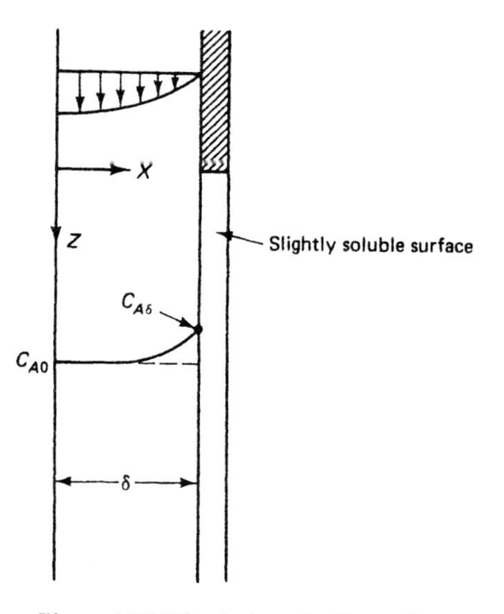

Figure 4-20 Dissolution of a flat surface

$$U_{Z,\max}\left(1 - \frac{X^2}{\delta^2}\right)\frac{\partial C_A}{\partial Z} = D_{AB}\frac{\partial^2 C_A}{\partial X^2} \tag{4-160}$$

and

$$\frac{\partial C_A}{\partial X} = 0 \quad \text{at } X = 0 \tag{4-161}$$

$$C_A = C_{A0} \quad \text{at all } X \quad \text{for } Z < 0 \tag{4-162}$$

$$C_A = C_{A\delta} \quad \text{at } X = \delta \quad \text{for } Z > 0 \tag{4-163}$$

Determine the pertinent dimensionless groups that describe the dissolution process.

Solution:

Define dimensionless and reference variables as follows:

$$Z = \bar{Z}Z_a \qquad X = \bar{X}X_a \qquad C_A = \bar{\theta}C_a + K_a \tag{4-164}$$

The terms $\bar{Z}$, $\bar{X}$, and $\bar{\theta}$ are dimensionless variables and the terms with the subscript a are arbitrary reference quantities. Introducing the dimensionless variables $\bar{Z} = Z/Z_a$, $\bar{X} = X/X_a$, and $\bar{\theta} = (C_A - K_a)/C_a$ into the differential equation and boundary conditions above, we obtain

$$\frac{U_{Z,\max}X_a^2}{Z_aD_{AB}}\left[1 - \bar{X}^2\left(\frac{X_a}{\delta}\right)^2\right]\frac{\partial\bar{\theta}}{\partial\bar{Z}} = \frac{\partial^2\bar{\theta}}{\partial\bar{X}^2} \tag{4-165}$$

and

$$\frac{\partial\bar{\theta}}{\partial\bar{X}} = 0 \quad \text{at } \bar{X} = 0 \tag{4-166}$$

$$\bar{\theta}_0 = \frac{C_{A0} - K_a}{C_a} \quad \text{at all } \bar{X} \quad \text{for } \bar{Z} < 0 \tag{4-167}$$

$$\bar{\theta}_\delta = \frac{C_{A\delta} - K_a}{C_a} \quad \text{at } \bar{X} = \frac{\delta}{X_a} \quad \text{for } \bar{Z} > 0 \tag{4-168}$$

Thus the dimensionless concentration can be written in functional form as

$$\bar{\theta} = f\left[\bar{X}, \bar{Z}, \frac{X_a}{\delta}, \frac{C_{A0} - K_a}{C_a}, \frac{C_{A\delta} - K_a}{C_a}, \frac{U_{Z,\max} X_a^2}{Z_a D_{AB}}\right] \tag{4-169}$$

Setting X_a/δ, $(C_{A0} - K_a)/C_a$, and $U_{Z,\max}X_a^2/Z_a^2 D_{AB}$ equal to unity and $(C_{A\delta} - K_a)/C_a$ equal to zero, we find that

$$K_a = C_{A\delta} \qquad C_a = C_{A0} - C_{A\delta} \qquad X_a = \delta$$

and

$$Z_a = \frac{U_{Z,\max}\,\delta^2}{D_{AB}}$$

After carrying out the operation above, we can express the remaining terms in the functionality as

$$\bar{\theta} = \frac{C_A - C_{A\delta}}{C_{A0} - C_{A\delta}} = f[\bar{X}, \bar{Z}] = \left[\frac{X}{X_a}, \frac{Z}{Z_a}\right] \tag{4-170}$$

If we substitute for the arbitrary reference values in the equation above, we obtain the pertinent dimensionless groups

$$\frac{C_A - C_{A\delta}}{C_{A0} - C_{A\delta}} = f\left[\frac{X}{\delta}, \frac{Z D_{AB}}{\delta^2 U_{Z,\max}}\right] \tag{4-171}$$

The second term in the functionality can be shown to contain the frequently used Reynolds and Schmidt numbers. Multiplying and dividing the second term by ν gives

$$\frac{Z}{\delta}\,\frac{\nu}{\delta U_{Z,\max}}\,\frac{D_{AB}}{\nu} = \frac{Z}{\delta}\,\frac{1}{\mathrm{Re}}\,\frac{1}{\mathrm{Sc}} \tag{4-172}$$

Let us now consider a case in which the system is under determined and a similarity transformation is obtained.

Example 4.8

Resolve Example 4.7 and assume that the fluid velocity profile can be written as U_0. The differential equation that describes the dissolution process becomes

$$U_0 \frac{\partial C_A}{\partial Z} = D_{AB} \frac{\partial^2 C_A}{\partial X^2} \tag{4-173}$$

The boundary conditions are the same as those given in Example 4.7.

Solution:

Introducing the dimensionless variables $\bar{Z} = Z/Z_a$, $\bar{X} = X/X_a$, and $\bar{\theta} = (C_A - K_a)/C_a$ into Eq. (4-173) and Eqs. (4-161) through (4-163) gives

$$\frac{U_0 X_a^2}{Z_a D_{AB}} \frac{\partial \bar{\theta}}{\partial \bar{Z}} = \frac{\partial^2 \bar{\theta}}{\partial \bar{X}^2} \tag{4-174}$$

and Eqs. (4-166) through (4-168). When written in functional form, the dimensionless concentration is

$$\bar{\theta} = f\left[\bar{X}, \bar{Z}, \frac{C_{A0} - K_a}{C_a}, \frac{C_{A\delta} - K_a}{C_a}, \frac{U_0 X_a^2}{Z_a D_{AB}}\right] \tag{4-175}$$

Setting $(C_{A0} - K_a)/C_a$ and $U_0 X_a^2/Z_a D_{AB}$ equal to 1 and $(C_{A\delta} - K_a)/C_a$ equal to zero gives $K_a = C_{A\delta}$, $C_a = C_{A0} - C_{A\delta}$, and $Z_a = U_0 X_a^2/D_{AB}$. Since one arbitrary reference term is defined in terms of another, the system is under determined. Thus the variables can be combined to yield a similarity transformation as follows:

$$\bar{\theta} = \frac{C_A - C_{A\delta}}{C_{A0} - C_{A\delta}} = f\left[\frac{X}{X_a}, \frac{ZD_{AB}}{U_0 X_a^2}\right] \tag{4-176}$$

By dividing both terms in the functionality by X/X_a, we have

$$\frac{C_A - C_{A\delta}}{C_{A0} - C_{A\delta}} = f\left[\frac{ZD_{AB}}{U_0 X^2}\right] \tag{4-177}$$

The similarity transformation identified in Eq. (4-177) can be used to reduce Eq. (4-173) from a partial to an ordinary differential equation by using the method shown in Section 4.2.

REFERENCES

BIRKHOFF, G., *Hydrodynamics*, Princeton University Press, Princeton, N.J., 1961.

BOELTER, L. M. K., V. H. CHERRY, H. A. JOHNSON, and R. C. MARTINELLI, *Heat Transfer Notes*, McGraw-Hill, New York, 1965.

CARSLAW, H. S., and J. C. JAEGER, *Conduction of Heat in Solids*, Oxford University Press, Oxford, 1959.

CHURCHILL, R. V., *Operational Mathematics*, McGraw-Hill, New York, 1958.

CRANK, J., *The Mathematics of Diffusion*, Oxford University Press, Oxford, 1956.

FINLAYSON, B. A., *The Method of Weighted Residuals and Variational Principles*, Academic Press, New York, 1972.

HELLUMS, J. D., and S. W. CHURCHILL, *AIChE J.*, *10*, 110 (1964).

KANTOROVICH, L. V., and V. I. KRYLOV, *Approximate Methods of Higher Analysis*, Interscience, New York, 1958.

ROBERTS, G. E., and H. KAUFMAN, *Table of Laplace Transforms*, Saunders, Philadelphia, 1966.

SCHLICHTING, H., *Boundary Layer Theory*, McGraw-Hill, New York, 1968.

VILLADSEN, J., and M. L. MICHELSEN, *Solution of Differential Equation Models by Polynomial Approximation*, Prentice-Hall, Englewood Cliffs, N.J., 1978.

WICKE, E., *Kolloid Z.*, *86*, 167 (1939).

NOTATIONS

a, b = integration constants

A, B = arbitrary constants

C_A = molar concentration of A, mol/L^3

$C_{A,s}, C_{A1}$ = concentration of A in the liquid at the gas–liquid interface, mol/L^3

C_{A0} = initial concentration of A, mol/L^3

$C_{A\infty}$ = concentration of A in the bulk liquid, mol/L^3
C_A^* = concentration of A in the solid which is in equilibrium with the bulk concentration $C_{A\infty}$, mol/L^3
D = diffusion coefficient, L^2/t
D_{AB} = binary diffusivity for system A–B, L^2/t
f = dimensionless concentration defined by Eq. (4-13)
h = height of column, L
k_c = convective mass transfer coefficient, mol/tL^2 (concentration difference)
K = distribution coefficient
l = width of column, L
L = length of a slab, L
m = Henry's constant
M = molecular weight, M/mol
M_t = mass of substance adsorbed at time t, M
M_∞ = mass adsorbed at infinite time, M
N_{AX}, N_{AY}, N_{AZ} = molar flux of A with respect to a fixed reference frame in the X, Y, and Z directions, mol/L^2t
P = pressure, F/L^2
R = gas constant; radius of a tube or cylinder, L
t = time, t
T = temperature, T
U_X, U_Z = velocity in the X and Z directions, L/t
$U_{Z,\max}$ = maximum velocity in the Z direction, L/t
W = thickness of the liquid film, L
x_A = mole fraction of A

Greek Letters

δ = dimensionless width as defined by $\delta = Z/l$
ζ = special function of distance Z as defined by Eq. (4-42)
η = dimensionless group defined by Eq. (4-18)
θ = dimensionless group defined by Eq. (4-37)
λ = integration constant
ξ = dimensionless length as defined by $\xi = X/L$
τ = special function of time t as defined by Eq. (4-42)

PROBLEMS

4.1 Solute A is absorbed from an inert gas into liquid B in a 0.5-m-long wetted-wall column. The initial concentration of A in the liquid is 2 kgmol/m^3 and the partial pressure of A in the gas is 0.015 atm. The concentrations of A in the gas and liquid at the interface are related by the expression

$$C_A = 150P_A$$

where P_A has units of atm and C_A has units kgmol/m^3. If the maximum velocity of the liquid is 0.003 m/s, determine **(a)** the concentration of A at a point 0.01 m from the gas–liquid interface, and **(b)** the penetration distance of A in the liquid. Assume that the diffusion coefficient is constant and has a value of 9.0×10^{-8} m^2/s. State the assumptions used in solving this problem.

4.2 A porous membrane initially contains solute A with concentration C_{A0}. One side of the membrane is suddenly brought into contact with a gas and the concentration of A on the surface is raised to C_{A1} (see Figure 4-21). The concentration of A on the other surface is held constant at C_{A0}. Obtain an expression for the concentration profile of A in the membrane.

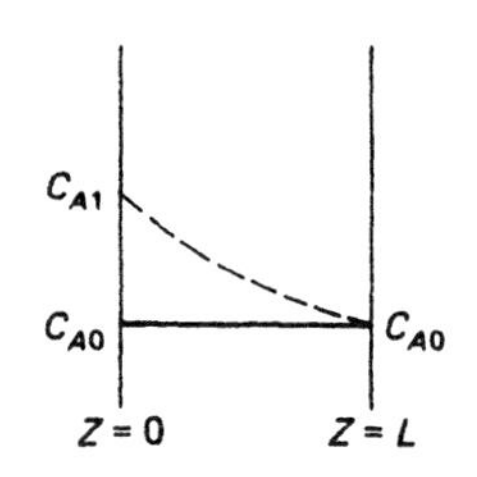

Figure 4-21 Problem 4.2: diffusion through a membrane

4.3 A 0.025-m-thick porous slab, initially containing 0.5 kg of solute A per kilogram of solid, is placed in a well-agitated tank of liquid B (see Figure 4-22). If the effective diffusivity of A through the porous slab is equal to 5.0×10^{-10} m^2/s, determine the time required for the solute content at the center of the slab to decrease to 0.005 kg A/kg solid. Neglect the counterdiffusion of B into the solid.

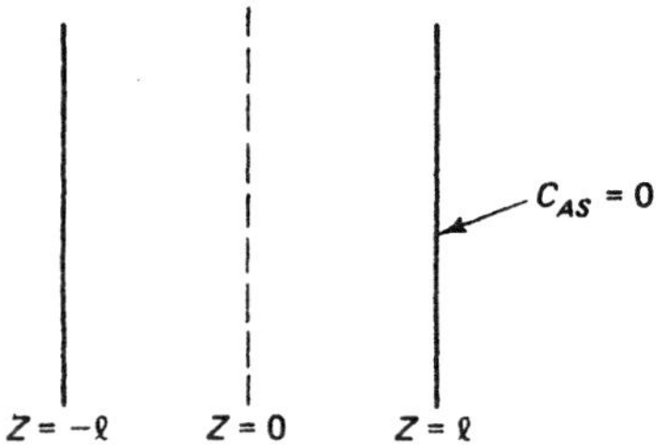

Figure 4-22 Problem 4.3: diffusion through a porous slab

4.4 Starting with Eq. (4-80) for diffusion into a sphere and the boundary conditions given by Eqs. (4-81) through (4-83), obtain **(a)** the concentration profile given by Eq. (4-85), and **(b)** the relative mass transfer given by Eq. (4-86).

4.5 A cylinder, initially at a concentration of C_{A0}, is placed in a large well-stirred tank that has a concentration of $C_{A\infty}$. Determine the unsteady-state concentration profile in the cylinder. Assume that the concentrations of the solution and the surface of the cylinder are constant and are related by the equilibrium expression $C_{AR} = (1/b)C_{A\infty}$. Neglect mass transfer through the ends of the cylinder.

4.6 In Problem 4.2 determine the concentration of A at $Z = L/2$ after 1 h if the thickness is $L = 0.1$ m and $D_A = 5.75 \times 10^{-8}$ m^2/s. The initial concentration is $C_{A0} = 2.45$ $kgmol/m^3$ and the imposed concentration at $Z = 0$ is $C_{A1} = 4.80$ $kgmol/m^3$. Check your answer against values taken from Figure 4-5 and comment on any differences.

4.7 The pores of a spherical anion resin are filled with a solution that contains a Cl^- concentration of 4.0×10^{-3} $kgmol/m^3$ solution. If the spherical resin is placed in a large stirred tank that contains a Cl^- concentration of 0.1 kgmol/m^3, **(a)** determine the time necessary to raise the Cl^- concentration to 8.0×10^{-3} kgmol/m^3 solution at a position $R/2$ in the sphere, and **(b)** compare your answer with values obtained from Figure 4-7. Particle radius = 0.002 m, effective diffusivity = 3.54×10^{-11} m^2/s.

4.8 Polyethylene terephthalate (PET) is a polymer used in making synthetic fabrics. If it is stored for any length of time prior to extruding, it will absorb moisture from the atmosphere and plug the extruder when processed. A shipment of 0.01-m-thick PET tile with a surface area of 0.3 m^2 has been stored in a warehouse at 32°C and 70% relative humidity for several months. It is desired to reduce the moisture content of the tile to 0.06 wt% before processing them in an extruder. This will be achieved by kiln drying at 160°C in dry air which has a dew point of 0°C. Determine the time required to dry the PET. The diffusion coefficient of water in the PET can be calculated from the Arrhenius equation

$$D_W = D_0 \exp\left(\frac{-E}{RT}\right)$$

where $D_0 = 2.6 \times 10^{-2}$ m^2/h and $E = 8.66$ kcal/mol. The equilibrium water content of the PET tile in moist air is given as

$$W_g = 0.8\left(\frac{\phi}{100}\right)$$

where ϕ is the percent relative humidity and W_g is the weight percent of water in the tile. Neglect mass transfer from the edges.

4.9 A water stream is to be chlorinated by contacting it with pure chlorine gas in a counter current spray tower. The spray contactor is operated at 15.5°C and 1.07 atm pressure. Assuming that the spray droplets are spherical with a diameter of 4.0×10^{-4} m and the contactor height is 1.83 m, calculate the average chlorine concentration of the water at the column exit. The concentration of chlorine on the surface of the droplet can be expressed in terms of the chlorine pressure as

$$C_A = 0.11 P_A$$

where C_A has units kgmol/m^3 and P is expressed in atmospheres. Since this problem is similar to diffusion into a sphere described by Eqs. (4-74) through (4-77), the concentration profile is given by Eq. (4-85). The average concentration is found by integrating Eq. (4-85) over the volume of the sphere.

4.10 A porous sphere saturated with water is placed in a wind tunnel in order to study the unsteady-state mass transfer. The water will diffuse radially to the surface of the sphere, where it will be removed by convective mass transfer into a stream of air. The initial concentration of water is equal to C_{A0} and the concentration of water in the airstream is $C_{A\infty}$. The diffusion coefficient of water through the porous sphere and the convective mass transfer coefficient at the surface are constant. Derive the concentration profile for the water in the sphere as a function of time and position. The concentration profile for this problem is given by Eq. (4-139).

4.11 In Section 4.5 the equation and associated boundary conditions that describe unsteady-state diffusion in an infinitely long cylinder with convection at the surface was presented. **(a)** Derive the concentration profile for species A inside the cylinder as a function of time and position. The concentration profile is given by

Eq. (4-136). **(b)** Obtain Eq. (4-138) by following the method presented in Eqs. (4-70) through (4-72).

4.12 An inert carrier gas containing a small quantity of species A flows axially through a circular tube at a velocity U_0. The inside of the tube is coated with a catalyst that promotes the conversion of A to B according to the first-order reaction

$$A \xrightarrow{k_1} B$$

(a) Assuming plug flow in the tube, derive the steady-state partial differential equation and associated boundary conditions that describe the concentration of A as a function of position in the r and Z directions. **(b)** Obtain the concentration profile as a function of position. The concentration of A entering the tube at $Z = 0$ is C_{A0}. At the inside surface of the cylinder, $r = R$, the boundary condition is

$$-D_A \frac{\partial C_A}{\partial r} = k_1 C_A$$

4.13 Consider the problem in which gas A diffuses into the semi-infinite porous solid shown in Figure 4-23. Initially, the concentration of A in the solid is C_{A0}. If the concentration of A at the surface of the plate, $X = 0$, is suddenly increased to $C_A = C_{A0} + 10/\sqrt{t}$, obtain the concentration profile of A in the solid. Use the Laplace transform method.

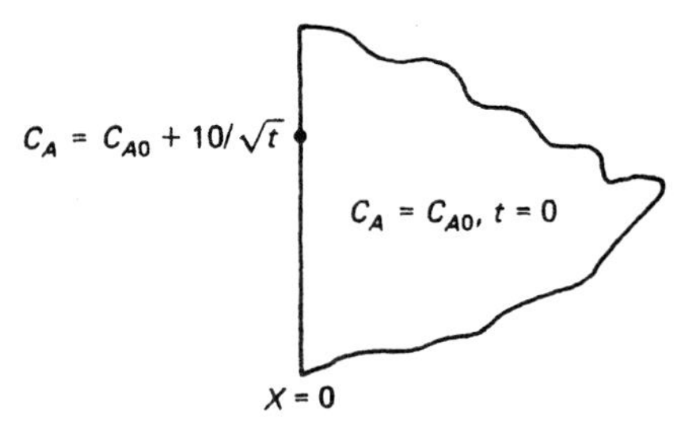

Figure 4-23 Problem 4.13: diffusion into a semi-infinite solid

4.14 A large well-stirred tank containing a dilute solution of species A is separated from a long circular conduit by a solid membrane as shown in Figure 4-24. The circular tube contains a porous catalyst. When the solid membrane is ruptured, species A diffuses into the conduit where it reacts according to the first-order reaction

$$A \xrightarrow{k_1} B$$

Initially, the concentration of A in the solid is zero. However, upon rupturing the membrane the concentration of A at the entrance is constant at C_{A0}. Assum-

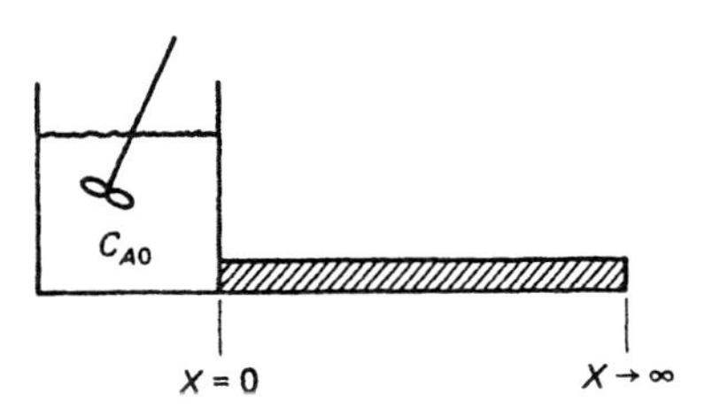

Figure 4-24 Problem 4.14: diffusion with chemical reaction in a semi-infinite porous solid

ing that the circular conduit is very long, use the Laplace transform method to calculate the concentration profile of A in the porous solid as a function of time and position. Assume that the concentration of A in the stirred tank is constant.

4.15 A 1.0-m-thick porous plate initially contains 0.5 kgmol A/m^3 solid. One side of the plate is suddenly exposed to a gas stream which contains 0.02 kgmol of A/m^3. The other side of the plate is sealed so that mass will not be transferred from that surface. If the convective mass transfer coefficient is 5.6×10^{-3} m/s, determine **(a)** the time necessary to decrease the concentration of A at the center of the plate to 0.3 kgmol/m^3, and **(b)** the concentration of A on the sealed surface at the time calculated in part (a). The diffusion coefficient of A in the solid is $D_A = 7.5 \times 10^{-3}$ m^2/s. The concentrations of A in the solid and in the gas stream are related by the expression $C_A^* = 0.8C_{A,g}$.

4.16 In Problem 4.2 the concentration on one face was suddenly increased to C_{A1} while the concentration on the other face was kept constant at C_{A0}. Consider the case in which the concentration at $Z = 0$ is increased to C_{A1} but decays to C_{A0} for large times. The concentration at the surface is expressed by the equation

$$C_A = a + be^{-t}$$

Obtain the concentration profile of A in the wall for the boundary condition above. The solution to this problem can be found by starting with the solution obtained in Problem 4.2 and applying Duhamel's superposition integral for the time-varying boundary condition.

4.17 Consider the case of diffusion in a slab as shown in Figure 4-4. Initially, the concentration of A is C_{A0}. For times greater than zero, the concentrations on the surfaces of the slab are described by

$$C_A = C_{A0} + (C_{A1} - C_{A0})\frac{Dt}{8l^2} \quad \text{for } \frac{Dt}{4l^2} \leq 10$$

and $$C_A = C_{A0} \quad \text{for } \frac{Dt}{4l^2} > 10$$

Obtain an expression for the concentration profile inside the slab as a function of time and position.

4.18 Show that the solution given by Eq. (4-129) for the flat plate reduces to Eq. (4-99) for the case in which the solute concentration in the free stream, $C_{A\infty}$, is zero.

4.19 In Problem 4.5 the separation-of-variables method was used to obtain an exact solution. Using Galerkin's method, rework Problem 4.5 to obtain an approximate concentration profile.

Mass Transfer Coefficients 5

5.1 Introduction

In the preceding chapters we considered molecular transport and developed detailed models describing the concentration profile in a medium. From this we were able to calculate the mass flux of an individual component. Quite often we are not interested in a detailed description of the process and even though we might be, a detailed model may not be feasible. We are therefore encouraged to seek a convenient means of describing the mass transfer process. Since the flux is proportional to the concentration gradient, we define an empirical mass transfer coefficient in a manner analogous with heat transfer.

Let us consider the steady-state diffusion of solute A through a membrane as shown in Figure 5-1. After the solute diffuses through the membrane, it is swept from the external surface by a gas stream. A mass transfer coefficient for transfer of component A into the free stream is defined in terms of diffusion at the interface by the expression

$$k_c^* = \frac{{}_M J_A}{C_{A1} - C_{A\infty}} = -\frac{D(\partial C_A/\partial Z)_{Z=0}}{C_{A1} - C_{A\infty}} \tag{5-1}$$

where k_c^* is the mass transfer coefficient. The concentration, $C_{A\infty}$, is evaluated in the free stream flowing over the surface and C_{A1} is the concentration on the surface, but in the fluid phase.

Equation (5-1) can be written in the same form as Newton's law of cooling, where k_c^* is analogous to the convective heat transfer coefficient.

$$N_A = k_c^*(C_{A1} - C_{A\infty}) \tag{5-2}$$

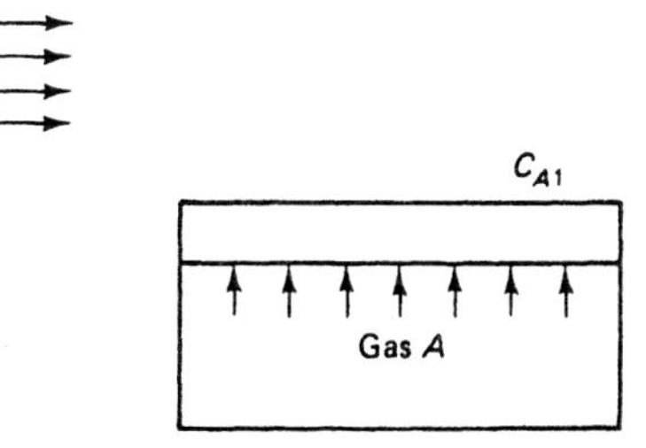

Figure 5-1 Definition of the mass transfer coefficient

The expression above is limited to low mass transfer rates in which the bulk flow is negligible and the concentration profile is not distorted. Obviously, the equation above does not take into account either the case in which the fluids in the free stream diffuse countercurrently to species A or the case of high flux rates. In this chapter Eq. (5-2) is extended to include a variety of physical situations, and theoretical models for the mass transfer coefficient will be presented.

5.2 Definitions of the Mass Transfer Coefficient

Simple forms of the convective mass transfer coefficient can be found by considering the steady-state evaporation of a single component A into a gas as shown in Figure 5-2. Although only a single component is evaporating, informa-

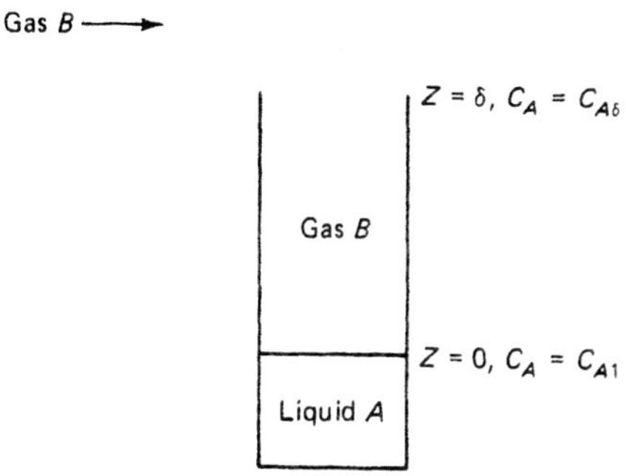

Gas B →

$Z = \delta,\ C_A = C_{A\delta}$

Gas B

$Z = 0,\ C_A = C_{A1}$

Liquid A

Figure 5-2 Evaporation of a single component

tion regarding the transfer of gas B must be known. Two cases that can be used to derive theoretical values for the mass transfer coefficient are equimolar counter diffusion and diffusion of a single component.

Equimolar counter diffusion

For equimolar counter diffusion, the flux of component A is equal to the negative of flux B (i.e., $N_{AZ} = -N_{BZ}$). Thus the bulk flow term of Eq. (1-34) is zero and the resulting flux of A is equal to transfer by diffusion only. This is represented as

$$N_{AZ} = -D_{AB}\frac{dC_A}{dZ} \tag{5-3}$$

For convenience, the direction of the flux will not be indicated in the remainder of the chapter. Separating the variables and assuming that the molar flux and diffusivity are constant, the equation above becomes

$$N_A \int_0^{\delta} dZ = -D_{AB}\int_{C_{A1}}^{C_{A\delta}} dC_A \tag{5-4}$$

Upon integrating, Eq. (5-4) can be written in terms of the molar flux as

$$N_A = \frac{D_{AB}}{\delta}(C_{A1} - C_{A\delta}) \tag{5-5}$$

where δ is defined as the length of the diffusion path. Comparing Eqs. (5-2) and (5-5), we see that the convective mass transfer coefficient is related to the diffusion coefficient and the length of the diffusion path. Equation (5-5) can be written more specifically in terms of the Colburn–Drew mass transfer coefficient, k'_c, as

$$N_A = k'_c(C_{A1} - C_{A\delta}) \tag{5-6}$$

where $k'_c = D_{AB}/\delta$. Thus for liquids

$$N_A = k'_c(C_{A1} - C_{A\delta}) = k'_L(C_{A1} - C_{A\delta}) \tag{5-7}$$

and

$$N_A = Ck'_L(x_{A1} - x_{A\delta}) = k'_x(x_{A1} - x_{A\delta}) \tag{5-8}$$

where k'_L has been substituted for k'_c in Eq. (5-7). For gases, the concentration can be written in terms of the ideal gas equation to give

$$N_A = \frac{k'_c}{RT}(\bar{P}_{A1} - \bar{P}_{A\delta}) = k'_G(\bar{P}_{A1} - \bar{P}_{A\delta}) \tag{5-9}$$

where $\bar{P}_A$ is the partial pressure. Since at low pressure the partial pressure can be related to the mole fraction and total pressure by Dalton's law (i.e., $\bar{P}_A = y_AP$), we have

$$N_A = Pk'_G(y_{A1} - y_{A\delta}) = k'_y(y_{A1} - y_{A\delta}) \tag{5-10}$$

Diffusion of *A* through a stagnant gas

For diffusion of A through a stagnant gas B, Eq. (1-34) can be reduced as shown below by setting the flux of B equal to zero. For this case we have

$$N_A = -D_{AB}\frac{dC_A}{dZ} + \frac{C_A}{C}N_A \tag{5-11}$$

or

$$N_A\left(1 - \frac{C_A}{C}\right) = -D_{AB}\frac{dC_A}{dZ} \tag{5-12}$$

By assuming that the molar flux and diffusivity are constant, the variables in Eq. (5-12) can be separated and written as integrals over the length of the diffusion path shown in Figure 5-2.

$$N_A \int_0^\delta dZ = -D_{AB} \int_{C_{A1}}^{C_{A\delta}} \frac{dC_A}{1 - C_A/C} \tag{5-13}$$

Following integration and substitution of the limits, Eq. (5-13) becomes

$$N_A = \frac{CD_{AB}}{\delta} \ln \frac{1 - C_{A\delta}/C}{1 - C_{A1}/C} \tag{5-14}$$

Since the flux is customarily written as a mass transfer coefficient times a linear driving force, Eq. (5-14) will be modified by multiplying and dividing it by $[(1 - C_{A\delta}/C) - (1 - C_{A1}/C)]$. The resulting equation is

$$N_A = \frac{D_{AB}}{\delta(1 - C_A/C)_M}(C_{A1} - C_{A\delta}) \tag{5-15}$$

where the log mean concentration difference, $(1 - C_A/C)_M$, is defined by the relationship

$$\left(1 - \frac{C_A}{C}\right)_M = \frac{(1 - C_{A\delta}/C) - (1 - C_{A1}/C)}{\ln \dfrac{1 - C_{A\delta}/C}{1 - C_{A1}/C}} = \left(\frac{C_B}{C}\right)_M \tag{5-16}$$

Introducing the Colburn–Drew mass transfer coefficient, Eq. (5-15) becomes

$$N_A = \frac{k'_c}{(1 - C_A/C)_M}(C_{A1} - C_{A\delta}) = k_c(C_{A1} - C_{A\delta}) \tag{5-17}$$

When written for liquids we have

$$N_A = k_c(C_{A1} - C_{A\delta}) = k_L(C_{A1} - C_{A\delta}) \tag{5-18}$$

or

$$N_A = Ck_L(x_{A1} - x_{A\delta}) = k_x(x_{A1} - x_{A\delta}) \tag{5-19}$$

By introducing the ideal gas equation and making use of Dalton's law, the molar flux can be written as

$$N_A = \frac{PD_{AB}}{RT\delta(\bar{P}_B)_M}(\bar{P}_{A1} - \bar{P}_{A\delta}) \tag{5-20}$$

Making use of the Colburn–Drew mass transfer coefficient, we have

$$N_A = \frac{Pk'_c}{RT(\bar{P}_B)_M}(\bar{P}_{A1} - \bar{P}_{A\delta}) = \frac{Pk'_G}{(\bar{P}_B)_M}(\bar{P}_{A1} - \bar{P}_{A\delta}) \tag{5-21}$$

or

$$N_A = k_G(\bar{P}_{A1} - \bar{P}_{A\delta}) = Pk_G(y_{A1} - y_{A\delta}) = k_y(y_{A1} - y_{A\delta}) \tag{5-22}$$

A summary of the mass transfer coefficients for equimolar counter diffusion and diffusion through a stagnant film is given in Table 5-1. A comparison of these equations shows the effect of introducing the counter diffusion of a second species. As shown, counter diffusion reduces the molar flux of A. For very dilute solutions, however, the log mean pressure difference is approximately equal to 1.0 and the mass transfer coefficient for transfer through a stagnant film can be used safely. The molar flux for cases in which we have neither of the limiting conditions given in Table 5-1 will be discussed later. To convert from one type of mass transfer coefficient to another, we can use the following definitions:

TABLE 5-1. MASS TRANSFER COEFFICIENTS

Equimolar Counter Diffusion		*Diffusion through Stagnant Film*		
Flux	*Mass Transfer Coefficient*	*Flux*	*Mass Transfer Coefficient*	*Unit of Mass Transfer Coefficients*
Gases:		Gases:		
$N_A = k'_c(C_{A1} - C_{A\delta})$	$k'_c = \frac{D_{AB}}{\delta}$	$N_A = k_c(C_{A1} - C_{A\delta})$	$k_c = \frac{PD_{AB}}{\delta(\bar{P}_B)_M}$	$\frac{\text{mol}}{\text{(time)(area)(mol/vol.)}}$
$N_A = k'_G(\bar{P}_{A1} - \bar{P}_{A\delta})$	$k'_G = \frac{D_{AB}}{\delta RT}$	$N_A = k_G(\bar{P}_{A1} - \bar{P}_{A\delta})$	$k_G = \frac{PD_{AB}}{\delta RT(\bar{P}_B)_M}$	$\frac{\text{mol}}{\text{(time)(area)(pressure)}}$
$N_A = k'_y(y_{A1} - y_{A\delta})$	$k'_y = \frac{PD_{AB}}{\delta RT}$	$N_A = k_y(y_{A1} - y_{A\delta})$	$k_y = \frac{P^2 D_{AB}}{\delta RT(\bar{P}_B)_M}$	$\frac{\text{mol}}{\text{(time)(area)(mole fraction)}}$
Liquids:		Liquids:		
$N_A = k'_L(C_{A1} - C_{A\delta})$	$k'_L = \frac{D_{AB}}{\delta}$	$N_A = k_L(C_{A1} - C_{A\delta})$	$k_L = \frac{D_{AB}}{\delta(x_B)_M}$	$\frac{\text{mol}}{\text{(time)(area)(mol/vol.)}}$
$N_A = k'_x(x_{A1} - x_{A\delta})$	$k'_x = \frac{CD_{AB}}{\delta}$	$N_A = k_x(x_{A1} - x_{A\delta})$	$k_x = \frac{CD_{AB}}{\delta(x_B)_M}$	$\frac{\text{mol}}{\text{(time)(area)(mole fraction)}}$

Liquids:

$$k'_x = Ck'_L = \frac{\rho}{M}k'_L = k_x(x_B)_M = Ck_L(x_B)_M \tag{5-23}$$

$$(C_B)_M = C(x_B)_M$$

Gases:

$$\frac{Pk'_c}{RT} = Pk'_G = k'_y = (\bar{P}_B)_M k_G = \frac{(\bar{P}_B)_M k_y}{P} \tag{5-24}$$

Example 5.1

Consider the case in which liquid A evaporates and diffuses through stagnant gas B as shown in Figure 5-2. Assuming that the process is carried out isothermally at 25°C and the total pressure is 1 atm, calculate the molar flux and convective mass transfer coefficient when the level of liquid A is 0.05 m below the top of the container. Assume that $D_{AB} = 1.0 \times 10^{-5}$ m²/s. The mole fraction of A at the gas–liquid interface is 0.2. At the top of the container, the mole fraction is 0.001.

Solution:

Starting with Eq. (5-20), we have

$$N_A = \frac{PD_{AB}}{RT\delta(\bar{P}_B)_M}(\bar{P}_{A1} - \bar{P}_{A\delta})$$

$$\bar{P}_A = y_A P \qquad \bar{P}_B = y_B P$$

$$\bar{P}_{A1} = 0.2 \text{ atm} \qquad \bar{P}_{B1} = 0.8 \text{ atm}$$

$$\bar{P}_{A\delta} = 0.001 \text{ atm} \qquad \bar{P}_{B\delta} = 0.999 \text{ atm}$$

$$(\bar{P}_B)_M = \frac{\bar{P}_{B\delta} - \bar{P}_{B1}}{\ln(\bar{P}_{B\delta}/\bar{P}_{B1})}$$

$$(\bar{P}_B)_M = \frac{0.999 - 0.8}{\ln(0.999/0.8)} = 0.896 \text{ atm}$$

$$N_A = \frac{(1)(1.0 \times 10^{-5})(0.2 - 0.001)}{(82.06 \times 10^{-3})(298)(0.05)(0.896)} = 1.82 \times 10^{-6} \frac{\text{kgmol}}{\text{m}^2\cdot\text{s}}$$

The mass transfer coefficient is calculated by using the definitions given in Table 5-1.

$$k_G = \frac{PD_{AB}}{RT\delta(\bar{P}_B)_M} = \frac{1.0 \times 10^{-5}}{(82.05 \times 10^{-3})(298)(0.05)(0.896)}$$

$$= 9.13 \times 10^{-6} \frac{\text{kgmol}}{\text{m}^2\cdot\text{s}\cdot\text{atm}}$$

Coefficients for arbitrary flux rates and eddy diffusion

The use of the theoretically derived mass transfer coefficients is generally quite limited due to difficulty in defining the length of the diffusion path. Also the assumptions involved in their derivation are not often found in industrial operations. Thus a more general relationship must be used to account for situations in which we cannot assume either equimolar counter diffusion or transfer through a stagnant film. In addition, when the fluid flow that we previously considered is turbulent, eddy diffusion will also contribute to the overall flux. This contribution is expressed by the equation

$$N_A = -E_D \frac{dC_A}{dZ} \tag{5-25}$$

where E_D is the eddy diffusivity. The eddy diffusivity is not a molecular property of the fluid but depends on the level of turbulence in the system. Although it can be related to the mixing length and mean eddy velocity, it is difficult to evaluate experimentally.

The eddy diffusivity can be combined with the molecular diffusivity in Eq. (1-34) to give

$$N_A = -(D_{AB} + E_D)\frac{dC_A}{dZ} + \frac{C_A}{C}(N_A + N_B) \tag{5-26}$$

A general form for the molar flux can be written in terms of the convective mass transfer coefficient by integrating Eq. (5-26) over the diffusion path shown in Figure 5-2.

$$N_A = \frac{N_A}{N_A + N_B} \frac{C(D_{AB} + E_D)}{\delta} \ln \frac{\dfrac{N_A}{N_A + N_B} - \dfrac{C_{A\delta}}{C}}{\dfrac{N_A}{N_A + N_B} - \dfrac{C_{A1}}{C}} \tag{5-27}$$

In the equation above, the flux and the molecular and eddy diffusivities are assumed to be constant. Introducing the log mean driving force as shown in Eq. (5-15), we can express the equation above as

$$N_A = \frac{k_c'}{\beta}(C_{A1} - C_{A\delta}) \tag{5-28}$$

where the Colburn–Drew mass transfer coefficient includes both the molecular and eddy diffusivities and β is the bulk flow correction factor. It is given by the expression

$$\beta = \frac{\left(\dfrac{N_A}{N_A + N_B} - \dfrac{C_{A\delta}}{C}\right) - \left(\dfrac{N_A}{N_A + N_B} - \dfrac{C_{A1}}{C}\right)}{\dfrac{N_A}{N_A + N_B} \ln \dfrac{\dfrac{N_A}{N_A + N_B} - \dfrac{C_{A\delta}}{C}}{\dfrac{N_A}{N_A + N_B} - \dfrac{C_{A1}}{C}}} \tag{5-29}$$

Comparing Eq. (5-28) to Eq. (5-6), we see that $\beta = 1$ for equimolar counter diffusion. For diffusion of A through stagnant B, we note that $\beta = (x_B)_M$.

5.3 Theoretical Models for Mass Transfer at a Fluid–Fluid Interface

Film theory

Several models have been developed to describe mass transfer at a fluid phase boundary. The earliest and simplest of these is the film theory proposed by Whitman (1923). This model is based on the assumption that for a fluid flowing turbulently over a solid, the entire resistance to mass transfer resides in a stagnant film in the fluid next to the surface. The thickness of the film, which is greater than the laminar sublayer, is such that it provides the same resistance to mass transfer by molecular diffusion as exists for the actual convective process.

Consider the case of a fluid flowing in a cylinder in which a solid dissolves from the walls of the cylinder into the fluid. Since the film is very thin, a shell balance can be made using rectangular coordinates (see Figure 5-3). For steady-

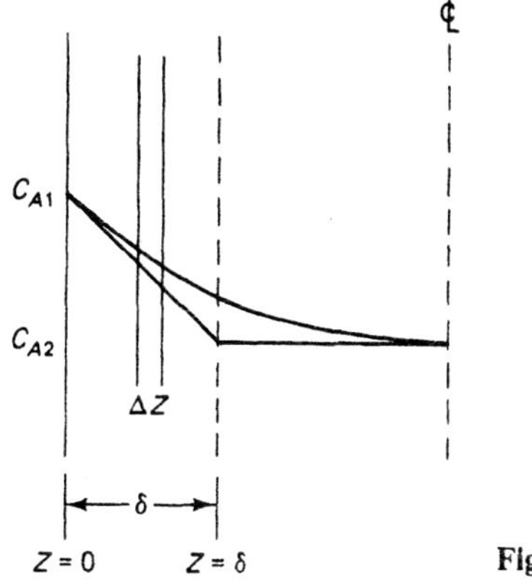

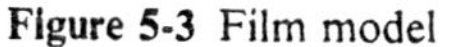

Figure 5-3 Film model

state equimolar counter diffusion with no chemical reaction, a shell balance over the differential element gives

$$\frac{d^2C_A}{dZ^2} = 0 \tag{5-30}$$

The boundary conditions for this process are

$$\text{BC1:} \quad C_A = C_{A1} \quad \text{at } Z = 0 \tag{5-31}$$

$$\text{BC2:} \quad C_A = C_{A2} \quad \text{at } Z = \delta \tag{5-32}$$

After integrating and evaluating the constants of integration, we obtain

$$C_A = \left(\frac{C_{A2} - C_{A1}}{\delta}\right)Z + C_{A1} \tag{5-33}$$

But for equimolar counter diffusion the flux expression becomes

$$N_A = -D_{AB}\frac{dC_A}{dZ} \tag{5-34}$$

Therefore, by taking the derivative of the concentration profile, we obtain the molar flux in terms of a theoretical mass transfer coefficient. This is given as

$$N_A = \frac{D_{AB}}{\delta}(C_{A1} - C_{A2}) = k_c'(C_{A1} - C_{A2}) \tag{5-35}$$

where δ is the thickness of the stagnant film. From a comparison of Eqs. (5-6) and (5-35) we see that the Colburn–Drew mass transfer coefficients are based on the film theory. The film theory predicts that the mass transfer coefficient is proportional to $D_{AB}^{1.0}$. Experimental data show, however, that k_c' is more nearly proportional to $D_{AB}^{2/3}$. The inability to predict the film thickness a priori next to a solid surface limits the usefulness of this theory as a method for calculating mass transfer coefficients.

Example 5.2

Ammonia is absorbed from an air stream into a water film flowing down an inclined plane at 20°C and 1 atm as shown in Figure 5-4. The flow rate of the water is 5×10^{-3} kg/s per meter width of the plate. If laminar flow is assumed,

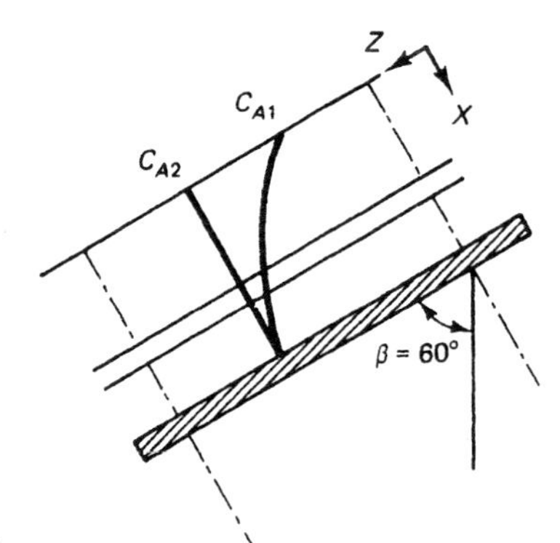

Figure 5-4 Absorption of ammonia

the thickness of the film can be calculated from the expression

$$\delta = \left(\frac{3\mu\Gamma}{\rho^2 g \cos\beta}\right)^{1/3}$$

where Γ is defined as the mass rate of flow per unit width of the plate. Assuming that the film thickness calculated from the expression above is equal to the thickness of the stagnant film for the film theory, estimate the mass transfer coefficient for the absorption of ammonia into the water.

Solution:

From Figure A-1 in Appendix A,

$$\mu_{H_2O} = 10.5 \times 10^{-4} \text{ kg/m·s}$$

$$\Gamma = 5 \times 10^{-3} \text{ kg/m·s}$$

$$\rho = 998 \text{ kg/m}^3 \qquad \beta = 60°$$

$$D_{NH_3\text{-}H_2O} = 1.76 \times 10^{-9} \text{ m}^2/\text{s}$$

$$\delta = \left[\frac{(3)(8.75 \times 10^{-4})(5 \times 10^{-3})}{(998)^2(9.806)(0.5)}\right]^{1/3} = 1.39 \times 10^{-4} \text{ m}$$

For the film theory $k_c' = D_{AB}/\delta$. Therefore,

$$k_c' = \frac{1.76 \times 10^{-9} \text{ m}^2/\text{s}}{1.39 \times 10^{-4} \text{ m}}$$

$$= 1.27 \times 10^{-5} \text{ m/s}$$

Penetration theory

Although the presence of a laminar sublayer is often used in fluid mechanics, the concept of a stagnant film is unrealistic on the basis of the unstableness of a fluid–fluid interface. In order to describe the physical process more accurately, Higbie (1935) suggested a penetration theory for transfer across a gas–liquid interface. This theory assumes that the liquid surface consists of small fluid elements that contact the gas phase for an average time, after which they penetrate into the bulk liquid. Each element is then replaced by another element from the bulk liquid phase. For turbulent flow, the penetration theory provides a reasonable mechanism for describing mass transfer. This model can be described as the unsteady-state diffusion of a solute into a liquid phase of infinite thickness.

Consider Figure 5-5 in which solute A diffuses with the element from the gas–liquid interface into the bulk liquid. For dilute concentrations of A in the

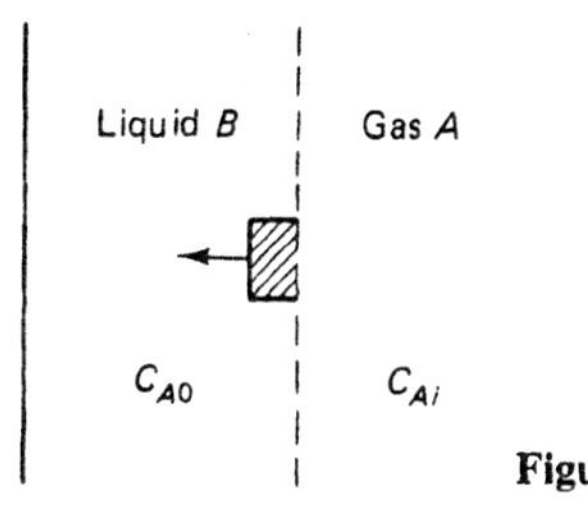

Figure 5-5 Penetration model

liquid the diffusion process is described by the equation

$$D_{AB}\frac{\partial^2 C_A}{\partial X^2} = \frac{\partial C_A}{\partial t} \tag{5-36}$$

with boundary conditions

$$\text{BC1:} \quad C_A = C_{A0} \quad \text{at } t = 0 \tag{5-37}$$

$$\text{BC2:} \quad C_A = C_{Ai} \quad \text{at } X = 0 \tag{5-38}$$

$$\text{BC3:} \quad C_A = C_{A0} \quad \text{at } X = \infty \tag{5-39}$$

The solution to the equation above is the same as that found for two-dimensional mass transfer in a wetted-wall column. This is readily shown by simply writing $t = Z/U_{Z,max}$ in Eq. (4-12) and rewriting the boundary condition given by Eq. (4-3) as the initial condition. The concentration profile is expressed as

$$\frac{C_A - C_{A0}}{C_{Ai} - C_{A0}} = 1 - \operatorname{erf}\left(\frac{X}{\sqrt{4D_{AB}t}}\right) = \operatorname{erfc}\left(\frac{X}{\sqrt{4D_{AB}t}}\right) \tag{5-40}$$

where erfc is defined as the complementary error function.

The molar flux is given by

$$N_A|_{X=0} = -D_{AB}\left(\frac{\partial C_A}{\partial X}\right)_{X=0} \tag{5-41}$$

The concentration gradient is found by differentiating Eq. (5-40) as shown.

$$\frac{\partial C_A}{\partial X} = (C_{Ai} - C_{A0})\frac{\partial}{\partial X}\left(-\frac{2}{\sqrt{\pi}}\int_0^{X/\sqrt{4D_{AB}t}} e^{-X^2/4D_{AB}t}\, d\frac{X}{\sqrt{4D_{AB}t}}\right) \tag{5-42}$$

$$= (C_{Ai} - C_{A0})\left(-\frac{2}{\sqrt{\pi}}e^{-X^2/4D_{AB}t}\right)\frac{1}{\sqrt{4D_{AB}t}} \tag{5-43}$$

$$= -(C_{Ai} - C_{A0})\frac{1}{\sqrt{\pi D_{AB}t}}e^{-X^2/4D_{AB}t} \tag{5-44}$$

Therefore, the instantaneous flux rate at the interface is given by

$$N_A|_{X=0} = \sqrt{\frac{D_{AB}}{\pi t}}(C_{Ai} - C_{A0}) \tag{5-45}$$

The average flux rate is found by integrating over the time during which the element has been exposed at the surface. Thus

$$N_A|_{avg} = \frac{1}{t_s}(C_{Ai} - C_{A0})\sqrt{\frac{D_{AB}}{\pi}}\int_0^{t_s}\frac{dt}{t^{1/2}} \tag{5-46}$$

$$= \sqrt{\frac{4D_{AB}}{\pi t_s}}(C_{Ai} - C_{A0}) \tag{5-47}$$

where t_s is the exposure time of the fluid element. The penetration theory predicts that the mass transfer coefficient is proportional to $\sqrt{D_{AB}}$:

$$k_c' = 2\sqrt{\frac{D_{AB}}{\pi t_s}} \tag{5-48}$$

Surface renewal theory

In order to improve the penetration theory, Danckwerts (1951) suggested that the constant exposure time be replaced by an average exposure time determined from an assumed time distribution. He assumed that the chance of an element being replaced on the surface was independent of the time during which it had been exposed. For the surface renewal model the average flux rate is given by

$$N_A|_{avg} = (C_{Ai} - C_{A0})\sqrt{\frac{D_{AB}}{\pi}} \int_0^\infty \frac{\tau(t)}{t^{1/2}}\, dt \tag{5-49}$$

where $\tau(t)\, dt$ represents the fraction of the surface consisting of elements with an age in the time span t to $t + dt$. The sum of the fractions is equal to 1.0:

$$\int_0^\infty \tau(t)\, dt = 1.0 \tag{5-50}$$

Danckwerts presented an analytical form for the age distribution function by assuming that the rate of disappearance of surface elements of a certain age was proportional to the number of elements of that same age. Thus

$$-\frac{d\tau}{dt} = s\tau \tag{5-51}$$

where s is the rate of surface renewal and is equal to the reciprocal of the exposure time of the elements. After integration, the equation above becomes

$$\tau = Ke^{-st} \tag{5-52}$$

The constant of integration can be found by integrating Eq. (5-52) over the total surface to give

$$\int_0^\infty Ke^{-st}\, dt = 1.0 \tag{5-53}$$

This gives the constant of integration equal to the rate of surface renewal. The age distribution function is thus given as

$$\tau = se^{-st} \tag{5-54}$$

The average flux can be obtained by integrating the product of the instantaneous flux and the age distribution over all times.

$$N_A|_{avg} = (C_{Ai} - C_{A0})\sqrt{\frac{D_{AB}}{\pi}} \int_0^\infty \frac{se^{-st}}{t^{1/2}}\, dt \tag{5-55}$$

$$= \sqrt{D_{AB}s}\,(C_{Ai} - C_{A0}) \tag{5-56}$$

For this model the mass transfer coefficient is equal to $\sqrt{D_{AB}s}$. Although values for s are not generally known, they can be determined experimentally by measuring the mass transfer coefficient.

Harriott (1962) modified the surface renewal model to account for the effect of eddies not penetrating down to the wall itself. He assumed that the eddies arrived at the interface at random times and then penetrated random distances into the liquid.

Film penetration theory

In addition to the theories above, a number of other models have been developed by modifying or combining the film and penetration theories. Toor and Marchello (1958) proposed a model that combined the film theory with the penetration model. They proposed that the total resistance to mass transfer took place in a laminar film at the fluid interface, as in the film theory, but that the mass transfer is an unsteady-state process. The molar flux at the interface for this model is expressed by the relationship

$$N_A|_{x=0} = (C_{Ai} - C_{A0})\frac{D_{AB}}{\delta}\left[1 + 2\sum_{n=1}^{\infty} \exp\left(-\frac{n^2\pi^2 D_{AB}t}{\delta^2}\right)\right] \tag{5-57}$$

In the range $\pi \leq \delta^2/D_{AB}t < \infty$, the equation above can be written in a form similar to the penetration theory. For $0 < \delta^2/D_{AB}t \leq \pi$, Eq. (5-57) can be expressed in a form related to the film theory. These are

$$N_A|_{x=0} = (C_{Ai} - C_{A0})\sqrt{\frac{D_{AB}}{\pi t}}\left[1 + 2\exp\left(-\frac{\delta^2}{D_{AB}t}\right)\right] \qquad \pi \leq \frac{\delta^2}{D_{AB}t} < \infty \tag{5-58}$$

and

$$N_A|_{x=0} = (C_{Ai} - C_{A0})\frac{D_{AB}}{\delta}\left[1 + 2\exp\left(-\frac{\pi^2 D_{AB}t}{\delta^2}\right)\right] \qquad 0 < \frac{\delta^2}{D_{AB}t} \leq \pi \tag{5-59}$$

The exponential terms in the equations above contribute no more than 8.64% to the molar flux. If they are neglected, Eqs. (5-58) and (5-59) reduce to the penetration model and film model, respectively.

5.4 Overall Mass Transfer Coefficients

Most important mass transfer processes involve the contact of two or more phases and the transfer of a solute across a phase boundary. In gas absorption, for example, a solute diffuses through the carrier gas phase to a liquid surface, where it dissolves and is transferred by diffusion or convective mixing into the liquid phase. The rate at which the solute is transferred depends on a number of factors such as the solubility of the solute in the liquid and its displacement from equilibrium. The physical conditions under which the transfer takes place also are extremely important to the rate at which the transfer occurs. Consequently, the design of mass transfer equipment includes a consideration of interfacial contact area, distribution of both phases throughout the column, and the temperature and pressure of the process.

A number of models have been proposed to describe the transfer across a phase boundary. A two-resistance theory, developed by Lewis and Whitman (1924), described interphase transfer as being confined to two thin stagnant films on either side of the gas–liquid interface. This model assumes that all the resistance to mass transfer is contained in the two films and none at the interface. Consequently, the interfacial concentrations are in equilibrium and can be determined from the equilibrium distribution curve expressed simply as $y_{Ai} = f(x_{Ai})$. A graphical description of the two-resistance model is shown in Figure 5-6 for gas–liquid contact. The concentrations y_A and x_A are bulk phase composi-

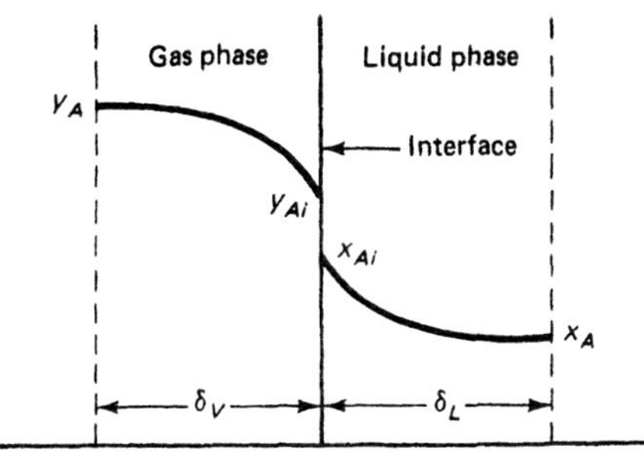

Figure 5-6 Concentration profiles across an interface

tions, and y_{Ai} and x_{Ai} are compositions at the gas–liquid interface. The equilibrium relationship for dilute solutions is expressed in terms of Henry's law by the expression

$$\bar{P}_{Ai} = mC_{Ai} \tag{5-60}$$

where $\bar{P}_{Ai}$ is the equilibrium partial pressure of component A in the gas phase at the interface, m is the Henry's law constant, and C_{Ai} is the equilibrium composition of A at the interface in the liquid phase. Equation (5-60) may be expressed in terms of mole fractions by

$$y_{Ai} = mx_{Ai} \tag{5-61}$$

In addition to the interfacial compositions, equilibrium compositions, as indicated with an asterisk, are expressed in terms of the Henry's law constant as

$$y_A^* = mx_A \tag{5-62}$$

and

$$y_A = mx_A^* \tag{5-63}$$

Now let us consider the case of steady-state transfer of solute A from a gas to a liquid phase. At steady-state the flux of A through one phase must equal the flux in the other phase. By starting with Eq. (5-28), which contains the bulk flow correction factor and the Colburn–Drew mass transfer coefficient, the flux of A can be written in terms of a mole fraction driving force for each single phase as

$$N_A = \frac{k'_y}{\beta_{V-i}}(y_A - y_{Ai}) = \frac{k'_x}{\beta_{i-L}}(x_{Ai} - x_A) \tag{5-64}$$

where β is defined by Eq. (5-29) and applies in the region of interest. For the left side of Eq. (5-64), β_{V-i} applies from the bulk gas composition to the interfacial gas composition. The bulk correction factor β_{i-L} applies from the liquid-phase interfacial composition to the bulk liquid composition.

Interfacial compositions can be obtained by solving Eq. (5-64).

$$-\frac{k'_x/\beta_{i-L}}{k'_y/\beta_{V-i}} = \frac{y_A - y_{Ai}}{x_A - x_{Ai}} \tag{5-65}$$

By knowing the local coefficients we can, in principle, determine the interfacial compositions. However, the bulk correction terms also contain interfacial compositions. The solution is thus obtained by a trial-and-error procedure in which either the gas or liquid interfacial composition is assumed.

Although it is neither practical nor usually possible to determine interfacial concentrations in an experimental apparatus, the bulk concentrations of the phases can be readily found. Thus we find it convenient to introduce an overall mass transfer coefficient which relates the flux to the concentration difference of solute A in both the liquid and gas phases. For the gas phase this becomes

$$N_A = \frac{K'_y}{\beta_{V-V^*}}(y_A - y_A^*) \tag{5-66}$$

where β_{V-V^*} = bulk flow correction factor related to the composition of the bulk phase and the equilibrium composition,
y_A = mole fraction of A in the bulk gas phase,
y_A^* = mole fraction of A in the gas phase in equilibirum with the concentration of A in the bulk liquid phase,
K'_y = overall mass transfer coefficient for the entire diffusional resistance in both phases but related to the gas-phase driving force.

For the liquid phase we have

$$N_A = \frac{K'_x}{\beta_{L^*-L}}(x_A^* - x_A) \tag{5-67}$$

where β_{L^*-L} = bulk flow correction factor related to the equilibrium composition and the bulk liquid composition,
x_A^* = mole fraction of A in the liquid phase in equilibrium with A in the bulk gas phase,
x_A = mole fraction of A in the bulk liquid phase,
K'_x = overall liquid-phase mass transfer coefficient related to the liquid-phase driving force.

The relationship between the local mass transfer coefficients and the equilibrium curve is shown in Figure 5-7. From a further consideration of Figure 5-7, we can determine the amount of resistance to mass transfer that occurs in both the gas and liquid phases. Since the resistance in the gas phase is proportional to the concentration difference in that phase and the total resistance can be related to the overall gas phase concentration difference, we have

$$\frac{\text{gas-phase resistance}}{\text{overall resistance}} = \frac{y_A - y_{Ai}}{y_A - y_A^*} = \frac{N_A/(k'_y/\beta_{V-i})}{N_A/(K'_y/\beta_{V-V^*})} \tag{5-68}$$

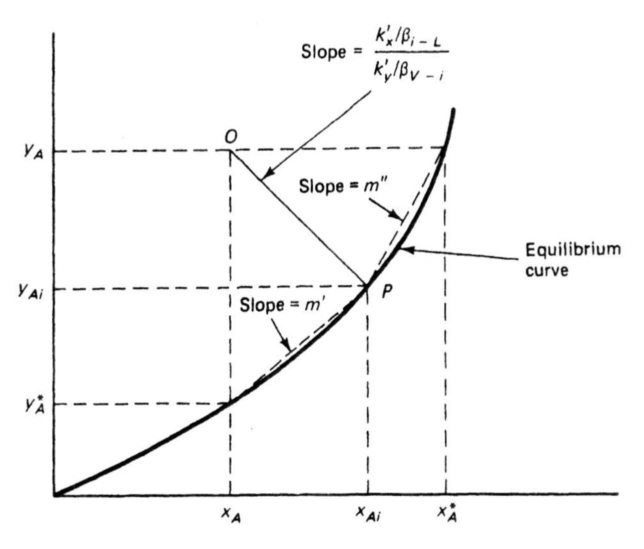

Figure 5-7 Concentration driving forces

which gives the fractional resistance in the gas phase as

$$\text{fractional gas resistance} = \frac{1/(k'_y/\beta_{V-i})}{1/(K'_y/\beta_{V-V^*})} = \frac{K'_y/\beta_{V-V^*}}{k'_y/\beta_{V-i}} \tag{5-69}$$

In analogous fashion we find the fractional resistance in the liquid phase to be

$$\text{fractional liquid resistance} = \frac{1/(k'_x/\beta_{i-L})}{1/(K'_x/\beta_{L^*-L})} = \frac{K'_x/\beta_{L^*-L}}{k'_x/\beta_{i-L}} \tag{5-70}$$

Relationships between the local and overall mass transfer coefficients can be readily established for dilute solutions. Starting with the molar flux equation written in terms of the overall mass transfer coefficient, we have on rearranging

$$\frac{1}{K'_y/\beta_{V-V^*}} = \frac{y_A - y_A^*}{N_A} \tag{5-71}$$

As noted in Figure 5-7,

$$y_A - y_A^* = y_A - y_{Ai} + y_{Ai} - y_A^* \tag{5-72}$$

Thus we can write Eq. (5-71) as

$$\frac{1}{K'_y/\beta_{V-V^*}} = \frac{y_A - y_{Ai}}{N_A} + \frac{y_{Ai} - y_A^*}{N_A} \tag{5-73}$$

When we introduce Eqs. (5-61) and (5-62) for the region from y_{Ai} to y_A^*, Eq. (5-73) becomes

$$\frac{1}{K'_y/\beta_{V-V^*}} = \frac{1}{k'_y/\beta_{V-i}} + \frac{m'(x_{Ai} - x_A)}{N_A} \tag{5-74}$$

Using Eq. (5-64), we have

$$\frac{1}{K'_y/\beta_{V-V^*}} = \frac{1}{k'_y/\beta_{V-i}} + \frac{m'}{k'_x/\beta_{i-L}} \tag{5-75}$$

Thus the overall resistance is shown to equal the sum of the gas-phase and liquid-phase resistances.

Now consider the relationship between the overall liquid-phase resistance and the individual gas- and liquid-phase resistances. Starting with the definition of the overall liquid-phase coefficient and noting that

$$x_A^* - x_A = x_A^* - x_{Ai} + x_{Ai} - x_A \tag{5-76}$$

we have

$$\frac{1}{K'_x/\beta_{L^*-L}} = \frac{x_A^* - x_{Ai}}{N_A} + \frac{x_{Ai} - x_A}{N_A} \tag{5-77}$$

Using the Henry's law constant in the region x_{Ai} to x_A^* and introducing Eq. (5-64), we can write Eq. (5-77) as

$$\frac{1}{K'_x/\beta_{L^*-L}} = \frac{1}{m''k'_y/\beta_{V-i}} + \frac{1}{k'_x/\beta_{i-L}} \tag{5-78}$$

For a system in which the solute is highly soluble in the liquid phase, the Henry's law constant is very small. Thus for the case in which the liquid- and gas-phase resistances are approximately equal, Eq. (5-75) reduces to

$$\frac{1}{K'_y/\beta_{V-V^*}} \approx \frac{1}{k'_y/\beta_{V-i}} \tag{5-79}$$

Therefore, when the solute is very soluble in the liquid, the overall gas-phase coefficient can be approximated with the local gas-phase coefficient.

Conversely, if the solute is relatively insoluble in the liquid, the Henry's law constant is large. We then note that for the case in which the local coefficients are about equal, $1/(m''k'_y) \ll 1/k'_x$. Equation (5-78) becomes

$$\frac{1}{K'_x/\beta_{L^*-L}} \approx \frac{1}{k'_x/\beta_{i-L}} \tag{5-80}$$

For dilute solutions in which $m' = m''$, we can solve Eqs. (5-75) and (5-78) simultaneously to obtain an expression relating the overall mass transfer coefficients.

$$\frac{K'_x}{\beta_{L^*-L}} = \frac{mK'_y}{\beta_{V-V^*}} \tag{5-81}$$

In the previous developments it can be seen that the overall coefficient depends on both local coefficients in addition to the slope of the equilibrium curve over the concentration range of interest. If we consider mass transfer in a wetted-wall column, the bulk compositions are a function of position in the column. Because the slope of the equilibrium curve depends on bulk compositions inside the column, the overall mass transfer coefficient is also dependent on the bulk compositions. Hence, when using experimental overall coefficients we must be certain that the experimental values apply in the region of interest. For dilute solutions in which the equilibrium curve is straight and the liquid and gas flow rates do not change over the length of the column, the overall coefficients are constant. Ideally, local coefficients should be used to calculate the overall coefficients when available.

Equimolar counter transfer

For equimolar counter transfer, the bulk flow correction factor is equal to 1. The flux equations can be reduced to

$$N_A = k'_y(y_A - y_{Ai}) \tag{5-82a}$$

$$N_A = k'_x(x_{Ai} - x_A) \tag{5-82b}$$

$$N_A = K'_y(y_A - y_A^*) \tag{5-82c}$$

$$N_A = K'_x(x_A^* - x_A) \tag{5-82d}$$

The slope of line OP in Figure 5-7 is then equal to

$$-\frac{k'_x}{k'_y} = \frac{y_A - y_{Ai}}{x_A - x_{Ai}} \tag{5-83}$$

The relationships between the local and overall mass transfer coefficients given by Eqs. (5-75) and (5-78) are

$$\frac{1}{K'_y} = \frac{1}{k'_y} + \frac{m'}{k'_x} \tag{5-84}$$

and

$$\frac{1}{K'_x} = \frac{1}{m''k'_y} + \frac{1}{k'_x} \tag{5-85}$$

To demonstrate the use of the equations above we will consider the following example.

Example 5.3

In the dilute concentration region, equilibrium data for SO_2 distributed between air and water can be approximated by

$$\bar{P}_A = 25x_A$$

where the partial pressure of SO_2 in the vapor is expressed in atmospheres. For an absorption column operating at 10 atm, the bulk vapor and liquid concentrations at one point in the column are $y_A = 0.01$ and $x_A = 0.0$. The mass transfer coefficients for this process are

$$k'_x = 10 \text{ kgmol/(m}^2\cdot\text{h)(mole fraction)}$$

$$k'_y = 8 \text{ kgmol/(m}^2\cdot\text{h)(mole fraction)}$$

Assuming equimolar counter transfer, (a) find K'_x. (b) Determine the interfacial compositions, x_{Ai} and y_{Ai}, and (c) calculate the molar flux, N_A.

Solution:

(a) $\bar{P}_A = 25x_A$ but $y_A = \bar{P}_A/P$. Therefore, $y_A = 2.5x_A$. For equimolar counter transfer, the bulk flow correction factors are equal to 1. From Eq. (5-84),

$$\frac{1}{K'_x} = \frac{1}{m''k'_y} + \frac{1}{k'_x}$$

Upon substituting $m'' = 2.5$ and the mass transfer coefficients above into Eq. (5-84), we obtain

$$\frac{1}{K'_x} = \frac{1}{(2.5)(8)} + \frac{1}{10}$$

$$K'_x = 6.67 \text{ kgmol/(m}^2\cdot\text{h)(mole fraction)}$$

(b) Using Eq. (5-83), we obtain

$$-\frac{k'_x}{k'_y} = \frac{y_A - y_{Ai}}{x_A - x_{Ai}} = -\frac{10}{8} = -1.25$$

The interfacial compositions are found by constructing a line with slope -1.25 from $y_A = 0.01$ and $x_A = 0.0$ to intersect the equilibrium curve as shown in Figure 5-8. From the figure we find $y_{Ai} = 0.0067$ and $x_{Ai} = 0.00267$. The inter-

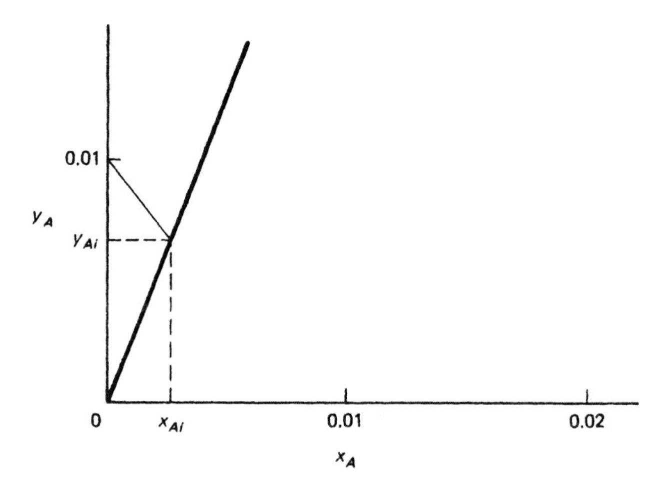

Figure 5-8 Interfacial concentrations

facial compositions can also be found by expressing the line through the bulk compositions as

$$y_A = -1.25x_A + 0.01$$

and solving this equation simultaneously with the equation representing the equilibrium curve.

(c) The mass flux can be determined from Eq. (5-64). Thus

$$\begin{aligned} N_A &= k'_x(x_{Ai} - x_A) \\ &= 10(0.00267 - 0) \\ &= 0.0267 \text{ kgmol/m}^2\cdot\text{h} \end{aligned}$$

This value can be checked by using the local gas-phase coefficient as shown:

$$\begin{aligned} N_A &= k'_y(y_A - y_{Ai}) \\ &= 8(0.01 - 0.0067) \\ &= 0.0264 \text{ kgmol/m}^2\cdot\text{h} \end{aligned}$$

The difference in the calculated values results from errors introduced by reading the interfacial compositions from the figure.

Transfer of component A through a stagnant film

In absorption, the assumption of solute transfer through a stagnant film is frequently made. For this case the bulk flow correction term is replaced by the log mean concentration difference. The flux equations can then be expressed as

$$N_A = \frac{k'_y}{(1 - y_A)_{iM}}(y_A - y_{Ai}) \tag{5-86a}$$

$$N_A = \frac{k'_x}{(1 - x_A)_{iM}}(x_{Ai} - x_A) \tag{5-86b}$$

$$N_A = \frac{K'_y}{(1 - y_A)_{*M}}(y_A - y^*_A) \tag{5-86c}$$

$$N_A = \frac{K'_x}{(1 - x_A)_{*M}}(x^*_A - x_A) \tag{5-86d}$$

The slope of line OP in Figure 5-7 is now given by

$$-\frac{k'_x/(1 - x_A)_{iM}}{k'_y/(1 - y_A)_{iM}} = \frac{y_A - y_{Ai}}{x_A - x_{Ai}} \tag{5-87}$$

The mass transfer coefficients are related as follows:

$$\frac{1}{K'_y/(1 - y_A)_{*M}} = \frac{1}{k'_y/(1 - y_A)_{iM}} + \frac{m'}{k'_x/(1 - x_A)_{iM}} \tag{5-88}$$

and

$$\frac{1}{K'_x/(1 - x_A)_{*M}} = \frac{1}{m''k'_y/(1 - y_A)_{iM}} + \frac{1}{k'_x/(1 - x_A)_{iM}} \tag{5-89}$$

where

$$(1 - y_A)_{iM} = \frac{(1 - y_{Ai}) - (1 - y_A)}{\ln[(1 - y_{Ai})/(1 - y_A)]} \tag{5-90a}$$

$$(1 - x_A)_{iM} = \frac{(1 - x_A) - (1 - x_{Ai})}{\ln[(1 - x_A)/(1 - x_{Ai})]} \tag{5-90b}$$

and

$$(1 - y_A)_{*M} = \frac{(1 - y^*_A) - (1 - y_A)}{\ln[(1 - y^*_A)/(1 - y_A)]} \tag{5-90c}$$

$$(1 - x_A)_{*M} = \frac{(1 - x_A) - (1 - x^*_A)}{\ln[(1 - x_A)/(1 - x^*_A)]} \tag{5-90d}$$

As noted in Eq. (5-87), the slope of line OP in Figure 5-7 identifies the interfacial compositions. To obtain values for x_{Ai} and y_{Ai}, we must first guess a composition and by a trial-and-error procedure determine the correct value. This is demonstrated in the following example.

Example 5.4

Repeat part (b) of Example 5.3 for bulk concentrations $y_A = 0.04$ and $x_A = 0.01$. Assume transfer of component A through a stagnant film.

Solution:

The determination of interfacial compositions for transfer through a stagnant film requires that a trial-and-error procedure be used. To begin, we use Eq. (5-83) and the interfacial compositions as found by initially assuming equimolar counter transfer.

$$-\frac{k'_x}{k'_y} = \frac{y_A - y_{Ai}}{x_A - x_{Ai}}$$

$$-\frac{10}{8} = \frac{y_A - y_{Ai}}{x_A - x_{Ai}} = -1.25$$

$$y_{Ai} = 1.25(x_A - x_{Ai}) + y_A$$

$$y_{Ai} = -1.25x_{Ai} + 0.0525$$

Solving the equation above simultaneously with the equation of the equilibrium curve

$$y_{Ai} = 2.5x_{Ai}$$

we obtain $x_{Ai} = 0.014$ and $y_{Ai} = 0.035$

Introducing Eq. (5-87), we obtain

$$-\frac{k'_x/(1-x_A)_{iM}}{k'_y/(1-y_A)_{iM}} = \frac{y_A - y_{Ai}}{x_A - x_{Ai}}$$

$$(1-x_A)_{iM} = \frac{(1-x_{Ai})-(1-x_A)}{\ln[(1-x_{Ai})/(1-x_A)]} = \frac{(1-0.014)-(1-0.01)}{\ln[(1-0.014)/(1-0.01)]} = 0.988$$

$$(1-y_A)_{iM} = \frac{(1-y_{Ai})-(1-y_A)}{\ln[(1-y_{Ai})/(1-y_A)]} = \frac{(1-0.035)-(1-0.04)}{\ln[(1-0.035)/(1-0.04)]} = 0.962$$

Therefore, $$-\frac{k'_x/(1-x_A)_{iM}}{k'_y/(1-y_A)_{iM}} = -\frac{10/0.988}{8/0.962} = -1.217$$

Constructing a line from the bulk compositions with a slope equal to -1.217 to intersect with the equilibrium curve, as shown in Figure 5-9, we obtain a

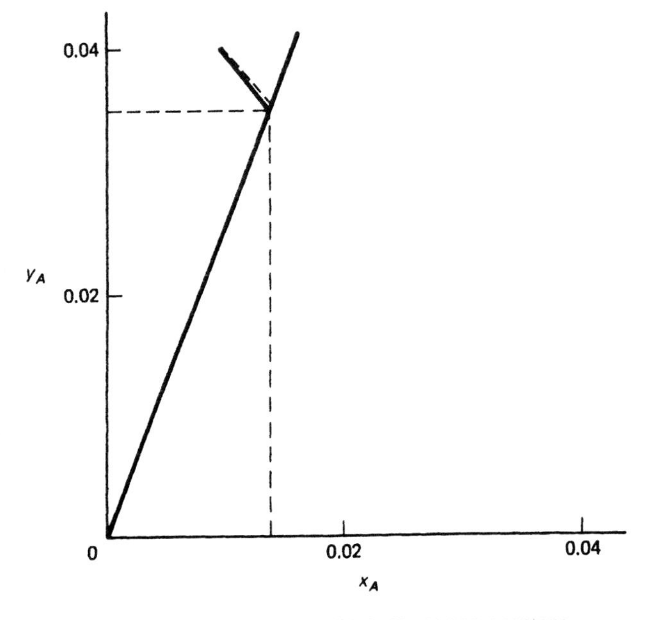

Figure 5-9 Determination of interfacial compositions

better estimate of the interfacial compositions. The new values are $x_{Ai} = 0.01405$ and $y_{Ai} = 0.0364$. Calculating new values for the log mean concentration differences we find that $(1 - x_A)_{iM} = 0.988$ and $(1 - y_A)_{iM} = 0.962$. Consequently, the interfacial compositions are $x_{Ai} = 0.01405$ and $y_{Ai} = 0.0364$, as indicated by the intersection of the dashed line with the equilibrium curve. These values are only slightly different from the values found by assuming equimolar counter transfer. For more concentrated solutions, however, more trials would be required to obtain the correct interfacial compositions.

REFERENCES

DANCKWERTS, P. V., *Ind. Eng. Chem.*, *43*, 1460 (1951).

HARRIOTT, P., *Chem. Eng. Sci.*, *17*, 149 (1962).

HIGBIE, R., *Trans. AIChE*, *31*, 365 (1935).

LEWIS, W. K., and W. G. WHITMAN, *Ind. Eng. Chem.*, *16*, 1215 (1924).

TOOR, H. L., and J. M. MARCHELLO, *AIChE J.*, *4*, 97 (1958).

WHITMAN, W. G., *Chem. Met. Eng.*, *29*, 146 (1923).

NOTATIONS

C = total molar concentration, mol/L^3

C_{Ai}, C_{A1} = molar concentration of A in the liquid at gas-liquid interface, mol/L^3

$C_{A\infty}$ = molar concentration of A in the free stream, mol/L^3

$C_{A\delta}$ = molar concentration of A in the bulk fluid, mol/L^3

E_D = eddy diffusivity, L^2/t

K = integration constant

K'_x = overall liquid-phase mass transfer coefficient, mol/L^2t (concentration difference)

K'_y = overall gas-phase mass transfer coefficient, mol/L^2t (concentration difference)

P = total pressure of the system, F/L^2

$\bar{P}_{Ai}, \bar{P}_{A1}$ = partial pressure of A at interface, F/L^2

$\bar{P}_{A\delta}$ = partial pressure of A in the bulk fluid, F/L^2

$(\bar{P}_B)_M$ = logarithmic mean pressure of B, F/L^2

R = gas constant; radius of a tube or a cylinder, L

s = rate of surface renewal, t^{-1}

t_s = exposure time of a fluid element, t

T = temperature of the system, T

x_{Ai}, x_{A1} = mole fraction of A in the liquid at the gas–liquid interface

$x_{A\delta}$ = mole fraction of A in the bulk liquid

$(x_B)_M$ = logarithmic mean mole fraction of B

x_A^* = mole fraction of A in the liquid phase in equilibrium with A in the bulk gas phase

y_{Ai}, y_{A1} = mole fraction of A in the gas at the gas–liquid interface

$y_{A\delta}$ = mole fraction of A in the bulk gas phase

y_A^* = mole fraction of A in the gas phase in equilibrium with the concentration of A in the bulk liquid phase

Greek Letters

β_{i-L} = bulk flow correction factor related to the liquid phase interfacial composition and the bulk liquid composition

β_{L^*-i} = bulk flow correction factor related to the equilibrium composition and the bulk liquid composition

β_{V-i} = bulk flow correction factor related to the bulk gas-phase composition and the interfacial gas composition

β_{V-V^*} = bulk flow correction factor related to the bulk gas composition and the equilibrium gas composition

δ = diffusion path, L; film thickness, L

$\tau(t)$ = fraction of the surface consisting of elements with an age in the time span t to $t + dt$

PROBLEMS

5.1 Carbon dioxide is absorbed in water at 20°C and 1 atm in a packed column. For this system it may be assumed that 98% of the resistance to mass transfer occurs in the liquid film and the overall coefficient is $K'_G = 0.95$ kgmol/m²·h·atm. For a dilute solution of carbon dioxide in water, the equilibrium curve can be expressed in terms of the Henry's law constant by

$$\bar{P}_A = 1.42 \times 10^3 x_A$$

where $\bar{P}_A$ is the partial pressure of A in the gas in atm, x_A is the mole fraction in the liquid, and the Henry's law constant has units, atm/mole fraction. Determine **(a)** K'_y, **(b)** k'_G, **(c)** k'_y, **(d)** k'_x, and **(e)** K'_x.

5.2 In Example 5.1, liquid A evaporated and diffused through stagnant gas B. If for each mole of liquid A diffusing into gas B an equal amount of gas B dissolves and diffuses into the liquid A, the molar flux will change. Calculate the molar flux and convective mass transfer coefficient for the case of equimolar counter diffusion for A under the same conditions as stated in Example 5.1.

5.3 Various methods have been investigated to reduce the evaporation of water from large reservoirs in semiarid regions. One method that has been tested is to spray a nonvolatile chemical on the reservoir surface. In an attempt to determine the effectiveness of this method, the surface of a 1 m × 5 m rectangular container filled with water was covered with a 0.002-m-thick layer of the chemical. Since the water and the chemical are only slightly miscible the rate of water evaporation can be determined by calculating the diffusion of water through the stagnant chemical film. Using the data given below, calculate the mass transfer coefficient k_L and the rate at which water evaporates. The diffusion coefficient for water through the film is 2.30×10^{-9} m²/s. The concentration of water in the chemical at the liquid–liquid interface is $C_{W1} = 0.3$ kgmol/m³. The concentration of water in the chemical at the gas–liquid interface is $C_{W\delta} = 0.05$ kgmol/m³. Assume that the molar density of the chemical is 0.35 kgmol/m³.

5.4 The penetration theory is not applicable for cases in which the pentration distance of a solute diffusing into a solution is greater than the thickness of the liquid film into which it diffuses. To determine if the penetration theory is applicable to the absorption of ammonia into water as described in Example 5.2, calculate the penetration distance and compare your answer with the film thickness. Assume that the gas stream above the water contains 2 mol% ammonia and the water initially contains 0.01 mol% ammonia. The plate is 2 m long and very wide.

5.5 Using the two-resistance theory discussed in Section 5.4, show that the overall mass transfer coefficient K'_x is related to fundamental transport properties and film thicknesses by the expression

$$K'_x = \frac{m''PD_{AV}CD_{AL}\beta_{L^\bullet-L}}{\delta_V RT\beta_{V-i}CD_{AL} + \delta_L\beta_{i-L}m''PD_{AV}}$$

where D_{AL} and D_{AV} are the diffusion coefficients of species A through the liquid and vapor phases, respectively.

5.6 At one point in an absorption column the bulk compositions were found to be $x_A = 0.0$ and $y_A = 0.08$. The corresponding interfacial compositions were estimated to be $x_{Ai} = 0.025$ and $y_{Ai} = 0.04$. If the overall mass transfer coefficient for the liquid phase is 50 kgmol/m$^2\cdot$h$\cdot$(mole fraction), determine the percentage resistance to mass transfer for the gas phase. Assume that the equilibrium relationship for the gas and liquid phases can be described by Henry's law.

5.7 Most important mass transport processes involve the contact of two phases. In a gas–liquid transport process, the overall mass transfer coefficients were expressed by Eqs. (5-84) and (5-85). Develop a similar equation to relate the overall mass transfer coefficient to the local mass transfer coefficients in a liquid–liquid transport process. Assume that equilibrium at the liquid–liquid interface can be written as $\gamma^{\text{I}}_{Ai}x^{\text{I}}_{Ai}f^{\circ\text{I}}_{Ai} = \gamma^{\text{II}}_{Ai}x^{\text{II}}_{Ai}f^{\circ\text{II}}_{Ai}$. The fluxes for the two liquid phases are as follows:

Liquid phase I: $N_A = k^{\text{I}}_x(x^{\text{I}}_A - x^{\text{I}}_{Ai})$

Liquid phase II: $N_A = k^{\text{II}}_x(x^{\text{II}}_A - x^{\text{II}}_{Ai})$

5.8 In an apparatus used for studying the absorption of sulfur dioxide by water, the overall mass transfer coefficient K_G was found to be 0.3 kgmol SO_2/m$^2\cdot$h$\cdot$atm. Assuming that 40% of the resistance to mass transfer is in the gas phase, calculate the overall mass transfer coefficient based on the liquid concentration. The system temperature was 30°C and the pressure was maintained at 1 atm.

Convective Mass Transfer 6

6.1 Introduction

In Chapter 5 mass transfer at a surface was described in terms of a convective mass transfer coefficient. Except for special models describing transfer across a fluid–fluid interface, consideration was not given to the actual mechanism of convective mass transfer from a surface. Most problems of an applied nature, such as the design of packed bed absorption, distillation, and extraction columns, require information about the mass transfer mechanism and the use of a convective mass transfer coefficient.

In this chapter models will be developed that describe mass transfer from surfaces of different geometry. Particular emphasis will be given to forced convective mass transfer inside tubes and from flat surfaces for both laminar and turbulent flow regimes.

6.2 Dimensional Analysis: Buckingham Pi Method

When attempting to determine mass transfer coefficients experimentally, we must identify and investigate the pertinent variables that describe the physical process. For example, consider the case of forced convection mass transfer that results when a fluid flows over a solid surface. The obvious variables that contribute to the overall transfer are velocity U, viscosity μ, density ρ, mass diffusivity D_{AB}, a convective mass transfer coefficient from the surface k_c, and a characteristic length l. In functional form the mass transfer can be expressed as

$$f(U, \mu, \rho, D_{AB}, k_c, l) = 0 \tag{6-1}$$

Experimental determinations of how each of the variables above affect the mass transfer would be extremely difficult and time consuming. However, the variables above can be arranged into pertinent dimensionless groups that will reduce the amount of data necessary to describe the process.

The dimensionless groups describing convective mass transfer can be determined by using the *Buckingham pi theorem*, which states that the number of independent dimensionless groups that can be formed by combining the physical variables of the problem is equal to the total number of physical variables minus the number of primary dimensions required to express the physical quantities. Expressing the dimensionless groups as π_1, π_2, and so on, the relationship for these groups can be expressed as $f(\pi_1, \pi_2, \text{etc.}) = 0$. If we select a process that is described by seven variables and the number of primary dimensions is four, the functional group is given as either $f(\pi_1, \pi_2, \pi_3) = 0$ or as $\pi_1 = g(\pi_2, \pi_3)$.

The use of the Buckingham pi theorem can be demonstrated by considering the convective mass transfer of a solute from a solid surface to a fluid.

Example 6.1

A sheet of naphthalene is placed in a conduit, where it undergoes sublimation into a flowing gas stream. Using the Buckingham pi method, determine the minimum number of dimensionless groups necessary to characterize this process and identify each group. The important variables identified for this case with their dimensions are presented in Table 6-1.

TABLE 6-1. VARIABLES RELATED TO FORCED CONVECTIVE MASS TRANSFER

Parameter	*Dimensions*
U	$[L/t]$
μ	$[M/Lt]$
ρ	$[M/L^3]$
D_{AB}	$[L^2/t]$
k_c	$[L/t]$
l	$[L]$

Solution:

Since there are three primary dimensions, L = length, t = time, and M = mass, for all six of the physical variables, we can identify three dimensionless groups or π terms as $f(\pi_1, \pi_2, \pi_3) = 0$. These groups can be found if we write

π as the product of the variables each raised to an unknown power. Thus

$$\pi = U^a \mu^b \rho^c D_{AB}^d k_c^e l^f \tag{6-2}$$

or

$$\pi = [Lt^{-1}]^a[ML^{-1}t^{-1}]^b[ML^{-3}]^c[L^2t^{-1}]^d[Lt^{-1}]^e[L]^f \tag{6-3}$$

The term π is dimensionless only if the sum of the exponents for each primary dimension is equal to zero. Thus by equating the sum of the exponents for each primary dimension, we have

$$\begin{aligned} [L]&: \quad a - b - 3c + 2d + e + f = 0 \\ [t]&: \quad -a - b - d - e = 0 \\ [M]&: \quad b + c = 0 \end{aligned}$$

Because there are six variables but only three equations, we must choose values for the exponents for three of the variables with the restriction that the selected exponents are independent. This can be done by selecting three repeating variables and calculating the various π terms. For this example, the selected repeating variables are U, μ, and ρ. Therefore, the three π groups are

$$\begin{aligned} \pi_1 &= U^a \mu^b \rho^c D_{AB} \\ &= [Lt^{-1}]^a[ML^{-1}t^{-1}]^b[ML^{-3}]^c[L^2t^{-1}]^1 = L^0M^0t^0 \end{aligned} \tag{6-4}$$

$$\begin{aligned} \pi_2 &= U^a \mu^b \rho^c k_c \\ &= [Lt^{-1}]^a[ML^{-1}t^{-1}]^b[ML^{-3}]^c[Lt^{-1}]^1 = L^0M^0t^0 \end{aligned} \tag{6-5}$$

$$\begin{aligned} \pi_3 &= U^a \mu^b \rho^c l \\ &= [Lt^{-1}]^a[ML^{-1}t^{-1}]^b[ML^{-3}]^c[L]^1 = L^0M^0t^0 \end{aligned} \tag{6-6}$$

Equating exponents for π_1 for each primary dimension gives us

$$\begin{aligned} [L]&: \quad a - b - 3c + 2 = 0 \\ [M]&: \quad b + c = 0 \\ [t]&: \quad -a - b - 1 = 0 \end{aligned}$$

Solving the system of equations above gives $a = 0$, $b = -1$, and $c = 1$. Therefore, the dimensionless group represented by π_1 is given by the relationship

$$\pi_1 = \frac{D_{AB}}{\mu/\rho} = \frac{D_{AB}}{\nu} = \frac{1}{\nu/D_{AB}} \tag{6-7}$$

The term ν/D_{AB} is defined as the *Schmidt number*, Sc, and gives the ratio of the momentum diffusivity to the mass diffusivity. Equating exponents for π_2 for the primary dimensions as previously shown gives us three equations and three unknown variables.

$$\begin{aligned} [L]&: \quad a - b - 3c + 1 = 0 \\ [M]&: \quad b + c = 0 \\ [t]&: \quad -a - b - 1 = 0 \end{aligned}$$

Solving for a, b, and c gives

$$\pi_2 = \frac{k_c}{U} \tag{6-8}$$

For π_3 we have

$$\pi_3 = \frac{U\rho l}{\mu} = \frac{Ul}{\nu} \tag{6-9}$$

which is defined as the *Reynolds number*. Other dimensionless groups can be obtained by multiplying or dividing one group by another as shown. Thus

$$\frac{\pi_2\pi_3}{\pi_1} = \frac{(k_c/U)(Ul/\nu)}{D_{AB}/\nu} = \frac{k_c l}{D_{AB}} \tag{6-10}$$

The dimensionless group above is defined as the *Sherwood number* or the *Nusselt number* for mass transfer and will be represented here as $\text{Sh} = k_c l/D_{AB}$.

The resulting functional equation that describes convective mass transfer is

$$f(\text{Sh}, \text{Re}, \text{Sc}) = 0 \tag{6-11}$$

If we want to determine the mass transfer coefficient, which is included in the Sherwood number, we can write the functionality as

$$\text{Sh} = g(\text{Re}, \text{Sc}) \tag{6-12}$$

Experimental data can now be correlated in terms of the groupings above instead of the original six variables. Although dimensional analysis does not provide specific information about the nature of the function, it does identify pertinent parameters that can be studied to reduce experimental efforts.

Convective mass transfer also takes place by natural convection for systems in which a density gradient is present. An example of natural convective mass transfer is a heated vertical tube coated with naphthalene. The important parameters governing the rate of mass transfer are given in Table 6-2. Application of the Buckingham pi theorem to the system variables in Table 6-2 shows that the Sherwood number is a function of the Schmidt number and *Grashof number*, Gr_{AB}. The Grashof number is defined by the expression

$$\text{Gr}_{AB} = \frac{L^3 g\, \Delta\rho_A}{\mu\nu} \tag{6-13}$$

It accounts for the transfer of mass due to a difference in densities.

TABLE 6-2. VARIABLES RELATED TO NATURAL CONVECTION

System Variables		*Typical Units*
Tube diameter	d	m
Diffusion coefficient	D_{AB}	m^2/s
Fluid viscosity	μ	kg/m·s
Fluid density	ρ	kg/m^3
Mass transfer coefficient	k_c	m/s
Bouyancy force	$g\,\Delta\rho$	$(m/s^2)(kg/m^3)$

6.3 Analogies between Mass, Momentum, and Heat Transfer

Comparison of the equations of change

From studies of fluid mechanics and heat transfer it has been shown that the Sherwood number for mass transfer is analogous to the Nusselt number used in convective heat transfer. Further, the Schmidt number bears the same relationship to momentum and mass transfer as the Prandtl number does to momentum and heat transfer. The *Prandtl* and *Schmidt numbers* are defined by

$$\mathrm{Pr} = \frac{\mu/\rho}{k/\rho c} = \frac{\nu}{\alpha} = \frac{\text{momentum diffusivity}}{\text{thermal diffusivity}} \tag{6-14}$$

and

$$\mathrm{Sc} = \frac{\nu}{D_{AB}} = \frac{\text{momentum diffusivity}}{\text{mass diffusivity}} \tag{6-15}$$

Another dimensionless grouping that finds application in mass transfer problems can be obtained by dividing the Schmidt number by the Prandtl number. This grouping, identified as the *Lewis number*, relates the thermal and mass diffusivities as follows:

$$\mathrm{Le} = \frac{\mathrm{Sc}}{\mathrm{Pr}} = \frac{\text{thermal diffusivity}}{\text{mass diffusivity}} = \frac{\alpha}{D_{AB}} \tag{6-16}$$

A comparison of mass, momentum, and heat transfer can be made by considering the equations of change. Following a procedure similar to that used in deriving Eq. (3-8), we can also obtain the equation of energy. Neglecting heat generation due to viscous forces and assuming that the density and thermal conductivity of the fluid are constant, we can write the energy equation as

$$\frac{\partial T}{\partial t} + \left(U_X\frac{\partial T}{\partial X} + U_Y\frac{\partial T}{\partial Y} + U_Z\frac{\partial T}{\partial Z}\right) = \alpha\left(\frac{\partial^2 T}{\partial X^2} + \frac{\partial^2 T}{\partial Y^2} + \frac{\partial^2 T}{\partial Z^2}\right) + \frac{\dot{q}}{\rho c} \tag{6-17}$$

where $\dot{q}$ is the volumetric rate of heat generation, excluding viscous heat dissipation.

For fluid flow at constant density and viscosity the Navier–Stokes equation for momentum transfer in the X direction can be written in a form similar to Eq. (6-17) as

$$\begin{aligned} \frac{\partial U_X}{\partial t} &+ \left(U_X\frac{\partial U_X}{\partial X} + U_Y\frac{\partial U_X}{\partial Y} + U_Z\frac{\partial U_X}{\partial Z}\right) \\ &= \nu\left(\frac{\partial^2 U_X}{\partial X^2} + \frac{\partial^2 U_X}{\partial Y^2} + \frac{\partial^2 U_X}{\partial Z^2}\right) - \frac{1}{\rho}\frac{\partial P}{\partial X} + g_X \end{aligned} \tag{6-18}$$

The sum of the pressure gradient and gravity is analogous to energy generation.

The corresponding mass transfer equation for a dilute solution for constant mass concentration and mass diffusivity is

$$\frac{\partial C_A}{\partial t} + \left(U_X \frac{\partial C_A}{\partial X} + U_Y \frac{\partial C_A}{\partial Y} + U_Z \frac{\partial C_A}{\partial Z}\right) = D_{AB}\left(\frac{\partial^2 C_A}{\partial X^2} + \frac{\partial^2 C_A}{\partial Y^2} + \frac{\partial^2 C_A}{\partial Z^2}\right) + \dot{R}_A^v \tag{6-19}$$

A comparison of the equations above suggests that by writing each equation in dimensionless form, one solution can be obtained that is valid for either heat, mass, or momentum transfer. If we neglect the heat generation term of Eq. (6-17), the pressure and gravity terms of Eq. (6-18), and the chemical reaction term in the expression above, the resulting equations can be written in dimensionless form by introducing the groups of variables

$$\phi = \frac{U_1 - U_X}{U_1 - U_\infty} = \frac{T - T_1}{T_\infty - T_1} = \frac{C_A - C_{A1}}{C_{A\infty} - C_{A1}} \tag{6-20}$$

where U_1, T_1, and C_{A1} represent the values at the surface and U_∞, T_∞, and $C_{A\infty}$ are the values at some distance from the surface. The resulting equation is

$$\frac{\partial \phi}{\partial t} + \left(U_X \frac{\partial \phi}{\partial X} + U_Y \frac{\partial \phi}{\partial Y} + U_Z \frac{\partial \phi}{\partial Z}\right) = \mathbb{D}\left(\frac{\partial^2 \phi}{\partial X^2} + \frac{\partial^2 \phi}{\partial Y^2} + \frac{\partial^2 \phi}{\partial Z^2}\right) \tag{6-21}$$

where $\mathbb{D}$ represents the appropriate diffusivity.

Turbulent transport

On the basis of experimental studies of turbulent flow in tubes, it has been suggested that momentum, heat, and mass are transferred by elements of fluid moving from one region to another. Reynolds (1901) proposed that for turbulent flow, equations similar to the molecular equations of Newton, Fourier, and Fick could be used to predict the flux rates by introducing turbulent or "eddy" diffusivities, denoted as E_V, E_H, and E_D, respectively. In contrast to the molecular diffusivities, which are considered to be properties of the fluid, turbulent diffusivities depend on the type of fluid as well as on the type of flow.

The relative magnitude of the molecular and eddy diffusivities depends on the position of the fluid relative to the wall. In the region near the wall, the flow approaches laminar motion and the turbulent fluctuations are equal to zero. This region is appropriately called the *laminar sublayer*. In the region far removed from the wall, described as the *turbulent core*, the turbulent fluctuations in the fluid are more pronounced. Because of the high transfer rate due to the movement of fluid elements in this region, molecular diffusion contributes very little to the overall transport process and can be neglected. The region between the turbulent core and the laminar sublayer is described as the transition region or *buffer layer*. Here the eddy and molecular diffusivities are both important and must be included when predicting the transfer rates. The different flow regimes are shown in Figure 6-1.

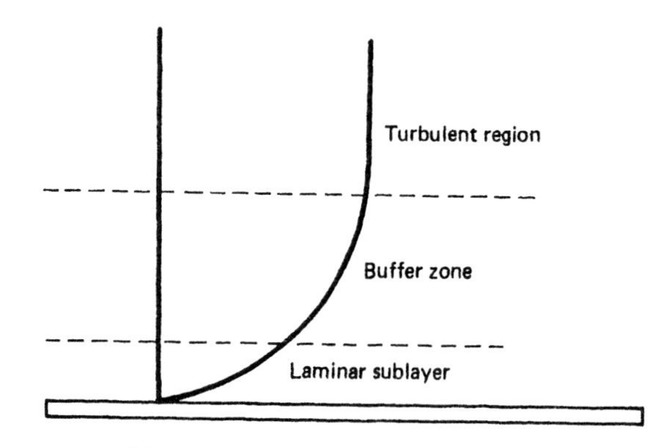

Figure 6-1 Flow regimes near a surface

The general transfer equations are written to include both molecular and eddy diffusivities as

Mass transfer:

$$N_A = -(D_{AB} + E_D)\frac{\partial C_A}{\partial Y} \tag{6-22}$$

Heat transfer:

$$q = -(\alpha + E_H)\frac{\partial(\rho c T)}{\partial Y} \tag{6-23}$$

Momentum transfer:

$$\tau g_c = -(\nu + E_V)\frac{\partial(\rho U_X)}{\partial Y} \tag{6-24}$$

where U_X is the velocity in the direction of flow at a given point. From the form of the equations above we see that the eddy diffusivities have units of length^2/time.

Notter and Sleicher (1971) conducted an extensive evaluation of experimental data for mass and heat transfer and obtained the following expression for the eddy diffusivity over the range $0 < y^* < 45$:

$$\frac{E_D}{\nu} = \frac{E_H}{\nu} = \frac{0.00090 y^{*3}}{(1 + 0.0067 y^{*2})^{0.5}} \tag{6-25}$$

where y^* is a modified Reynolds number defined by Eqs. (6-26) and (6-27). The analysis of momentum, heat, and mass transfer in turbulent flowing systems is very difficult because of a lack of information a priori on the unsteady motion of the fluid elements. For this reason semiempirical theories of turbulent flow have been developed.

Many investigators have measured velocity profiles for fully developed flow in smooth pipes over a wide range of flow conditions. These data have been correlated successfully in terms of the two dimensionless groups

$$\frac{U_X}{\sqrt{\tau_0 g_c/\rho}} \quad \text{and} \quad \frac{Y\sqrt{\tau_0 g_c/\rho}}{\nu} \tag{6-26}$$

where τ_0 is the shear stress at the wall. The term $\sqrt{\tau_0 g_c/\rho}$ is called the *friction velocity* and is often written as U_τ. Introducing the friction velocity into Eq. (6-26) gives

$$u^* = \frac{U_X}{U_\tau} \quad \text{and} \quad y^* = \frac{U_\tau Y}{\nu} \tag{6-27}$$

The graph shown in Figure 6-2 gives a reasonably accurate representation of flow data in smooth pipes. The velocity profile for flow near a flat plate can also be approximated by Figure 6-2 with acceptable accuracy.

As proposed by von Kármán (1939), the graph in Figure 6-2 can be divided into three separate zones: the laminar sublayer ($0 < y^* < 5$), the buffer layer

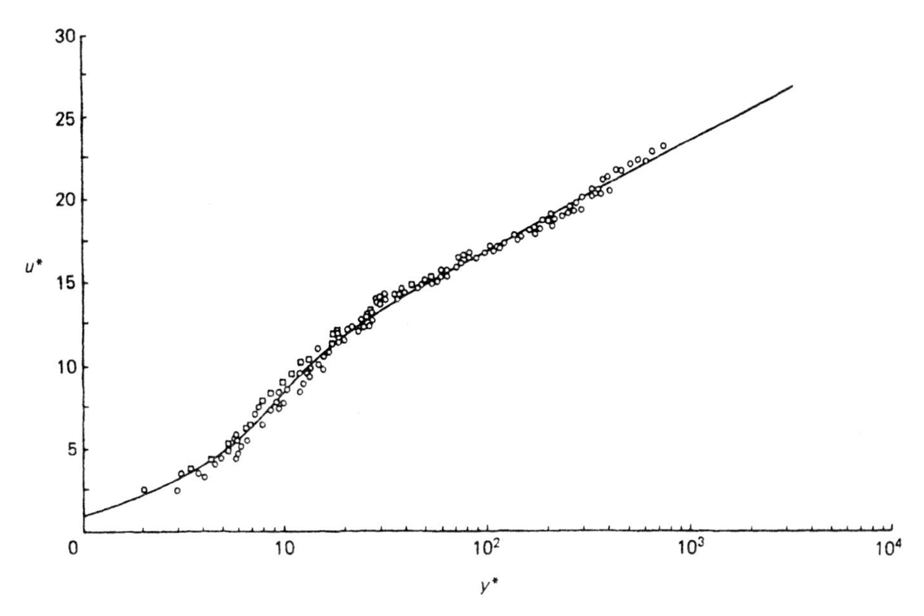

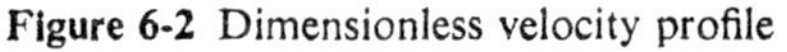

Figure 6-2 Dimensionless velocity profile

($5 < y^* < 30$), and the turbulent core ($y^* > 30$). Equations for the three regions are as follows:

$$u^* = y^* \quad \text{for } 0 < y^* < 5 \tag{6-28}$$

$$u^* = 5.0 \ln y^* - 3.05 \quad \text{for } 5 < y^* < 30 \tag{6-29}$$

$$u^* = 2.5 \ln y^* + 5.5 \quad \text{for } y^* > 30 \tag{6-30}$$

These equations are identified as the universal velocity profile. Various values of y^* are used in the literature to indicate the positions of the buffer and turbulent zones. Several analogies between heat, mass, and momentum transfer have been developed by using the universal velocity profile given by Eqs. (6-28) through (6-30). Selected analogies are presented later.

Reynolds analogy

On the basis of experimental observations, Reynolds proposed that for turbulent flow, both momentum and heat transfer occur by analogous mechanisms. This observation is readily extended to include mass transfer. A demonstration of this analogy is seen by comparing Fick's first law with Fourier's law of heat conduction and Newton's law of viscosity. The eddy diffusivities are included to provide a general interpretation of the analogy.

Mass transfer:

$$N_A = -(D_{AB} + E_D)\frac{\partial C_A}{\partial Y}\bigg|_{Y=0} = k_c^*(C_{A1} - C_{Am}) \tag{6-31}$$

Heat transfer:

$$q = -(\alpha + E_H)\frac{\partial(\rho c T)}{\partial Y}\bigg|_{Y=0} = h(T_1 - T_m) \tag{6-32}$$

Momentum transfer:

$$\frac{\tau g_c}{U_m} = -(\nu + E_V)\frac{\partial(\rho U_X/U_m)}{\partial Y}\bigg|_{Y=0} = \frac{f\rho}{2}(U_m - U_1) \tag{6-33}$$

In the equations above, h is the convective heat transfer coefficient with typical units of J/s·m²·K, τ is the shear stress with units N/m², and f is a dimensionless quantity defined as the *Fanning friction factor*. The grouping $f\rho/2$ can be defined as a convective momentum transfer coefficient. Equations (6-31) through (6-33) relate the fluxes to a driving force and a diffusivity. Since we are considering transfer from the wall of a tube, the driving force is equal to the difference between the temperature or concentration at the wall and the mean or bulk value across the tube. The mean values across the tube are denoted as C_{Am}, T_m, and U_m. The velocity U_X is in the direction of flow. A summary of the remaining terms presented in the equations above is given in Table 6-3.

TABLE 6-3. DEFINITION OF TERMS USED TO DESCRIBE THE ANALOGY BETWEEN HEAT, MASS, AND MOMENTUM TRANSFER

	Mass Transfer	*Momentum Transfer*	*Heat Transfer*
Flux	$N_A\left(\frac{\text{mol}}{\text{time-area}}\right)$	$\frac{\tau g_c}{U_m}\left(\frac{\text{mass}}{\text{time-area}}\right)$	$q\left(\frac{\text{energy}}{\text{time-area}}\right)$
Molecular diffusivity	$D_{AB}\left(\frac{\text{area}}{\text{time}}\right)$	$\nu\left(\frac{\text{area}}{\text{time}}\right)$	$\alpha\left(\frac{\text{area}}{\text{time}}\right)$
Eddy diffusivity	$E_D\left(\frac{\text{area}}{\text{time}}\right)$	$E_V\left(\frac{\text{area}}{\text{time}}\right)$	$E_H\left(\frac{\text{area}}{\text{time}}\right)$
Differential driving force	$C_A\left(\frac{\text{mol}}{\text{vol.}}\right)$	$\rho\frac{U_X}{U_m}\left(\frac{\text{mass}}{\text{vol.}}\right)$	$\rho c T\left(\frac{\text{energy}}{\text{vol.}}\right)$
Convective coefficient	$k_c^*\left[\frac{\text{mol}}{\text{time-area(mol/vol.)}}\right]$	$\frac{f\rho}{2}\left(\frac{\text{mass}}{\text{vol.}}\right)$	$h\left(\frac{\text{energy}}{\text{time-area-temp.}}\right)$

The Reynolds analogy can be obtained readily by writing the equations above in dimensionless form. After replacing U_∞, T_∞, and $C_{A\infty}$ in Eq. (6-20) by the bulk values, the dimensionless variables defined in Eq. (6-20) can be substituted into Eqs. (6-31) through (6-33) to give

Mass transfer:

$$\left.\frac{\partial\phi}{\partial Y}\right|_{Y=0} = \frac{k_c^*}{D_{AB} + E_D} \tag{6-34}$$

Heat transfer:

$$\left.\frac{\partial\phi}{\partial Y}\right|_{Y=0} = \frac{h}{\rho c(\alpha + E_H)} \tag{6-35}$$

Momentum transfer:

$$\left.\frac{\partial\phi}{\partial Y}\right|_{Y=0} = \frac{fU_m}{2(\nu + E_V)} \tag{6-36}$$

The analogy between momentum and heat transfer is obtained by combining Eqs. (6-35) and (6-36) to give

$$\frac{fU_m}{2(\nu + E_V)} = \frac{h}{\rho c(\alpha + E_H)} \tag{6-37}$$

In the turbulent region the eddy diffusivities are much larger than the molecular diffusivities. If the assumption is made that the eddy diffusivities for momentum and heat transfer are approximately equal, and ν and α can be neglected, the Reynolds analogy results.

$$\frac{f}{2} = \frac{h}{\rho c U_m} \equiv \mathrm{St} \quad \text{(Stanton number)} \tag{6-38}$$

The Reynolds analogy for mass transfer is obtained by combining Eqs. (6-34) and (6-36). Thus

$$\frac{f}{2} = \frac{k_c^*}{U_m}\left(\frac{\nu + E_V}{D_{AB} + E_D}\right) \tag{6-39}$$

For laminar flow the eddy diffusivities are equal to zero. Thus we have

$$\frac{f}{2} = \frac{k_c^*}{U_m}\,\mathrm{Sc} \tag{6-40}$$

In turbulent flow the molecular diffusivities can be neglected since they are much less than the eddy terms. Following the assumption that mass, momentum, and heat are transferred by the same mechanism for turbulent flow, and assuming that the eddy diffusivities are approximately equal, we obtain the Reynolds analogy given as

$$\mathrm{St}^* \equiv \frac{k_c^*}{U_m} = \frac{h}{\rho c U_m} = \frac{f}{2} \tag{6-41}$$

The ratio of the mass transfer coefficient to the average or free stream velocity is defined as the Stanton number as shown above. This can be written as Sh/ReSc.

The expression above can be used to predict turbulent heat and mass transfer data for fluids which have a Prandtl or Schmidt number close to unity. However, for Schmidt numbers significantly different from 1 the Reynolds analogy predicts values substantially different from those observed in practice. An examination of Eqs. (6-39) and (6-40) suggests that the dependence of experimental data on the Schmidt number is related to the flow regime of the fluid. If the molecular and eddy diffusivities are nearly equal, which is the case near a solid surface, a dependence on the Schmidt number should be expected.

Prandtl analogy

Prandtl (1910) suggested that the flow be divided into two separate regions. One is near the wall, where flow is laminar and transport occurs by a molecular mechanism, and the other is the turbulent region, which can be described by the Reynolds analogy. Prandtl's analogy takes into account the variation of experimental data with the Schmidt number:

$$\frac{k_c^*}{U_m} = \frac{f}{2}\left[\frac{1}{1 + 5\sqrt{f/2}(\mathrm{Sc} - 1)}\right] \tag{6-42}$$

The equation above is based on the assumption that the laminar sublayer can be described by Eq. (6-28). For a Schmidt number of 1, Eq. (6-42) reduces to the Reynolds analogy. Although the Prandtl analogy provides some improvement, it is unsatisfactory for large Schmidt numbers.

von Kármán analogy

The Prandtl analogy was extended by von Kármán (1939) to include the existence of a buffer zone between the laminar sublayer and the turbulent core. The von Kármán analogy is based on the universal velocity profile for which the laminar sublayer extends from $y^* = 0$ to $y^* = 5$, the buffer zone extends from $y^* = 5$ to $y^* = 30$, and the turbulent region begins at $y^* = 30$. It can be applied readily to mass transfer from the wall of a smooth tube. The flux expressions for mass and momentum transfer can be expressed respectively as

$$N_A\Big|_{Y=0} = N_{A,0} = -(D_{AB} + E_D)\frac{dC_A}{dY} \tag{6-43}$$

and

$$\tau g_c\Big|_{Y=0} = \tau_0 g_c = (\nu + E_V)\frac{d(\rho U_X)}{dY} \tag{6-44}$$

In the equations above, $Y = R - r$, where R is the radius of the tube. The positive sign is used in Eq. (6-44) since the velocity increases with increasing Y and the shear stress of the fluid acts on the tube wall.

In the laminar sublayer the eddy diffusivities are assumed to be equal to zero. Thus for transfer at the surface, Eqs. (6-43) and (6-44) become

$$N_A\Big|_{Y=0} = N_{A,0} = -D_{AB}\frac{dC_A}{dY} \tag{6-45}$$

and

$$\tau g_c\Big|_{Y=0} = \tau_0 g_c = \nu\frac{d(\rho U_X)}{dY} \tag{6-46}$$

Upon introducing Eq. (6-27), we can write Eq. (6-45) as

$$N_{A,0} = -\frac{U_\tau D_{AB}}{\nu}\frac{dC_A}{dy^*} \tag{6-47}$$

Separating the variables in the equation above and integrating from $y^* = 0$ to $y^* = 5$ gives

$$C_{A,0} - C_{A,5} = \frac{5\nu N_{A,0}}{U_\tau D_{AB}} \tag{6-48}$$

Both molecular and eddy diffusion are considered to be important in the buffer zone. Beginning with Eq. (6-29), we have

$$\frac{du^*}{dy^*} = \frac{5}{y^*} \tag{6-49}$$

From Eq. (6-27), $du^* = dU_X/U_\tau$ and $dy^* = U_\tau dY/\nu$. Substituting for the derivatives in Eq. (6-49) gives

$$\frac{dU_X}{dY} = \frac{5U_\tau^2}{y^*\nu} = \frac{5\tau_0 g_c/\rho}{y^*\mu/\rho} \tag{6-50}$$

Introducing the results above into Eq. (6-44) and rearranging gives us an expression for the eddy momentum diffusivity.

$$E_V = \nu\left(\frac{y^*}{5} - 1\right) \tag{6-51}$$

The mass flux thus becomes

$$N_{A,0} = -\left[D_{AB} + \nu\left(\frac{y^*}{5} - 1\right)\right]\frac{dC_A}{dY} \tag{6-52}$$

Inserting the relationship $dy^* = U_\tau dY/\nu$ into Eq. (6-52) and rearranging gives

$$dC_A = -\frac{N_{A,0}\mathrm{Sc}\,dy^*}{U_\tau[\mathrm{Sc}\,y^*/5 - (\mathrm{Sc} - 1)]} \tag{6-53}$$

Integrating the expression above and applying the limits that correspond to the buffer zone gives the concentration profile from $y^* = 5$ to $y^* = 30$.

$$C_{A,5} - C_{A,30} = \frac{5N_{A,0}}{U_\tau}\ln(5\mathrm{Sc} + 1) \tag{6-54}$$

Transport in the turbulent core takes place primarily by eddy mixing. Thus, if the eddy momentum and eddy mass diffusivities are equal, we have

$$-\frac{\rho N_{A,0}}{\tau_0 g_c} = \frac{dC_A}{dU_X} \tag{6-55}$$

By integrating from the outer edge of the buffer zone to the bulk values inside the turbulent core, we obtain

$$-\frac{N_{A,0}}{\tau_0 g_c/\rho} = \frac{C_{Am} - C_{A,30}}{U_m - U_{y^*=30}} \tag{6-56}$$

The velocity at the edge of the turbulent core is

$$U_{y^*=30} = U_\tau(5.0 \ln 30 - 3.05) \tag{6-57}$$

Substituting Eq. (6-57) into Eq. (6-56) and rearranging gives us the concentration difference between the buffer zone and turbulent core.

$$C_{A,30} - C_{Am} = \frac{N_{A,0}}{\tau_0 g_c/\rho}[U_m - U_\tau(5.0 \ln 30 - 3.05)] \tag{6-58}$$

$$= \frac{N_{A,0}}{U_\tau}\left[\frac{U_m}{U_\tau} - 5(\ln 6 + 1)\right] \tag{6-59}$$

In the equation above, the following substitutions were made: $\tau_0 g_c/\rho = U_\tau^2$ and $3.05 = 5 \ln 5 - 5$.

The total concentration difference between the solid surface and the turbulent core is found by adding Eqs. (6-48), (6-54), and (6-59).

$$C_{A,0} - C_{Am} = \frac{N_{A,0}}{U_\tau}\left\{5\,\mathrm{Sc} + 5 \ln(5\,\mathrm{Sc} + 1) + \left[\frac{U_m}{U_\tau} - 5(\ln 6 + 1)\right]\right\} \tag{6-60}$$

After rearranging and introducing the definition of the mass transfer coefficient, we obtain

$$\frac{N_{A,0}}{C_{A,0} - C_{Am}} = k_c^* = \frac{U_\tau}{5\,\mathrm{Sc} + 5 \ln(5\,\mathrm{Sc} + 1) + U_m/U_\tau - 5 \ln 6 - 5} \tag{6-61}$$

Introducing $U_\tau = \sqrt{\tau_0 g_c/\rho}$ and using the relation between the shear stress and friction factor given as $f/2 = \tau_0 g_c/\rho U_m^2$, we obtain

$$\frac{k_c^*}{U_m} = \frac{f/2}{1 + 5\sqrt{f/2}\left[\mathrm{Sc} + \ln\left(\frac{5\,\mathrm{Sc} + 1}{6}\right) - 1\right]} \tag{6-62}$$

Reasonable results are obtained for cases in which the Schmidt number is less than 25. The von Kármán analogy reduces to the analogy proposed by Reynolds for a Schmidt number equal to 1.0.

Chilton–Colburn *j*-factor analogy

Chilton and Colburn (1934) observed that by replacing the term $1 + 5\sqrt{f/2}(\mathrm{Sc} - 1)$ of the Prandtl analogy by $\mathrm{Pr}^{2/3}$ that an improvement could be obtained in the correlation of heat transfer data. This can be extended to mass transfer by replacing the Prandtl number with the Schmidt number. Thus the analogy becomes

$$\frac{k_c^*}{U_m}\mathrm{Sc}^{2/3} = \frac{f}{2} \tag{6-63}$$

where the first term is defined as the Chilton–Colburn *j*-factor, j_D. Thus we have

$$j_D = \frac{k_c^*}{U_m}\mathrm{Sc}^{2/3} = \frac{\mathrm{Sh}}{\mathrm{Re}\mathrm{Sc}^{1/3}} = \frac{f}{2} \tag{6-64}$$

The Chilton and Colburn *j*-factor analogy provides reasonable correlation of experimental data for flow past flat plates and in pipes.

6.4 Mass Transfer in Cylindrical Tubes

A number of important applications can be cited of mass transfer in cylindrical tubes. Included in these are: reverse osmosis, dialysis, ultrafiltration during flow through tubular synthetic membranes, oxygenation of blood in the capillaries, and mass transfer in wetted-wall columns.

The differential equation describing mass transfer in a cylinder is obtained from the generalized equation E given in Table 3-2. Considering steady-state mass transfer with no chemical reaction and assuming that the concentration is symmetric in the θ-direction, we obtain

$$U_r^m \frac{\partial C_A}{\partial r} + U_X^m \frac{\partial C_A}{\partial X} = D_{AB}\left[\frac{1}{r}\frac{\partial}{\partial r}\left(r\frac{\partial C_A}{\partial r}\right) + \frac{\partial^2 C_A}{\partial X^2}\right] \tag{6-65}$$

If we further restrict our discussion to fully developed flow, then $U_r^m = 0$. In addition, diffusion in the axial direction, as represented by the term $\partial^2 C_A/\partial X^2$, is frequently much smaller than bulk transfer and the concentration of A in the solution is low. For these limiting conditions, Eq. (6-65) reduces to

$$U_X \frac{\partial C_A}{\partial X} = D_{AB}\left[\frac{1}{r}\frac{\partial}{\partial r}\left(r\frac{\partial C_A}{\partial r}\right)\right] \tag{6-66}$$

The equation above has been used to obtain mass transfer coefficients for a variety of conditions. A number of limiting cases will be presented here.

Fully developed velocity profile and developing concentration profile

The discussion here will be limited to low mass transfer rates. Several combinations of velocity and concentration profile development have been presented by Skelland (1974). The heat transfer equation analogous to Eq. (6-66) was solved by Graetz in 1883 for the case of unheated entrance length followed by a section in which the wall temperature was constant. A plug flow velocity profile was used in this early study. Semiempirical relationships for the Nusselt number were developed by Hausen (1943) for a developing temperature profile with both a developing velocity profile and a fully developed parabolic velocity profile. The expressions applied to mass transfer are given below.

1. Constant wall concentration and fully developed parabolic velocity profile:

$$\frac{k_c^* d}{D_{AB}} = \mathrm{Sh}_d^* = 3.66 + \frac{0.0668[(d/L)\,\mathrm{Re}_d\mathrm{Sc}]}{1 + 0.04[(d/L)\,\mathrm{Re}_d\mathrm{Sc}]^{2/3}} \tag{6-67}$$

2. Constant mass input to the stream and fully developed parabolic velocity profile:

$$\frac{k_c^* d}{D_{AB}} = \mathrm{Sh}_d^* = 4.36 + \frac{0.023[(d/L)\,\mathrm{Re}_d\mathrm{Sc}]}{1 + 0.0012[(d/L)\,\mathrm{Re}_d\mathrm{Sc}]} \tag{6-68}$$

In the expressions above, the Sherwood and Reynolds' numbers are based on the tube diameter.

Fully developed velocity and concentration profiles: constant mass input at the surface

Convective mass transfer coefficients for the case of fully developed velocity and concentration profiles can be readily calculated if we assume that a nondimensional concentration profile can be used over the entire length of the tube. Contrary to mass transfer from a flat plate, the mass flux from a tube wall is related to the difference between the concentration at the wall and the average or bulk concentration at some position along the tube axis. The mass transfer coefficient is given by the expression

$$k_c^* = \frac{D_{AB}(\partial C_A/\partial r)_{r=R}}{C_{AR} - C_{Am}} = \frac{N_A}{C_{AR} - C_{Am}} \tag{6-69}$$

where C_{AR} is the concentration of species A at the tube wall and C_{Am}, the bulk concentration, is defined as

$$C_{Am} = \frac{\int_0^R 2\pi r U C_A \, dr}{\int_0^R 2\pi r U \, dr} \tag{6-70}$$

A nondimensional concentration profile can be defined in terms of the bulk concentration as $\theta = (C_{AR} - C_A)/(C_{AR} - C_{Am})$. If the nondimensional profile is invariant in the axial direction, we have

$$\frac{\partial}{\partial X}\left(\frac{C_{AR} - C_A}{C_{AR} - C_{Am}}\right) = 0 \tag{6-71}$$

Expanding Eq. (6-71) gives

$$\frac{\partial C_A}{\partial X} = \frac{dC_{AR}}{dX} - \frac{C_{AR} - C_A}{C_{AR} - C_{Am}}\frac{dC_{AR}}{dX} + \frac{C_{AR} - C_A}{C_{AR} - C_{Am}}\frac{dC_{Am}}{dX} \tag{6-72}$$

If the mass flux is constant, $dC_{AR}/dX = dC_{Am}/dX$. Thus

$$\frac{\partial C_A}{\partial X} = \frac{dC_{AR}}{dX} = \frac{dC_{Am}}{dX} \tag{6-73}$$

Equation (6-66) can then be reduced to

$$\frac{1}{r}\frac{\partial}{\partial r}\left(r\frac{\partial C_A}{\partial r}\right) = \frac{U_X}{D_{AB}}\frac{dC_{Am}}{dX} \tag{6-74}$$

with boundary conditions

$$C_A = C_{AR} \quad \text{at } r = R$$

$$\frac{\partial C_A}{\partial r} = 0 \quad \text{at } r = 0$$

The fully developed velocity profile for laminar flow in a tube is given in terms of the average velocity by the expression

$$U_X = 2\bar{U}\left(1 - \frac{r^2}{R^2}\right) \tag{6-75}$$

where $\bar{U}$ is the average velocity. Equation (6-75) can be substituted into Eq. (6-74) and integrated. The concentration profile is found by applying the boundary conditions above. Thus

$$C_A = C_{AR} - \frac{2\bar{U}}{D_{AB}}\left(\frac{dC_{Am}}{dX}\right)\left(\frac{3}{16}R^2 + \frac{r^4}{16R^2} - \frac{r^2}{4}\right) \tag{6-76}$$

The bulk concentration is found by substituting Eq. (6-76) into Eq. (6-70) and integrating.

$$C_{Am} = C_{AR} - \frac{2\bar{U}}{D_{AB}}\left(\frac{11}{96}R^2\right)\frac{dC_{Am}}{dX} \tag{6-77}$$

The mass transfer coefficient is obtained by evaluating the derivative of the concentration profile at the surface of the tube and substituting this value and Eq. (6-77) into Eq. (6-69). This gives

$$k_c^* = \frac{24D_{AB}}{11R} = \frac{48D_{AB}}{11d} \tag{6-78}$$

or

$$\frac{k_c^* d}{D_{AB}} = \mathrm{Sh}_d^* = 4.36 \tag{6-79}$$

The mass transfer coefficient depends only on the tube diameter and the diffusion coefficient for the case of laminar flow, where both the velocity and concentration profiles are fully developed. As noted in Eqs. (6-67) and (6-68), the mass transfer coefficients depend on the Reynolds and Schmidt numbers for developing concentration profiles.

Fully developed velocity and concentration profiles: constant wall concentration

If the concentration at the wall is constant, Eq. (6-72) reduces to

$$\frac{\partial C_A}{\partial X} = \frac{C_{AR} - C_A}{C_{AR} - C_{Am}}\frac{dC_{Am}}{dX} \tag{6-80}$$

The differential equation describing this process and the boundary conditions are

$$\frac{1}{r}\frac{\partial}{\partial r}\left(r\frac{\partial C_A}{\partial r}\right) = \frac{U_X}{D_{AB}}\frac{C_{AR} - C_A}{C_{AR} - C_{Am}}\left(\frac{dC_{Am}}{dX}\right) \tag{6-81}$$

and

$$C_A = C_{AR} \quad \text{at } r = R$$

$$\frac{\partial C_A}{\partial r} = 0 \quad \text{at } r = 0$$

A solution to the equations above can be found by the method of successive approximations. As a first approximation, we make use of the solution obtained for the case of constant flux at the wall. The dimensionless concentration profile is thus found by combining Eqs. (6-76) and (6-77).

$$\frac{C_{AR} - C_A}{C_{AR} - C_{Am}} = \frac{6}{11}\left(3 + \frac{r^4}{R^4} - 4\frac{r^2}{R^2}\right) \tag{6-82}$$

Upon substituting the expression above into Eq. (6-81) and integrating, we obtain

$$C_A = C_{AR} - \frac{12\bar{U}}{11D_{AB}}\left(\frac{dC_{Am}}{dX}\right)\left(\frac{251}{576}R^2 - \frac{3}{4}r^2 + \frac{7}{16}\frac{r^4}{R^2} - \frac{5}{36}\frac{r^6}{R^4} + \frac{r^8}{64R^6}\right) \tag{6-83}$$

Continuing as in the preceding example, we ultimately obtain a value for the Sherwood number.

$$\text{Sh}_d^* = \frac{k_c^* d}{D_{AB}} = 3.658 \tag{6-84}$$

For large values of X, Eqs. (6-67) and (6-68) reduce to Eqs. (6-84) and (6-79), respectively. This should be expected since both the velocity and concentration profiles will be fully developed for large distances in the tube.

Turbulent flow in a smooth tube

The friction factor for flow in circular tubes is defined in terms of the shear stress at the tube wall and the bulk velocity by the expression

$$\frac{f}{2} = \frac{\tau g_c}{\rho U_m^2} \tag{6-85}$$

Petukhov (1970) proposed an empirical equation for the friction factor inside smooth tubes. His expression, which provides a good fit to experimental data over the turbulent range $10^4 < \text{Re} < 5 \times 10^6$, is

$$\frac{f}{2} = (2.236 \ln \text{Re} - 4.639)^{-2} \tag{6-86}$$

where the Reynolds number is defined in terms of the tube diameter and mean velocity as $\text{Re} = dU_m/\nu$. Substitution of Eq. (6-86) for the friction factor into any of the analogies previously discussed provides us with a means of predicting the mass transfer coefficient for turbulent flow in a smooth tube.

A slightly simpler and more commonly employed empirical equation for obtaining the friction factor is

$$\frac{f}{2} = 0.023\text{Re}^{-0.2} \tag{6-87}$$

The friction factor above provides a good fit to experimental data over the Reynolds' number range $3 \times 10^4 < \text{Re} < 10^6$.

Pinczewski and Sideman (1974) developed a model to predict mass transfer coefficients at moderate and high Schmidt numbers for constant surface concentration. They considered the region near the tube wall to be a mosaic of periodically replaced developing boundary layers. Their model, which utilizes the definition of the friction factor given by Eq. (6-87), results in good agreement with available heat and mass transfer data. Three equations are proposed for a wide range of Schmidt numbers. These are

$0.5 < \mathrm{Sc} < 10$:

$$\mathrm{Sh}^* = 0.0097\mathrm{Re}^{9/10}\mathrm{Sc}^{1/2}(1.10 + 0.44\mathrm{Sc}^{-1/3} - 0.70\mathrm{Sc}^{-1/6}) \tag{6-88}$$

$10 < \mathrm{Sc} < 1000$:

$$\mathrm{Sh}^* = \frac{0.0097\mathrm{Re}^{9/10}\mathrm{Sc}^{1/2}(1.10 + 0.44\mathrm{Sc}^{-1/3} - 0.70\mathrm{Sc}^{-1/6})}{1 + 0.064\mathrm{Sc}^{1/2}(1.10 + 0.44\mathrm{Sc}^{-1/3} - 0.70\mathrm{Sc}^{-1/6})} \tag{6-89}$$

$\mathrm{Sc} > 1000$:

$$\mathrm{Sh}^* = 0.0102\mathrm{Re}^{9/10}\mathrm{Sc}^{1/3} \tag{6-90}$$

A comparison of the equations above with experimental data is made in Figure 6-3.

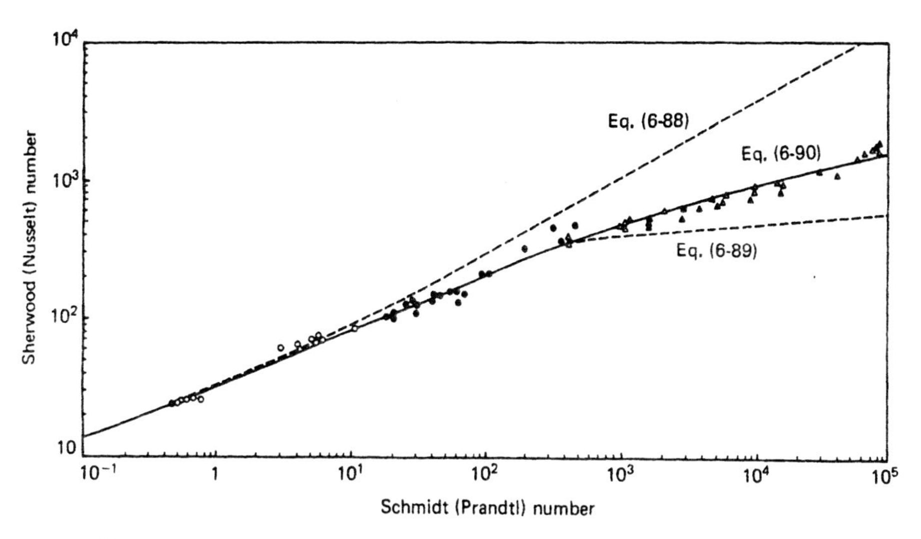

Figure 6-3 Comparison of Pinczewski and Sideman (1974) model with experimental data

In the region $\mathrm{Sc} < 10$, Einstein and Li (1956) and Hanratty (1956) developed models for the Sherwood number by utilizing Danckwerts' surface renewal theory. For the same region, Ruckenstein (1963) developed an equation by using a quasi-steady flow model. The Einstein and Li model overestimates the transfer coefficient, whereas the Ruckenstein model underestimates the transfer coefficients.

A large portion of the experimental mass transfer data inside circular tubes has been obtained by using a wetted-wall column. Mass transfer coefficients were obtained by Gilliland and Sherwood (1934) for the evaporation of nine different liquids into air. Their results were correlated by the empirical equation

$$\mathrm{Sh}_M = \frac{k_c d}{D_{AB}} \frac{\bar{P}_{B,M}}{P} = 0.023\mathrm{Re}^{0.83}\mathrm{Sc}^{0.44} \tag{6-91}$$

where k_c = mass transfer coefficient based on the logarithmic mean driving force at the ends of the column,
D_{AB} = diffusion coefficient for the vapor into the gas,
Sc = Schmidt number for vapor in the gas,
Re = Reynolds number of the gas relative to the column,
d = diameter of the column,
$\bar{P}_{B,M}$ = log mean composition of the carrier gas evaluated between the surface and bulk compositions.

Equation (6-91) is valid for the range $2000 \leq \mathrm{Re} \leq 35{,}000$ and $0.6 \leq \mathrm{Sc} \leq 2.5$ at column pressures from 0.1 to 3 atm. In a later study Linton and Sherwood (1950) investigated the flow of water through circular tubes made of benzoic acid, cinnamic acid, and β-napthol. On the basis of their results, they modified the correlation of Gilliland and Sherwood as shown:

$$\mathrm{Sh}_M = 0.023\mathrm{Re}_d^{0.83}\mathrm{Sc}^{1/3} \tag{6-92}$$

The equation above is valid over the range $2000 \leq \mathrm{Re} \leq 70{,}000$ and $0.6 \leq \mathrm{Sc} \leq 2500$. Equations (6-91) and (6-92) are written for species A diffusing through stagnant B. They can be expressed in terms of the Colburn–Drew coefficient by simply replacing $k_c\bar{P}_{B,M}/P$ by k'_c. For dilute concentrations, however, k_c and k'_c can be written as k_c^*. The j-factor representation is given by the expression

$$j_D = \frac{\mathrm{Sh}^*}{(\mathrm{Re})(\mathrm{Sc})^{1/3}} = 0.023\mathrm{Re}^{-0.17} \tag{6-93}$$

If j_D is equated to $f/2$, where f is the Fanning friction factor, the correlation above gives mass transfer coefficients that are about 20% lower than the experimental values.

Example 6.2

Calculate the convective mass transfer coefficient in a wetted-wall column for the transfer of benzene into carbon dioxide at 0°C. The gas velocity through the column is 1.25 m/s. The diameter of the column is 0.16 m. Assume that the benzene concentration in carbon dioxide is very low.

Solution:

For a dilute mixture $\bar{P}_{B,M} \cong P$. Thus using Eq. (6-91),

$$\mathrm{Sh} = \frac{k_c d}{D_{AB}} = 0.023\mathrm{Re}^{0.83}\mathrm{Sc}^{0.44}$$

$$\rho_{CO_2} = 1.966\ \mathrm{kg/m^3}$$

From Figure A-1 in Appendix A, $\mu = 1.35 \times 10^{-5}$ kg/m·s. Thus

$$\nu = \frac{\mu}{\rho} = \frac{1.35 \times 10^{-5}}{1.966} = 6.87 \times 10^{-6} \text{ m}^2/\text{s}$$

From the *Chemical Engineers' Handbook*, $D_{AB} = 5.28 \times 10^{-6}$ m²/s.

$$\text{Sc} = \frac{\nu}{D_{AB}} = \frac{6.87 \times 10^{-6} \text{ m}^2/\text{s}}{5.28 \times 10^{-6} \text{ m}^2/\text{s}} = 1.3$$

$$\text{Re} = \frac{\rho\, dU}{\mu} = \frac{dU}{\nu} = \frac{(0.16 \text{ m})(1.25 \text{ m/s})}{6.87 \times 10^{-6} \text{ m}^2/\text{s}} = 29{,}112$$

Therefore, $$\text{Sh} = \frac{k_c d}{D_{AB}} = 0.023(29{,}112)^{0.83}(1.3)^{0.44} = 130.9$$

and $$k_c = \frac{\text{Sh} D_{AB}}{d} = \frac{(130.9)(5.28 \times 10^{-6} \text{ m}^2/\text{s})}{0.16 \text{ m}} = 4.3 \times 10^{-3} \text{ m/s}$$

It is difficult to maintain a smooth liquid surface in a wetted-wall column without adding surfactants. For conditions in which ripples are present in the liquid film, the empirical equation suggested by Kafesjian et al. (1961) is recommended. Their equation, which introduces an additional correction term for the Reynolds number, is expressed as

$$\frac{k_c d}{D_{AB}} \frac{\bar{P}_{B,M}}{P} = 0.00814 \text{Re}^{0.83} \text{Sc}^{0.44} \left(\frac{4\Gamma}{\mu}\right)^{0.15} \tag{6-94}$$

where Γ is the flow rate per unit width of column surface. At a Reynolds number of 1000 (i.e., $4\Gamma/\mu = 1000$), the equation above reduces to that proposed by Gilliland and Sherwood (1934).

6.5 Exact Boundary Layer Analysis for Laminar Flow across a Flat Plate

Boundary layer theory has found wide application in fluid mechanics and provides a method for gaining a better understanding of convective mass transfer from a surface. Prandtl (1904) introduced the concept of a thin boundary layer next to a solid surface in which the viscous effects that retard flow were concentrated. Outside the boundary layer he considered the viscous effects to be negligible. In this section exact solutions of laminar boundary layer models will be discussed for flow over exterior surfaces.

Consider the laminar flow of an incompressible fluid flowing over a flat plate with a free-stream velocity U_∞. When the fluid reaches the leading edge of the plate, the fluid next to the surface is brought to rest as a result of shear stresses between the fluid and solid. Shear stresses in the fluid subsequently retard the flow of fluid over a very small thickness defined as the boundary layer (see Figure 6-4). The laminar boundary layer continues to thicken for a distance along the length of the plate until a point defined as the critical distance, X_c,

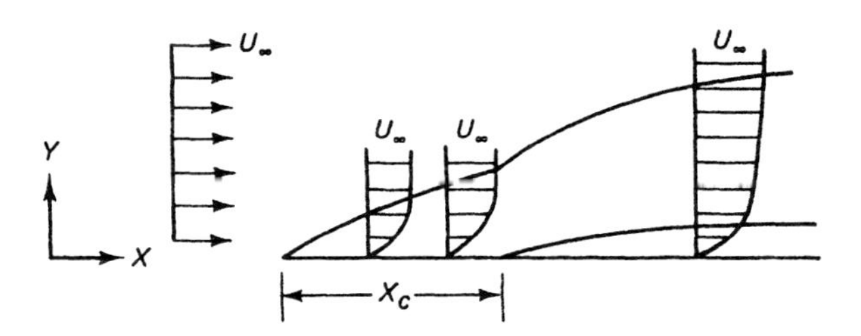

Figure 6-4 Boundary layer model

is reached. Past that point the boundary layer flow becomes unstable and a turbulent boundary layer forms. Because of the low velocities and the large shear stresses near the surface of the plate, it is assumed that a laminar sublayer forms between the plate and the turbulent boundary layer. The thickness of the boundary layer is defined such that the velocity at the edge of the boundary layer equals 99% of the free-stream velocity. Typical applications that yield to a boundary layer analysis are sublimation of a solid or evaporation of a liquid from a flat surface such as in drying.

The boundary layer equations that describe mass, momentum, and continuity can be found for the X and Y directions by using an order of magnitude analysis as shown by Schlichting (1968). Let us consider the case in which a flowing fluid approaches a flat plate at a steady uniform velocity. If the properties of the fluid are constant and there are no pressure gradients, the boundary layer equations are

Continuity:
$$\frac{\partial U_X}{\partial X} + \frac{\partial U_Y}{\partial Y} = 0 \tag{6-95}$$

Momentum:
$$U_X \frac{\partial U_X}{\partial X} + U_Y \frac{\partial U_X}{\partial Y} = \nu \frac{\partial^2 U_X}{\partial Y^2} \tag{6-96}$$

Mass:
$$U_X \frac{\partial C_A}{\partial X} + U_Y \frac{\partial C_A}{\partial Y} = D_{AB} \frac{\partial^2 C_A}{\partial Y^2} \tag{6-97}$$

The velocity of the fluid along the surface of the plate in the X direction is equal to zero and at the outer edge of the boundary layer the velocity is equal to the free-stream velocity, U_∞. Concentrations of A in the fluid at the solid surface and in the free stream are C_{A1} and $C_{A\infty}$, respectively. In the absence of flow through the solid, the fluid velocity in the Y direction is equal to zero at the solid surface. The boundary conditions for this case are

$$\text{BC1:} \quad Y = 0, \quad U_X = 0, \quad C_A = C_{A1}, \quad U_Y = 0 \tag{6-98}$$

$$\text{BC2:} \quad Y = \infty, \quad U_X = U_\infty, \quad C_A = C_{A\infty} \tag{6-99}$$

For either suction or injection of the fluid through the solid surface, the velocity boundary condition given by Eq. (6-98) becomes $U_Y = U_0$.

If chemical reaction in the boundary layer is neglected, the boundary layer equations can be reduced to ordinary differential equations by using the appropriate similarity transformation. Concentration and velocity profiles are given in Figure 6-5 as a function of the blowing or suction parameter, $U_0\sqrt{\mathrm{Re}_X}/U_\infty$, for Schmidt numbers of 0.7 and 1.0. Positive values for the blowing parameter indicate that mass is transferred from the solid into the fluid, whereas a negative value is used for suction or mass transfer toward the wall.

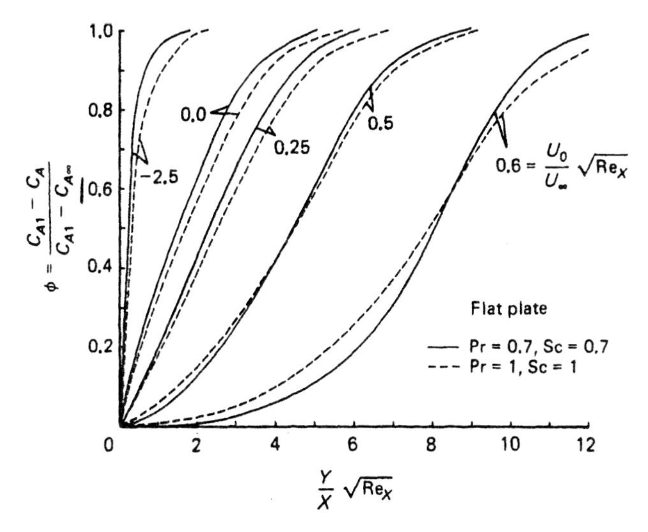

Figure 6-5 Dimensionless concentration and velocity profiles for laminar flow over a flat plate (from Hartnett and Eckert, 1957)

The velocity gradient and the nondimensional concentration gradient evaluated at the solid surface for Sc = 1 and $U_0 = 0$ is expressed in terms of the local Reynolds number as

$$\frac{\partial}{\partial Y}\left(\frac{U_X}{U_\infty}\right)_{Y=0} = \frac{\partial}{\partial Y}\left(\frac{C_{A1} - C_A}{C_{A1} - C_{A\infty}}\right)_{Y=0} = 0.332\frac{\sqrt{\mathrm{Re}_X}}{X} \tag{6-100}$$

If the mass transfer rate from the surface is small and does not alter the velocity profile, the flux can be written as

$$N_A = -D_{AB}\left(\frac{\partial C_A}{\partial Y}\right)_{Y=0} = k_c^*(C_{A1} - C_{A\infty}) \tag{6-101}$$

where the bulk flow correction term has been neglected. Combining Eqs. (6-100) and (6-101) gives the mass transfer coefficient in terms of the Reynolds number.

$$k_c^* = 0.332 D_{AB}\frac{\sqrt{\mathrm{Re}_X}}{X} \tag{6-102}$$

The equation above can be written in terms of the local Sherwood number by multiplying each side by distance and dividing by the diffusivity.

$$\mathrm{Sh}_X^* = \frac{k_c^* X}{D_{AB}} = 0.332(\mathrm{Re}_X)^{1/2} \tag{6-103}$$

For cases in which the Schmidt number is not equal to unity, the Sherwood number can be found by extending the Pohlhausen (1921) solution to mass transfer. For mass transfer in the absence of flow through the solid surface, the dimensionless concentration profiles for Sc > 0.6 can be reduced to a single curve when plotted versus $(Y/X)(\text{Re}_X)^{1/2}(\text{Sc})^{1/3}$. The local Sherwood number thus becomes

$$\text{Sh}_X^* = \frac{k_c^* X}{D_{AB}} = 0.332(\text{Re}_X)^{1/2}(\text{Sc})^{1/3} \tag{6-104}$$

where the Schmidt number is the ratio of the momentum and concentration boundary layer thicknesses.

A general relationship for the local Sherwood number can be written in terms of the slope of the concentration profile evaluated at the surface for any value of the blowing parameter.

$$\text{Sh}_X^* = (\text{slope})_{Y=0}(\text{Re}_X)^{1/2}(\text{Sc})^{1/3} \tag{6-105}$$

Approximate values of the slopes are shown in Table 6-4 as a function of the blowing parameter.

TABLE 6-4. SLOPES OF THE CONCENTRATION PROFILE (SISSOM AND PITTS, 1972)

$\frac{U_0}{U_\infty}\sqrt{\text{Re}_X}$	0.6	0.50	0.25	0.0	−2.5
$(\text{Slope})_{Y=0}$	0.01	0.06	0.17	0.332	1.64

The average mass transfer coefficient over the total length of the plate can be found by integrating the local value over the length as shown.

$$k_{c,\text{avg}}^* = \frac{\int_0^L k_c^*\, dX}{\int_0^L dX} \tag{6-106}$$

Thus the average mass transfer coefficient for flow over a flat plate is

$$k_{c,\text{avg}}^* = 0.332 D_{AB}(\text{Sc})^{1/3} \frac{\int_0^L [(\rho U_X X/\mu)^{1/2}(1/X)]\, dX}{\int_0^L dX} \tag{6-107}$$

$$= 0.664 \frac{D_{AB}(\text{Sc})^{1/3}}{L}\left(\frac{\rho U_X L}{\mu}\right)^{1/2} \tag{6-108}$$

The average value of the Sherwood number thus becomes

$$\text{Sh}_{\text{avg}}^* = 0.664(\text{Re}_L)^{1/2}(\text{Sc})^{1/3} \tag{6-109}$$

6.6 Approximate Integral Analysis for Flow across a Flat Plate

Integral balance

The exact boundary layer analysis presented in the preceding section is limited to the case of laminar flow and generally is quite complicated. An approximate but accurate integral method was introduced by von Kármán (1921) that can be used for either laminar or turbulent flow. The von Kármán integral method can be developed by writing an integral balance for the control volume shown in Figure 6-6. The hydrodynamic and concentration boundary layer thicknesses are designated as δ and δ', respectively.

In Figure 6-6 the plate will be assumed to be both porous with injection through the surface and soluble in the fluid. For steady-state flow the molar

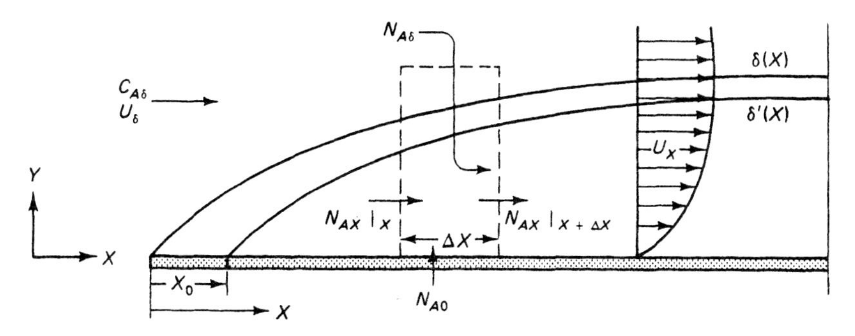

Figure 6-6 Integral boundary layer model

balance for species A over a unit depth in the Z direction is

$$(1)\int_0^{\delta'} (N_{AX}|_X)\,dY + 1(\Delta X)N_{A\delta} + 1(\Delta X)N_{A0} + 1(\Delta X)\int_0^{\delta'} \dot{R}_A^v\,dY = (1)\int_0^{\delta'} (N_{AX}|_{X+\Delta X})\,dY \tag{6-110}$$

Dividing the equation above by the area, $1(\Delta X)$, and applying the limiting process gives

$$\frac{d}{dx}\int_0^{\delta'} N_{AX}\,dY = N_{A\delta} + N_{A0} + \int_0^{\delta'} \dot{R}_A^v\,dY \tag{6-111}$$

The integral above is evaluated only over the concentration boundary thickness since we will assume that it is always thinner than the hydrodynamic boundary layer.

Mass transfer in the X direction through the control volume is primarily by bulk flow. Thus the molar flux rate is written as

$$N_{AX} = C_A U_X \tag{6-112}$$

Since the mass transferred through the top of the control volume is primarily by bulk flow, the molar flux becomes

$$N_{A\delta} = C_{A\delta'}U_\delta - D_{AB}\left(\frac{dC_A}{dY}\right)_{Y=\delta'} \approx C_{A\delta'}U_\delta \tag{6-113}$$

The rate of flow from the solid surface depends on whether species A is dissolved from the surface into the boundary layer or is injected through the solid. Dissolution of the surface is a diffusional process, whereas if A is injected through the surface, diffusion may be small compared to bulk flow. The case in which both transport mechanisms are equally important is used here. Thus

$$N_{A0} = C_{A0}U_0 - D_{AB}\left(\frac{dC_A}{dY}\right)_{Y=0} \tag{6-114}$$

Substituting Eqs. (6-112) through (6-114) into Eq. (6-111) gives

$$\frac{d}{dx}\int_0^{\delta'} C_A U_X \, dY = C_{A\delta'}U_\delta + C_{A0}U_0 - D_{AB}\left(\frac{dC_A}{dY}\right)_{Y=0} + \int_0^{\delta'} \dot{R}_A^v \, dY \tag{6-115}$$

Now let us write the equation above for the case in which there is no chemical reaction and the properties of the fluid remain constant across the boundary layer (i.e., $C_{\delta'} \approx C_0$). Thus we have

$$\frac{d}{dX}\int_0^{\delta} U_X \, dY = U_\delta + U_0 \tag{6-116}$$

Combining Eqs. (6-115) and (6-116) gives the integral equation for the concentration profile across the boundary layer.

$$\frac{d}{dX}\left[\int_0^{\delta'} (C_{A\delta'} - C_A)U_X \, dY\right] = U_0(C_{A\delta'} - C_{A0}) + D_{AB}\left(\frac{dC_A}{dY}\right)_{Y=0} - \int_0^{\delta'} \dot{R}_A^v \, dY \tag{6-117}$$

Calculation of the concentration boundary layer thickness requires that an expression be known for the momentum boundary layer thickness as a function of the distance along the surface of the plate. The momentum integral equation may be obtained by a procedure similar to the one described above. For the case of injection or suction through the plate's surface, the momentum integral is

$$\frac{d}{dX}\left[\int_0^{\delta} (U_\delta - U_X)U_X \, dY\right] = \nu\left(\frac{dU_X}{dY}\right)_{Y=0} + U_0 U_\delta \tag{6-118}$$

If we want to calculate the mass transfer from the wall, we must first find an expression for the concentration boundary layer thickness so that a value for the mass transfer coefficient can be determined. However, a solution to the integral equations above requires that velocity and concentration profiles be known. Approximate profiles can be obtained by using general polynomial equations fitted to the boundary conditions for the boundary layer. The accuracy with which the mass transfer coefficient can be determined depends on the ingenuity of the individual in selecting the appropriate form for the approximate profiles.

The approximate profile for the velocity should satisfy the boundary conditions given by Eqs. (6-98) and (6-99). The free-stream velocity at the edge of the boundary layer will be denoted as U_δ in this development instead of U_∞ as previously shown. From Figure 6-6, we also see that the velocity gradient is zero at the outer edge of the boundary layer. In the absence of injection through the solid surface a fourth boundary condition can be found by evaluating Eq. (6-96) at the solid surface. This condition is $\partial^2 U_X/\partial Y^2 = 0$ at $Y = 0$. On the basis of the similarity between momentum and mass transfer, analogous boundary conditions exist for mass transfer in the boundary layer. Velocity and concentration boundary conditions are summarized below.

$$\text{BC1:} \quad U_X = 0 \qquad C_A = C_{A1} \quad \text{at } Y = 0 \tag{6-119a}$$

$$\text{BC2:} \quad U_X = U_\delta \qquad C_A = C_{A\delta'} \quad \text{at } Y = \delta(\delta') \tag{6-119b}$$

$$\text{BC3:} \quad \frac{\partial U_X}{\partial Y} = 0 \qquad \frac{\partial C_A}{\partial Y} = 0 \quad \text{at } Y = \delta(\delta') \tag{6-119c}$$

$$\text{BC4:} \quad \frac{\partial^2 U_X}{\partial Y^2} = 0 \qquad \frac{\partial^2 C_A}{\partial Y^2} = 0 \quad \text{at } Y = 0 \tag{6-119d}$$

Laminar flow across a flat plate

The integral method is readily demonstrated by using the first two boundary conditions and a polynomial with two arbitrary constants, $U_X = a + bY$. Applying boundary conditions (6-119a) and (6-119b) gives the velocity profile

$$\frac{U_X}{U_\delta} = \frac{Y}{\delta} \tag{6-120}$$

In the absence of injection or suction through the surface we can substitute Eq. (6-120) into Eq. (6-118) to obtain

$$U_\delta^2 \frac{d}{dX}\left[\int_0^\delta \left(1 - \frac{Y}{\delta}\right)\frac{Y}{\delta}\, dY\right] = \nu \frac{U_\delta}{\delta} \tag{6-121}$$

After integration, the equation above becomes

$$\frac{U_\delta^2}{6}\frac{d\delta}{dX} = \nu \frac{U_\delta}{\delta} \tag{6-122}$$

or

$$\delta\, d\delta = \frac{6\nu}{U_\delta}\, dX \tag{6-123}$$

Upon integrating Eq. (6-123), we obtain

$$\delta^2 = \frac{12\nu X}{U_\delta} + C \tag{6-124}$$

As shown in Figure 6-6 the boundary layer starts to form at the leading edge of the plate. Since $\delta = 0$ at $X = 0$, the constant of integration is zero. Therefore,

$$\delta = 3.46\sqrt{\frac{\nu X}{U_\delta}} \tag{6-125}$$

or

$$\frac{\delta}{X} = \frac{3.46}{\sqrt{\mathrm{Re}_X}} \tag{6-126}$$

where Re_X is the local Reynolds number defined in terms of distance along the plate. The linear velocity profile given by Eq. (6-120) does not give a smooth transition for the velocity at the edge of the boundary layer. An improved approximate velocity profile can be obtained by using higher-order polynomials with additional boundary conditions.

Example 6.3

Consider the dissolution of a solid from a nonporous flat plate as shown in Figure 6-7, in which the development of the momentum boundary layer begins at the leading edge of the plate. Dissolution of the plate and the subsequent

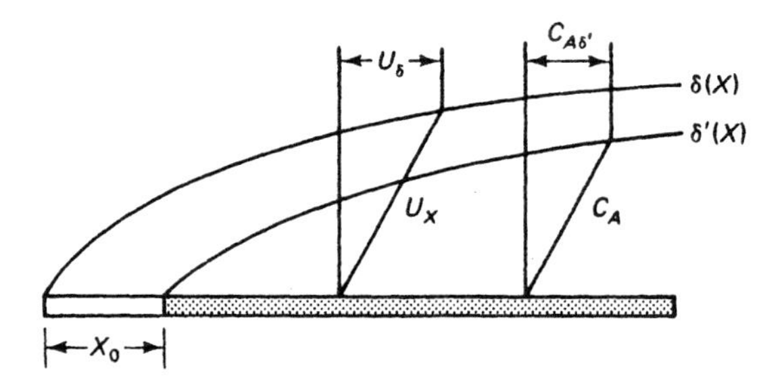

Figure 6-7 Dissolution of a flat plate

development of the concentration boundary layer begins at $X = X_0$. Using a first-order polynomial to represent the concentration boundary layer, and the approximate velocity profile given by Eq. (6-120), obtain an expression for the concentration boundary layer thickness as a function of distance along the plate.

Solution:

In the absence of chemical reaction with no transfer through the surface, Eq. (6-117) can be written as

$$U_\delta(C_{A\delta'} - C_{A1})\frac{d}{dX}\left[\int_0^{\delta'}\left(1 - \frac{C_A - C_{A1}}{C_{A\delta'} - C_{A1}}\right)\frac{U_X}{U_\delta}\,dY\right] = D_{AB}\left(\frac{dC_A}{dY}\right)_{Y=0} \tag{6-127}$$

An approximate profile for concentration inside the boundary layer is found by using a first-order polynomial written in terms of a concentration difference as

$$C_A - C_{A1} = a + bY$$

After applying the boundary conditions given by Eqs. (6-119a) and (6-119b), we have

$$\frac{C_A - C_{A1}}{C_{A\delta'} - C_{A1}} = \frac{Y}{\delta'}$$

The linear concentration profile above and the approximate velocity profile given by Eq. (6-120) are substituted into Eq. (6-127) to give

$$U_\delta(C_{A\delta'} - C_{A1})\frac{d}{dX}\left[\int_0^{\delta'}\left(1 - \frac{Y}{\delta'}\right)\frac{Y}{\delta}\,dY\right] = \frac{D_{AB}(C_{A\delta'} - C_{A1})}{\delta'}$$

If we assume that the concentration boundary layer is thinner than the hydrodynamic boundary layer, we need to integrate only to δ'. The integral is equal to zero for $Y > \delta'$. After integrating we have

$$\frac{1}{6}\frac{d}{dX}\left(\frac{\delta'^2}{\delta}\right) = \frac{D_{AB}}{\delta' U_\delta}$$

Making the substitution $\eta = \delta'/\delta$, the equation above becomes

$$\frac{d}{dX}(\eta^2\delta) = \frac{6D_{AB}}{\eta\delta U_\delta}$$

Differentiating gives

$$\eta^2\frac{d\delta}{dX} + 2\delta\eta\frac{d\eta}{dX} = \frac{6D_{AB}}{\eta\delta U_\delta}$$

or

$$\eta^3\delta\frac{d\delta}{dX} + 2\delta^2\eta^2\frac{d\eta}{dX} = \frac{6D_{AB}}{U_\delta}$$

But from Eqs. (6-123) and (6-125),

$$\delta\frac{d\delta}{dX} = \frac{6\nu}{U_\delta}$$

and

$$\delta^2 = \frac{12\nu X}{U_\delta}$$

Thus we have

$$\eta^3 + 4X\eta^2\frac{d\eta}{dX} = \frac{D_{AB}}{\nu} \tag{6-128}$$

The expression above is a first-order linear differential equation in terms of η^3. The solution is

$$\eta^3 = \frac{D_{AB}}{\nu} + \frac{C}{X^{3/4}}$$

At $X = X_0$, $\delta' = 0$, which gives $\eta = 0$. Therefore,

$$\eta^3 = \frac{D_{AB}}{\nu}\left[1 - \left(\frac{X_0}{X}\right)^{3/4}\right]$$

Thus we obtain

$$\delta' = \delta(\text{Sc})^{-1/3}\left[1 - \left(\frac{X_0}{X}\right)^{3/4}\right]^{1/3} \tag{6-129}$$

where the Schmidt number has been introduced.

The mass transfer coefficient at the surface can be found from Eq. (5-1).

$$k_c^* = -\frac{D_{AB}(dC_A/dY)_{Y=0}}{C_{A1} - C_{A\delta'}} = \frac{D_{AB}}{\delta'}$$

Thus introducing Eqs. (6-125) and (6-129) into the equation above, we obtain

$$k_c^* = 0.289(\text{Sc})^{1/3}(\text{Re}_X)^{1/2}\frac{D_{AB}}{X}\left[1 - \left(\frac{X_0}{X}\right)^{3/4}\right]^{-1/3} \tag{6-130}$$

Rearranging the expression above in terms of the local Sherwood number gives

$$\text{Sh}_X^* = \frac{k_c^* X}{D_{AB}} = 0.289(\text{Sc})^{1/3}(\text{Re}_X)^{1/2}\left[1 - \left(\frac{X_0}{X}\right)^{3/4}\right]^{-1/3} \tag{6-131}$$

If the plate is soluble over its entire length, $X_0 = 0$ and

$$\text{Sh}_X^* = 0.289(\text{Sc})^{1/3}(\text{Re}_X)^{1/2} \tag{6-132}$$

The average mass transfer coefficient and Sherwood number are found by integrating over the length of the plate to give

$$k_{c,\text{avg}}^* = 2k_{c,X=L}^* \tag{6-133}$$

and

$$\text{Sh}_{\text{avg}}^* = 2\text{Sh}_{X=L}^* \tag{6-134}$$

Thus

$$\text{Sh}_{\text{avg}}^* = 0.578(\text{Sc})^{1/3}(\text{Re}_L)^{1/2} \tag{6-135}$$

The use of a first-order polynomial gives a value for the Sherwood number only slightly less than the value obtained from the exact solution, Eq. (6-109). A third-order polynomial gives approximately the same value for the Sherwood number as the exact solution.

Turbulent flow across a flat plate

The analysis for turbulent flow across a flat plate can be made in much the same way as that shown previously for laminar flow. The approximate velocity and concentration profiles are assumed to obey the following expressions:

$$\frac{U}{U_\delta} = \left(\frac{Y}{\delta}\right)^{1/7} \tag{6-136}$$

and

$$\frac{C_A - C_{A1}}{C_{A\delta'} - C_{A1}} = \left(\frac{Y}{\delta'}\right)^{1/7} \tag{6-137}$$

For turbulent flow the thickness of the momentum boundary layer can be obtained by introducing an expression for the turbulent shear stress at the surface into the momentum integral equation given by Eq. (6-118). For turbulent flow, $\nu(dU_X/dY)_{Y=0} = 0.0228U_\delta^2(\nu/U_\delta\delta)^{1/4}$. Neglecting injection through the surface of the plate, we can write Eq. (6-118) as

$$\frac{d}{dX}\left[\int_0^\delta (U_\delta - U_X)U_X\,dY\right] = 0.0228U_\delta^2\left(\frac{\nu}{U_\delta\delta}\right)^{1/4} \tag{6-138}$$

Substituting for the value of U_X given by Eq. (6-136) and integrating gives an expression that relates the boundary layer thickness to the distance from the leading edge of the plate and the local Reynolds number.

$$\frac{\delta}{X} = 0.376(\text{Re}_X)^{-1/5} \tag{6-139}$$

Equation (6-139) is valid for local Reynolds numbers ranging from 5×10^5 to 10^7.

In the absence of blowing or suction through the plate surface and for fluids with Schmidt numbers close to unity, Eqs. (6-136), (6-137), and (6-139) can be combined with Eq. (6-117) to give an expression for the local mass transfer coefficient and Sherwood number. Thus

$$k_c^* = 0.0292U_\delta(\text{Re}_X)^{-1/5} \tag{6-140}$$

and

$$\text{Sh}_X^* = \frac{k_c^* X}{D_{AB}} = 0.0292(\text{Re}_X)^{4/5}\text{Sc} \tag{6-141}$$

If the flow is turbulent over the entire length of the plate, the average Sherwood number is

$$\text{Sh}_{\text{avg}}^* = \frac{k_c^* L}{D_{AB}} = 0.0365(\text{Re}_L)^{4/5}\text{Sc} \tag{6-142}$$

Since the momentum and concentration boundary layer thicknesses were assumed to be equal, the Schmidt numbers in Eqs. (6-141) and (6-142) are equal to unity.

Example 6.4

Typically, laminar flow exists over a portion of the plate. The distance from the leading edge of the plate to the point at which flow becomes turbulent is described as the critical length and corresponds to a Reynolds number of 5×10^5. Find the average mass transfer coefficient for the case in which both laminar and turbulent flow exist on the surface by integrating the individual coefficients over the regions of interest.

Solution:

The average mass transfer coefficient is expressed by

$$k_{c,\text{avg}}^* = \frac{D_{AB}}{L}\left[0.332\left(\frac{U_\delta}{\nu}\right)^{1/2}\int_0^{X_c} X^{-1/2}\,dX + 0.0292\left(\frac{U_\delta}{\nu}\right)^{4/5}\int_{X_c}^{L} X^{-1/5}\,dX\right] \tag{6-143}$$

After integrating we obtain

$$k_{c,\text{avg}}^* = \frac{D_{AB}}{L}\{0.664(\text{Re}_{X_c})^{1/2} + 0.0365[(\text{Re}_L)^{4/5} - (\text{Re}_{X_c})^{4/5})]\} \tag{6-144}$$

For a critical Reynolds number of 5×10^5 the average Sherwood number becomes

$$\text{Sh}_{\text{avg}}^* = 0.0365(\text{Re}_L)^{4/5} - 853 \tag{6-145}$$

j-factor for a flat plate

As previously noted, experimental mass transfer data can be accurately and conveniently correlated in terms of the Chilton and Colburn j-factor. By replacing the bulk velocity for flow in a tube with the free stream velocity for flow across a flat plate, we have

$$j_D = \frac{k_c^*}{U_\infty}(\text{Sc})^{2/3} \tag{6-146}$$

Correlation of data for evaporation from a flat plate is shown in Figure 6-8.

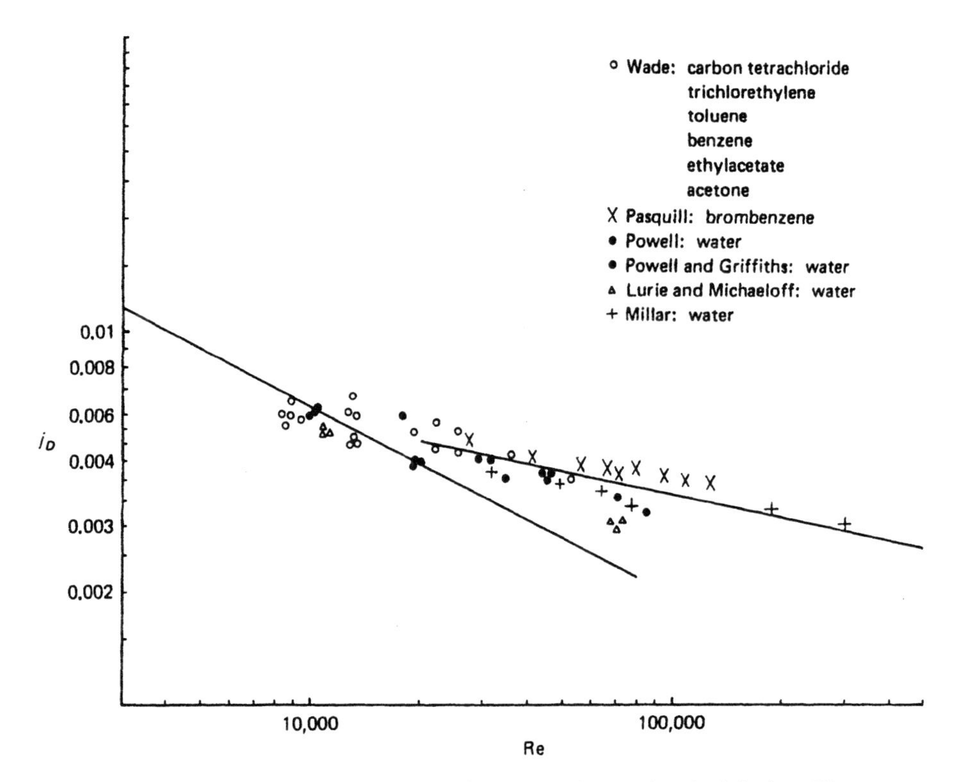

Figure 6-8 Evaporation from a flat plate (from Sherwood and Pigford, 1952)

In the laminar flow region, the j-factor based on the Reynolds number and calculated for the total plate length is

$$j_D = \frac{0.664}{\sqrt{\text{Re}_L}} \tag{6-147}$$

In the turbulent flow region over a flat plate, 10,000 < Re < 300,000, experimental data can be represented by the relationship

$$j_D = \frac{0.037}{(\text{Re}_L)^{0.2}} \tag{6-148}$$

For a flat plate the skin friction coefficient is used in place of the Fanning friction factor. Equations (6-147) and (6-148) are compared with experimental data in Figure 6-8. When predicting mass transfer coefficients from the correlations above, the best results are obtained by using properties evaluated at the film conditions. For flow over a surface this is defined as the average of the free stream and the interface.

6.7 Mass Transfer Coefficients for Spheres and Cylinders

Transfer from single spheres

Mass transfer either to or from single spheres can be correlated with the Frössling (1938) equation.

$$\mathrm{Sh} = 2 + 0.552\mathrm{Re}_d^{0.5}\mathrm{Sc}^{1/3} \tag{6-149}$$

Although the equation above provides good correlation of data at low Reynolds numbers, Sherwood numbers calculated for high Reynolds numbers are lower than values found experimentally. Steinberger and Treybal (1960) analyzed the data obtained by several investigators and developed a correlation that included natural convection. Their equation is

$$\mathrm{Sh}^* = \mathrm{Sh}^*_{\mathrm{NC}} + 0.347(\mathrm{Re}\mathrm{Sc}^{1/2})^{0.62} \tag{6-150}$$

where $\mathrm{Sh}^*_{\mathrm{NC}}$ is the correction term for natural convection. This is expressed by

$$\mathrm{Sh}^*_{\mathrm{NC}} = 2.0 + 0.569(\mathrm{GrSc})^{0.25} \quad \text{for GrSc} < 10^8 \tag{6-151}$$

and

$$\mathrm{Sh}^*_{\mathrm{NC}} = 2.0 + 0.0254(\mathrm{GrSc})^{1/3}\mathrm{Sc}^{0.244} \quad \text{for GrSc} > 10^8 \tag{6-152}$$

Equation (6-150) is valid for the range $1 \leq \mathrm{Re} \leq 3 \times 10^4$ and $0.6 \leq \mathrm{Sc} \leq 3200$. The average deviation is 12.7%. For Reynolds numbers greater than 2000, the Sherwood number can be predicted to within 15% by the expression

$$\mathrm{Sh}^* = 0.347\mathrm{Re}^{0.62}\mathrm{Sc}^{0.31} \tag{6-153}$$

Mass transfer coefficients in packed and fluidized beds

Mass transfer in packed beds has been studied for a number of gases and liquids. Yoshida et al. (1962) presented correlations defined in terms of the j-factor as

$$j_D = 0.91\psi\mathrm{Re}^{-0.51} \quad (\mathrm{Re} < 50) \tag{6-154}$$

and

$$j_D = 0.61\psi\mathrm{Re}^{-0.41} \quad (\mathrm{Re} > 50) \tag{6-155}$$

where the geometry of the packing in the bed is characterized by the shape factor ψ. These are given in Table 6-5.

TABLE 6-5. SHAPE FACTORS FOR PACKED BEDS (GAMSON, 1951)

Particle Shape	ψ
Spheres	1.0
Cylinders	0.91
Flakes	0.86

The Reynolds number for Eqs. (6-154) and (6-155) is expressed as

$$\mathrm{Re} = \frac{U'}{\nu\psi a} \tag{6-156}$$

where U' is the superficial velocity and is based on the empty column cross section. It can be written in terms of the bed void fraction and interstitial velocity as $U' = \epsilon U_X$. As shown, the Reynolds number is defined in terms of the parameter a, which gives the surface area of the packing per unit volume of bed. This is defined as $a = 6(1 - \epsilon)/d_p$, where d_p is the diameter of the particle. For spheres d_p is equal to the diameter. For nonspherical geometries the diameter of the particle can be expressed in terms of its surface area by the expression

$$d_p = 0.567\sqrt{A_p} \tag{6-157}$$

where A_p is the surface area of the particle. The j-factor for a packed bed is related to the superficial velocity by the expression

$$j_D = \frac{k_c^*}{U'}(\mathrm{Sc})^{2/3} \tag{6-158}$$

Wilson and Geankoplis (1966) investigated mass transfer between liquids and spheres in packed beds. They correlated their data in the range $0.0016 < \mathrm{Re} < 55$, $165 < \mathrm{Sc} < 70{,}600$, and $0.35 < \epsilon < 0.75$ with the expression

$$\epsilon j_D = 1.09\mathrm{Re}^{-2/3} \tag{6-159}$$

In the range $55 < \mathrm{Re} < 1500$ and $165 < \mathrm{Sc} < 10{,}690$, they proposed the equation

$$\epsilon j_D = 0.25\mathrm{Re}^{-0.31} \tag{6-160}$$

The Reynolds number is defined in terms of the diameter of the spheres, d_p, and the superficial velocity of the liquid as $\mathrm{Re} = \rho U' d_p/\mu$.

Mass transfer coefficients for gases in packed beds were correlated in the range $90 < \mathrm{Re} < 4000$ by Gupta and Thodos (1963, 1964). For beds of packed spheres their correlation is given by

$$\epsilon j_D = 2.06\mathrm{Re}^{-0.575} \tag{6-161}$$

where the Reynolds number is the same as that used in the equation above. For packings other than spheres, the correlation above can be modified by using the geometry correction factors given in Table 6-6.

TABLE 6-6. GEOMETRY CORRECTION FACTORS FOR EQ. (6-161) (GEANKOPLIS, 1972)

Geometry	*Correction Factor* $\epsilon j_D/\epsilon j_D$ *(Sphere)*
Sphere	1.0
Cylinder (length = diameter)	0.79
Cube	0.71

Most of the experimental data obtained prior to 1977 for mass transfer between particles and fluids in fixed and fluidized beds were reanalyzed and correlated by Dwivedi and Upadhyay (1977). They proposed a single correlation for both gases and liquids in packed and fluidized beds in terms of the j-factor.

$$\epsilon j_D = \frac{0.765}{\text{Re}^{0.82}} + \frac{0.365}{\text{Re}^{0.386}} \tag{6-162}$$

The Reynolds number in the equation above is expressed in terms of the superficial velocity as $\text{Re} = \rho U' d_p/\mu$. Their equation, which is valid over the range $0.01 \leq \text{Re} \leq 15{,}000$, correlates experimental data with an average deviation of 17.95%. Equation (6-162) is recommended for preliminary design purposes, except for gases for which the Reynolds number is less than 10.

Example 6.5

Air at 75°C flows through a 0.05-m-long packed tube of naphthalene spheres at an interstitial velocity of 0.6 m/s. The diameter of each sphere is 0.01 m and the bed porosity is 0.5. The convective mass transfer coefficient for this process was found to be 0.057 m/s. (a) Calculate the bulk concentration of naphthalene in air at the exit of the tube. (b) Determine the percent saturation of the exiting air. The entering air is free of naphthalene. The vapor pressure of naphthalene at 75°C is 5.61 mmHg (7.47×10^2 N/m^2).

Solution:

(a) A shell balance over a differential section of the packed bed gives

$$\frac{dN_A}{dZ} = k_c^* a(C_{Ai} - C_A)$$

where C_{Ai} is the solute concentration at the solid interface and C_A is the solute concentration in the bulk fluid at any position in the tube. Since mass transfer in the axial direction is due primarily to bulk flow, the molar flux can be expressed as

$$N_A \cong C_A U'$$

If plug flow is assumed and U' is constant, then

$$U'\frac{dC_A}{dZ} = k_c^* a(C_{Ai} - C_A)$$

Separating the variables and integrating over the tube length gives

$$\int_{C_{A1}}^{C_{A2}} \frac{dC_A}{C_{Ai} - C_A} = \int_0^L \frac{k_c^* a}{U'} dZ$$

or

$$\ln \frac{C_{Ai} - C_{A1}}{C_{Ai} - C_{A2}} = \frac{k_c^* a L}{U'}$$

where

$$U' = \epsilon U_Z = 0.5(0.6) = 0.3 \text{ m/s}$$

$$a = \frac{6(1-\epsilon)}{d_p} = \frac{6(1-0.5)}{0.01} = 300 \text{ m}^2/\text{m}^3$$

$$C_{A1} = 0$$

$$C_{Ai} = \frac{P_{Ai}}{RT} = \frac{7.47 \times 10^2}{8314(348)} = 2.58 \times 10^{-4} \text{ kgmol/m}^3$$

Solving for C_{A2} gives

$$C_{A2} = C_{Ai} + (C_{A1} - C_{Ai}) \exp\left(\frac{-k_c^* a L}{U'}\right)$$

$$= 2.58 \times 10^{-4} + (0 - 2.58 \times 10^{-4}) \exp\left[\frac{-(0.057)(300)(0.05)}{0.3}\right]$$

$$= 2.43 \times 10^{-4} \text{ kgmol/m}^3$$

(b) The percent saturation is related to the concentration at the solid interface. Thus

$$\% \text{ saturation} = \frac{C_{A2}}{C_{Ai}} \times 100$$

$$= \frac{2.43 \times 10^{-4}}{2.58 \times 10^{-4}} \times 100 = 94.2\%$$

REFERENCES

CHILTON, T. H., and A. P. COLBURN, *Ind. Eng. Chem.*, *26*, 1183 (1934).

DWIVEDI, P. N., and S. N. UPADHYAY, *Ind. Eng. Chem., Process Des. Dev.*, *16*, 157 (1977).

EINSTEIN, H. A., and H. LI, *Structural Div. J. Am. Soc. Civil Eng.*, *82*, 293 (1956).

FRÖSSLING, N., *Beitr. Geophys.*, *52*, 170 (1938).

GAMSON, B. W., *Chem. Eng. Prog.*, *47*, 19 (1951).

GEANKOPLIS, C. J., *Mass Transport Phenomena*, Holt, Rinehart and Winston, New York, 1972.

GILLILAND, E. R., and T. K. SHERWOOD, *Ind. Eng. Chem.*, *26*, 516 (1934).

GUPTA, A. S., and G. THODOS, *AIChE J.*, *9*, 751 (1963).

GUPTA, A. S., and G. THODOS, *Ind. Eng. Chem. Fundam.*, *3*, 218 (1964).

HANRATTY, T. J., *AIChE J.*, *2*, 359 (1956).

HARTNETT, J. P., and E. R. G. ECKERT, *Trans. ASME*, *79*, 247 (1957).

HAUSEN, H., *Z. VDI Beih. Verfarenstech.*, *4*, 91 (1943).

KAFESJIAN, R., C. A. PLANK, and E. R. GERHARD, *AIChE J.*, *7*, 463 (1961).

LINTON, W. H., and T. K. SHERWOOD, *Chem. Eng. Prog.*, *46*, 258 (1950).

NOTTER, R. H., and C. A. SLEICHER, *Chem. Eng. Sci.*, *26*, 161 (1971).

PETUKHOV, B. S., in *Advances in Heat Transfer*, Vol. 6, p. 503, Academic Press, New York, 1970.

PINCZEWSKI, W. V., and S. SIDEMAN, *Chem. Eng. Sci.*, *29*, 1969 (1974).

POHLHAUSEN, E., *Z. Angew. Math. Mech.*, *1*, 115 (1921).

PRANDTL, L., *Phys. Z.*, *11*, 1072 (1910).

PRANDTL, L., Proc. 3rd Int. Math. Congr., Heidelberg, 1904; reprinted in *Natl. Advisory Comm. Aeronaut., Tech. Mem.*, 452 (1928).

REYNOLDS, O., *Scientific Papers of Osborne Reynolds*, Vol. 2, Cambridge University Press, London, 1901.

RUCKENSTEIN, E., *Chem. Eng. Sci.*, *18*, 233 (1963).

SCHLICHTING, H., *Boundary Layer Theory*, McGraw-Hill, New York, 1968.

SHERWOOD, T. K., and R. L. PIGFORD, *Absorption and Extraction*, 2nd ed., McGraw-Hill, New York, 1952.

SISSOM, L. E., and D. R. PITTS, *Elements of Transport Phenomena*, McGraw-Hill, New York, 1972.

SKELLAND, A. H. P., *Diffusional Mass Transfer*, Wiley, New York, 1974.

STEINBERGER, R. L., and R. E. TREYBAL, *AIChE J.*, *6*, 227 (1960).

VON KÁRMÁN, T., *Trans. ASME*, *61*, 705 (1939).

VON KÁRMÁN, T., *Z. Angew. Math. Mech.*, *1*, 233 (1921); reprinted in *Natl. Advisory Comm. Aeronaut., Tech. Mem.*, 1092 (1946).

WILSON, E. J., and C. J. GEANKOPLIS, *Ind. Eng. Chem. Fundam.*, *5*, 9 (1966).

YOSHIDA, F., D. RAMASWAMI, and O. A. HOUGEN, *AIChE J.*, *8*, 5 (1962).

NOTATIONS

a = surface area of packing per unit volume of bed, L^2/L^3
A_p = surface area of a particle, L^2
C_{Ai} = molar concentration of A in the liquid at the gas–liquid interface, mol/L^3
C_{Am} = bulk concentration of A, mol/L^3
C_{AR} = molar concentration of A at the tube wall, mol/L^3
$C_{A\infty}$ = molar concentration of A in the free stream, mol/L^3
c = heat capacity, E/mol T
d = diameter, L
d_p = diameter of particle, L
D_{AB} = binary diffusivity for system A–B, L^2/t
$\mathbf{D}$ = diffusivity, L^2/t
E_D = eddy diffusivity for mass transfer, L^2/t
E_H = eddy diffusivity for heat transfer, L^2/t
E_V = eddy diffusivity for momentum transfer, L^2/t
f = Fanning friction factor

Gr = Grashof number
h = heat transfer coefficient, E/TL^2
j_D = Chilton and Colburn j-factor
k = thermal conductivity, E/tLT
$k_c, k_c^*, k_{c,\mathrm{avg}}^*$ = mass transfer coefficients, mol/tL^2 (concentration difference)
l = characteristic length, L
L = length of the plate or column
Le = Lewis number
M = mass or molecular weight, M/mol
N_A = molar flux of A with respect to a fixed reference frame, mol/tL^2
$N_{A\delta}$ = molar flux of A in bulk fluid with respect to a fixed reference frame, mol/L^2t
P = pressure of the system, F/L^2
$(\bar{P}_B)_M$ = logarithmic mean pressure of B evaluated between the surface and bulk compositions, F/L^2
Pr = Prandtl number
$\dot{q}$ = heat flux, E/tL^2
$\dot{r}_A^v$ = net mass rate of production of A, M/tL^3
R = tube diameter, L
$\dot{R}_A^v$ = net molar rate of production of A, mol/tL^3
Re = Reynolds number
Re_d = Reynolds number based on tube diameter
Re_L = Reynolds number based on tube length
Re_{X_c} = Reynolds number at critical point where flow becomes turbulent
Sc = Schmidt number
Sh = Sherwood number
$\mathrm{Sh}_{\mathrm{avg}}^*$ = average Sherwood number
Sh_d^* = Sherwood number based on tube diameter
$\mathrm{Sh}_{\mathrm{NC}}^*$ = Sherwood number for natural convection
St = Stanton number
St^* = Stanton number used for turbulent flow
t = time, t
T_1 = temperature at gas-liquid interface, T
T_m = temperature in bulk fluid, T
U_X, U_Y, U_Z = velocity in X, Y, and Z directions, L/t
U_τ = friction velocity, L/t
$\bar{U}$ = average velocity, L/t
U' = superficial velocity, L/t
y^* = modified Reynolds number defined by Eq. (6-88)

Greek Letters

α = thermal diffusivity, L^2/t
δ = boundary layer thickness for momentum transfer, L
δ' = boundary layer thickness for mass transfer, L
ϵ = porosity

θ = dimensionless bulk concentration
μ = viscosity, M/Lt
ν = kinematic viscosity, L^2/t
$\pi_1, \pi_2, \ldots$ = dimensionless groups
ρ = density, M/L^3
τ = shear stress, F/L^2
τ_0 = shear stress at wall, F/L^2
ϕ = dimensionless velocity
ψ = shape factor

PROBLEMS

6.1 Apply the Buckingham pi method to the case of mass transfer due to natural convection. The variables are as follows:

Parameter	*Dimensions*
Tube diameter, d	L
Diffusion coefficient, D_{AB}	L^2t^{-1}
Fluid viscosity, μ	$ML^{-1}t^{-1}$
Fluid density, ρ	ML^{-3}
Mass transfer coefficient, k_c	Lt^{-1}
Tube length, l	L
Buoyancy force, $g\,\Delta\rho_A$	$ML^{-2}t^{-2}$

6.2 Air flows through a cylindrical tube made of naphthalene at a velocity of 5 m/s. The diameter of the tube is 0.1 m and the temperature of the air is 20°C. Using the correlation proposed by Linton and Sherwood, **(a)** calculate the mass transfer coefficient for the transfer of naphthalene to air. Compare the result with values obtained by using the following: **(b)** Reynolds analogy, **(c)** Prandtl analogy, **(d)** von Kármán analogy, and **(e)** the Chilton–Colburn j-factor correlation.

$$\text{Viscosity of air} = 1.8 \times 10^{-5}\ \text{kg/m}\cdot\text{s}$$

$$\text{Density of air} = 1.2\ \text{kg/m}^3$$

$$\text{Effective diffusivity} = 4.24 \times 10^{-6}\ \text{m}^2/\text{s}$$

6.3 Water is stored in a spherical earthen pot that has an inside diameter 0.3 m. The temperature of the water is 25°C. Calculate the time needed to cool the water from 25°C to 22°C. Assume that air at 25°C and a relative humidity of 20% flows over the outside surface of the pot at a velocity of 0.1 m/s. The viscosity of air is 1.8×10^{-5} kg/m·s; other data are as follows: $\rho_{air} = 1.18$ kg/m^3, $D_{H_2O\text{-}air} = 2.56 \times 10^{-5}$ m^2/s, ΔH_v(water) = 560 kcal/kg.

6.4 An N_2–O_2 mixture flows through a 0.15-m-inside diameter (I.D.) wetted-wall column at a velocity of 1.5 m/s. The column temperature is 20°C and the pressure is 20 atm. At some point in the column, the partial pressure of O_2 in N_2 is 2.0 atm and the concentration of O_2 in the water at the gas–liquid interface is 0.01 mol %. Calculate the mass transfer coefficient and the molar flux of oxygen into the water for the conditions given.

6.5 Nitrogen flows through a porous tube at a velocity of 0.3 m/s and a pressure of 2.7 atm as shown in Figure 6-9. The tube is 0.003 m I.D. and 0.9 m long. A small amount of benzene is injected uniformly through the walls of the tube at a constant rate, where it is carried away by the nitrogen. Determine the mass transfer coefficient of benzene into the nitrogen if the tube temperature is 90°C. Assume that a fully developed parabolic velocity profile is maintained in the tube.

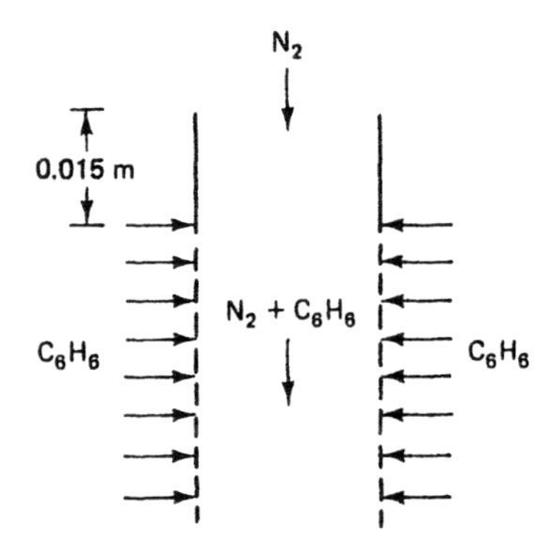

Figure 6-9 Problem 6.5: mass transfer in a porous tube

6.6 A fluid flows across a flat plate that is coated with solute A at a free stream velocity of 0.35 m/s. The concentration of A in the free stream is 0.2 kgmol/m³, and the concentration of A in the fluid at the surface is 0.5 kgmol/m³. Obtain the concentration of A at the point $X = 0.15$ m and $Y = 0.3$ m in the boundary layer as shown in Figure 6-10. Assume that the velocity in the boundary layer at the point above is 0.275 m/s. The kinematic viscosity of the fluid is 0.093 m²/s. Pr = 0.7, Sc = 0.7 for this case.

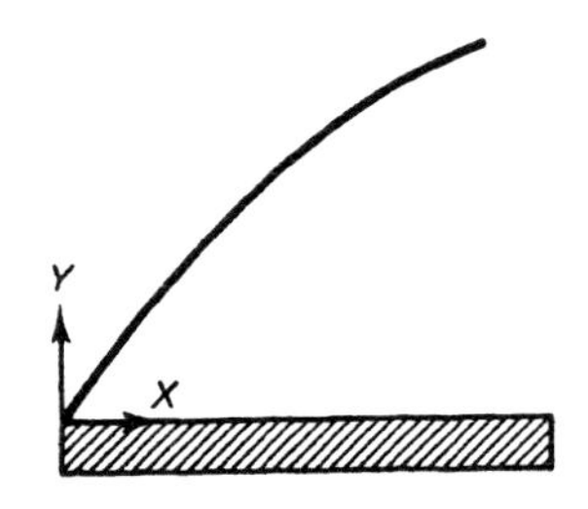

Figure 6-10 Problem 6.6: mass transfer from a flat plate

6.7 Fluid B flows over a 1-m-long porous flat plate at a free-stream velocity of 0.2 m/s as shown in Figure 6-11. A gas is forced through the surface of the plate at a velocity U_0. The diffusivity of the gas in the fluid is 5.0×10^{-9} m²/s and the Schmidt number is 100. **(a)** Calculate k_c^* at a point 0.5 m from the leading edge of the plate and **(b)** determine $k_{c,\text{avg}}^*$ for the entire plate. The gas velocity through the plate can be related to the free-stream velocity by the equation $U_0/U_\infty = 0.000707X^{-1/2}$, where X is expressed in meters.

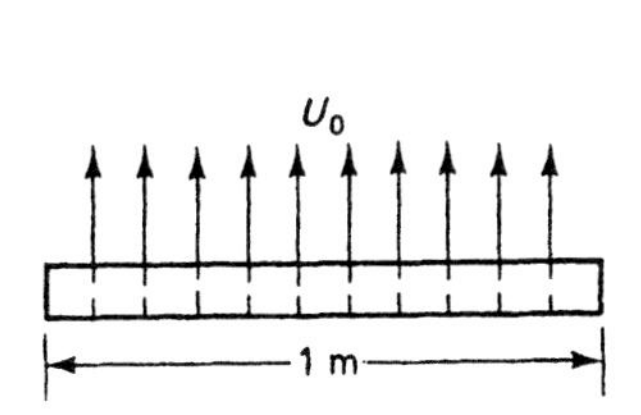

Figure 6-11 Problem 6.7: injection through a porous plate

6.8 Derive the momentum integral equation given by Eq. (6-118) for flow across a flat plate with injection or suction through the plate's surface.

6.9 In solving Eq. (6-118) for the momentum boundary layer thickness, the concentration and velocity profiles must be known. **(a)** Assuming that the velocity and concentration profiles can be expressed as

$$U_X = a \sin bY + c \cos dY$$

and

$$C_A - C_{A1} = a \sin bY + c \cos dY$$

evaluate the constants a, b, c, and d in the two equations above for a laminar boundary layer. **(b)** Calculate the momentum boundary layer thickness δ.

6.10 In Example 6.3, a first-order polynomial was used to represent the concentration and momentum boundary layers. Rework Example 6.3 by using the third-order polynomials given below to describe the concentration and velocity profiles.

$$C_A - C_{A1} = a + bY + cY^2 + dY^3$$

$$U_X = a + bY + cY^2 + dY^3$$

6.11 Consider the case in which a flat plate coated with solute A dissolves into liquid B as shown in Figure 6-12. If both the concentration and velocity profiles can be expressed as first-order polynomials, calculate k_c^* at $X = 1$ m for **(a)** $X_0 = 0$, and **(b)** $X_0 = 0.5$ m. **(c)** Calculate k_c^* at $X = 0.5$ m for $X_0 = 0$ and discuss the physical significance for the variation of k_c^* at different positions along the plate. Physical properties are as follows:

$$D_{AB} = 2 \times 10^{-9}\ \text{m}^2/\text{s}$$

$$U_\infty = 0.1\ \text{m/s}$$

$$\nu = 2 \times 10^{-8}\ \text{m}^2/\text{s}$$

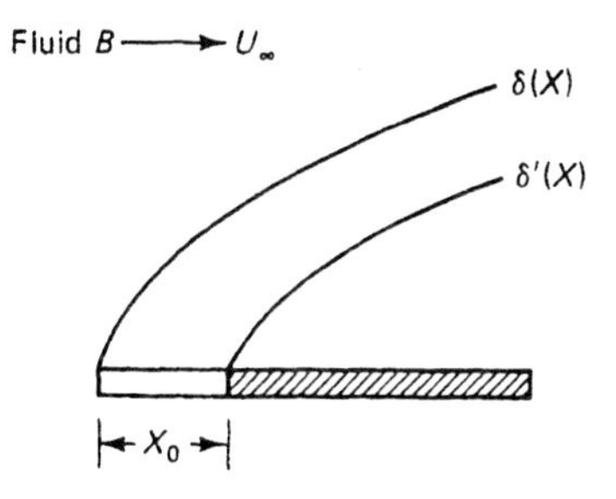

Figure 6-12 Problem 6.11: dissolution of a flat plate

6.12 During the winter a highway surface is covered by a 1×10^{-4} m layer of ice as shown in Figure 6-13. The temperature of the surface and the air is -1.0°C. The relative humidity of the air is 10%. Find the time required to completely remove the ice by sublimation at a point 0.3 m from the edge of the highway if the wind blows across the surface at a velocity of 6.7 m/s.

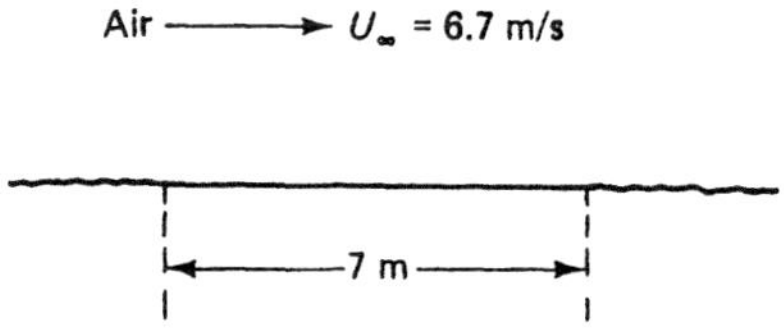

Air → $U_\infty = 6.7$ m/s

7 m

Figure 6-13 Problem 6.12

6.13 Air flows over an open rectangular container that is filled with benzene as shown in Figure 6-14. The benzene evaporates and is carried away by the air. The temperature of the ambient air is 25°C and the velocity is 0.2 m/s. Determine the convective mass transfer coefficient. Assume that the process is isothermal.

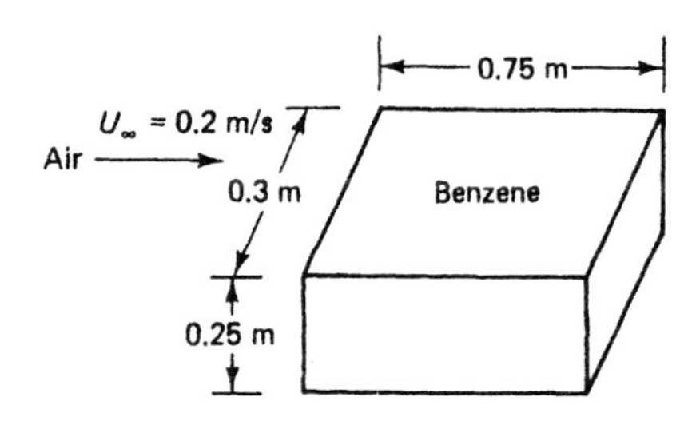

Figure 6-14 Problem 6.13

6.14 Air at 32°C is humidified by flowing over a 1.2-m-long container filled with water. The interfacial temperature is 20°C. If the initial humidity of the air is 25% and its velocity is 0.15 m/s, calculate **(a)** the convective mass transfer coefficient, and **(b)** the amount of water evaporated per unit width of the container.

6.15 The velocity of air in Problem 6.14 is increased to 1.5 m/s. If the flow becomes turbulent when the Reynolds number exceeds 20,000, calculate the convective mass transfer coefficient.

6.16 A 0.04-m-diameter naphthalene sphere is placed in an air duct, where it undergoes sublimation into a 25°C air stream that flows at 0.6 m/s. Calculate the convective mass transfer coefficient. $D_{n\text{-air}} = 5.85 \times 10^{-6}$ m²/s.

6.17 Calculate the convective mass transfer coefficient in a packed bed for naphthalene spheres undergoing sublimation into air. Use the equation of Yoshida, Ramaswami, and Hougen and compare your result to the value predicted from the equation of Gupta and Thodos. The spheres have a diameter of 0.04 m and the bed porosity is 0.4. The superficial velocity of the air is 0.6 m/s and the temperature is 25°C.

6.18 Air at 25°C flows through a bed filled with cylindrical pellets of naphthalene. Both the length and diameter of the pellets are 6.35×10^{-3} m. Using the correlation of Gupta and Thodos, calculate the convective mass transfer coefficient for the sublimation of naphthalene. The superficial velocity of the air is 0.75 m/s and the bed porosity is 0.4.

6.19 The average heat transfer coefficient for natural convection from a single sphere in a large body of fluid is given by

$$\frac{h_m d}{k} = 2 + 0.6\left(\frac{d^3 \rho^2 g \beta \Delta T}{\mu^2}\right)^{1/4}\left(\frac{c\mu}{k}\right)^{1/3} \quad \text{for} \quad \mathrm{Gr}^{1/4}\,\mathrm{Pr}^{1/3} < 200$$

where d is the diameter of the sphere and the fluid properties are evaluated at the mean temperature of the sphere and bulk fluid. Using the analogy between mass and heat transfer, calculate the instantaneous rate of sublimation at the surface of a naphthalene sphere in air at 145.5°C and 1 atm. The vapor pressure of naphthalene at 145.5°C is 0.13 atm. Assume that $D_{n\text{-}air} = 5.85 \times 10^{-6}$ m²/s. The diameter of the sphere is 7.5×10^{-2} m.

Phase Equilibrium 7

7.1 Introduction

The quantity and reliability of data available for equilibrium systems vary widely. Perhaps the most widely and completely investigated type of system is the vapor–liquid system containing paraffin hydrocarbons. Because of the large sums of money that depend on predicting reservoir behavior and natural gas processing conditions, the behavior of these systems has been thoroughly studied. For other vapor–liquid systems, such as oxygenated or chlorinated hydrocarbons, the information available is less complete. Generalizations in this area are difficult and the designer is wise to insist on experimental data for the particular system on hand. For liquid–liquid systems little has been reported in the way of generalized methods for predicting thermodynamic properties. The designer is required to have experimental data on the specific system of interest. For vapor–solid and liquid–solid systems in which the vapor or liquid is adsorbed onto the solid, almost no generalized procedures exist. Without experimental data for the specific system the designer has no way to proceed.

7.2 Thermodynamic Background

The thermodynamics of phase distribution can become very complex and there is no intent here to reproduce the large volume of literature that exists in this area. However, some thermodynamic definitions will be discussed in order to establish proper ground rules for understanding the material that follows.

An ideal gas is, by definition, one that follows the ideal gas law:

$$PV = nRT \tag{7-1}$$

where P = absolute pressure,
V = total volume,
n = number of moles of gas,
R = ideal gas constant, appropriate units,
T = absolute temperature.

The conditions under which a given component or mixture approaches ideal behavior depend somewhat on the critical temperature and critical pressure. Generally, however, the assumption of ideal vapor-phase behavior is a good approximation below pressures of approximately 50 psia (344.8 kPa).

Two principles of gas behavior are *Dalton's law of additive pressures* and *Amagat's law of additive volumes*. These are

$$\pi = \bar{P}_A + \bar{P}_B + \cdots \tag{7-2}$$

and

$$V_t = V_A + V_B + \cdots \tag{7-3}$$

where π = total system pressure,
V_t = total volume of system,
$\bar{P}$ = partial pressure,
V = pure component volume,
A, B = components.

Both Dalton's law and Amagat's law are correct when conditions are such that each of the components and the mixture obey the ideal gas law. There are instances in which either Dalton's law or Amagat's law can be assumed to apply even though the mixture does not follow the ideal gas law. However, both apply simultaneously only for ideal gases.

The rate of mass transfer was shown previously to be governed by the displacement from equilibrium as indicated by the difference in chemical potential within a single phase. Using thermodynamics the equilibrium between phases can be related to the equality of the chemical potential for each species in all phases. This is expressed mathematically as

$$\mu_i^\delta = \mu_i^\beta \tag{7-4}$$

where μ_i is the chemical potential. The chemical potential is defined in terms of the Gibbs free energy by the expression

$$\mu_i = \left[\frac{\partial G}{\partial n_i}\right]_{T,P,n_j} = \bar{G}_i \tag{7-5}$$

where $\bar{G}_i$ is the partial molar Gibbs free energy.

Using the thermodynamic relationship between Gibbs free energy, temperature, and pressure, we obtain, for the case of constant temperature

$$d\mathcal{G} = \mathcal{V}\, dP \tag{7-6}$$

where $\mathcal{G}$ is the molar Gibbs free energy and $\mathcal{V}$ is the molar volume. Considering pure fluid i and introducing the ideal gas law, we can write the expression above as

$$d\mathcal{G} = RT\frac{dP}{P} = RT\,d\ln P \tag{7-7}$$

Prediction of the free energy from the equation above is restricted to ideal gases but can be modified to include real fluids by introducing a new function defined as the fugacity. Equation (7-7) becomes

$$d\mathcal{G}_i = RT\,d\ln f_i \tag{7-8}$$

where f_i is the fugacity of pure i with units of pressure. To complete this definition, the fugacity becomes equal to pressure as the pressure approaches zero. Thus

$$\lim_{P\to 0}\frac{f_i}{P} = 1.0 \tag{7-9}$$

For a single component in a mixture at constant temperature the fugacity is defined by the expressions

$$d\bar{G}_i = RT\,d\ln\bar{f}_i \tag{7-10}$$

and

$$\lim_{P\to 0}\frac{\bar{f}_i}{y_iP} = 1.0 \tag{7-11}$$

where $\bar{f}_i$ is the fugacity of i in solution. By fixing the reference state at zero pressure, the numerical value of the fugacity for any component is also fixed. The choice of zero pressure is purely arbitrary but is by far the most widely used choice for fixing the relationship between pressure and fugacity. Integration of Eq. (7-10) gives

$$\mu_i = RT\ln\bar{f}_i + \psi_i \tag{7-12}$$

where the constant of integration ψ_i is a function of temperature only. Since the criteria for equilibrium also require that the temperature and pressure of the existing phases be equal, we can combine Eqs. (7-4) and (7-12) to give the criterion for equilibrium in terms of fugacities.

$$\bar{f}_i^{\delta} = \bar{f}_i^{\beta} \quad (i = 1, 2, \ldots, m) \tag{7-13}$$

The criterion for equilibrium may also be expressed in terms of the activity, which is the ratio of the fugacity of i in the mixture to a standard-state fugacity. By integrating Equation (7-10) from a standard state to the state of i in the mixture, we obtain

$$\mu_i - \mu_i^{\circ} = RT\ln\frac{\bar{f}_i}{f_i^{\circ}} = RT\ln a_i \tag{7-14}$$

where a_i is the activity and f_i° is the standard-state value of the fugacity. The criterion for equilibrium expressed in terms of the activity thus becomes

$$a_i^{\delta} f_i^{\circ\delta} = a_i^{\beta} f_i^{\circ\beta} \tag{7-15}$$

The difference between the standard state and the reference state must be clearly understood. The reference state refers to that part of the definition of fugacity where the fugacity is assumed to be equal numerically to pressure. The standard state is used in the definition of activity. The standard state is fixed as a matter of convenience and the specific state chosen will differ in different instances.

Deviation from ideality is not uncommon for liquid and solid solutions. An indication of the nonideality is described in terms of the activity coefficient. For component i in a mixture this is defined by

$$\gamma_i = \frac{a_i}{z_i} = \frac{\bar{f}_i}{z_i f_i^\circ} \tag{7-16}$$

where γ_i = activity coefficient of component i,
z_i = mole fraction of component i in the mixture.

At equilibrium we have

$$(\gamma_i z_i f_i^\circ)^\delta = (\gamma_i z_i f_i^\circ)^\beta \tag{7-17}$$

Nonideality in the gas phase is usually accounted for by using the fugacity coefficient instead of the activity coefficient.

7.3 Selectivity and Distribution Coefficients

The ultimate extent of separation for any phase separation process can be described in terms of a separation factor or selectivity. The selectivity, which is related to the compositions in the different phases, can be expressed by the relationship

$$\alpha_{ij} = \frac{z_i^\delta / z_j^\delta}{z_i^\beta / z_j^\beta} = \frac{z_i^\delta / z_i^\beta}{z_j^\delta / z_j^\beta} \tag{7-18}$$

where the components are denoted by i and j and the phases are given by δ and β. For a separation process in which $\alpha_{ij} = 1$, the ratio of i to j would be the same in both phases. Obviously, separation is not possible when this occurs since the driving force for mass transfer is zero. If $\alpha_{ij} > 1$, more component i would accumulate in phase δ than in phase β and the concentration of j in phase β would be greater than in phase δ. Clearly, we may conclude that the magnitude of the selectivity relates to the driving force for mass transfer and is a measure of the ease of separation for the process. From the above we see that the selectivity can be rearranged to give the ratio of each component in each phase. This ratio, denoted as the *equilibrium distribution ratio*, is defined as

$$K_i = \frac{z_i^\delta}{z_i^\beta} \tag{7-19}$$

If we introduce Eq. (7-17) into the expression above, we obtain the distribution between two equilibrium phases in terms of the activity coefficients and the standard-state fugacities as shown:

$$K_i = \frac{z_i^{\delta}}{z_i^{\beta}} = \frac{\gamma_i^{\beta} f_i^{\circ\beta}}{\gamma_i^{\delta} f_i^{\circ\delta}} \tag{7-20}$$

The calculational model used to predict the behavior for most of the equilibrium separation operations is the equilibrium or theoretical stage. A theoretical stage is defined as one in which the two streams leaving the stage are in thermodynamic equilibrium. The streams may be different in composition, but the relationship between these compositions can be predicted from the distribution coefficients that apply at equilibrium. For this reason, thermodynamic data for the system being considered must be available before we can predict the phase behavior and complete the design of the stagewise operation.

Selecting the proper thermodynamic data to be used in the calculation is not always easy. Also, the predicted performance can vary greatly when different data are used. Stocking et al. (1960) made a study to evaluate the effects of various sources of thermodynamic data on the predicted performance of a distillation column—a high-pressure depropanizer. Different data caused the prediction of significantly different flow rates and product compositions. Table 7-1 shows the different compositions of overhead product from the column caused by using different equilibrium distribution coefficients or K values. The compositions shown in Table 7-1 clearly demonstrate that the values used for distribution coefficients have a pronounced effect on the composition of the product streams from the equilibrium stagewise unit. This should also make clear that distribution coefficients differ among various sources. The selection of the proper values for distribution coefficients and other thermodynamic properties is one of the major concerns of the engineer charged with designing an equilibrium stagewise operation unit. The choices among various alternatives are seldom obvious and never "cut and dried." They depend in large part on experience, judgment, and hints regarding the choice of thermodynamic data that will most accurately describe the degree of separation that can be obtained with actual plant equipment.

TABLE 7-1. PRODUCT COMPOSITIONS CALCULATED USING DISTRIBUTION COEFFICIENTS FROM DIFFERENT SOURCES (STOCKING ET AL., 1960)

	Mole Fraction				
Component	K_1	K_2	K_3	K_4	K_5
C_2	0.010	0.010	0.010	0.010	0.010
C_3	0.918	0.982	0.918	0.949	0.955
iC_4	0.046	0.003	0.050	0.031	0.021
nC_4	0.026	0.005	0.022	0.010	0.008
	1.000	1.000	1.000	1.000	1.000

7.4 Miscible Liquid Behavior

Two liquids are described as being miscible if they can be mixed in all proportions and only one liquid phase will exist. In this section both ideal and nonideal systems will be considered.

Ideal solutions

An ideal liquid or gas solution is one for which the activity coefficient is unity. There are very few liquid systems that behave ideally at any pressure. The assumption of ideal liquid behavior is generally correct only for members of an homologous series that are close together in molecular weight. However, the assumption of ideal liquid behavior is good from a practical engineering standpoint at dilute solute concentrations.

At low pressure the vapor phase of a mixture approaches ideal behavior and follows the ideal gas law. The pressure for this assumption varies but generally is in the range 30 to 50 psia (207 to 345 kPa). For these conditions, we can equate the fugacity with the partial pressure, and the standard-state fugacity of the liquid becomes equal to the vapor pressure.

Figure 7-1 shows a temperature–composition diagram at constant pressure for a miscible binary mixture. The X axis is the composition axis. The accepted convention is to plot the most volatile or lower boiling constituent increasing in composition in the positive X direction. Since only two components are present, the compositions of both components are known if we specify the composition of one. Curve AC on Figure 7-1 represents the locus of the bubble point temperatures for all mixtures of components A and B. Curve BD represents

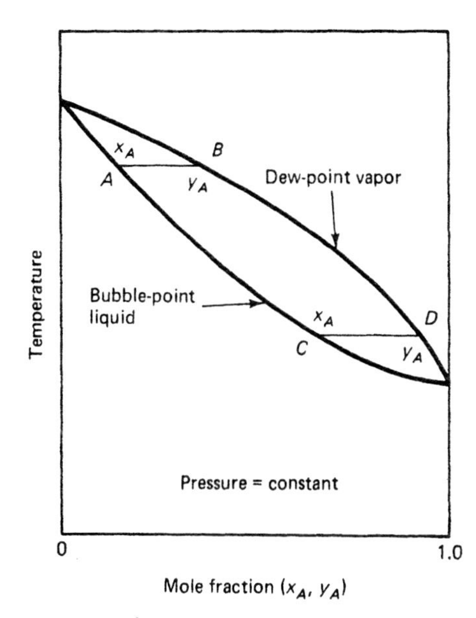

Figure 7-1 Diagram for a "normal" system

the locus of the vapor dew-point temperatures for all mixtures of components A and B. Both of these curves originate at the boiling temperature of one constituent and terminate at the boiling temperature of the other constituent. For a liquid of composition x_A the boiling-point temperature can be determined if we draw a vertical line from the point representing x_A to the line AC and then draw a horizontal line to the Y axis to read the temperature. The vapor in equilibrium with the liquid can be determined if we draw a horizontal line from the intersection of x_A and the line AC to line BD and then draw a vertical line to the X axis. In this way we can use Figure 7-1 to relate temperature to the liquid and vapor compositions for a miscible binary mixture at constant pressure.

Figure 7-2 illustrates another use for the temperature–composition diagram. Assume that a mixture of composition z_A exists at a temperature represented by point E on Figure 7-2. As this mixture is heated it will remain all liquid until point D is reached, where the first small bubble of vapor will form. If heating of the mixture is continued, and the vapor formed is maintained in contact with the liquid, the mixture will eventually reach a point C on the diagram where both vapor and liquid exist. The overall composition of the total mixture remains z_A, but the liquid portion will have a composition x_A and the vapor portion will have a composition y_A. The vapor and liquid in equilibrium (x_A and y_A) are connected by the equilibrium tie line ab. The relative amounts

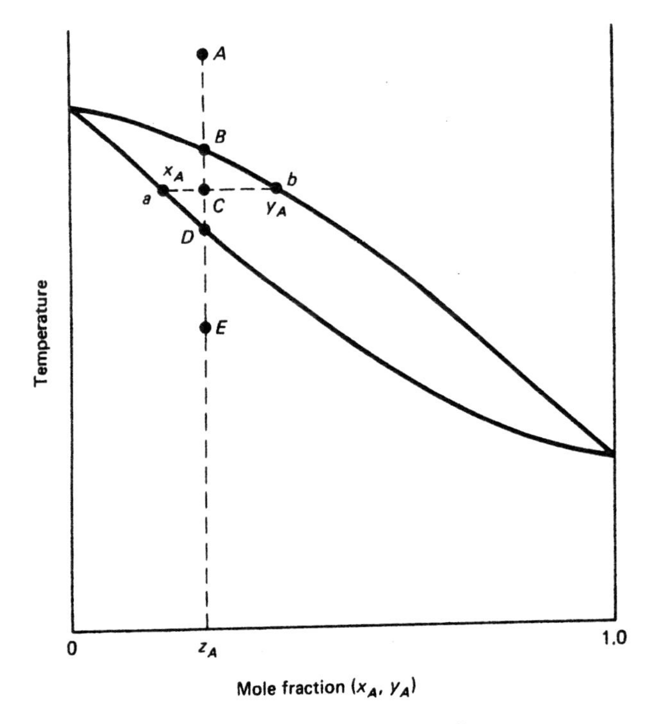

Figure 7-2 Illustration of a T–x, y diagram

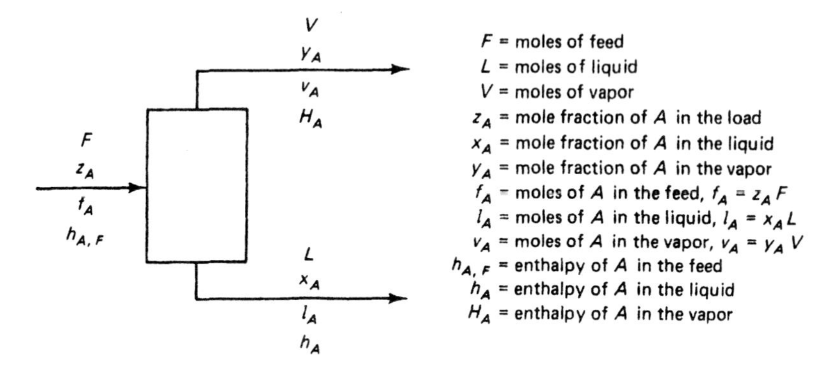

Figure 7-3 Schematic of an equilibrium flash separation

of the vapor and liquid present can be determined by a series of material balances around an equilibrium flash separator as shown in Figure 7-3. For component A

$$Fz_A = Lx_A + Vy_A \tag{7-21}$$

where z_A = mole fraction composition of component A in the feed mixture,
F = total moles of feed,
L = moles of liquid formed,
x_A = composition of liquid,
y_A = composition of vapor,
V = moles of vapor formed.

We can write an overall material balance as

$$F = V + L \tag{7-22}$$

If we eliminate V from Eqs. (7-21) and (7-22), we get

$$Fz_A = Lx_A + (F - L)y_A$$

$$F(z_A - y_A) = L(x_A - y_A) \tag{7-23}$$

$$\frac{L}{F} = \frac{z_A - y_A}{x_A - y_A}$$

By a similar procedure, only eliminating L, we obtain

$$\frac{V}{F} = \frac{x_A - z_A}{x_A - y_A} \tag{7-24}$$

If we had eliminated F, then

$$\frac{L}{V} = \frac{z_A - y_A}{x_A - z_A} \tag{7-25}$$

Equations (7-23) through (7-25) are the basis for the *inverse lever rule principle.* In Eq. (7-25) the value of $(z_A - y_A)$ is the equivalent of the length of the line from C to b on Figure 7-2. The value of $(x_A - y_A)$ is the equivalent of the length of line from a to b. Since the point x_A on the diagram represents the composition of L, the total amount of L is equivalent to the length of the line opposite its composition, hence the inverse. Equations (7-23) through (7-25) can be used to determine the composition and/or amount of vapor and liquid existing in a two-phase/two-component mixture.

If the mixture of composition z_A is heated beyond point C, eventually a temperature represented by B on the diagram will be reached. At this point all of the feed mixture has been vaporized and exists as a dew-point vapor. Further heating will cause the vapor to become superheated as indicated by point A on Figure 7-2.

Figure 7-2 is drawn at constant pressure. A similar diagram can be drawn at a constant temperature. For the constant-temperature diagram the pressure is changed in order to accomplish vaporization or condensation. A typical pressure–composition diagram is shown in Figure 7-4. The interpretation of

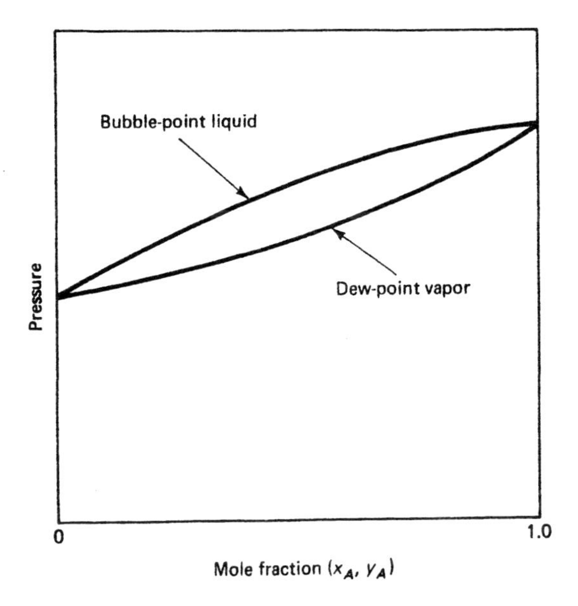

Figure 7-4 Typical P–x, y diagram

Figure 7-4 is exactly the same as for Figure 7-2, except that pressure is increased or decreased in order to change the condition of the mixture under consideration.

Another type of pressure–composition diagram is shown in Figure 7-5. In this case the mixture is assumed to follow Raoult's law, and the partial pressure of each component over the entire liquid composition range is a straight line. The total pressure over the liquid mixture is the sum of the two partial pressures. Under these conditions the distribution coefficient for the system can be predicted by combining Raoult's and Dalton's laws to give

$$\pi y_i = P_i^\circ x_i \tag{7-26}$$

or

$$K_i = \frac{y_i}{x_i} = \frac{P_i^\circ}{\pi} \tag{7-27}$$

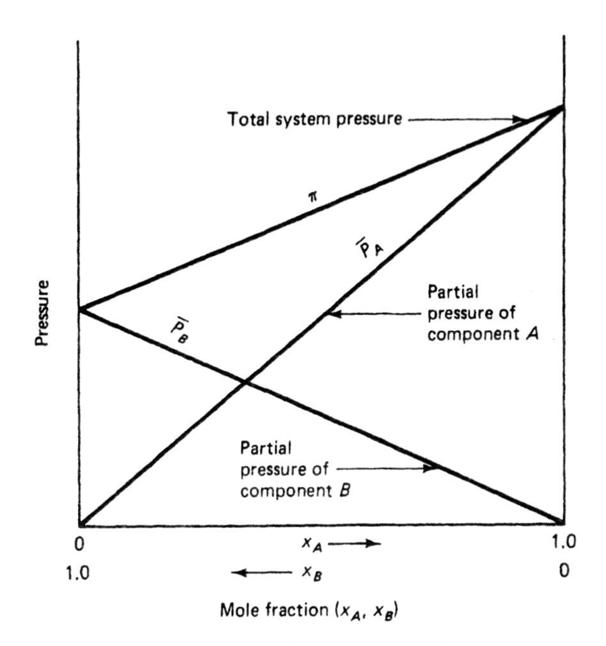

Figure 7-5 Typical P–x diagram for an ideal solution

where P_i° is the pure component vapor pressure. Vapor pressures can be obtained from a number of sources. Figure 7-6 shows a typical vapor pressure chart for a number of components.

Mixtures of gases form ideal solutions at pressures well above the pressure at which the same mixture will behave as an ideal gas. Dalton's law can be expected to hold up to pressures in the range 250 psi (1724 kPa) even though the mixture does not follow the ideal gas law at this pressure. At higher pressures vapor–liquid distribution coefficients become functions of temperature, pressure, and composition. Correlation and extrapolation of the coefficients under these conditions require the calculation of liquid and vapor fugacities for nonideal gases and liquid mixtures. These calculations require a complex equation of state and are beyond the scope of this treatment. The literature and thermodynamic textbooks are replete with methods and techniques for these procedures.

One simple graphical procedure that attempts the inclusion of temperature, pressure, and composition effects for paraffin hydrocarbons is the convergence pressure concept. This was originally presented by Hadden (1953) and does quite well for estimating purposes. Figure 7-7 shows a chart that permits us to estimate the convergence pressure for a hydrocarbon mixture. To use the chart, we consider the mixture to be a binary of the lightest component and the approximate weight average of the rest of the mixture in the liquid phase. We then read the convergence pressure at the temperature of concern from the curve corresponding to the two pseudo components. At pressures lower than 0.4 times the

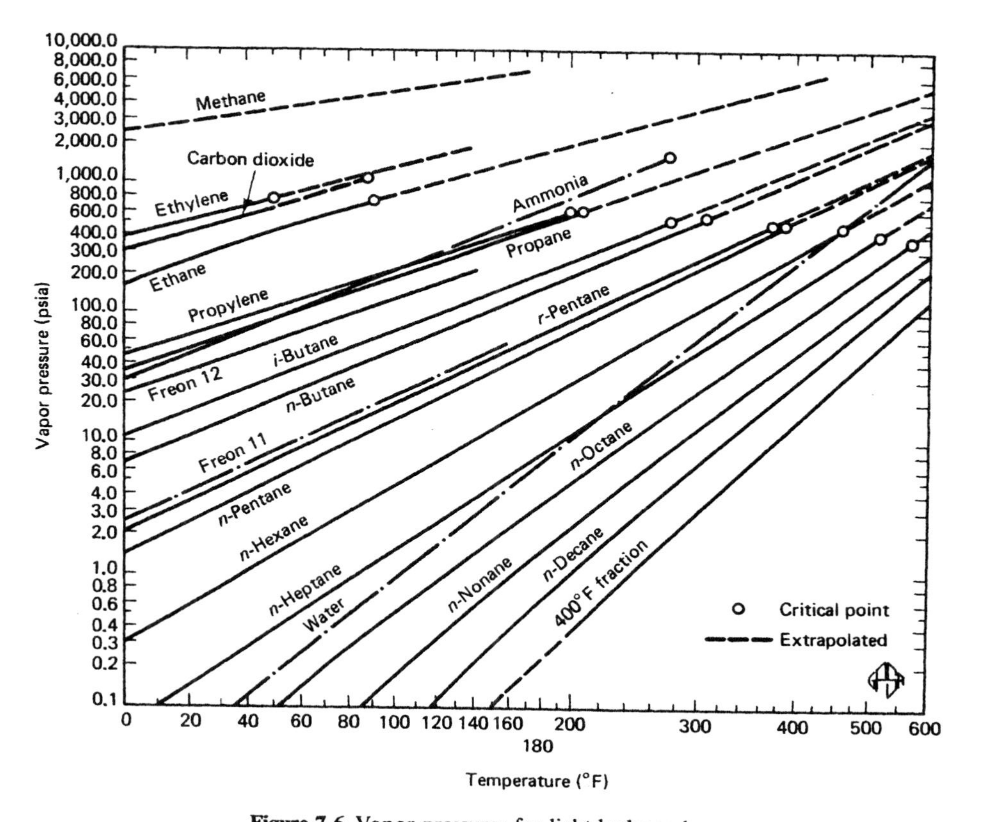

Figure 7-6 Vapor pressures for light hydrocarbons

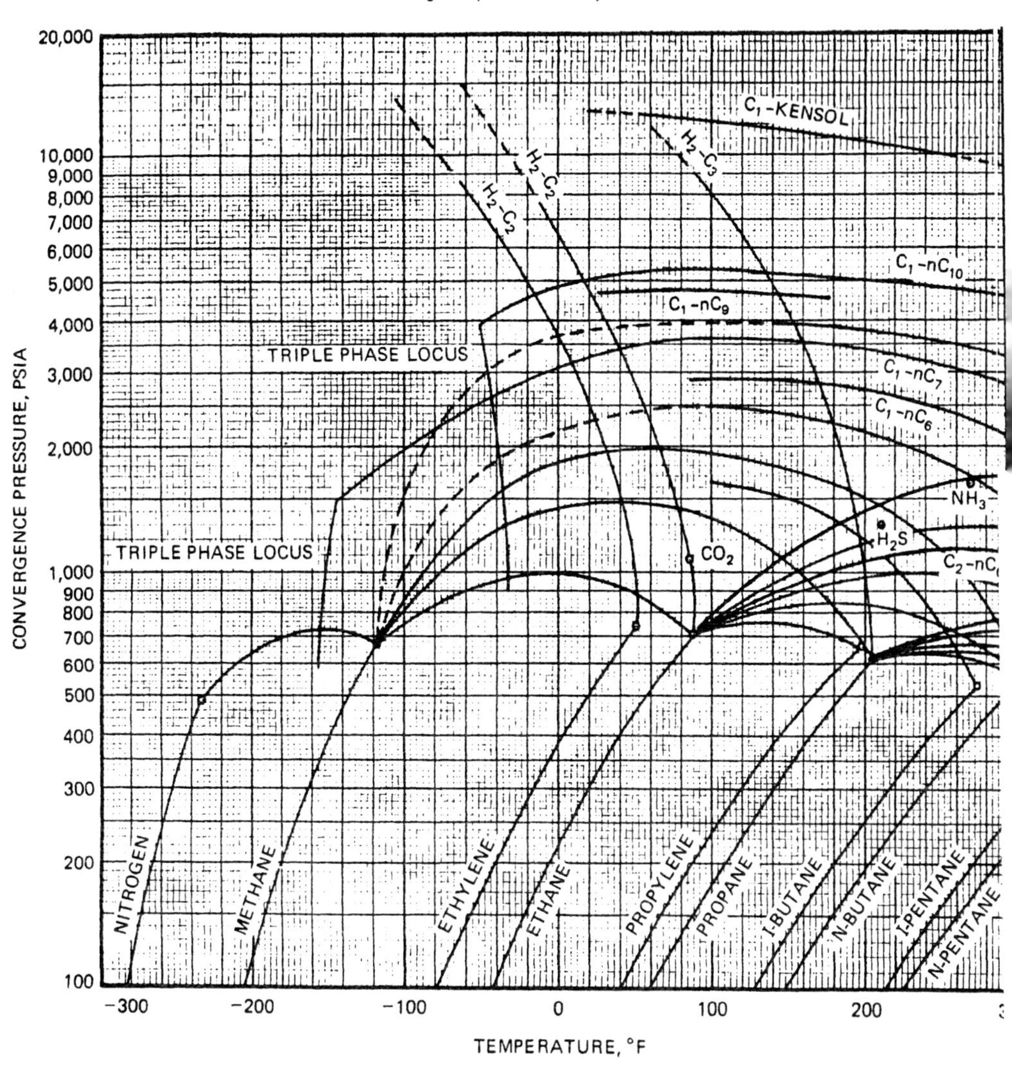

Figure 7-7 Convergence pressures for hydrocarbons

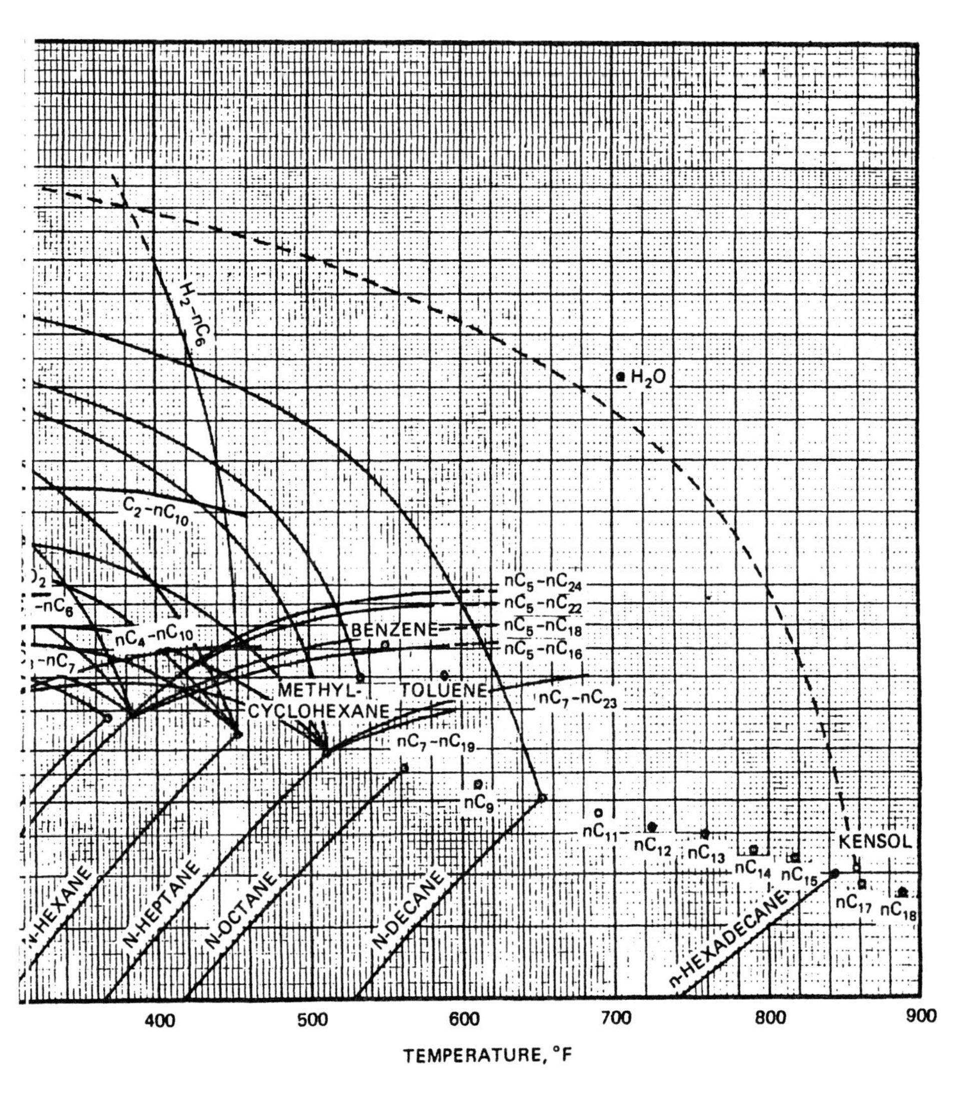

Figure 7-7 Continued

convergence pressure, the K values are essentially independent of the convergence pressure. For this reason, only an estimation of the convergence pressure is required. Figures C-1 through C-12 in Appendix C show vapor–liquid distribution coefficients for the paraffin hydrocarbons at a convergence pressure of 3000 psi (20.68 MPa).

Many other correlations for hydrocarbon vapor–liquid distribution ratios have been presented over the years. One of the easiest and most convenient to use is that presented by Hadden and Grayson (1961). The charts from the Hadden and Grayson work are shown in Figures 7-8 and 7-9. They are easily and readily used for determination of vapor–liquid distribution coefficients. The convergence pressure for the Hadden and Grayson charts is approximately 5000 psi (34.5 MPa). The original publication contains techniques for adjusting to other convergence pressures over the range 200 to 10,000 psi (1.38 to 68.95 MPa).

Most often today, the engineer will evaluate equilibrium constants and enthalpies for the paraffin hydrocarbons from an equation of state. As stated earlier, calculations of this nature are beyond the scope of this text. The generalized charts and curves presented in the appendices should check reasonably well with values for the light hydrocarbons calculated from an equation of state. For constituents other than the paraffin hydrocarbons, generalized procedures are not readily available for estimating equilibrium constants.

Example 7.1

Determine the equilibrium distribution coefficient at 100°F (38.1°C) and 1000 psia (6.89 MPa) for *n*-hexane. Use Figure C-10. From the equilibrium distribution coefficient, determine an apparent vapor pressure and compare this with the value determined from Figure 7-6. Repeat the calculation at 200°F (93.7°C) and 1000 psia (6.89 MPa).

Solution:

The K values, or equilibrium constants, as determined from Figure C-10 are:

Component	*Temperature* [°F (°C)]	*Pressure* [*psia* (*MPa*)]	K
n-Hexane	100 (38.1)	1000 (6.89)	0.0345
	200 (93.7)	1000 (6.89)	0.111

The apparent vapor pressure under these conditions is determined as follows:

At 100°F

$$K \simeq \frac{P^\circ}{\pi}$$

where P° = apparent vapor pressure,

$P^\circ_{100} = 1000(0.0345) = 34.5$ psia (237.9 kPa).

At 200°F $\quad P^\circ_{200} = 1000(0.111) = 111$ psia (165.5 kPa)

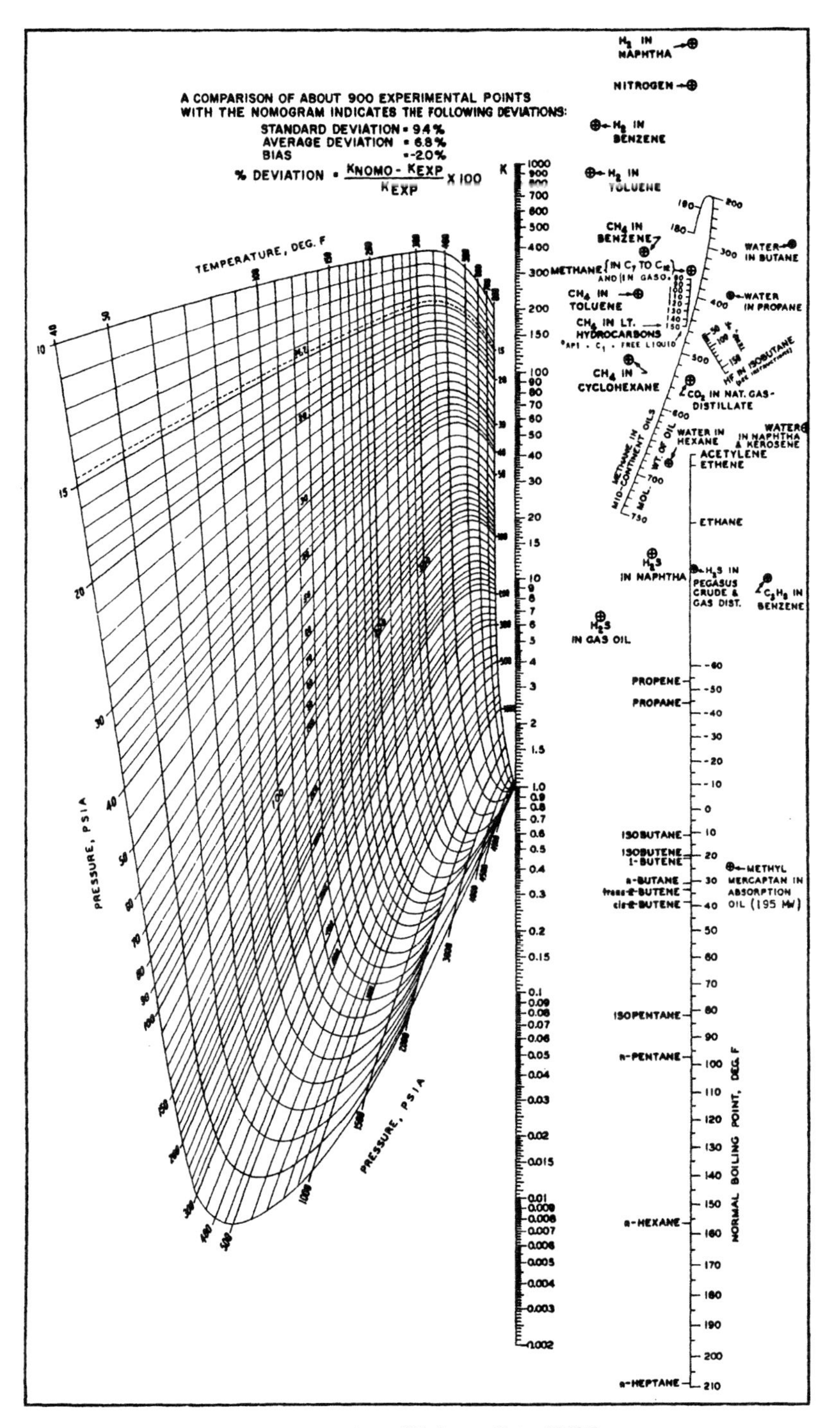

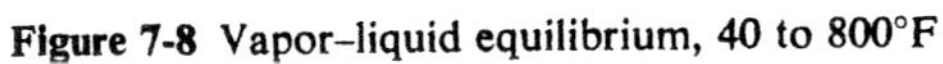
Figure 7-8 Vapor–liquid equilibrium, 40 to 800°F

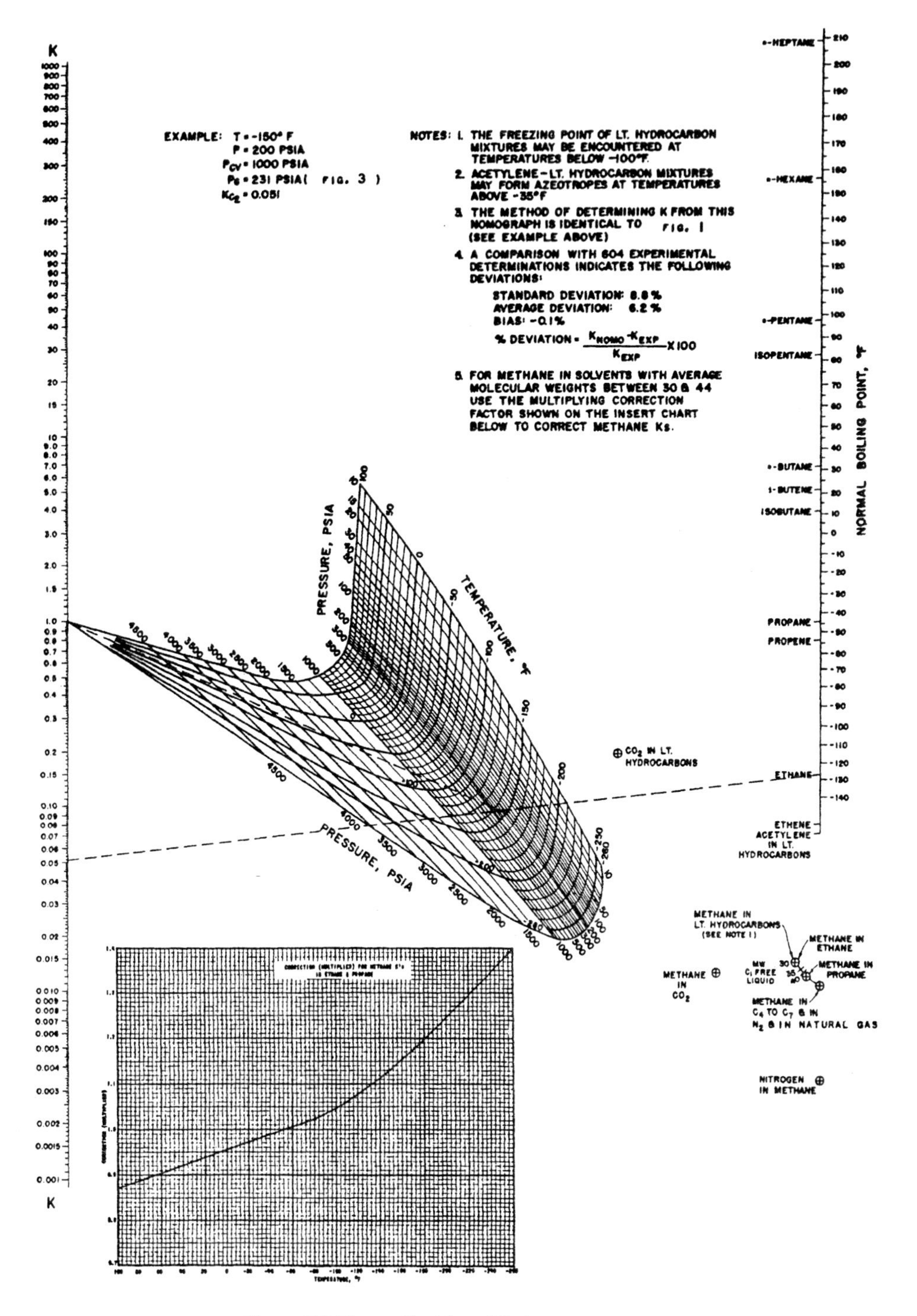

Figure 7-9 Vapor–liquid equilibrium, −260 to +100°F

Using Figure 7-6 the vapor pressures for n-hexane are

$$P^\circ_{100} = 5.0 \text{ psia (34.5 kPa)}$$

$$P^\circ_{200} = 30.5 \text{ psia (210.3 kPa)}$$

The vapor pressures read from Figure 7-6 are for the pure component at that temperature. The only pressure on the system is the vapor pressure of the component. The equilibrium constant charts (Figure C-10), on the other hand, consider the component to be a part of a mixture existing at the temperature and pressure specified. The larger apparent vapor pressures for n-hexane indicate that the component vapor pressure is not purely a function of temperature but is also affected (in this case increased) by a total system pressure which is greater than the vapor pressure of the pure component.

Example 7.2

Construct a temperature–composition diagram for the n-pentane/n-octane binary system at 1 atm pressure.

Solution:

The components involved are from an homologous series and form a nearly perfect solution in both the vapor and liquid phases. Because of the low pressure, vapor pressures may be used to determine K values by applying Raoult's law. The vapor pressure of n-pentane is 1 atm at 95°F (35.3°C). The vapor pressure of n-octane is 1 atm at 260°F (127°C). The temperature range to be covered by the temperature–composition diagram is 95 to 260°F. Vapor pressures for the two components may be determined from Figure 7-6. These vapor pressures may be converted to equilibrium distribution constants by

$$K_i = \frac{P_i^\circ}{\pi}$$

These calculations and the ones used to determine the points on the temperature–composition diagram are summarized in the following table.

T (°F)	$P^\circ_{nC_5}$ (psia)	K_{nC_5}	$P^\circ_{nC_8}$ (psia)	K_{nC_8}	$1 - K_{nC_8}$	$K_{nC_5} - K_{nC_8}$	x_{nC_5}	y_{nC_5}
130	27	1.84	1.3	0.088	0.912	1.752	0.521	0.958
165	48	3.27	2.6	0.177	0.823	3.093	0.266	0.870
200	76	5.17	5.2	0.353	0.647	4.817	0.134	0.694
235	110	7.48	8.0	0.544	0.456	6.936	0.066	0.492

The temperature–composition diagram for the n-pentane/n-octane system is shown in Figure 7-10. Additional points on the bubble-point and dew-point curves were calculated in order to find the proper shape of the curve.

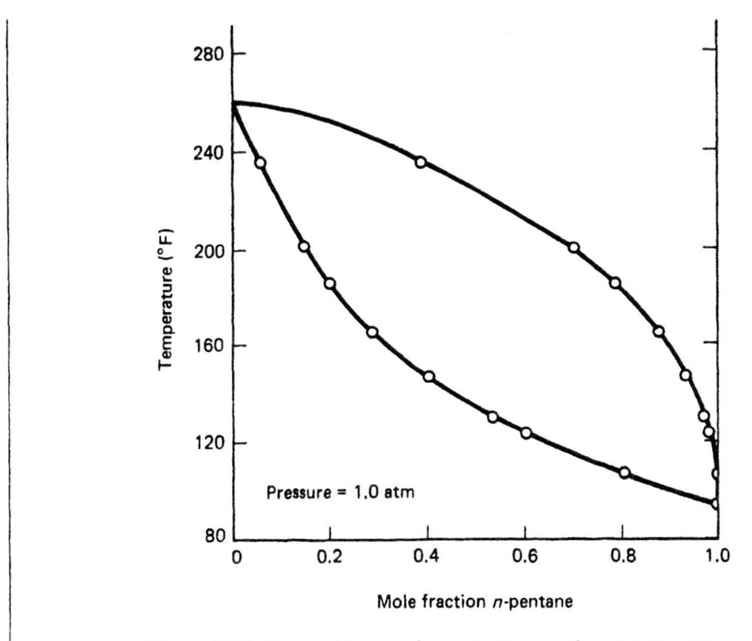

Figure 7-10 T–x, y diagram for n-pentane and n-octane system

Another type of equilibrium diagram that is frequently used in vapor–liquid separations is the so-called x–y diagram. This is developed from the temperature–composition or pressure–composition diagram by plotting liquid compositions on the x axis and the composition of the vapor that would be in equilibrium on the y axis. Figure 7-11 shows an x–y diagram for a typical binary mixture that exhibits essentially ideal behavior in both the liquid and vapor phases. Figure 7-11 can be used to determine the compositions of vapor and liquid in equilibrium.

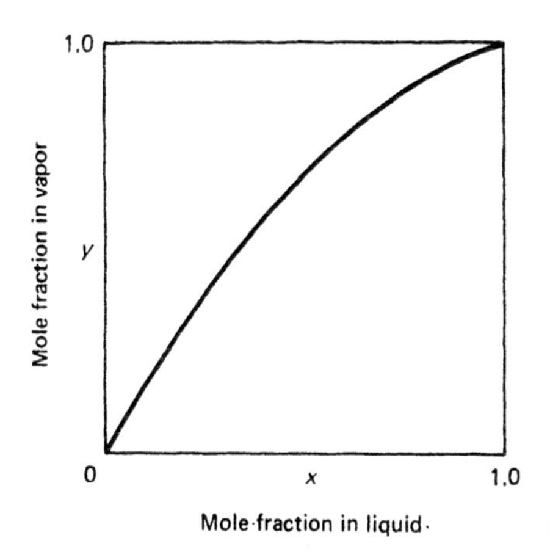

Figure 7-11 x–y diagram for a "normal" system

Nonideal behavior

Vapor-phase nonidealities are fairly regular and are relatively easily predicted on a generalized basis. Liquid-phase nonidealities are less regular and are much more difficult to predict. In extreme cases liquid-phase nonideality can be such as to reverse the volatility of the two components. When this occurs the system is referred to as *azeotropic* and the composition at which the volatility reversal occurs is referred to as the *azeotrope* or *azeotropic composition.*

Figure 7-12 shows the temperature–composition diagram at constant pressure for a binary system which forms a maximum boiling azeotrope. There

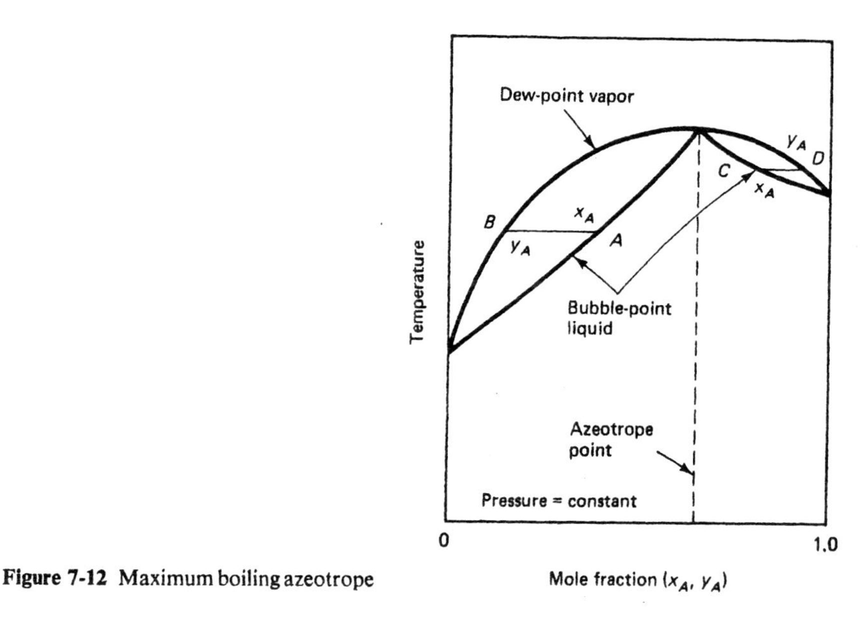

Figure 7-12 Maximum boiling azeotrope

are two vapor–liquid envelopes which meet at the azeotropic composition. Within each envelope, vapor–liquid equilibrium can be predicted from a horizontal tie line as in the case for the system exhibiting ideal behavior shown in Figure 7-11. However, one terminus of each envelope is at the azeotropic composition. The mixture is called a *maximum boiling* azeotrope because the boiling point of the constant composition mixture is higher than the boiling point of either pure component. At the azeotrope, or constant-composition point, both the vapor and the liquid in equilibrium have the same composition. Figure 7-13 shows the x–y diagram for the system and clearly indicates the volatility reversal at the azeotropic composition.

Figures 7-14 and 7-15 show the temperature–composition and x–y diagrams, respectively, for a system that exhibits a minimum boiling azeotrope. The behavior of this system is identical with that of the maximum boiling

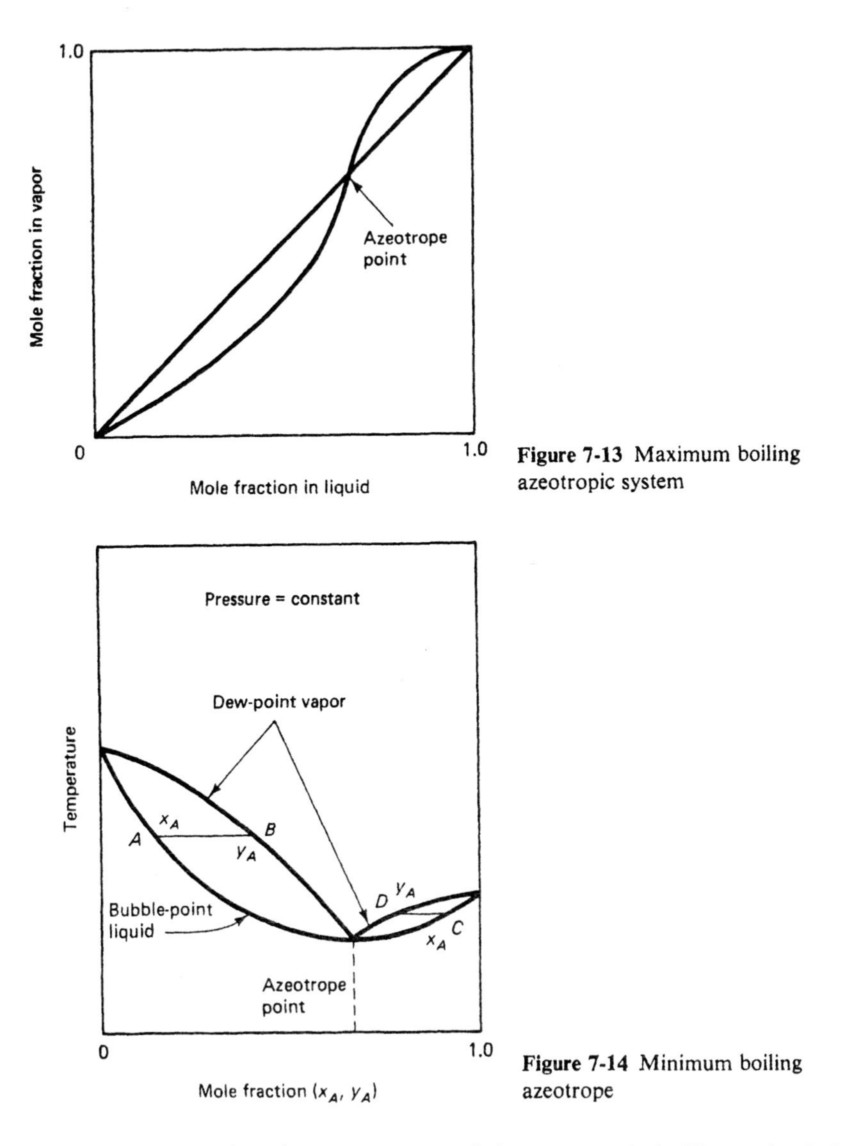

Figure 7-13 Maximum boiling azeotropic system

Figure 7-14 Minimum boiling azeotrope

azeotrope except that the temperature of the azeotropic boiling point is below the boiling point of either of the pure components.

Diagrams similar to Figures 7-12 and 7-14 can be developed at constant temperature using pressure as a variable. However, the temperature–composition diagram at constant pressure is by far the most common; the primary reason probably is that most separation operations are carried out at essentially constant pressure. Also, changing the pressure on a system may, in some cases, prohibit formation of an azeotrope.

There is no generally accepted, effective technique for correlating and/or

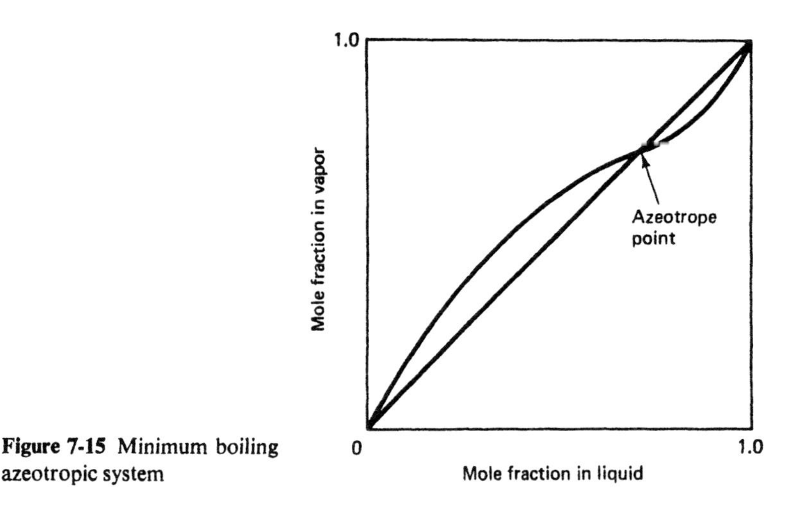

Figure 7-15 Minimum boiling azeotropic system

predicting distribution coefficients when the liquid phase is nonideal. The complexities raised when $\gamma \neq 1$ present a distinct and unsolved challenge to the solution thermodynamicist. Many industrial applications of stagewise operations are carried out at low pressures where the vapor phase may be assumed to be ideal in behavior. Even under these conditions, however, nonideal liquid behavior is common. For this reason, most of the attention directed to nonideal behavior has been directed toward solution behavior and the prediction of activity coefficients.

Techniques are available for extrapolating experimental data and/or testing data for thermodynamic consistency. One of the first procedures developed was that proposed by van Laar (1929). He assumed that the van der Waals equation of state applied to the pure components and mixture as both vapor and liquid, that the excess partial entropy of mixing of a component is zero, and that the partial molar volume change upon mixing is also zero. Under these restrictions he developed the following equations

$$T(\ln \gamma_1) = \frac{B}{[1 + A(x_1/x_2)]^2} \tag{7-28}$$

$$T(\ln \gamma_2) = \frac{AB}{[A + (x_2/x_1)]^2} \tag{7-29}$$

where A = the van Laar constant = b_1/b_2,
B = the van Laar constant = $(b_1/R)(a_1^{0.5}/b_1 - a_2^{0.5}/b_2)^2$,
$a = (27R^2T_{c,i}^2)/64P_{c,i}$,
$b = (RT_{c,i})/8P_{c,i}$.

The van Laar equations can be used in either of two ways: with experimental data to develop values for A and B and extrapolate or smooth the experimentally determined data points, or with critical temperatures and pressures of the components in the mixture to develop values for a and b so that values for A and B can be determined and used to predict distribution coefficients for the mixture. The first procedure is preferred. The second procedure should only be considered as a gross approximation and generally is not satisfactory for engineering calculations.

The van Laar equations are restricted in application to binary systems. Over the years many different equations of state have been used with essentially the same assumptions made by van Laar in order to develop similar types of equations. They all suffer from the same weakness—the inability to describe satisfactorily the behavior of the liquid state. Methods are available for extending extrapolation and consistency testing techniques for ternary and/or multicomponent systems. These techniques are complex in nature and are beyond the scope of this text.

7.5 Immiscible and Partially Miscible Systems

In the strictest sense of interpretation, immiscible liquid systems seldom exist. However, there are many instances in which the solubility of one constituent in the other is so small that, for practical purposes, the two liquids may be considered immiscible. Immiscible liquid systems do have practical significance and an understanding of their equilibrium behavior is necessary.

Figure 7-16 shows a pressure–temperature plot for two immiscible liquids. In the region of the plot where both components exist as liquids, each will exert its full pure-component vapor pressure. This means that the total pressure exerted by the system will be the sum of the two pure-component vapor pressures. This behavior is shown in that region of the graph from A to B. As the temperature increases the vapor pressure of each component increases, as does the total pressure exerted by the immiscible system. If the system is open so that the vapors that are formed can be removed when the total pressure exerted by the system

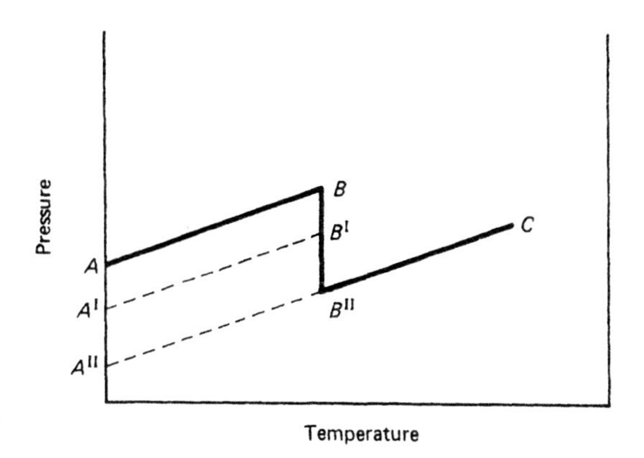

Figure 7-16 P–T plot for two immiscible liquids

becomes equal to the pressure of the surroundings, a temperature will eventually be reached at which vaporization of the two liquids begins to occur. At this point on the diagram, the pressure and temperature will remain constant until all of one or the other of the components (component I in this case) has been vaporized. When this occurs the pressure exerted by the system will immediately drop to the level of the vapor pressure of the remaining liquid constituent (section B-B^{II} of the solid line in Figure 7-16). If heat input to the system is continued, the temperature will then rise until the vapor pressure of the remaining constituent becomes equal to the total pressure on the system. At this temperature the remaining liquid component will begin to vaporize, as shown by point C in Figure 7-16.

Composition determinations in immiscible liquid systems are simple, once properly understood. The pressure exerted by one constituent is fixed at its vapor pressure value as long as there is any of that constituent present. The relative amount of that component present in the vapor formed can be determined from

$$y_A = \frac{P_A^\circ}{\pi} = \frac{P_A^\circ}{P_A^\circ + P_B^\circ} \tag{7-30}$$

where y_A = mole fraction of component A in the vapor,
P_A° = pure component vapor pressure of A,
P_B° = pure component vapor pressure of B,
π = total system pressure.

Because the two liquids are immiscible the vaporization behavior will be the same whether the system is considered to be open, such that vaporized material can flow from the system, or whether the system is considered to be in a closed container fitted with a movable piston. This is different from the behavior exhibited by miscible and partially miscible systems.

Example 7.3

A mixture of 75 kg of n-butane and 25 kg of water is heated in a closed container. (a) At what temperature will the vapor pressure exerted by the liquid be equal to 14.7 psia (101.35 kPa)? (b) At what temperature will the vapor pressure exerted by the liquid be equal to 50 psia (344.74 kPa)?

Solution:

Since n-butane and water are immiscible in the liquid phase, the solution to the problem consists of finding the temperature at which the sum of the vapor pressures of the two components is equal to the total system pressure.

(a) Total pressure = 14.7 psia. Assume that T = 30°F.

$$P_{nC_4}^\circ = 14.7 \text{ psia}$$

$$P_{H_2O}^\circ \cong 0$$

The temperature at which the pressure over the mixture will be 14.7 psia is 30°F.

(b) Total pressure = 50 psia. Assume that $T = 90°F$.

$$P^{\circ}_{nC_4} = 45 \text{ psia}$$

$$P^{\circ}_{H_2O} = 0.7 \text{ psia}$$

$$P^{\circ}_{nC_4} + P^{\circ}_{H_2O} = 45.7 \text{ psia} \neq 50 \text{ psia}$$

Assume that $T = 96°F$.

$$P^{\circ}_{nC_4} = 49 \text{ psia}$$

$$P^{\circ}_{H_2O} = 0.9 \text{ psia}$$

$$P^{\circ}_{nC_4} + P^{\circ}_{H_2O} = 49.9 \text{ psia} \approx 50 \text{ psia}$$

The temperature at which the pressure exerted by the mixture will be 50 psia is 96°F.

The condensation behavior of a vapor composed of two constituents, which when condensed form immiscible liquids, is shown in Figure 7-17.

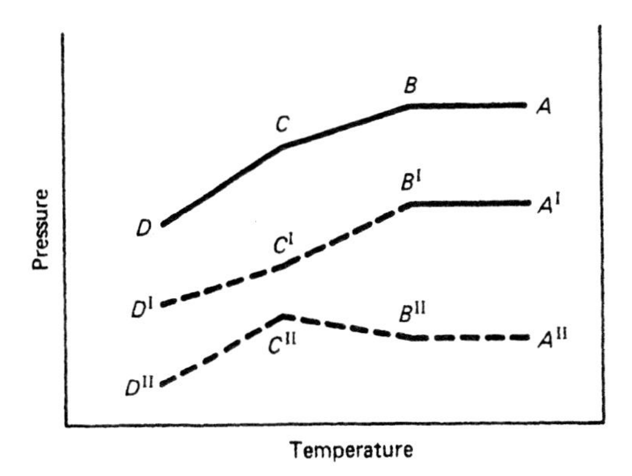

Figure 7-17 *P–T* plot for condensation of vapor of two immiscible liquids

Beginning at point A the two constituents are completely vaporized. Each exerts its partial pressure in the vapor phase, which is dependent on the relative amount of that component in the vapor. Thus

$$\bar{P}_A = y_A \pi \tag{7-31}$$

where $\bar{P}_A$ is the partial pressure of component A in the gas phase. As the temperature is lowered the pressure on the system will remain constant until some temperature is reached at which one of the constituents will have a partial pressure in the gas phase equal to its pure component vapor pressure. At this point that constituent will begin to condense (point B). As that constituent begins to condense its relative concentration in the vapor phase will decrease. This causes a decrease in the partial pressure of that constituent. Condensation of the one component will continue until a point C is reached, where the partial pressure of the other component also becomes equal to its pure component vapor pressure. At this temperature all of the remaining vapor will condense.

Once the vapor is totally condensed, the system pressure will then decrease with temperature as the two vapor pressures decrease (point C to D).

Example 7.4

If heat input to the n–butane–water mixture discussed in Example 7.3 is continued and the mixture is allowed to vaporize, which component will be the first to be completely vaporized at (a) 14.7 psia (101.35 kPa), and (b) at 50 psia (344.74 kPa)?

Solution:

$$\text{mole weight } nC_4 = 58; \quad nC_4 = \frac{75}{58} = 1.29 \text{ kgmol}$$

$$\text{mole weight } H_2O = 18; \quad H_2O = \frac{25}{18} = \underline{1.38} \text{ kgmol}$$

$$\text{Total } 2.67$$

$$\text{mole fraction } nC_4 = \frac{1.29}{2.67} = 0.483$$

$$\text{mole fraction } H_2O = \frac{1.38}{2.67} = 0.517$$

(a) At 30°F (the temperature at which the pressure of the mixture is equal to 14.7) the vapor is essentially all n-butane. This means that the n-butane will be the first component to vaporize completely.

(b) Pressure = 50 psia. The vapor generated has the composition

$$\text{mole fraction } nC_4 = \frac{49.1}{50} = 0.982$$

$$\text{mole fraction } H_2O = \frac{0.9}{50} = 0.018$$

Since the vapor generated over the solution is much richer in n-butane than is the combined liquid composition, the n-butane will also be the first component completely vaporized at 50 psia.

Example 7.5

Assume that the n–butane–water mixture is completely vaporized. At what temperature will condensation begin, and what will be the composition of the first drop of liquid formed (a) at 14.7 psia (101.35 kPa) and, (b) at 50 psia (344.74 kPa)?

Solution:

(a) The partial pressures of the two components in the vapor phase at 14.7 psia total pressure are

$$\bar{P}_{nC_4} = 0.483 \times 14.7 = 7.1 \text{ psia}$$

$$\bar{P}_{H_2O} = 0.517 \times 14.7 = 7.6 \text{ psia}$$

From Figure 7-5,

$$P^\circ_{nC_4} = 7.1 \text{ psia at } 0°\text{F}$$

$$P^\circ_{H_2O} = 7.6 \text{ psia at } 180°\text{F}$$

Condensation will begin at 180°F. Water will condense first and the first drop will be 100% water.

(b) The partial pressures at a total pressure of 50 psia are

$$\bar{P}_{nC_4} = 0.483 \times 50 = 24.2 \text{ psia}$$

$$\bar{P}_{H_2O} = 0.517 \times 50 = 25.8 \text{ psia}$$

From Figure 7-6,

$$P^\circ_{nC_4} = 24.21 \text{ psia at } 34°\text{F}$$

$$P^\circ_{H_2O} = 25.8 \text{ psia at } 238°\text{F}$$

Condensation will begin at 238°F. Water will be the first component to condense and the first drop will be 100% water.

Variations of Figures 7-16 and 7-17 can be prepared when the pressure varies and the temperature is held constant or when the temperature varies and the pressure is held constant. Also, cases can be encountered in which one of the immiscible phases contains more than one constituent. These different permutations and combinations of immiscible liquid behavior can be easily and quickly assimilated if the foregoing discussion of the behaviors illustrated in Figures 7-16 and 7-17 is clearly understood.

A number of binary systems exist which exhibit partial miscibility. Over a portion of the composition range the two components are miscible with each other and we can predict their vapor–liquid composition behavior in the same way as for any miscible system. In another portion of the composition range the two components are immiscible and in this range the laws of immiscible liquids are followed. Figures 7-18 and 7-19 are the temperature–composition and x–y diagrams for a system exhibiting partial miscibility at both ends of the composition range. In Figure 7-18 the two equilibrium envelopes indicate the region in which the laws of equilibrium for miscible mixtures apply. A horizontal line in either of these regions represents equilibrium between the indicated liquid and vapor compositions. In the region of liquid composition between R and T on Figure 7-18 the two liquids are immiscible. Over this range of liquid compositions the vapor formed is dependent only on the ratio of the vapor pressures of the two liquid components and is indicated in this case by the dashed line $\overline{UV}$. The constancy of vapor composition with varying liquid composition is graphically demonstrated on the x–y diagram in Figure 7-19.

In general, the liquid–liquid systems utilized in separation processes exhibit partially miscible behavior. The presentation of the equilibrium compositions is different from those shown earlier for liquid–vapor systems because the distribution is between two liquid phases. The general shape of a temperature–composition diagram in a partially miscible liquid–liquid system is shown in

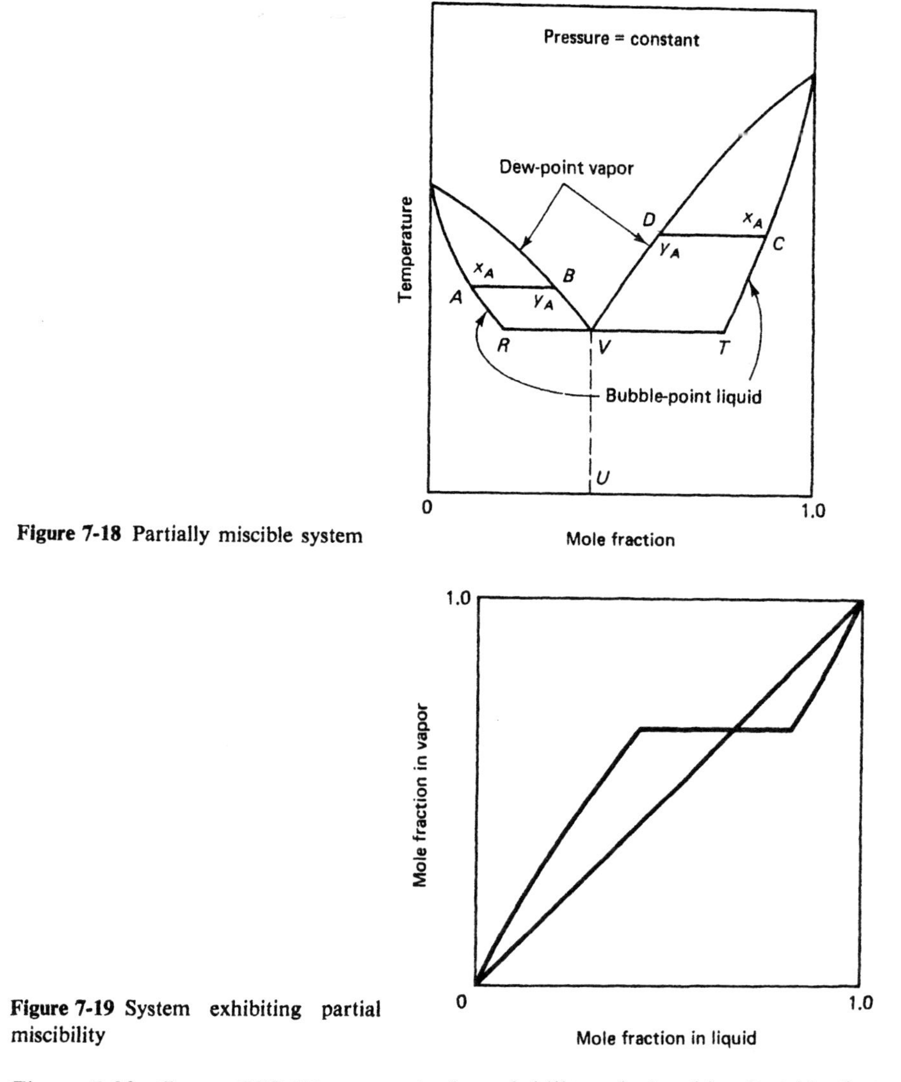

Figure 7-18 Partially miscible system

Figure 7-19 System exhibiting partial miscibility

Figure 7-20. Curve *CGDJE* represents the solubility relationship. Outside that curve the components are soluble in all proportions and completely miscible one with the other. Inside the curve two phases are formed with equilibrium compositions represented by a horizontal tie line *GJ*. The point at *D* is termed the *critical solution temperature*. The system in Figure 7-20 demonstrates a maximum critical solution temperature. Other liquid systems may exhibit a minimum critical solution temperature as shown in Figure 7-21. In some systems both a lower and an upper critical solution temperature exist, as shown in Figure 7-22.

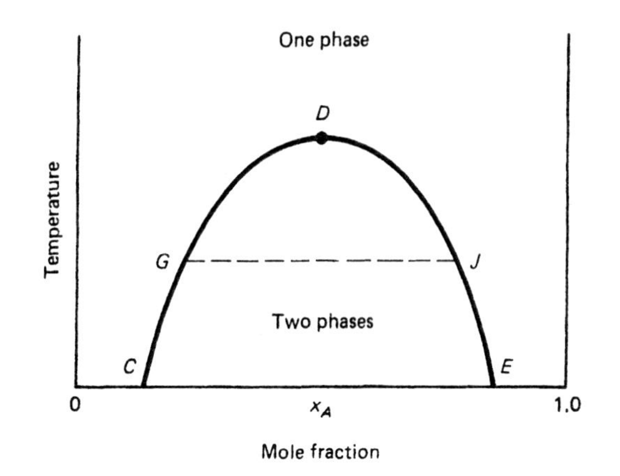

Figure 7-20 T–x plot for two partially miscible liquids exhibiting a maximum critical solution temperature

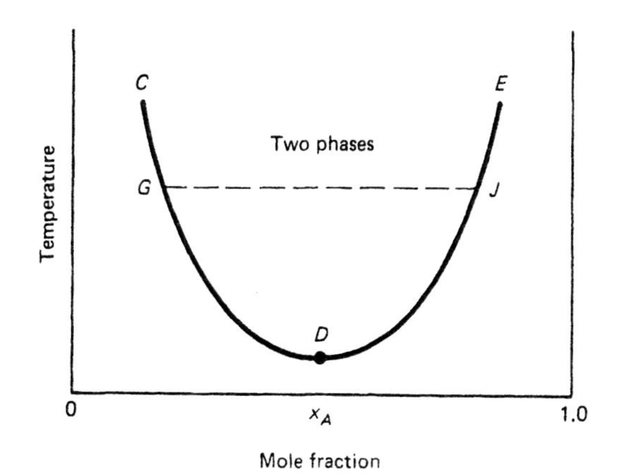

Figure 7-21 T–x plot for two partially miscible liquids exhibiting a minimum critical solution temperature

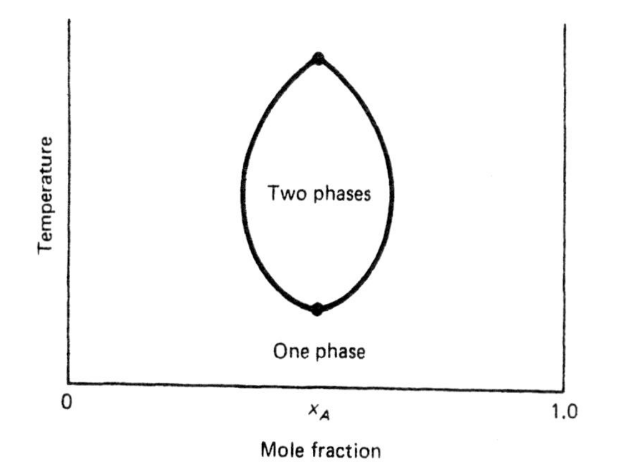

Figure 7-22 T–x plot for two partially miscible liquids exhibiting a minimum and maximim solution temperature

The liquid systems of interest in separation processes are usually three component systems. Two constituents will exhibit some mutual solubility as shown in Figures 7-20 through 7-22. In addition, a solute that is being transferred from one liquid phase to the other will be soluble in both liquids. In order to accommodate these three composition parameters the phase diagrams are ordinarily plotted on triangular coordinates as shown in Figure 7-23. Each apex

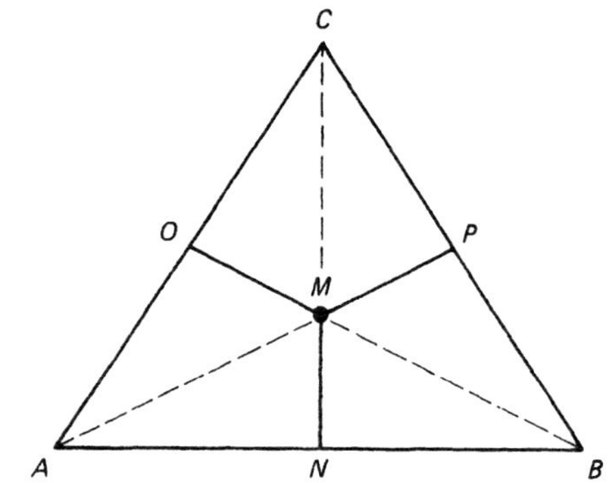

Figure 7-23 Determining compositions on a triangular diagram

of the triangular plot represents the pure component indicated. The side opposite the apex represents zero concentration of that component or a binary mixture of the two components connected by the baseline. At point A, for example, the solution would be 100% component A; the line connecting C and B represents 0% A, and a binary mixture of B and C. To determine the composition represented by a point such as point M, we draw perpendiculars to each side of the diagram. The amount of component C in Figure 7-23 is determined by drawing a line segment from point M to line AB (line $\overline{MN}$); the percentage of C in mixture M is $\overline{NM}/\overline{NC}$. In similar fashion we may determine the percentage of B in the mixture from the line segment $\overline{MO}$ and the percentage of A from the line segment $\overline{MP}$. Figure 7-24 shows an equilibrium phase diagram for a three-component mixture plotted on triangular graph paper. Within the envelope a

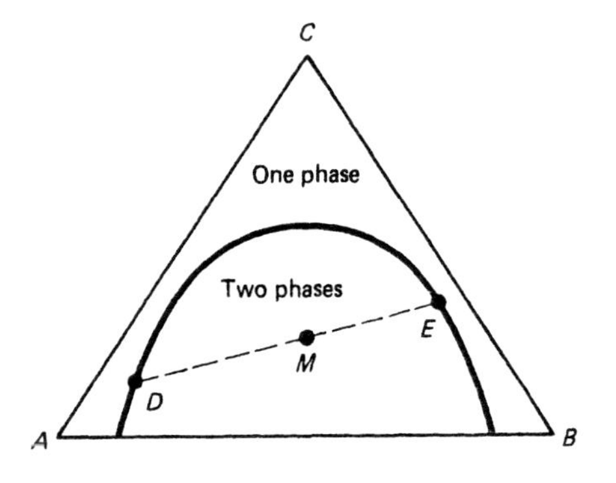

Figure 7-24 Typical two-phase region on a triangular diagram

mixture of composition M would separate into two phases, D and E, with the line $\overline{DE}$ representing the equilibrium tie line for the system.

Liquid–liquid equilibrium data for ternary systems can also be plotted on rectangular coordinate graph paper. A typical plot for a partially miscible ternary system is shown in Figure 7-25. The phase envelope and the equilibrium tie line $\overline{DE}$ have the same significance as on the equilateral triangular plot. The difference is that the third component composition is represented by lines parallel to the hypotenuse of the triangle as shown.

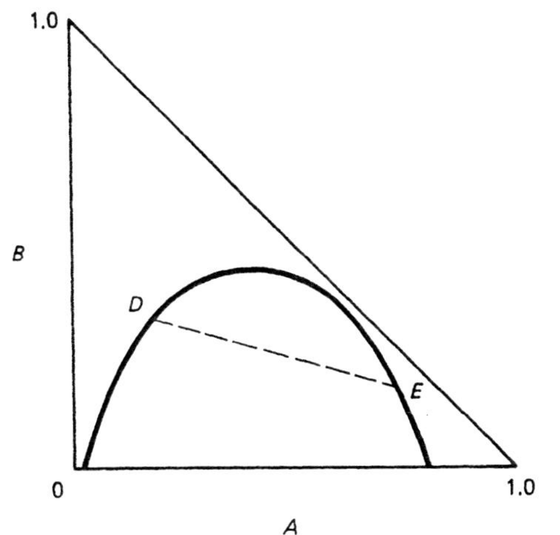

Figure 7-25 Typical two-phase region on a rectangular coordinate diagram

7.6 Enthalpy–Composition Diagrams

In many separation operations enthalpy is involved as an important variable. An enthalpy–composition diagram can be generated in essentially the same way as a temperature–composition diagram. The compositions of the coexisting equilibrium liquid and vapor phases must be calculated. The enthalpy of each phase is determined and the resulting values are plotted as shown in Figure 7-26. The composition axis is for both vapor and liquid. The enthalpies of the vapor and liquid are different because of the latent heat of vaporization.

The inverse lever rule can be shown to apply to the enthalpy–composition diagram as it does to the temperature–composition diagram. First, we write an overall material balance for the separator shown in Figure 7-2.

$$F = V + L \tag{7-22}$$

The overall heat balance is

$$h_F F = HV + hL \tag{7-32}$$

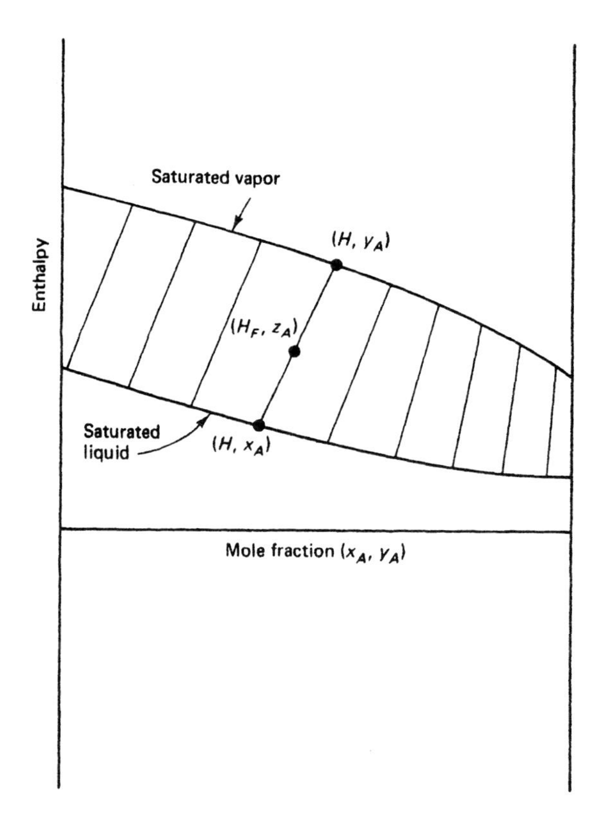

Figure 7-26 Typical enthalpy–composition diagram

where h_F = feed enthalpy,
H = vapor enthalpy,
h = liquid enthalpy.

The enthalpies and flow rates in the balance equations must be written with consistent units. If flow rates are in moles, enthalpies must be in cal/mole.

Eliminating one of the three variables F, V, or L between the two equations, we obtain

$$h_F F = H(F - L) + hL \tag{7-33}$$

or

$$\frac{L}{F} = \frac{h_F - H}{h - H} \tag{7-34}$$

In a fashion similar to that used for the derivation of Eq. (7-23), we can also derive a composition relationship for L and F. Equations (7-23) and (7-34) show that the inverse lever rule applies to measurements on the enthalpy–composition diagram in the same way that it does on the temperature–composition diagram.

Relationships for V/F and L/V can be derived in terms of the same basic variables depending on which variable is eliminated between the overall material balance and the overall heat balance.

7.7 Equilibrium Calculations for Vapor–Liquid Systems

Relatively few multicomponent systems exist for which completely generalized equilibrium data are available. The most widely available data are those for vapor–liquid systems and these are frequently referred to as vapor–liquid equilibrium distribution coefficients or K values. Since the K values vary considerably with temperature and pressure, the selectivity that is equal to the ratio of the K values is used. For vapor–liquid systems this is typically referred to as the *relative volatility* and is expressed for a binary system as

$$\frac{K_A}{K_B} = \alpha = \frac{y_A x_B}{x_A y_B} \tag{7-35}$$

The relative volatility is a useful term because it tends to change less with temperature than does the equilibrium constant. If we write Eq. (7-35) in terms of the more volatile component A, then

$$y_A = \frac{\alpha x_A}{1 + (\alpha - 1)x_A} \tag{7-36}$$

This equation relates equilibrium compositions y_A and x_A in terms of the relative volatility. If the assumption is made that α is independent of temperature and composition, then Eq. (7-36) becomes the equation of the equilibrium line or y–x curve, as shown in Figure 7-11. In the discussion that follows, most of the examples utilized will be from vapor–liquid systems, although the concepts are more widely applicable. The universal tools employed are the material balance and the composition distribution at equilibrium.

Bubble point calculation

The bubble point of a liquid mixture is defined as that combination of temperature and pressure at which the first bubble of vapor forms. The amount of vapor formed is assumed to be so small as to have no effect on the liquid composition. The equation for the bubble point composition of a binary mixture was developed in Section 7.4 without further explanation. For a binary system we have

$$x_1 = \frac{1 - K_2}{K_1 - K_2} \tag{7-37}$$

and

$$x_2 = 1.0 - x_1$$

In many instances there will be more than two components in the liquid mixture. For this case no direct solution such as Eq. (7-37) exists.

The basic definition of the equilibrium coefficient is

$$y_i = K_i x_i \tag{7-38}$$

The condition (by definition) is that in any stable system the mole fractions must sum to unity. Thus

$$\sum y_i = 1.0 = \sum K_i x_i \tag{7-39}$$

This becomes the basic criterion for establishing the bubble-point temperature and pressure. In the procedure for using Eq. (7-39), we assume that either the temperature or the pressure is fixed. We then vary the other parameter (this changes the value for the equilibrium distribution coefficient) until the criterion for a stable system is satisfied. If the summation of the calculated vapor compositions is different from unity, we must change the combination of temperature and pressure. There is no known direct procedure that will permit a reasonable estimation of the amount of change required. There are, however, techniques for minimizing the number of trials that we must make.

Figure 7-27 shows one technique for minimizing the number of trial calculations necessary to determine the bubble point. Pressure is assumed constant, with temperature being the assumed variable for determining the distribution coefficient. Three levels of temperature are assumed, which bracket the bubble-point temperature. A smooth curve is then drawn through these three points and the interpolated value for the bubble-point temperature is used in the final calculation. If the first two assumed temperatures bracket the bubble-point temperature, Figure 7-27 can be based on two assumed values. The line connecting the points on Figure 7-27 is not a straight line, so the temperature points must be reasonably close together to provide a good estimate of the bubble point.

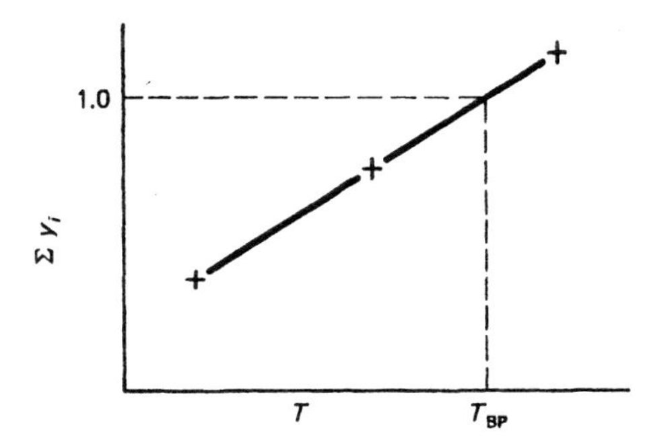

Figure 7-27 Estimating temperature for a bubble-point or dew-point calculation

Example 7.6

For a mixture with the composition shown below, calculate the bubble-point temperature at 100 psia (689.6 kPa). Use the extrapolation–interpolation method shown in Figure 7-27.

Component	*Mol%*
C_3	22.0
iC_4	21.6
nC_4	24.7
iC_5	19.2
nC_5	12.7

Solution:

The criterion for the bubble point is that $\sum y_i = \sum K_i x_i = 1.0$. To determine the bubble point, a temperature must be assumed, the K values determined, and the summation of the vapor mole fractions checked against the criterion of unity.

Assume a temperature of 140°F (60°C).

Component	x_i	K_i	$x_i K_i = y_i$
C_3	0.220	2.75	0.605
iC_4	0.216	1.23	0.266
nC_4	0.247	0.94	0.232
iC_5	0.192	0.44	0.0845
nC_5	0.127	0.35	0.0445
			1.2320

This temperature is obviously too high, so we assume a lower temperature, say 120°F (48.9°C).

Component	x_i	K_i	$x_i K_i = y_i$
C_3	0.220	2.2	0.484
iC_4	0.216	0.98	0.212
nC_4	0.247	0.72	0.178
iC_5	0.192	0.328	0.063
nC_5	0.127	0.26	0.033
			0.970

Since these two values bracket the bubble-point temperature, a graph similar to Figure 7-27 can be constructed to assist in the determination of the correct bubble-point temperature. A temperature of 122°F (50°C) is indicated in Figure 7-28. Using this as a basis to determine a better estimate for the K values, we make the following bubble-point calculation.

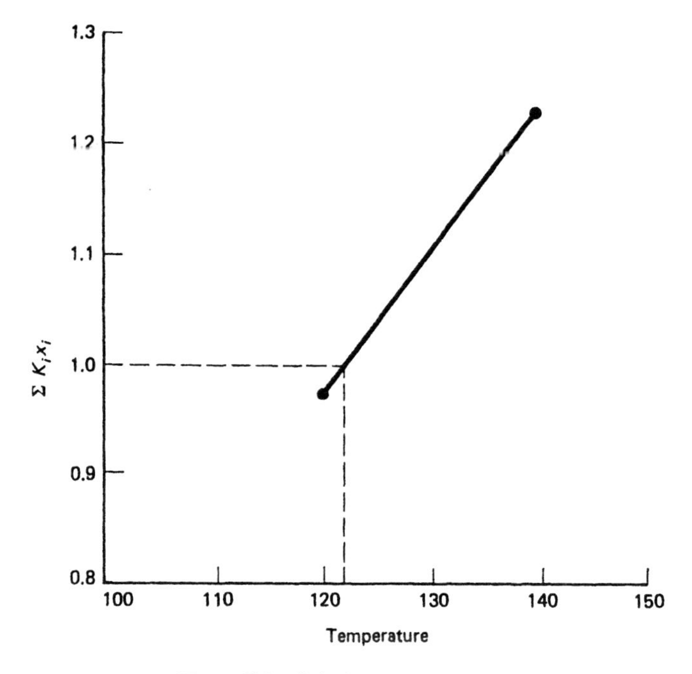

Figure 7-28 Solution to Example 7.6

Component	x_i	K_i	$x_iK_i = y_i$
C_3	0.220	2.26	0.497
iC_4	0.216	1.01	0.218
nC_4	0.247	0.744	0.1838
iC_5	0.192	0.346	0.0661
nC_5	0.127	0.276	0.0351
			1.0000

The bubble-point temperature is 122°F (50°C) since $\sum x_iK_i = 1.0$.

Another technique for minimizing the number of trials is one suggested by Dodge (1944). His procedure is based on the assumption that the calculated vapor-phase composition, when normalized, will represent a good estimate of the vapor composition at the correct bubble point. The vapor composition is normalized as follows:

$$y_i' = \frac{y_i}{\sum y_i} \tag{7-40}$$

where y_i = mole fraction of component i in the vapor phase as calculated at the trial bubble-point condition,

y_i' = normalized vapor composition ($\sum y_i' = 1.0$).

After the normalized compositions have been calculated, we select a reference component and the K value at the bubble point for that component is estimated. The criteria for selecting the reference component are as follows:

1. The component represents a substantial contribution to the vapor-phase composition.
2. The rate of change of the K value of the reference component with respect to temperature or pressure should be large.

These criteria will usually lead us to selecting one of the lighter components of the mixture as the reference component. The corrected K value is then calculated by

$$K_i' = \frac{y_i'}{x_i} \tag{7-41}$$

where y_i' = corrected mole fraction calculated by Eq. (7-40),
x_i = composition of the original liquid mixture whose bubble point is sought.

After we calculate the corrected K value for the reference component, we will use charts such as Figures C–1 through C–12 to determine the temperature (or pressure) that will give this K value for the component at the fixed pressure (or temperature). K values for the other components in the mixture are determined at the temperature (or pressure). The procedure recommended by Dodge will usually close to a satisfactory bubble-point temperature (or pressure) in about four trials. The procedure will be clarified by studying the following example.

Example 7.7

At 200 psia (1.38 MPa), calculate the bubble-point temperature of the following mixture using the procedure recommended by Dodge.

Component	*Mol%*
iC_4	18.2
nC_4	23.8
iC_5	33.7
nC_5	12.1
C_6	12.2

Solution:

Following the procedure outlined, we assume a temperature and then calculate the bubble point. In the case of the first temperature assumption of 260°F (126.7°C), the temperature is too high. *i*-pentane is taken as the reference component and the mole fraction of i-pentane in the vapor phase is normalized to what the projected value would be if the sum of the mole fractions was unity. The equilibrium constant under these conditions is determined and the temperature is found for conditions at which the K value for i-pentane is the value indicated. This procedure is repeated until the final calculation shows a temperature of 237°F (113.9°C) as the bubble-point temperature for the mixture. This value is found after four trials.

		Trial 1, T = 260°F		*Trial 2, T = 240°F*		*Trial 3, T = 235°F*		*Trial 4, T = 237°F*	
Component	*Mole Fraction*	K_i	$y_i = x_iK_i$	K_i	$y_i = x_iK_i$	K_i	$y_i = x_iK_i$	K_i	$y_i = x_iK_i$
iC_4	0.182	1.92	0.3495	1.69	0.3075	1.62	0.2945	1.65	0.300
nC_4	0.238	1.58	0.376	1.39	0.331	1.35	0.3215	1.35	0.3215
iC_5	0.337	0.93	0.3135	0.78	0.263	0.76	0.256	0.77	0.2595
nC_5	0.121	0.81	0.0980	0.66	0.0799	0.64	0.0775	0.64	0.0775
C_6	0.122	0.42	0.0513	0.33	0.0403	0.315	0.0385	0.32	0.0391
			1.1883		1.0217		0.9880		0.9976

$y'_{iC_5} = \frac{0.3135}{1.1883} = 0.264$, $K'_{iC_5} = 0.784$

$y'_{iC_5} = \frac{0.0263}{1.0217} = 0.257$, $K'_{iC_5} = 0.763$

$y'_{iC_5} = \frac{0.256}{0.988} = 0.259$, $K'_{iC_5} = 0.77$

Since $\sum y_i = \sum x_iK_i \approx 1.0$, the bubble-point temperature is 237°F.

When calculating either the bubble point or the dew point, the question naturally arises as to how close the sum of the calculated mole fractions should approach unity. The answer to this varies somewhat with conditions. However, the following criterion is usually satisfactory for hand calculations in which K values were obtained from the charts or figures.

$$0.995 \leq \sum K_ix_i \leq 1.005 \tag{7-42}$$

Dew-point calculation

The dew point of a vapor is defined as that combination of temperature and pressure at which the first drop of liquid condenses. The first drop of liquid can be interpreted as the formation of a differential amount of liquid such that the overall composition of the remaining vapor is unchanged. The dew-point criterion is similar in principle and in interpretation to the bubble-point criterion. The dew-point criterion is

$$\sum x_i = 1.0 = \sum \frac{y_i}{K_i} \tag{7-43}$$

For binary mixtures we can develop an equation similar to Eq. (7-38). Thus

$$y_1 = \frac{K_1(1 - K_2)}{K_1 - K_2} \qquad (7\text{-}44)$$

and

$$y_2 = 1 - y_1$$

These two equations can be used to calculate directly the composition of a binary vapor that will begin to condense at a given temperature and pressure.

The calculation for systems containing more than two components is similar to that presented for calculating bubble-point conditions, the sole difference being that a liquid composition is calculated and tested for stability. A plot similar to Figure 7-27 can also be used for estimating dew-point conditions.

Example 7.8

Calculate the dew point of the mixture in Example 7.7 if the pressure is 50 psia (344.8 kPa).

Solution:

This is a trial-and-error solution which involves assuming a temperature, calculating the composition of the liquid, and checking against the dew-point criterion $\sum x_i = \sum y_i/K_i = 1.0$. The dew-point temperature is reached when the criterion is satisfied.

Assume a temperature of 100°F (37.8°C).

Component	y_i	K_i	$y_i/K_i = x_i$
C_3	0.22	3.35	0.0657
iC_4	0.216	1.40	0.1543
nC_4	0.247	1.02	0.2420
iC_5	0.192	0.445	0.431
nC_5	0.127	0.335	0.379
			1.2720

This temperature is too low, so we assume a temperature of 120°F (48.9°C).

Component	y_i	K_i	$y_i/K_i = x_i$
C_3	0.22	4.10	0.0537
iC_4	0.216	1.82	0.1187
nC_4	0.247	1.33	0.1857
iC_5	0.192	0.59	0.3254
nC_5	0.127	0.47	0.2702
			0.9537

Plotting as before and interpolating, a temperature of 117°F (47.2°C) is used for the third trial. Thus we have

Component	y_i	K_i	$y_i/K_i = x_i$
C_3	0.22	4.0	0.055
iC_4	0.216	1.75	0.1234
nC_4	0.247	1.28	0.1930
iC_5	0.192	0.56	0.3429
nC_5	0.127	0.45	0.2822
			0.9965

Since the sum of the mole fractions for x_i is close to unity, the dew-point temperature of the mixture at 50 psia is 117°F.

The procedure recommended by Dodge can also be adapted to dew-point calculations. In this case, however, we normalize the calculated liquid-phase composition and assume this to be a good estimate of the final dew-point liquid composition.

$$\sum x_i = \sum \frac{y_i}{K_i}$$

and

$$x'_i = \frac{x_i}{\sum x_i} \tag{7-45}$$

where y_i = composition of original vapor,
x_i = liquid composition calculated at assumed temperature and pressure,
x'_i = normalized liquid composition.

A single reference component is selected according to the following criteria:

1. The selected component represents a substantial contribution to the calculated liquid-phase composition.
2. The rate of change of the K value of the component with respect to temperature or pressure should be large.

The pseudo K value for the component is then calculated according to

$$K'_i = \frac{y_i}{x'_i} \tag{7-46}$$

where K'_i is the K value based on the original vapor composition and normalized liquid composition. After the corrected K value has been calculated, we will use K-value charts to determine the temperature (or pressure) that will give this value for K for the pseudocomponent at the fixed pressure (or temperature). The use of this procedure is demonstrated in the following example.

Example 7.9

At 100 psia (689.6 kPa) calculate the dew-point temperature of the mixture in Example 7.8 using the procedure recommended by Dodge.

Solution:

As before, the solution is trial and error. *i*-pentane will be the reference component chosen and its composition will be normalized. We will estimate the K value as before. After two trials, the temperature of the dew point is found to be 208°F at 100 psia.

		Trial 1, $T = 200°F$		Trial 2, $T = 208°F$	
Component	y_i	K_i	$x_i = y_i/K_i$	K_i	$x_i = y_i/K_i$
iC_4	0.182	2.20	0.0827	2.40	0.0758
nC_4	0.238	1.76	0.1351	1.91	0.1246
iC_5	0.337	0.92	0.3663	1.02	0.3303
nC_5	0.121	0.765	0.1581	0.84	0.1440
C_6	0.122	0.33	0.3700	0.380	0.3210
			1.1122		0.9957
			$x'_{iC_5} = \frac{0.3663}{1.1122}$		
			$= 0.3293$		
			$K'_{iC_5} = 1.02$		

Thus the dew-point temperature is 208°F. The criteria for satisfactory solution at the dew point are the same as those expressed in Eq. (7-42) for the bubble point.

$$0.995 \leq \sum \frac{y_i}{K_i} \leq 1.005 \tag{7-47}$$

Differential flash vaporization

Flash vaporization can be conducted in two ways, by removing the vapor from contact with the liquid as it is formed or by keeping the vapor formed in intimate contact with the liquid so that the total vapor formed is in equilibrium with the residual liquid. The first of these operations is referred to as *differential vaporization* and the second as *equilibrium vaporization* or *equilibrium flash vaporization*.

In a differential vaporization process the liquid loses material of a given component and the vapor formed gains the same quantity of material. Thus

$$dV = -dL$$

We can write a component material balance for the differential mass transferred from the liquid to the vapor phase as

$$d(yV) = -d(Lx) \tag{7-48}$$

$$y\,dV + V\,dy = -x\,dL - L\,dx \tag{7-49}$$

Since $V\,dy \cong 0$, we can separate the variables and integrate to give

$$\int_{L_2}^{L_1} \frac{dL}{L} = \int_{x_2}^{x_1} \frac{dx}{y - x} \tag{7-50}$$

or

$$\ln \frac{L_1}{L_2} = \int_{x_2}^{x_1} \frac{1}{y - x}\,dx \tag{7-51}$$

where x = mole fraction of the more volatile component in the liquid at any instant,
y = mole fraction of the more volatile component in the vapor in equilibrium with the liquid at any instant,
x_1 = mole fraction of the more volatile component in the liquid initially,
x_2 = mole fraction of the more volatile component in the liquid after vaporization,
L_1 = initial number of total moles of liquid,
L_2 = total moles of liquid remaining after vaporization.

The right-hand side of Eq. (7-51) cannot be integrated directly since the relationship between y and x is not expressible analytically. The right-hand side of Eq. (7-51) can be integrated graphically through the use of a temperature-composition or x–y diagram such as that shown in Figure 7-2 or 7-11.

Example 7.10

A total of 100 kgmol of feed containing 70% heptane and 30% octane is to be separated. If the mixture is differentially vaporized, what will be the composition of the overhead when 80% of the total mixture has been vaporized? The system pressure is 20 psia.

Solution:

The differential vaporization equation is

$$\ln \frac{L_1}{L_2} = \int_{x_2}^{x_1} \frac{1}{y - x}\,dx$$

Graphical integration is necessary and this requires the development of a temperature–composition curve and a curve of y–x as a function of x (see Figure 7-29). The calculations are tabulated below.

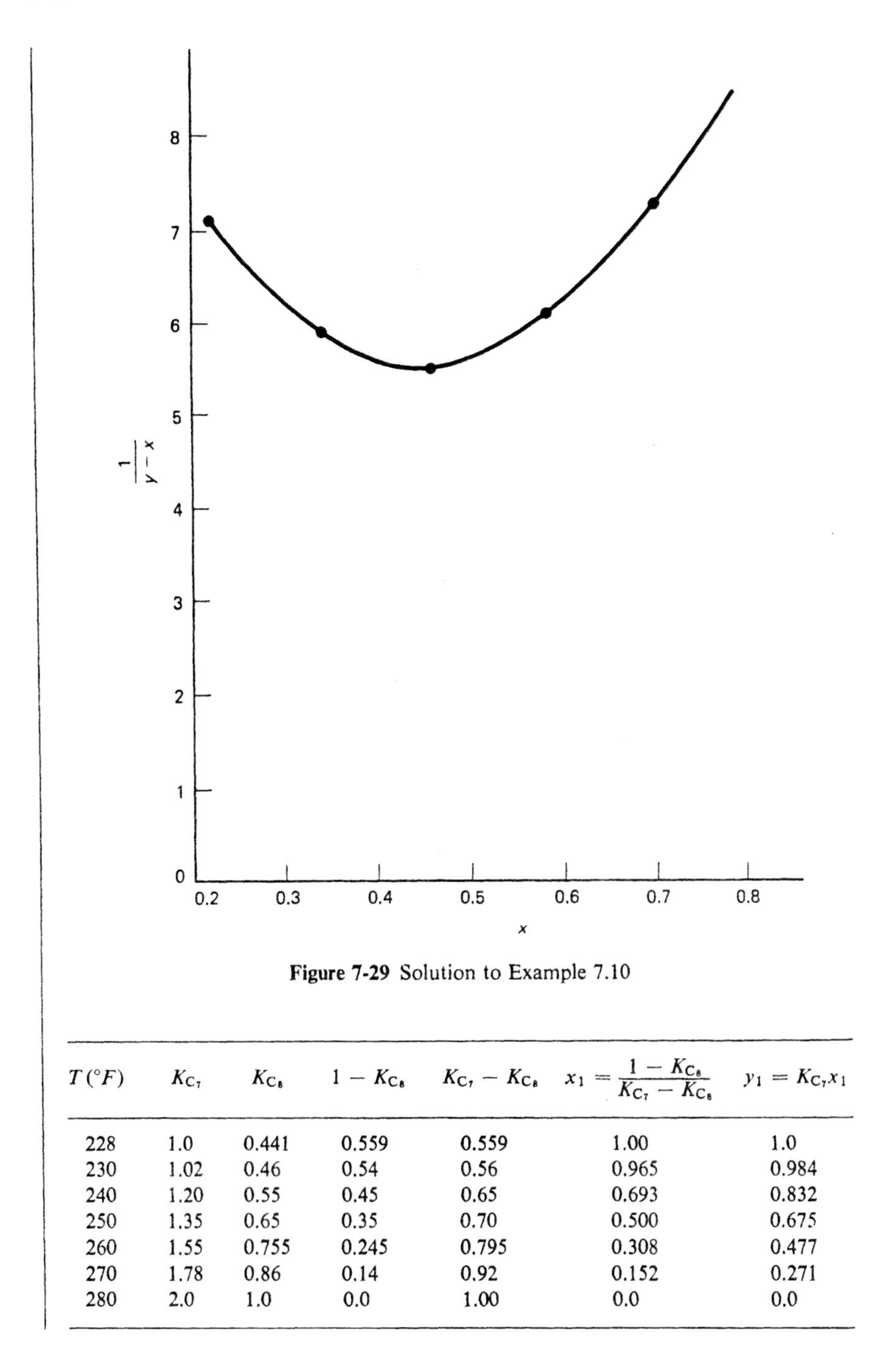

Figure 7-29 Solution to Example 7.10

T (°F)	K_{C_7}	K_{C_8}	$1 - K_{C_8}$	$K_{C_7} - K_{C_8}$	$x_1 = \dfrac{1 - K_{C_8}}{K_{C_7} - K_{C_8}}$	$y_1 = K_{C_7} x_1$
228	1.0	0.441	0.559	0.559	1.00	1.0
230	1.02	0.46	0.54	0.56	0.965	0.984
240	1.20	0.55	0.45	0.65	0.693	0.832
250	1.35	0.65	0.35	0.70	0.500	0.675
260	1.55	0.755	0.245	0.795	0.308	0.477
270	1.78	0.86	0.14	0.92	0.152	0.271
280	2.0	1.0	0.0	1.00	0.0	0.0

Since the solution is trial and error, we assume a value for x_2 (mole fraction of heptane remaining in the container) and determine the value of the integral graphically. The equality sought is

$$\ln \frac{100}{20} = 1.61 = \int_{x_2}^{0.7} \frac{1}{y - x}\, dx$$

For a value of $x_2 = 0.44$, the area under the curve is equal to 1.60 and the equality is satisfied. The liquid remaining in the container has the composition

$$\text{mole fraction } C_8 = 0.56$$

and

$$\text{mole fraction } C_7 = 0.44$$

The overhead product contains:

$$C_7\!: \quad 70 - (0.44 \times 20) = 61.2 \text{ kgmol}, \quad y_{C_7} = 0.765$$

$$C_8\!: \quad 30 - (0.56 \times 20) = 18.8 \text{ kgmol}, \quad y_{C_8} = 0.235$$

We can derive an equation to predict the amount and composition of liquid formed from differential condensation of vapor in exactly the same manner as was used to derive Eq. (7-51). The equation for differential condensation is

$$\ln \frac{V_1}{V_2} = \int_{y_2}^{y_1} \frac{1}{x - y}\, dy \tag{7-52}$$

where y_1 = initial composition of the vapor before condensation,
y_2 = final composition of vapor after condensation,
V_1 = initial total moles of vapor present,
V_2 = final moles of vapor present after condensation.

Equation (7-52) is solved and handled in exactly the same way as Eq. (7-51).

For the case of binary flash vaporization under conditions such that the relative volatility is constant, Eq. (7-51) becomes

$$\ln \frac{L_1}{L_2} = \frac{1}{\alpha - 1} \ln \frac{x_1(1 - x_2)}{x_2(1 - x_1)} + \ln \frac{1 - x_2}{1 - x_1} \tag{7-53}$$

This allows us to calculate the amount of liquid remaining after vaporization if the initial and final compositions are known. Since the relative volatility was nearly constant in Example 7.10, we can use Eq. (7-53) to predict the value for x_2.

Equilibrium vaporization

Figure 7-3 shows schematically the process used to visualize equilibrium flash vaporization. The feed F with composition z is separated into a vapor V of composition y and a liquid L of composition x, such that the vapor and liquid are in equilibrium. Development of the flash equation involves using the overall material balance given by Eq. (7-22).

$$F = L + V \tag{7-22}$$

A component material balance can be written as

$$Fz_i = Lx_i + Vy_i \tag{7-21}$$

or

$$f_i = l_i + v_i \tag{7-54}$$

The equilibrium distribution relationship is

$$y_i = K_i x_i$$

It can be written as

$$\frac{v_i}{V} = K_i \frac{l_i}{L} \tag{7-55}$$

If we substitute the component material balance for v_i, we obtain

$$\frac{f_i - l_i}{V} = K_i \frac{l_i}{L} \tag{7-56}$$

Solving for l_i gives

$$l_i = \frac{f_i}{K_i(V/L) + 1} \tag{7-57}$$

where

$$L = \sum l_i \tag{7-58}$$

Since Eq. (7-57) cannot be solved directly, a trial-and-error solution is necessary. A value for the total liquid (or vapor) formed in the equilibrium flash vaporization must be assumed. This assumption is then checked by calculation. If the assumed and calculated values agree, the calculation is correct. In general, more than one trial will be necessary to obtain agreement between the assumed and calculated values.

An equation to predict the amount and composition of the vapor formed in the flash vaporization can be derived in the same fashion. Thus for the vapor we obtain

$$v_i = \frac{K_i f_i}{(L/V) + K_i} \tag{7-59}$$

where

$$V = \sum v_i \tag{7-60}$$

Example 7.11

The following mixture is to be subjected to equilibrium flash vaporization at 300°F (148.9°C) and 280 psia (1.93 MPa). Determine the amount and composition of the vapor and liquid phases resulting from this equilibrium flash.

Component	*kgmol/h*
C_3	35.00
nC_4	30.00
nC_5	30.00
nC_6	10.00
nC_8	20.00
	125.00

Solution:

The first step in any equilibrium flash calculation is to make certain that the mixture is in the two-phase region. This is accomplished by determining whether or not the mixture is above its bubble point and below its dew point at the temperature and pressure of the flash.

Component	*kgmol/h*	K_i	$f_i K_i$	f_i/K_i
C_3	35.00	2.93	102.55	11.94
nC_4	30.00	1.55	46.5	19.35
nC_5	30.00	0.87	26.1	34.48
nC_6	10.00	0.49	4.9	20.41
nC_8	20.00	0.138	2.76	144.93
	125.00		182.81	231.11

From the calculations above we see that the mixture is in the two-phase region. The calculation of the flash is trial and error and requires an assumption of the quantity of liquid (or vapor) existing after the flash, and a calculation of that stream rate in order to check the assumption. Equation (7-57) is to be used. Assume that 54.4 kgmol of liquid will be formed in the flash.

Component	*kgmol/h*	K_i	$1 + \frac{L}{V} K_i$	l_i	v_i
C_3	35.0	2.93	4.81	7.29	27.71
nC_4	30.0	1.55	3.015	9.96	20.04
nC_5	30.0	0.87	2.132	14.09	15.91
nC_6	10.0	0.49	1.637	6.11	3.89
nC_8	20.0	0.138	1.179	16.96	3.04
				54.41	70.59

Since the amount of liquid calculated is essentially the same as the amount assumed, further trials are not necessary. The amount of each component in the liquid and vapor streams resulting from the flash are shown in the last two columns of the calculation table.

Equations (7-57) and (7-59) represent only two of many forms of the equilibrium flash vaporization equations that exist. However, in the opinion of the authors, these represent the most convenient forms to use for hand calculations. Values for the next assumption can be developed by preparing a graph similar to Figure 7-27 and plotting assumed and calculated values. By interpolating we can find the point at which the two values are the same.

Two simple rules should be followed when using Eqs. (7-57) and (7-59) that will do much to minimize wasted effort and needless trials.

1. Always calculate that stream (vapor or liquid) which is expected to be present in the smallest total amount.

2. Assume an initial value for the stream that is being calculated. If the calculated number of moles of the stream differs from the assumed value, the second assumption should be even further in the direction of difference.

If these rules are followed, the calculations will, in general, converge quickly and easily to accurate values for the conditions of the flash vaporization.

REFERENCES

DODGE, B. F., *Chemical Engineering Thermodynamics*, McGraw-Hill, New York, 1944.

Gas Processors Suppliers Association Engineering Data Book, Tulsa, Okla., 1972.

HADDEN S. T., *Chem. Eng. Prog. Symp. Ser. 49*, No. 7, 53 (1953).

HADDEN, S. T., and H. G. GRAYSON, *Hydrocarbon Process., 40*, No. 9, 207 (1961).

STOCKING, M. L., J. H. ERBAR, and R. N. MADDOX, *Refining Eng.*, p. C-15 (Apr. 1960).

VAN LAAR, J. J., *Z. Phys. Chem., 185*, 35 (1929).

NOTATIONS

a, b = van Laar constants
a_i = activity of i
a_i^α, a_i^β = activity of i in the α and β phases
A, B = van Laar constants defined by Eq. (7-28)
f_i = fugacity of i, F/L^2; flow rate of i in the feed mixture, mol/t
f_i° = standard state fugacity of i, F/L^2
$\hat{f}_i$ = fugacity of i in solution, F/L^2
$\hat{f}_i^\alpha, \hat{f}_i^\beta$ = fugacity of i in the α and β phases, F/L^2
$\hat{f}_i^{\circ\alpha}, \hat{f}_i^{\circ\beta}$ = standard state fugacity of i in the α and β phases, F/L^2
F = total flow rate of feed, mol/t
$\mathcal{G}$ = molar Gibbs free energy, E/mol
$\bar{G}_i$ = partial molar Gibbs free energy of i, E/mol
h = liquid enthalpy per mole, E/mol
H = vapor enthalpy per mole, E/mol
K_i = equilibrium distribution ratio of i as defined by Eq. (7-19)
K_i' = equilibrium distribution coefficient of i based on the original vapor composition and normalized liquid composition
l_i = flow rate of i in the liquid phase, mol/t
L = total liquid flow rate, mol/t
n = moles of gas, mol
P = pressure of the system, F/L^2
$P_{c,i}$ = critical pressure of i, F/L^2

P_i° = vapor pressure of i, F/L^2
$\bar{P}_A, \bar{P}_B$ = partial pressure of A and B, F/L^2
R = gas constant
T = absolute temperature
$T_{c,i}$ = critical temperature of i
v_i = flow rate of i in the vapor phase, mol/t
V = total vapor flow rate, mol/t
V_t = total volume, L^3
$\mathcal{V}$ = molar volume, L^3/mol
x_A = mole fraction of A in the liquid phase
y_A = mole fraction of A in the vapor phase
y'_A = normalized mole fraction of A in the vapor phase
z_A = mole fraction of A in the feed

Greek Letters

α_{ij} = separation factor or selectivity defined by Eq. (7-18)
γ_i = activity coefficient of i
$\mu_i^\alpha, \mu_i^\beta$ = chemical potential of i in α and β phases
π = total system pressure, F/L^2
ψ_i = constant of integration

PROBLEMS

7.1 Prepare a T–x diagram for the hexane–octane system at 20 psia using vapor pressures from Figure 7-6.

7.2 A temperature–composition diagram was constructed in Example 7.2 by using equilibrium K values. Derive the equations used in this example.

7.3 Prepare a T–x diagram for the hexane–heptane binary system at 20 psia using equilibrium constants determined from Figures 7-8 and 7-9.

7.4 Using the T–x diagram constructed in Problem 7.1, **(a)** determine the temperature at which a mixture containing 30 mol% hexane would be 50% vaporized. **(b)** Find the compositions of the vapor and liquid. **(c)** The vapor in part (b) is cooled until 50% of the vapor is condensed. Determine the temperature at which this occurs. **(d)** What are the compositions of the equilibrium vapor and liquid?

7.5 Use the T–x diagram prepared in Problem 7.1, and plot the y–x diagram for the hexane–octane binary system.

7.6 Prepare a pressure–temperature diagram to show the vapor–liquid behavior of a mixture that contains 20 lbs of n-pentane and 15 lbs of water. Determine the bubble-point temperature and composition of the vapor phase if the system pressure is 1 atm.

7.7 Find the temperature at which condensation begins for the mixture in Problem 7.6. The system pressure is 1 atm.

7.8 A 30-lb mixture containing 50 mol% hexane and 50 mol% heptane is added to a system that contains 30 lbs of water. **(a)** Prepare a pressure–temperature diagram showing the vaporization of the mixture at a system pressure of 1 atm, **(b)**

determine the bubble-point composition and temperature, and (c) calculate the dew-point composition and temperature.

7.9 Prepare an x–y diagram and an enthalpy–concentration diagram for mixtures of n-pentane and n-hexane. Determine the enthalpies of the saturated vapor, the saturated liquid, and an overall mixture that contains 38 mol% n-pentane which is 50% vaporized. The system pressure is 1 atm.

7.10 Using both the graphical interpolation–extrapolation technique and the method suggested by Dodge, calculate the bubble-point and dew-point temperatures of a mixture that contains 22% n-pentane, 28% n-hexane, 33% n-heptane, and 17% n-octane. The bubble-point and dew-point temperatures are to be determined at a pressure of 50 psia.

7.11 A mixture containing 30% propane and 70% n-butane is submitted to a differential flash vaporization. Determine the composition of the vapor for this operation when the final composition of the remaining liquid is 15 mol% propane. The vaporization is carried out at a pressure of 50 psia.

7.12 If the vaporization in Problem 7.11 is carried out at a pressure of 300 psia, find the composition of the vapor when the liquid contains 15 mol% propane.

7.13 An equal molar mixture of propane, i-butane, n-butane, and i-pentane is submitted to an equilibrium flash vaporization at 100 psia. (a) Find the temperature at which 50% of the initial mixture is vaporized. (b) How near is this temperature to the average of the bubble-point and dew-point temperatures for the mixture?

Absorption 8

8.1 Introduction

Absorption is the separation process which involves the transfer of one or more materials from the gas phase to a liquid solvent. The material(s) with net transfer from the gas to the liquid phase is referred to as the solute(s). Absorption is a physical phenomenon and involves no change in the chemical species present in a system. It may involve use of a given portion of solvent only one time. Most frequently, however, the condensible vapor (solute) is separated from the solvent and the solvent is recirculated to the process. The operation of removing the absorbed solute from the solvent is normally referred to as *stripping*.

Generally speaking, three approaches have been employed to develop the equations used to predict the performance of absorbers and absorption equipment:

1. The approach using mass transfer coefficients that depends on the molecular and the eddy diffusivities of the solute for the equipment in which the operation is being carried out
2. The technique of graphical solution generally attributed to Lewis (1927)
3. The absorption factor or overall approach generally attributed to Kremser (1930)

The use of mass transfer coefficients is covered in Chapter 12. The graphical solution is simple, direct, and easy to use for one or two components. It has the advantage of giving the thinking user an explicit graphical presentation of the interrelationships of the variables and parameters in an absorption process. It has the disadvantage, however, that it becomes very tedious when several solutes are present and must be considered. The absorption factor approach can be utilized either for hand or computer calculations. The principles involved are the same, but the equations and solution techniques differ.

8.2 Single-Component Absorption

Material balances for countercurrent operations

Most absorption operations are carried out in countercurrent flow processes, in which the gas phase is introduced in the bottom of the absorber and the liquid solvent is introduced in the top of the tower. The contact tower may be either equipped with trays or filled with an inert packing. From the standpoint of the mathematical analysis the two are equivalent.

Figure 8-1 shows schematically a countercurrent absorption process. The overall material balance is

$$L_b + V_a = L_a + V_b \tag{8-1}$$

where V = moles of gas phase per unit time,
L = moles of liquid phase per unit time,
a, b = top and bottom of tower, respectively.

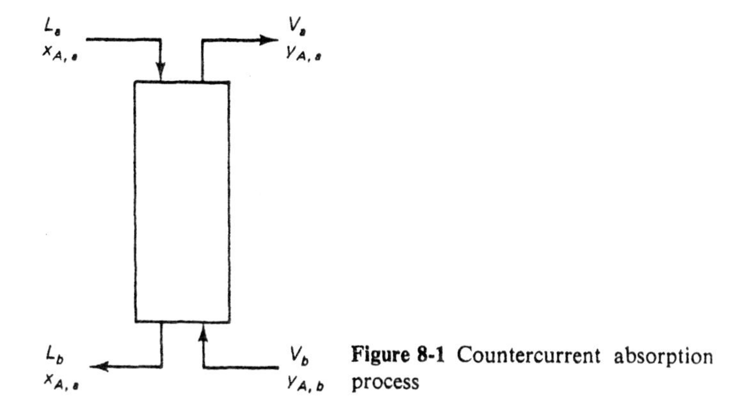

Figure 8-1 Countercurrent absorption process

If a single solute, A, is being transferred from the gas phase to the liquid phase, the component material balance is

$$L_b x_{A,b} + V_a y_{A,a} = L_a x_{A,a} + V_b y_{A,b} \tag{8-2}$$

where y_A = mole fraction of A in the vapor phase,
x_A = mole fraction of A in the liquid phase.

In many instances more convenient expressions can be derived for evaluating the absorption process if a *solute-free* basis is used for compositions rather than mole fractions. The solute-free concentration in the liquid phase is

$$\bar{X}_A = \frac{x_A}{1 - x_A} = \frac{\text{mole fraction of } A \text{ in the liquid}}{\text{mole fraction of non-}A \text{ components in the liquid}} \tag{8-3}$$

If the carrier gas is considered to be completely insoluble in the solvent and the solvent is considered to be completely nonvolatile, the carrier gas and solvent rates remain constant throughout the absorber. Using $\bar{L}$ to describe the flow rate of the nonvolatile solvent and $\bar{V}$ to describe the carrier gas flow rate, the material balance for solute A becomes

$$\bar{L}\bar{X}_{A,b} + \bar{V}\bar{Y}_{A,a} = \bar{L}\bar{X}_{A,a} + \bar{V}\bar{Y}_{A,b} \tag{8-4}$$

or

$$\bar{Y}_{A,a} = \frac{\bar{L}}{\bar{V}}\bar{X}_{A,a} + \left(\bar{Y}_{A,b} - \frac{\bar{L}\bar{X}_{A,b}}{\bar{V}}\right) \tag{8-5}$$

Obviously, Eqs. (8-4) and (8-5) are not restricted in application to the two column terminals but can be used to relate compositions and flow rates between any two points in the countercurrent absorber. When plotted on $\bar{X}_A$–$\bar{Y}_A$ coordinates, Eq. (8-5) gives a straight line with a slope of $\bar{L}/\bar{V}$ and a $\bar{Y}$ intercept of $(\bar{Y}_{A,b} - \bar{L}\bar{X}_{A,b}/\bar{V})$.

Gas solubilities in liquids are frequently given in terms of the Henry's law constants. *Henry's law* states that the quantity of gas that dissolves in a given quantity of solvent is directly proportional to its partial pressure over the solution. This is expressed by

$$\bar{P}_A = mx_A \tag{8-6}$$

where the Henry's law constant, m, has units of pressure/mole fraction. When the solubility is expressed in volumes of gas per volume of solution, the units will be pressure units/[(gas volume)/(liquid volume)]. In many cases the equilibrium between a gas and liquid will be expressed in terms of the equilibrium constant or vapor–liquid equilibrium ratio as

$$y_A = K_A x_A \tag{8-7}$$

Values of m for a number of gases in water, and a set of equilibrium constants for light-hydrocarbon gas mixtures are given in Appendix B.

If m in Eq. (8-6) or K in Eq. (8-7) may be assumed constant throughout the absorber, these two relationships can be plotted as straight lines on x_A–y_A coordinates.

Example 8.1

A solute A is to be recovered from an inert carrier gas B by absorption into a solvent. The gas entering into the absorber flows at a rate of 500 kgmol/h with $y_A = 0.3$. For the gas leaving the absorber $y_A = 0.01$. Solvent enters the absorber at the rate of 1500 kgmol/h with $x_A = 0.001$. The equilibrium relationship is $y_A = 2.8x_A$. The carrier gas may be considered nonsoluble in the solvent and the solvent may be considered nonvolatile. Construct the x–y plots for the equilibrium and operating lines using both mole fraction and solute-free coordinates.

Solution:

The first step is to determine the flow rates of the solvent and carrier gas. For the solvent

$$\bar{L} = L(1 - x_A) = 1500(1 - 0.001)$$
$$= 1498.5 \text{ kgmol/h}$$

For the gas

$$\bar{V} = V(1 - y_A) = 500(1 - 0.3)$$
$$= 350 \text{ kgmol/h}$$

The next step is to determine the concentration of A in the solvent stream leaving the absorber. The quantity of A leaving with the carrier gas is

$$y_{A,a} = 0.01 = \frac{\text{moles } A \text{ in } V_a}{\text{moles } A \text{ in } V_a + 350}$$

$$\text{moles } A \text{ in } V_a = 3.535$$

$$\text{moles } A \text{ in } L_b = (500 \times 0.3) - 3.535 + 1.5 = 147.965$$

$$x_{A,b} = \frac{147.965}{1498.5 + 147.965} = 0.0898$$

and

$$\bar{X}_{A,a} = 0.0010$$
$$\bar{X}_{A,b} = 0.0987$$
$$\bar{Y}_{A,a} = 0.0101$$
$$\bar{Y}_{A,b} = 0.4286$$

Points on the equilibrium curve may be calculated by assuming arbitrary values for x_A and calculating the corresponding equilibrium composition of the gas phase. For this absorber the calculations are summarized in the following table.

x_A	y_A	$\bar{X}_A$	$\bar{Y}_A$
0.001	0.0028	0.001001	0.00281
0.005	0.014	0.00503	0.0142
0.01	0.028	0.0101	0.0288
0.02	0.056	0.0204	0.0593
0.04	0.112	0.0417	0.126
0.06	0.168	0.0638	0.202
0.08	0.224	0.087	0.289
0.1	0.28	0.111	0.389

Points on the operating line may be calculated by using Eq. (8-5).

$$\bar{Y}_{A,a} = \frac{1498.5}{350} \frac{x_{A,a}}{1 - x_{A,a}} + \left(0.4286 - \frac{1498.5 \times 0.0987}{350}\right)$$

$$\bar{Y}_{A,a} = 4.28 \frac{x_{A,a}}{1 - x_{A,a}} + 0.00606$$

Points on the operating line are calculated by assuming arbitrary values for

$x_{A,a}$ and calculating the corresponding values of $\bar{Y}_{A,a}$. The results of the calculations are summarized below.

$x_{A,a}$	$\bar{X}_{A,a}$	$\bar{Y}_{A,a}$	$y_{A,a}$
0.001	0.001001	0.01034	0.01023
0.005	0.00503	0.0276	0.0268
0.01	0.0101	0.0493	0.0470
0.02	0.0204	0.0934	0.0854
0.04	0.0417	0.1845	0.1558
0.06	0.0638	0.279	0.218
0.08	0.087	0.378	0.275
0.0898	0.0987	0.4286	0.300

Figure 8-2 shows the equilibrium and operating lines plotted on mole fraction coordinates. The operating line is curved and the equilibrium line is straight. Normally, the compositions and/or temperatures will change from point to point in the absorber. These changes will bring about corresponding changes in

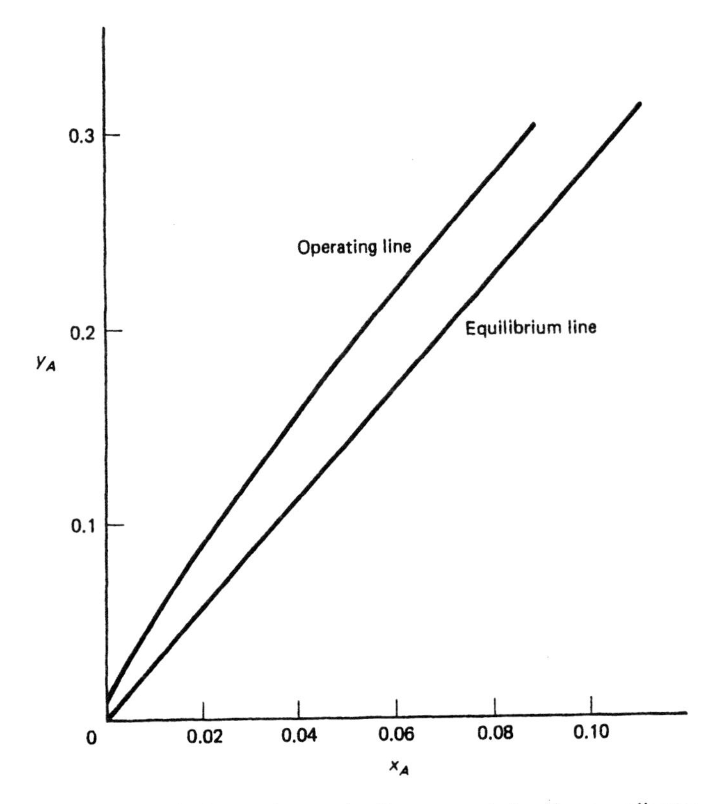

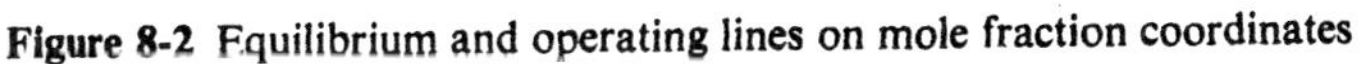
Figure 8-2 Equilibrium and operating lines on mole fraction coordinates

the equilibrium constant (Henry's law constant) and result in a curved equilibrium line.

Figure 8-3 shows the equilibrium and operating lines using solute-free coordinates. In this case the equilibrium line is also curved. Since the equilibrium line is going to be curved anyway, there is an advantage in using solute-free coordinates because the operating line will always be straight.

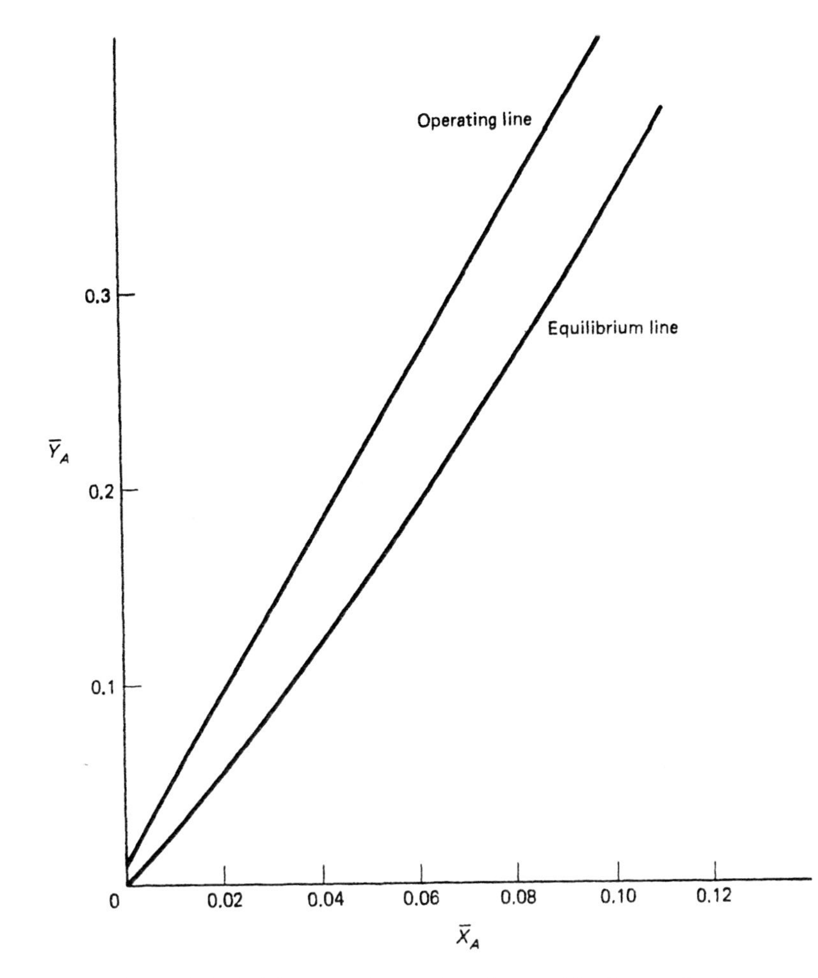

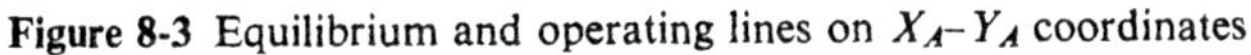

Figure 8-3 Equilibrium and operating lines on X_A–Y_A coordinates

Determining number of plates or contacts

Equation (8-7) in combination with Eq. (8-5) can be used to determine the number of theoretical or equilibrium plates required for a given absorption process. Equation (8-5) in combination with the inlet gas composition provides the composition of the liquid stream leaving the bottom tray of the absorber. If that tray operates as an equilibrium stage, the vapor phase in equilibrium with that stream can be determined from Eq. (8-7). The vapor stream calculated from Eq. (8-7) represents the vapor entering the next tray. Thus Eq. (8-5) can be used to calculate the composition of the liquid leaving that tray. This process could be continued until the required gas-phase composition leaving the absorber has been reached. This will give the number of theoretical or equilibrium trays required for the given absorption process. This same procedure can be carried out graphically as shown in Figure 8-4. The composition of the liquid stream leaving the absorber is determined by the vertical line from 1 to 2 when extended to the $\bar{X}_A$ axis. The horizontal line from 2 to 3 determines the composition of the vapor in equilibrium with that liquid. Consequently, steps 1–2–3 determine one theoretical or equilibrium plate. In the same manner steps 3–4–5 describe

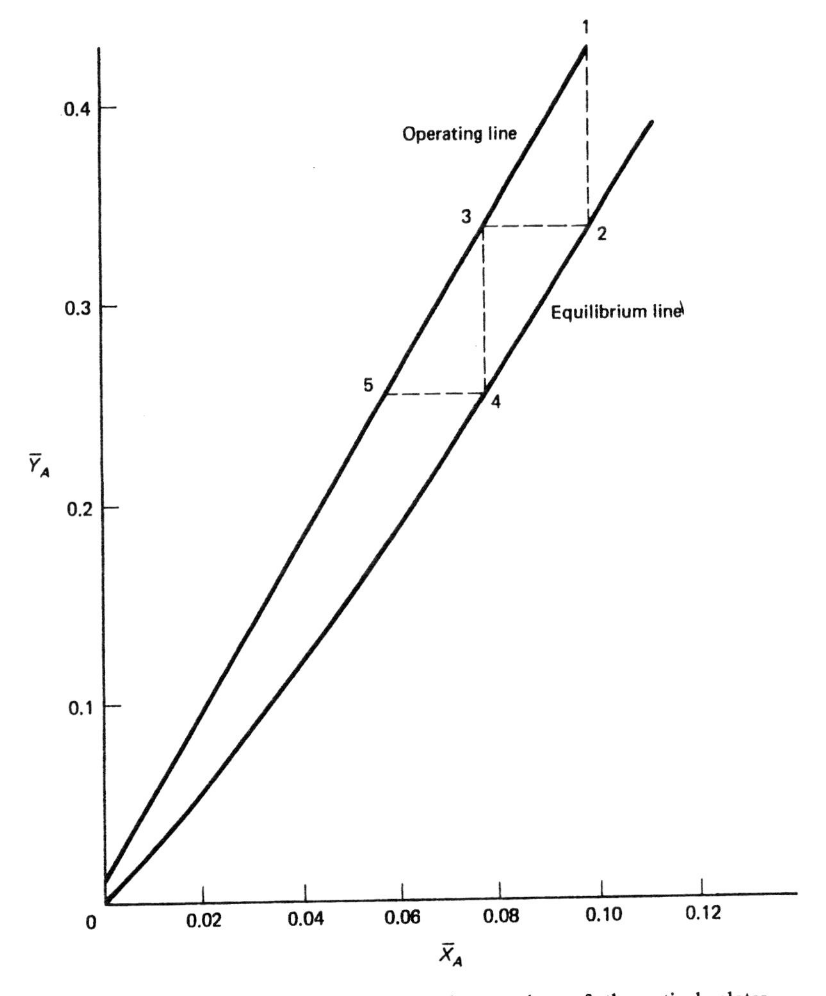

Figure 8-4 Graphical determination of the number of theoretical plates required

the separation on the second plate. If this procedure is continued until the composition of $\bar{Y}_{A,a}$ is reached, the number of theoretical plates required for the specified separation in Example 8.1 will have been determined.

Minimum solvent rates

A minimum solvent rate exists when the driving force for mass transfer from the vapor to the liquid phase becomes zero. This is indicated on the graph by the intersection of the operating and equilibrium lines. The various minimum rates are illustrated in Figures 8-5 through 8-7. Figures 8-5 and 8-6 have straight equilibrium and operating lines for simplicity. In Figure 8-5 the intersection of the equilibrium and operating lines occurs at the bottom of the absorber. This

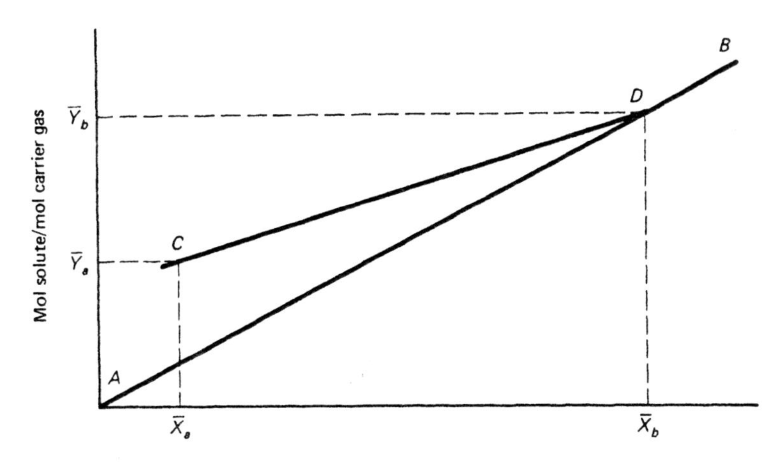

Figure 8-5 Moles solute/mole solvent graphical construction for determination of minimum solvent rate to recover specified amount of solute

condition defines the minimum solvent rate needed to recover a specified quantity of solute. In Figure 8-6 the intersection occurs at the top of the absorber. This condition represents the solvent rate required to remove the maximum possible amount of solute. For the case illustrated in Figure 8-6, in which a solvent denuded of solute enters the absorber, this solvent rate would be sufficient to remove all the solute from the carrier gas. If the equilibrium line is curved, the same two minima in solvent rates exist. The minimum solvent-to-vapor ratio for the case shown in Figure 8-5 can be calculated from the following expression:

$$\left(\frac{\bar{L}}{\bar{V}}\right)_{\min} = \frac{\bar{Y}_b - \bar{Y}_a}{\bar{X}_b - \bar{X}_a} \tag{8-8}$$

Figure 8-7 shows a curved equilibrium line that could become tangent to the operating line at a sufficiently low solvent rate. Since this is a condition that cannot be represented by straight lines, the slope of the tangent must be determined from the graph. The minimum liquid-to-vapor ratio for this case can be determined from Eq. (8-8) by replacing $\bar{Y}_b$ and $\bar{X}_b$ with $\bar{Y}_c$ and $\bar{X}_c$, respectively.

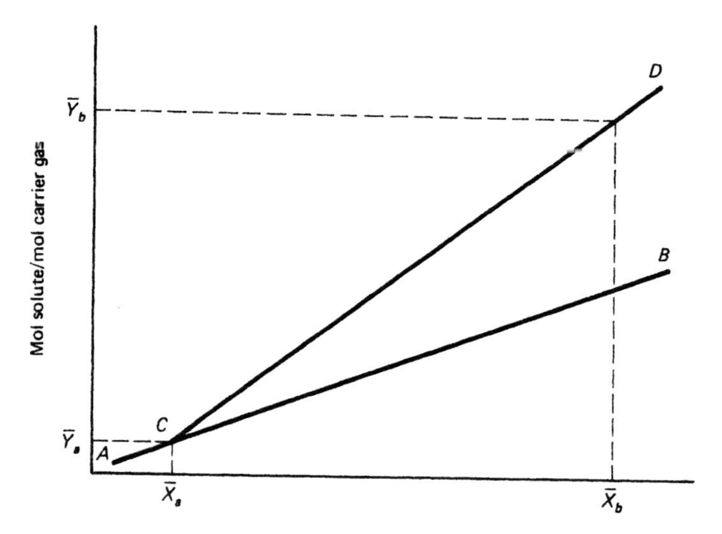

Figure 8-6 Moles solute/mole solvent graphical determination of minimum solvent rate to recover all of solute

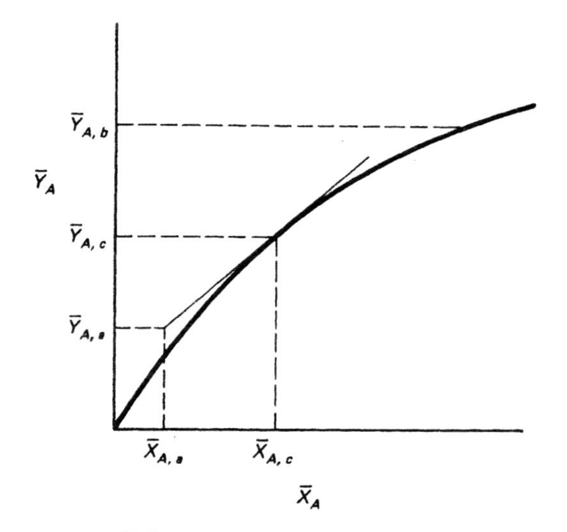

Figure 8-7 Minimum $\bar{L}/\bar{V}$ for equilibrium curve concaved away from operating line

Cocurrent flow

Although cocurrent mass transfer operations are encountered less frequently than countercurrent processes, cocurrent operations are used when contact between two phases involves one pure component such as in the saturation of a gas stream in a spray column. Also, cocurrent flow is sometimes used in a column when flooding is a problem. However, it is difficult to carry out physically, and is seldom used in practice.

For a steady-state cocurrent mass transfer process, such as that shown in Figure 8-8, the overall material balance is

$$L_b + V_b = L_a + V_a \tag{8-9}$$

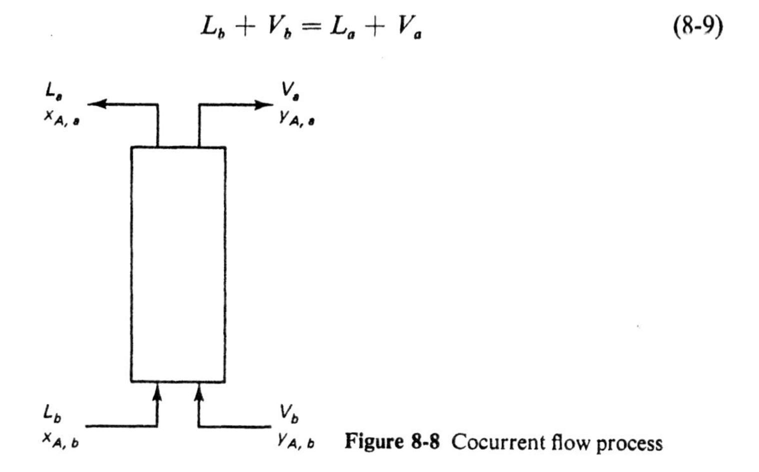

Figure 8-8 Cocurrent flow process

The overall material balance for component A is

$$L_b x_{A,b} + V_b y_{A,b} = L_a x_{A,a} + V_a y_{A,a} \tag{8-10}$$

Upon rewriting Eq. (8-10) in solute-free coordinates, we obtain

$$\bar{L}(\bar{X}_{A,b} - \bar{X}_{A,a}) = \bar{V}(\bar{Y}_{A,a} - \bar{Y}_{A,b}) \tag{8-11}$$

On $\bar{X}_A$-$\bar{Y}_A$ coordinates the expression above is the equation of a straight line with slope $-\bar{L}/\bar{V}$. The relationship between the operating line for cocurrent flow and the equilibrium curve is shown in Figure 8-9 for the case of transfer from the vapor to the liquid phase. The case of transfer from the liquid to vapor phase is shown in Figure 8-10.

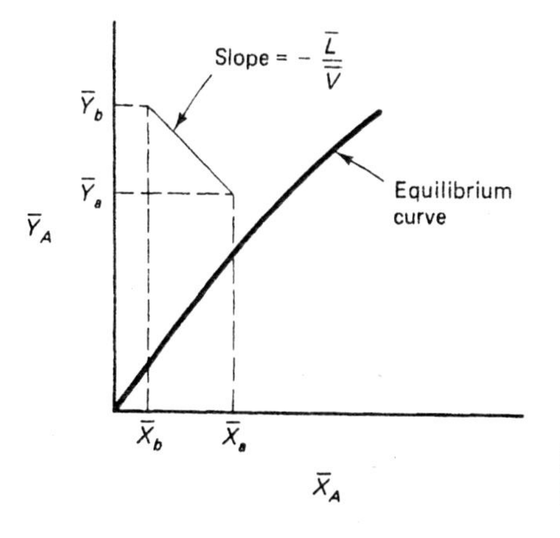

Figure 8-9 Cocurrent process: transfer from vapor to liquid

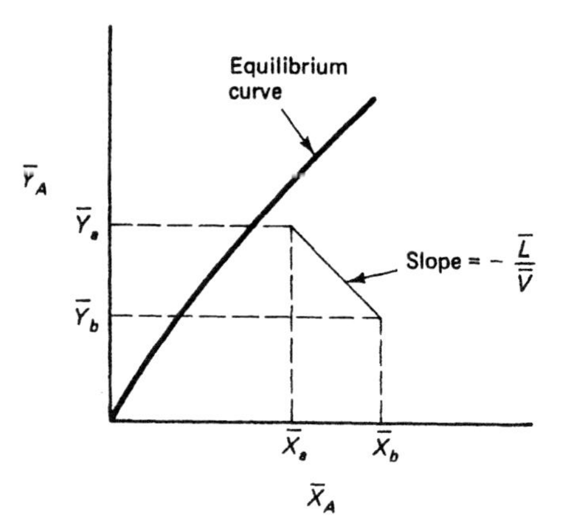

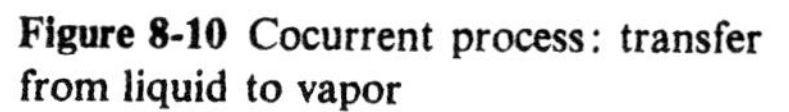
Figure 8-10 Cocurrent process: transfer from liquid to vapor

8.3 Multicomponent Absorption Using Shortcut Methods

Definition of the absorption factor

The equilibrium relationship between compositions in the vapor and the liquid phases for any equilibrium plate in the absorber may be expressed as

$$y_n = K_n x_n \tag{8-12}$$

where y = mole fraction of specified component in the vapor phase,
x = mole fraction of specified component in the liquid phase,
K = vaporization equilibrium constant for specified component on plate n,
n = arbitrary theoretical plate in the absorber.

Equation (8-12) can be rewritten in terms of the total molar flow rates of the vapor and liquid, and the flow rates for the specified component as

$$\frac{v_n}{V_n} = K_n \frac{l_n}{L_n} \tag{8-13}$$

where l_n = moles of specified component in the liquid stream leaving plate n,
v_n = moles of specified component in the vapor stream leaving plate n,
L_n = total moles of the liquid stream leaving plate n,
V_n = total moles of the vapor stream leaving plate n.

Solving for the individual molar flow rate of the specified component in the liquid leaving plate n, we obtain

$$l_n = \frac{L_n}{K_n V_n} v_n \tag{8-14}$$

The term $L_n/K_n V_n$ is referred to as the *absorption factor* and is expressed as

$$A_n = \frac{L_n}{K_n V_n} \tag{8-15}$$

where A_n is the absorption factor for any component on plate n in the absorber.

Absorption factor equations

General and rigorous equations involving the theoretical plate concept and the assumption of equilibrium between a gas and liquid on each theoretical plate can be derived by writing material balances around any plate n in the column. This can then be combined with the equilibrium expressions to give a generalized expression for the absorption process.

A material balance for any component around plate n of the absorber shown in Figure 8-11 is

$$L_{n-1}x_{n-1} + V_{n+1}y_{n+1} = L_n x_n + V_n y_n \tag{8-16}$$

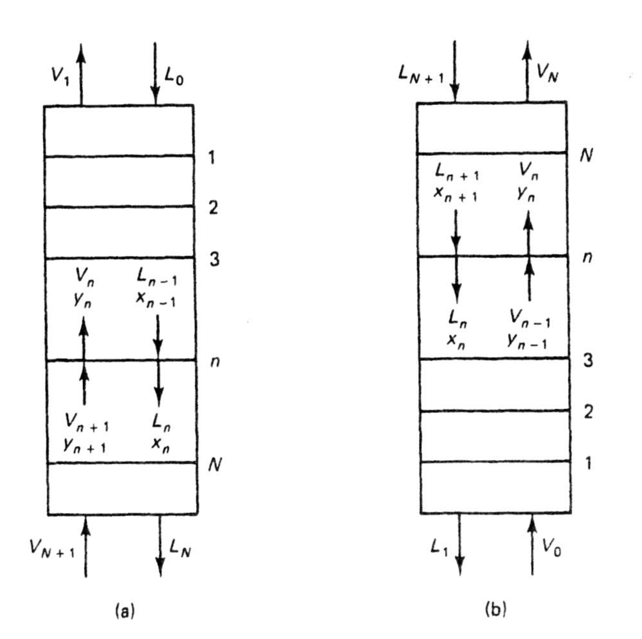

Figure 8-11 Schematic diagram of (a) an absorber, and (b) a stripper

However, the development of equations suitable for predicting component distributions in multicomponent absorption is more easily handled if compositions are placed on a slightly different basis from mole fractions. A convenient technique is to base all vapor stream compositions on the feed gas to the absorber and all liquid stream compositions on the lean solvent entering the absorber. We will denote these compositions by X' and Y', where

X' = moles of one component in the liquid stream per mole of lean solvent entering the absorber,

Y' = moles of one component in the gas stream leaving any plate per mole of rich gas entering the absorber.

For plate n these concentrations are defined as

$$X'_n = \frac{x_n L_n}{L_0} \tag{8-17}$$

and

$$Y'_n = \frac{y_n V_n}{V_{N+1}} \tag{8-18}$$

Using the composition parameters above, we can write the material balance around plate n shown in Figure 8-11 as

$$L_0 X'_{n-1} + V_{N+1} Y'_{n+1} = L_0 X'_n + V_{N+1} Y'_n \tag{8-19}$$

The material balance above can be written for the overall column as

$$Y'_{N+1} = \frac{L_0 X'_N}{V_{N+1}} + Y'_1 - \frac{L_0}{V_{N+1}} X'_0 \tag{8-20}$$

Equation (8-20) will plot as a straight line on Y'–X' coordinates with a slope of L_0/V_{N+1} and a Y' intercept of $Y'_1 - L_0 X'_0/V_{N+1}$. A comparison of Eqs. (8-5) and (8-20) should be considered. Although both equations can be plotted as straight lines, Eq. (8-5) is limited to systems in which the carrier gas is insoluble in the solvent and the solvent is nonvolatile. This greatly restricts its usefulness. Equation (8-20) is more general, however, and can be used for any absorption process.

The equilibrium relationship between the vapor and liquid on the tray, as represented by Eq. (8-12), can be described in terms of the new composition parameters. Equation (8-12) becomes

$$Y'_n = \frac{X'_n L_0/V_{N+1}}{L_n/K_n V_n} = \frac{X'_n L_0/V_{N+1}}{A_n} \tag{8-21}$$

By substituting the equation above into Eq. (8-19), we obtain

$$Y'_n = \frac{Y'_{n+1} + A_{n-1} Y'_{n-1}}{1 + A_n} \tag{8-22}$$

where $A_n = L_n/K_n V_n$ and $A_{n-1} = L_{n-1}/K_{n-1} V_{n-1}$. For a one-plate absorber Eq. (8-22) is

$$Y'_1 = \frac{Y'_2 + A_0 Y'_0}{1 + A_1} = \frac{Y'_2 + L_0 X'_0/V_{N+1}}{1 + A_1} \tag{8-23}$$

If the absorber has a second plate the material balance would be

$$Y'_2 = \frac{Y'_3 + A_1 Y'_1}{1 + A_2} \tag{8-24}$$

By combining Eqs. (8-23) and (8-24) we can write the material balance for a two-tray absorber as

$$Y'_2 = \frac{(A_1 + 1)Y'_3 + A_1 L_0 X'_0/V_{N+1}}{A_1 A_2 + A_2 + 1} \tag{8-25}$$

If the same procedure is followed for an absorber with three plates,

$$Y'_3 = \frac{(A_1 A_2 + A_2 + 1)Y'_4 + A_1 A_2 L_0 X'_0/V_{N+1}}{A_1 A_2 A_3 + A_2 A_3 + A_3 + 1} \tag{8-26}$$

Finally, for an absorber with N trays

$$Y'_N = \frac{(A_1A_2\cdots A_{N-1} + A_2A_3\cdots A_{N-1} + \cdots + A_{N-1} + 1)Y'_{N+1} + (A_1A_2\cdots A_{N-1})L_0X'_0/V_{N+1}}{A_1A_2\cdots A_N + A_2A_3\cdots A_N + A_3A_4\cdots A_N + \cdots + A_N + 1} \tag{8-27}$$

Since it is desirable to obtain an equation in terms of the absorber terminal conditions, Y'_N can be eliminated by combining Eq. (8-27) with an overall component material balance around the column. The overall component material balance is

$$L_0(X'_N - X'_0) = V_{N+1}(Y'_{N+1} - Y'_1) \tag{8-28}$$

An expression obtained from Eq. (8-21) for X'_N is

$$X'_N = \frac{A_N V_{N+1} Y'_N}{L_0} \tag{8-29}$$

If we replace X'_N in Eq. (8-28) by Eq. (8-29), we obtain

$$Y'_N = \frac{Y'_{N+1} - Y'_1 + L_0X'_0/V_{N+1}}{A_N} \tag{8-30}$$

Equation (8-27) can be written in terms of the terminal absorber conditions if we introduce Eq. (8-30). Thus

$$\frac{Y'_{N+1} - Y'_1}{Y'_{N+1}} = \frac{A_1A_2\cdots A_N + A_2A_3\cdots A_N + \cdots + A_N}{A_1A_2\cdots A_N + A_2A_3\cdots A_N + \cdots + A_N + 1}$$
$$- \frac{L_0X'_0}{V_{N+1}Y'_{N+1}}\,\frac{A_2A_3\cdots A_N + A_3A_4\cdots A_N + \cdots + A_N + 1}{A_1A_2\cdots A_N + A_2A_3\cdots A_N + \cdots + A_N + 1} \tag{8-31}$$

Equation (8-31) can be used to determine rigorously the terminal stream flow rates in a multicomponent absorber. The left side of Eq. (8-31) is the *fractional absorption* for any given component. Unfortunately, the use of Eq. (8-31) as shown above requires prior knowledge of the liquid and vapor flow rates for each tray in the column in addition to the temperature of each tray, since these are required to determine the absorption factor for the tray. For this reason further simplification is required if the equation is to be used.

Equation (8-31) can be simplified if certain restrictions are applied. If an average value of the absorption factor is assumed to be valid for each tray, the absorption factors in Eq. (8-31) can be written as

$$\frac{A^{N+1} - A}{A - 1} = A^N + A^{N-1} + \cdots + A^2 + A \tag{8-32}$$

and

$$\frac{A^{N+1} - 1}{A - 1} = A^N + A^{N-1} + \cdots + A^2 + A + 1 \tag{8-33}$$

This implies that $A = A_1 = A_2 = \cdots = A_N$, $A^2 = A_1A_2$, $A^3 = A_1A_2A_3$, and so on. By substituting the identities above into Eq. (8-31), we obtain the expression

$$\frac{Y'_{N+1} - Y'_1}{Y'_{N+1} - Y'_0} = \frac{A^{N+1} - A}{A^{N+1} - 1} = a \quad \text{(fraction absorbed)} \tag{8-34}$$

where A is the average absorption factor and Y'_0 is the moles of the component in the vapor in equilibrium with the lean solvent per mole of entering wet gas. Equation (8-34) is known as the Kremser (1930) and Brown (1932) equation. If the value of the average absorption factor for that component is known, the composition of the off-gas from the absorber and consequently the amount of material absorbed into the lean solvent may be readily calculated. One popular method of defining the average absorption factor for use in the equation above is in terms of the lean solvent and rich gas streams entering the absorber. Utilizing this technique, we can express the average absorption factor as

$$A_{avg} = \frac{L_0}{K_{avg} V_{N+1}} \tag{8-35}$$

The value of the average equilibrium constant to use in Eq. (8-35) is determined for each component at the average temperature and pressure in the absorber.

Graphical solution

Equation (8-34) may be solved either analytically or graphically, as shown in Figure 8-12. Three variables are identified in Figure 8-12: the absorption factor (plotted on the X axis), the value of the right (or left) side of Eq. (8-34) (plotted on the Y axis), and the number of theoretical plates in the absorber

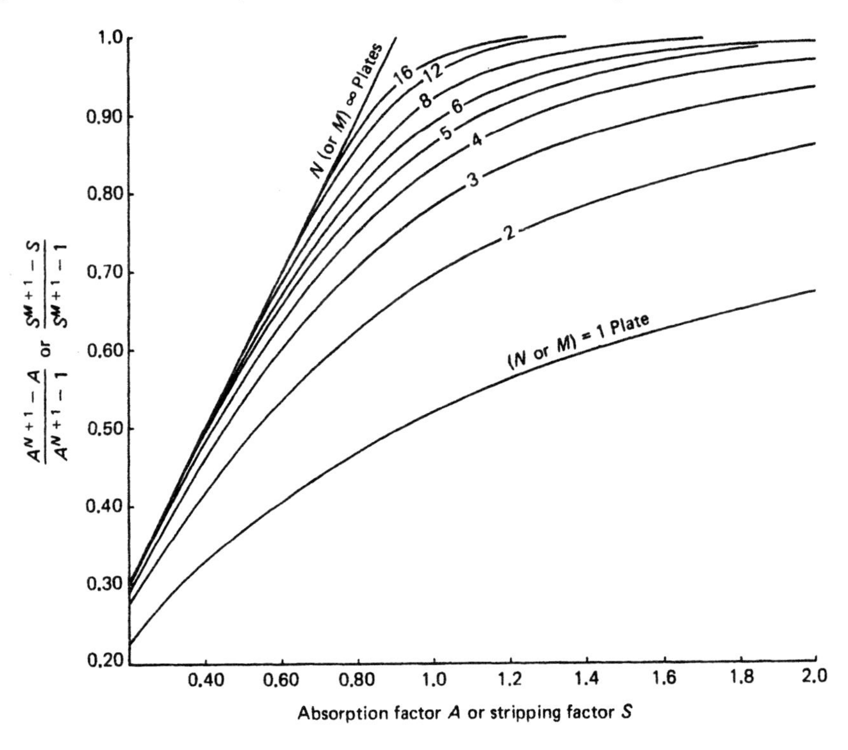

Figure 8-12 Graphical solution of Eq. (8-34)

(the parametric lines on Figure 8-12). Any one of these variables may be determined from the figure if the other two are known.

Example 8.2

An absorber containing six theoretical plates is used to process a natural gas stream having the composition shown below. The gas is available for processing at a pressure of 500 psia (3.45 MPa). If 65% of the propane in the gas stream is to be absorbed, what will be the composition of the gas stream leaving the absorber? Use the Kremser–Brown procedure. The average absorber temperature may be assumed to be 100°F (37.8°C).

Component	*Mol%*
C_1	76.524
C_2	13.110
C_3	4.938
nC_4	2.126
nC_5	2.09
nC_6	1.208

Solution:

Shortcut multicomponent absorption calculations are more conveniently carried out by creating a table as shown below. The sequence of steps for solution of the absorption problem is as follows:

1. Select a convenient basis of feed gas flow on which to base the absorption calculations. Convenient bases are 1 mol of feed gas, 100 mol of feed gas, or the number of moles of feed gas per hour entering the absorber. The first column in the calculation table will be the identification of the components in the gas stream and the second column will be the feed on which the calculations are to be based.
2. Obtain the equilibrium constants at the average temperature and pressure in the absorber.
3. Using Figure 8-12 with the specified fractional absorption of propane and the six theoretical plates specified for the column, determine the absorption factor for propane to be 0.675.
4. From Eq. (8-35) calculate the value of $L_0/V_{N+1} = 0.675 \times 0.49 = 0.331$.
5. Calculate the absorption factor for the other components, $A_i = 0.331/K_i$.
6. From Figure 8-12 determine the fractional absorption, a, for each component.
7. Calculate the number of moles of each component absorbed as [(a) × (moles in feed)].
8. Calculate the number of moles of each component in the off-gas from the absorber as [(moles in feed) − (moles absorbed)].

Component	Feed (mol%)	$K_{500}^{100°}$	Absorption Factor, A	Fraction Absorbed, a	Moles Absorbed	Moles Not Absorbed
C_1	76.524	6.3	0.0525	0.0525	4.018	72.506
C_2	13.110	1.38	0.240	0.240	3.146	9.964
C_3	4.938	0.49	0.675	0.65	3.210	1.728
nC_4	2.126	0.183	1.81	0.995	2.115	0.011
nC_5	2.09	0.070	4.73	1.00	2.094	—
nC_6	1.208	0.026	12.73	1.00	1.208	—
	100.000				15.791	84.209

Edmister procedure

Another technique often used for simplifying Eq. (8-34) was presented by Edmister (1934), who assumed that the absorber contained only two theoretical plates: the top plate and the bottom plate. He defined an *effective* absorption factor (A_e) for the column as

$$A_e = \sqrt{A_N(A_1 + 1) + 0.25} - 0.5 \tag{8-36}$$

The presence of solute in the lean solvent was taken into consideration by defining a modified absorption factor:

$$A' = \frac{A_N(A_1 + 1)}{A_N + 1} \tag{8-37}$$

Using these two definitions we can simplify Eq. (8-34) as follows:

$$\frac{Y'_{N+1} - Y'_1}{Y'_{N+1}} = \frac{A_e^{N+1} - A_e}{A_e^{N+1} - 1}\left(1 - \frac{L_0 X'_0}{A' V_{N+1} Y'_{N+1}}\right) \tag{8-38}$$

Equation (8-38) is rigorous only for an absorber with two theoretical plates. However, it represents a close approximation for many multicomponent absorbers. In a much later study based on rigorous tray calculations for multicomponent absorbers, Owens and Maddox (1968) concluded that 80% of the absorption occurs on the top and bottom trays with only about 20% occurring on other trays in the column.

Equation (8-38) requires knowledge of total stream flow rates and temperatures for the top and bottom trays in the column. Frequently, we can estimate these parameters rather reliably, but in many cases we find it difficult to estimate them.

Example 8.3

Solve Example 8.2 using the Edmister procedure. Assume that both the top and bottom trays in the absorber are at a temperature of 100°F.

Solution:

The following steps arc followed when applying the Edmister method.

1. In order to evaluate the absorption factors on the top and bottom trays for use in Eq. (8-36), the number of moles absorbed must be known. Using the Kremser–Brown calculation in Example 8.2 as a basis, assume that a total of 16 mol is absorbed per 100 mol of feed. This gives an absorption per tray of 16/6 = 2.67 mol.

2. The solvent rate to the tower must also be assumed. Using Example 8.2 as a basis, assume that 20 mol of solvent will be used for each 100 mol of gas. The absorption factors for propane can now be calculated. For the top tray

$$A_1 = \frac{20.0 + 2.67}{0.49(100.0 - 16.0)} = 0.551$$

For the bottom tray of the absorber

$$A_N = \frac{20.0 + 16.0}{0.49(100.0 - 2.67)} = 0.755$$

3. The effective absorption factor for propane may now be calculated using Eq. (8-36).

$$A_e = \sqrt{0.755(0.551 + 1.0) + 0.25} - 0.5 = 0.692$$

Using an absorption factor of 0.692 for propane in Figure 8-12, we find a fractional absorption for propane of 0.667. This is greater than the 0.65 that is specified. However, the 0.667 is based on assumed numbers that have not been proven. For this reason the calculation will continue.

4. The absorption factors on the top and bottom trays may now be calculated for the remaining components. For the top tray $L_1/V_1 = (20.0 + 2.67)/(100 - 16) = 0.2699$ and for the bottom tray $L_N/V_N = (20 + 16)/(100 - 2.67) = 0.3699$.

5. The effective absorption factors can now be calculated for each component and the calculation in the table can be completed. When this is done, the total number of moles absorbed is slightly greater than the number originally assumed. If greater accuracy is desired, the entire calculation procedure must be repeated until the assumed solvent rate produces a value for moles absorbed that agrees with the assumed value and, at the same time, matches the specified 65% absorption for the component.

Once again a tabular calculation of the absorption factor is most convenient.

Component	*Feed (mol%)*	$K_{500}^{100°}$	*Top Tray* A_1	*Bottom Tray* A_N	*Effective Absorption Factor,* A_e	*Fraction Absorbed,* a	*Moles Absorbed*
C_1	76.524	6.3	0.0428	0.0587	0.0579	0.0579	4.431
C_2	13.110	1.38	0.196	0.268	0.255	0.255	3.343
C_3	4.938	0.49	0.551	0.755	0.692	0.667	3.294
nC_4	2.126	0.183	1.477	2.021	1.793	0.987	2.098
nC_5	2.094	0.070	3.86	5.28	4.59	1.0	2.094
nC_6	1.208	0.026	10.38	14.23	12.24	1.0	1.208
							16.468

If only a small amount of solute is being absorbed, temperature changes in the column will not be great. For this condition, the assumption that the top tray is at the temperature of the lean solvent will be sufficiently accurate. The temperature of the bottom tray may be estimated by adding the heat of absorption for the solute to the lean solvent enthalpy and using that temperature for the bottom tray. Simple methods for estimating temperatures are not available for cases in which the solute absorption is large.

For cases in which the absorption is small, a good estimate for stream flow rates can be obtained by assuming that the same amount of solute is absorbed on each tray. At high levels of absorption, a reliable procedure for estimating stream flow rates is not available.

Both the Kremser–Brown and Edmister procedures are used for estimating purposes only. However, they provide reasonably good estimates of component absorption. The Kremser–Brown procedure gives solvent rates 15 to 25% too high. The required lean solvent flow estimated by Eq. (8-38) will be less than that found from Eq. (8-34) because Eq. (8-38) is based on absorption factors calculated at each terminal of the column. Only in rare cases should either calculation be used as a basis for design without first performing tray-to-tray calculations using techniques to be discussed later.

Stripping

Stripping is the reverse operation from absorption. The solute(s) absorbed in the solvent are removed by lowering the partial pressure of those components over the solution. This may be accomplished in several ways. The total pressure over the solution may be reduced, the rich solvent may be heated, or a noncondensible gas may be introduced.

An expression similar to Eq. (8-34) can be obtained for stripping by following the same procedure used in developing the absorption equations. The stripping equation, equivalent to Eq. (8-31), is

$$\frac{X'_{N+1} - X'_1}{X'_{N+1}} = \frac{S_1S_2 \cdots S_N + S_2S_3 \cdots S_N + \cdots + S_N}{S_1S_2 \cdots S_N + S_2S_3 \cdots S_N + \cdots + S_N + 1} - \frac{V_0Y'_0}{L_{N+1}X'_{N+1}} \frac{S_2S_3 \cdots S_N + S_3 \cdots S_N + \cdots + S_N + 1}{S_1S_2 \cdots S_N + S_2S_3 \cdots S_N + \cdots + S_N + 1} \tag{8-39}$$

where S is the stripping factor defined for tray n as

$$S_n = \frac{K_nV_n}{L_n} = \frac{1}{A_n} \tag{8-40}$$

Equation (8-39) can be simplified by applying the same assumptions and restrictions as those used for the Kremser–Brown equation.

$$\frac{X'_{N+1} - X'_1}{X'_{N+1} - X'_0} = \frac{S^{N+1} - S}{S^{N+1} - 1} \tag{8-41}$$

In the expression above S is the average stripping factor, defined by

$$S = \frac{K_{avg} V_0}{L_{N+1}} \tag{8-42}$$

The Edmister equations for stripping are

$$S_e = \sqrt{S_N(S_1 + 1) + 0.25} - 0.5 \tag{8-43}$$

and

$$S' = \frac{S_N(S_1 + 1)}{S_N + 1} \tag{8-44}$$

Thus

$$\frac{X'_{N+1} - X'_1}{X'_{N+1}} = \frac{S_e^{N+1} - S_e}{S_e^{N+1} - 1}\left(1 - \frac{V_0 Y'_0}{S' L_{N+1} X'_{N+1}}\right) \tag{8-45}$$

Over the years many different approaches have been developed in an effort to refine and make more accurate the calculated prediction of a multicomponent absorption process. Unfortunately, a multicomponent absorber is not amenable to the same degree of predictability as a distillation column. When we consider the absorber in Figure 8-11 we note that once the gas stream from which the solute(s) is to be absorbed has been fixed and the solvent to be used has been determined, only a few variables can be specified; the temperature of operation, the tower pressure, and the relative flow rate of solvent and gas are the available parameters. Once these have been determined or decided upon, the absorber performance has been fixed.

Estimating temperatures in multicomponent absorbers is difficult by using bubble-point and dew-point calculations. The wide range of component volatilities involved leads to difficulty in solution and instability in convergence. This is the primary reason that the shortcut absorber calculations have not involved the calculation of temperature and/or heat balances. The shortcut procedures outlined will provide a reasonable estimate of the amount of material absorbed and the composition of the gas leaving the absorber. The estimated solvent rate should be high, all other factors being the same. Shortcut methods generally provide good estimates of these parameters for use in a rigorous computer solution of the multicomponent absorption system.

8.4 Tray-by-Tray Calculations

The general approach taken in developing a procedure for rigorous tray-by-tray calculations includes material and heat balances. In this procedure, first developed by Sujata (1961), the temperature of each tray in the column is determined by an enthalpy balance, whereas liquid and vapor stream compositions and total flow rates are determined from equilibrium calculations involving the stripping factor defined previously. An absorption column with an external feed entering each tray is shown in Figure 8-13. The material balance around stage n is

$$F_n = V_n + L_n - L_{n-1} - V_{n+1} \tag{8-46}$$

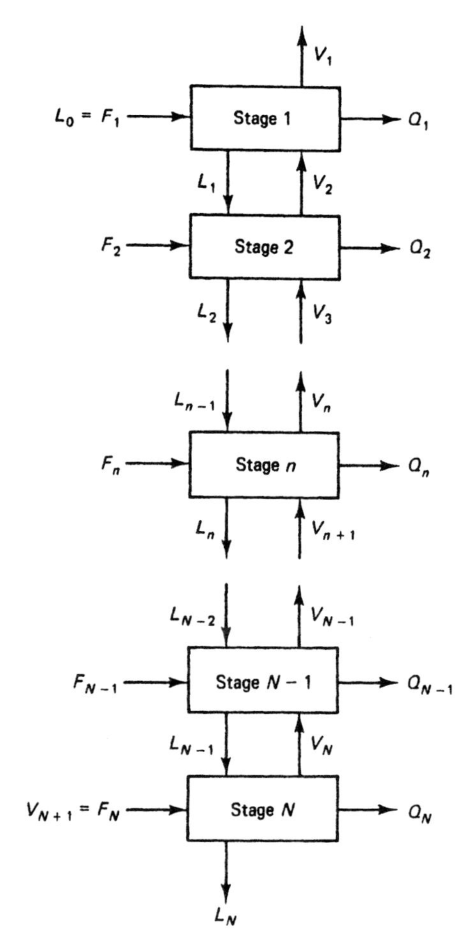

Figure 8-13 Nomenclature for an N-stage absorber

The material balance for one component is

$$f_n = v_n + l_n - l_{n-1} - v_{n+1} \tag{8-47}$$

where lower case letters are used to represent the flow of a single component. Utilizing the definition of the stripping factor and introducing Eq. (8-13), we can write Eq. (8-47) as

$$f_n = l_n(S_n + 1) - l_{n-1} - l_{n+1}S_{n+1} \tag{8-48}$$

For the top tray shown in Figure 8-13 and for a single-tray absorber we obtain

$$f_1 = l_1(S_1 + 1) - l_2S_2 \tag{8-49}$$

Note that for the one-tray absorber or for the top tray of an absorber with more than one theoretical plate, the lean solvent becomes the feed to the tray. Similarly, for the bottom tray the rich gas entering the absorber becomes the feed to that tray. If there is no external feed to an intermediate tray in the absorber, then $f = 0$. Writing Eq. (8-47) for each tray of a three-theoretical-tray absorber with no external feed to the middle tray, we obtain a set of linear equations in terms of the liquid flow rate which can be solved by using matrix techniques.

$$f_1 = l_1(S_1 + 1) - l_2 S_2 \tag{8-50}$$

$$f_2 = -l_1 + l_2(S_2 + 1) - l_3 S_3 \tag{8-51}$$

$$f_3 = -l_2 + l_3(S_3 + 1) \tag{8-52}$$

In order to obtain the stripping factor, estimates must be made of the liquid and vapor flow rate profiles and the temperature profile for the column. Once these estimates have been made and the material balance equation solved, new liquid and vapor rates can be estimated from

$$L_n = \sum^c l_n \qquad V_n = \sum^c v_n = \sum^c S_n l_n \tag{8-53}$$

The estimated tray temperature required for solution of Eqs. (8-50) through (8-52) must then be corrected by an enthalpy balance. Figure 8-14 shows the general tray schematic used for making heat balances. The total energy balance for all components around stage n is

$$F_n \bar{h}_n + \sum^c l_{n-1} h_{n-1} + \sum^c v_{n+1} H_{n+1} = \sum^c l_n h_n + \sum^c v_n H_n + Q_n \tag{8-54}$$

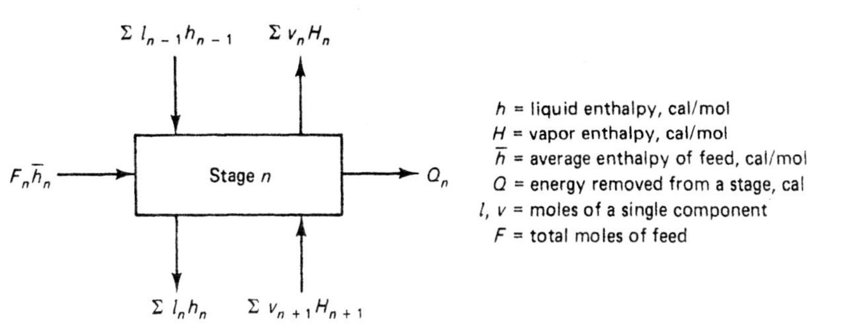

Figure 8-14 General absorber stage

The method of Sujata (1961) is based on the *sum rates* procedure, which uses the energy imbalance to correct the temperature profile. The energy imbalance is obtained by rearranging Eq. (8-54) as

$$E_n = Q_n + \sum^c v_n H_n + \sum^c l_n h_n - F_n \bar{h}_n - \sum^c l_{n-1} h_{n-1} - \sum^c v_{n+1} H_{n+1} \tag{8-55}$$

where E_n represents the energy imbalance or energy error for stage n. If the existing temperature for stage n is correct as obtained from the equilibrium calculation, E_n would be equal to zero. Since the energy imbalance is not zero, the temperature must be adjusted to correct for this. The adjustment procedure to be used assumes that any errors in the heat balance for the stage are functions of temperature only. Since the stage material balances probably are also not correct, the assumption is in error. The largest errors occur early in the iteration sequence, when individual stage vapor and liquid rates may change drastically. As convergence to the desired solution is approached, the assumption of energy balance errors being due to temperature only becomes more nearly correct. To develop a temperature correction, we make a first-order Taylor expansion of the energy balance error in terms of temperature. In general, the error in the heat balance on a given stage will depend on the temperature of that stage and the two adjacent stages. Thus for stage n,

$$dE_n = \left(\frac{\partial E_n}{\partial t_{n-1}}\right) dt_{n-1} + \left(\frac{\partial E_n}{\partial t_n}\right) dt_n + \left(\frac{\partial E_n}{\partial t_{n+1}}\right) dt_{n+1} \tag{8-56}$$

where

$$\frac{\partial E_n}{\partial t_{n-1}} = -\frac{\partial}{\partial t_{n-1}}\left(\sum^{c} l_{n-1} h_{n-1}\right) \tag{8-57a}$$

$$\frac{\partial E_n}{\partial t_n} = \frac{\partial}{\partial t_n}\left(\sum^{c} l_n h_n + \sum^{c} v_n H_n\right) \tag{8-57b}$$

$$\frac{\partial E_n}{\partial t_{n+1}} = -\frac{\partial}{\partial t_{n+1}}\left(\sum^{c} v_{n+1} H_{n+1}\right) \tag{8-57c}$$

If Eqs. (8-56) and (8-57) are written for each stage in the absorber, a set of n simultaneous equations results. Solution of these simultaneous equations yields the temperature change necessary for each tray to bring the energy imbalance for that tray into balance. The temperature of each tray is adjusted for the next iteration by algebraically adding the temperature and the temperature increment for each tray.

Obviously, the Sujata procedure as outlined above is suitable only for solution by digital computer. The calculations are so tedious and time consuming that no one would undertake to carry them out by hand or by pocket calculator. There is one characteristic of the Sujata equations that has been observed through their utilization in a number of absorber solutions. In nearly every case, errors in the heat balance on all trays tend to be in the same direction. This means that the engineer must very carefully assess the deviations from the energy balance closure that he or she will permit on a given tray. Since the energy balance errors tend to be additive, a satisfactory degree of closure for one stage may result in an unsatisfactory overall column balance closure.

8.5 Comparison of Shortcut and Tray-by-Tray Methods

A comparison of shortcut and tray-by-tray methods was recently carried out by Diab and Maddox (1982) for the absorption of propane from a multicomponent light-hydrocarbon gas stream. In one study they compared the effects of temperature on the quantity absorbed by applying the Kremser–Brown equation. They carried out the calculations for the case of 70% propane recovery at an

TABLE 8-1. TEMPERATURE EFFECTS ON ABSORPTION

		0°*F*		20°*F*		40°*F*	
Component	*Feed (mol/h)*	*Mol Absorbed*	*Off-Gas*	*Mol Absorbed*	*Off-Gas*	*Mol Absorbed*	*Off-Gas*
N_2	23.02	0.32	22.70	0.37	22.65	0.41	22.61
C_1	1,435.97	63.18	1,372.79	71.80	1,364.17	83.29	1,352.68
CO_2	125.59	13.31	112.28	14.57	111.02	15.95	109.64
C_2	183.73	47.40	136.33	49.06	134.67	51.08	132.65
C_3	60.49	42.34	18.15	42.34	18.15	42.34	18.15
iC_4	5.93	5.84	0.09	5.84	0.09	5.78	0.15
nC_4	8.40	8.40	—	8.40	—	8.40	—
iC_5	1.48	1.48	—	1.48	—	1.48	—
nC_5	1.02	1.02	—	1.02	—	1.02	—
C_6	0.42	0.42	—	0.42	—	0.42	—
C_{7+}	0.07	0.07	—	0.07	—	0.07	—
	1,846.12	183.78	1,662.34	195.37	1,650.75	210.24	1,635.88
		$\frac{L_0}{V_{N+1}} = 0.117 \frac{\text{mol oil}}{\text{mol gas}}$		$\frac{L_0}{V_{N+1}} = 0.142$		$\frac{L_0}{V_{N+1}} = 0.171$	
		216 mol oil/h		262 mol oil/h		316 mol oil/h	

absorption pressure of 1000 psia. Both the absorption oil and the gas stream were assumed to enter the absorber at 0°F. The results for a six-theoretical-tray absorber are shown in Table 8-1.

As noted in Table 8-1, temperature has very little effect on the absorption of components heavier than propane. The components lighter than propane are affected to a greater degree and, interestingly enough, are absorbed to a greater extent as the absorption temperature increases. This is a typical result, and another reason for using lower temperatures for the operation of multicomponent absorbers in the natural-gas industry. We should note that the amount of absorption oil increases with increasing temperature. However, the Kremser–Brown procedure always gives a high value for the solvent rate required. This is because the method essentially assumes that no absorption takes place, and uses the lowest oil circulation rate (the entering lean oil) and the highest gas flow rate (the entering gas stream) for evaluating the absorption factor. The procedure proposed by Edmister gives more accurate solvent rates, since this method takes into consideration the absorption on the top and bottom trays where the absorption factors are evaluated.

Diab and Maddox (1982) also made a comparison of the Edmister and Sujata tray-by-tray methods. In this comparison, the entering rich gas and lean oil temperature was 1°F, and the absorption pressure was 1000 psia. They used a lean oil circulation rate of 150 mol/h and a six-tray absorber. Their results are shown in Tables 8-2 and 8-3.

TABLE 8-2. RESULTS OF EDMISTER METHOD

	mol/h			
Component	*Rich Gas*	*Lean Oil*	*Lean Gas*	*Rich Oil*
Nitrogen	23.02	0.0	22.49	0.53
Methane	1,435.97	0.0	1,327.68	108.29
Carbon dioxide	125.59	0.0	105.98	19.61
Ethane	183.73	0.0	120.18	63.55
Propane	60.49	0.0	9.88	50.61
i-Butane	5.93	0.0	0.07	5.86
n-Butane	8.40	0.0	0.02	8.38
i-Pentane	1.48	0.0	0.00	1.48
n-Pentane	1.02	0.0	0.00	1.02
n-Hexane	0.42	0.0	0.00	0.42
Heptane plus	0.07	0.0	0.00	0.07
lean oil	0.00	150.0	0.00	150.00
Total	1,846.12	150.0	1,586.30	409.82
Temperature (°F)	1.00	1.00	26.00	4.18
Pressure (psia)	1,000	1,000	1,000	1,000

TABLE 8-3. RESULTS OF THE SUJATA METHOD

	mol/h			
Component	*Rich Gas*	*Lean Oil*	*Lean Gas*	*Rich Oil*
Nitrogen	23.02	0.0	22.09	0.93
Methane	1,435.97	0.0	1,267.23	168.74
Carbon dioxide	125.59	0.0	95.74	29.85
Ethane	183.73	0.0	96.80	86.93
Propane	60.49	0.0	4.99	55.50
i-Butane	5.93	0.0	0.02	5.91
n-Butane	8.40	0.0	0.01	8.39
i-Pentane	1.48	0.0	0.00	1.48
n-Pentane	1.02	0.0	0.00	1.02
n-Hexane	0.42	0.0	0.00	0.42
Heptane plus	0.07	0.0	0.00	0.07
lean oil	0.00	150.0	0.03	149.97
Total	1,846.12	150.0	1,486.91	509.24
Temperature (°F)	1.00	1.00	31.04	13.03
Pressure (psia)	1,000	1,000	1,000	1,000

Tray Number	*Liquid (mol/h)*	*Vapor (mol/h)*	*Temperature (°F)*	*Pressure (psia)*
1	509.20	1,730.87	13.03	1,000
2	393.96	1,689.73	20.10	1,000
3	352.82	1,666.21	25.07	1,000
4	329.30	1,648.36	28.83	1,000
5	311.45	1,628.59	31.34	1,000
6	291.68	1,486.91	31.04	1,000

In the Edmister method we must specify the temperature difference between the lean gas leaving the absorber and the entering lean oil. In the case above, this difference was assumed to be 25°F. As shown in the preceding tables, a lean oil rate of 150 mol/h, which is approximately 25% less than that predicted by the Kremser–Brown procedure, provides an absorption of almost 84% of the propane in the feed. The tray-by-tray method shows that 92% of the entering propane will be absorbed.

REFERENCES

BROWN, G. G., and M. SOUDERS, JR., *Ind. Eng. Chem.*, *24*, 519 (1932).

DIAB, S., and R. N. MADDOX, *Chem. Eng.*, *89*, No. 26, 38 (1982).

EDMISTER, W. C., *Petrol. Eng.*, *18*, No. 13 (1934).

KREMSER, A., *Natl. Petrol. News*, *22*, No. 21, 48 (1930).

LEWIS, W. K., *Trans. AIChE, 20,* 1 (1927).
OWENS, W. R., and R. N. MADDOX, *Ind. Eng. Chem., 60,* No. 12, 14 (1968).
SUJATA, A. D., *Hydrocarbon Process. Petrol. Refiner, 40,* No. 12, 137 (1961).

NOTATIONS

a = fraction absorbed
A_{avg} = average absorption factor defined by Eq. (8-35)
A_e = effective absorption factor defined by Eq. (8-36)
A_n = absorption factor for one component on plate n
A' = modified absorption factor defined by Eq. (8-37)
f_n = flow rate of one component in the feed stream entering plate n, mol/t
K_A = equilibrium constant of component A
K_n = equilibrium constant of any component on plate n
l_n = flow rate of any component in the liquid stream leaving plate n, mol/t
L = total flow rate of the liquid stream, mol/t
L_a, L_b = total flow rate of the liquid streams at the top and bottom of the tower, respectively, mol/t
L_n = total flow rate of the liquid stream leaving plate n, mol/t
$\bar{L}$ = flow rate of the solvent, mol/t
m = Henry's constant
$\bar{P}_A$ = partial pressure of A, F/L^2
S_e = effective stripping factor
S_n = stripping factor for one component on plate n
S' = modified stripping factor
v_n = flow rate of any component in the vapor stream leaving plate n, mol/t
V = total flow rate of the vapor stream, mol/t
V_a, V_b = total flow rate of the vapor streams at the top and bottom of the tower, respectively, mol/t
V_n = total flow rate of the vapor stream leaving plate n, mol/t
$\bar{V}$ = flow rate of insoluble carrier gas, mol/t
x_A = mole fraction of A in the liquid stream
$x_{A,a}$, $x_{A,b}$ = mole fractions of A in the liquid streams at the top and bottom of the tower, respectively
x_n = mole fraction of any component in the liquid stream leaving plate n
$\bar{X}_A$ = solute-free concentration of A in the liquid stream
$\bar{X}_{A,a}$, $\bar{X}_{A,b}$ = solute-free concentrations of A in the liquid at the top and bottom of the tower, respectively
X'_n = moles of one component in the liquid stream on plate n per mole of lean solvent entering the absorber
y_A = mole fraction of A in the vapor stream
$y_{A,a}$, $y_{A,b}$ = mole fractions of A in the vapor streams at the top and bottom of the tower, respectively

y_n = mole fraction of any component in the vapor stream leaving stage n

$\bar{Y}_A$ = solute-free concentration of A in the vapor stream

$\bar{Y}_{A,a}$, $\bar{Y}_{A,b}$ = solute-free concentrations of A in the vapor streams at the top and bottom of the tower, respectively

Y'_n = moles of one component in the vapor stream on plate n per mole of rich gas entering the absorber

PROBLEMS

8.1 Solute A is to be removed from an inert gas B in a multi-stage countercurrent absorption tower. The gas enters the tower at a rate of 200 kgmol/h and contains 25 mol% A. The solvent enters the tower at a rate of 800 kgmol/h and is initially free of solute. Determine **(a)** the concentration of the exiting gas stream and **(b)** the number of stages, if the exiting liquid stream contains 5.0 mol% A. The equilibrium relationship is $y_A = 4.0x_A$. Assume that the carrier gas is insoluble in the solvent and the solvent is nonvolatile.

8.2 Solute A is to be stripped from a liquid stream by contacting the liquid with a pure gas. The liquid enters the stripping tower at an A-free rate of 150 kgmol/h and contains 30 mol% A. The gas enters the column at a rate of 500 kgmol/h. Determine the number of stages required to reduce the concentration of A in the exiting liquid stream to 1.0 mol%. The distribution of A in the gas and liquid is expressed as $y_A = 0.4x_A$.

8.3 A countercurrent absorption process is to be used to recover solute A from an inert gas. The concentration of A is to be reduced from $\bar{Y}_A = 0.285$ to $\bar{Y}_A = 0.05$ by contacting the gas with a pure solvent. **(a)** Find the minimum liquid to vapor ratio on a solute-free basis, and **(b)** determine the number of equilibrium stages if the actual liquid rate is 1.2 times the minimum value. The equilibrium relationship can be expressed as $y_A = 4.0x_A$.

8.4 The liquid solution described in Problem 8.2 is fed to the column cocurrently with the pure gas. **(a)** If the exiting liquid solution is to contain no more than 10 mol% of A, find the maximum exiting gas concentration. **(b)** Determine the liquid to vapor ratio on a solute-free basis if the exiting gas concentration is 3.0 mol% A.

8.5 The gas stream described in Problem 8.3 is fed to the column cocurrently with a pure solvent. The concentration of A in the carrier gas is again reduced from $\bar{Y}_A = 0.285$ to $\bar{Y}_A = 0.05$ and the actual liquid rate is 1.2 times the minimum. Determine **(a)** $(\bar{L}/\bar{V})_{min}$, **(b)** the actual $(\bar{L}/\bar{V})$ ratio, and **(c)** the concentration of A in the exiting solvent solution.

8.6 Ammonia is to be absorbed from a gas mixture by water in a countercurrent absorption column. The absorption is to be carried out at 1 atm and 68°F (20°C). The entering gas mixture contains 30 mol% ammonia and the water entering the absorber may be considered completely free of ammonia. The water circulation rate will be 3.5 mol per mole of NH_3 in the gas mixture. **(a)** Determine the concentration of ammonia in the water leaving the absorber if the column has two theoretical stages, and **(b)** find the NH_3 concentration if the absorber has three theoretical stages.

8.7 A gas mixture contains 80% methane, 8% ethane, 6% propane, 2% *n*-butane, 2% *n*-pentane, and 2% *n*-hexane. The mixture is to be contacted by a solvent that is a mixture of heavy hydrocarbons. The absorption is to be carried out at 100°F and 500 psia. **(a)** How many moles of solvent are required per mole of gas to recover 80% of the propane if the absorber contains eight theoretical plates? **(b)** How many moles of oil would be required if the absorption were carried out at 0°F and 500 psia?

8.8 Evaluate the Edmister absorption factors for Problem 8.7 assuming equal absorption per tray and no change in temperature throughout the absorber.

8.9 If the solvent in Problem 8.7 initially contains 0.02 mole fraction of propane instead of being totally stripped, find the required circulation rate.

8.10 A physical solvent is to be used to remove H_2S and CO_2 from a natural gas stream that contains 10% H_2S, 5% CO_2, 80% C_1, 4.5% C_2, and 0.5% nC_5. For this process 85% of the H_2S in the gas is to be removed by the solvent. The absorption process will be carried out at 60°F and 1000 psia in an absorber that contains 10 theoretical stages. After leaving the absorber the solvent will flow to a flash tank that operates at 60°F and 250 psia. The total flow rate of the feed gas is 100 MMSCFD (measured at 60°F and 14.7 psia).

(a) Determine the moles of solvent required per mole of gas fed to the absorber. **(b)** Calculate the H_2S content of the flash gas that leaves the 250-psia separator. **(c)** If the solvent has a molecular weight of 99.09 and a specific gravity of 1.1329, find the solvent flow rate in gallons per minute.

Equilibrium constants at 60°F for the natural gas components in the solvent are shown in the following table.

Component	*K*(10 *psia*)	*K*(100 *psia*)	*K*(1000 *psia*)
C_1	2200	250	30
C_2	550	55	6.0
nC_5	32	4.5	0.65
H_2S	47	3.1	0.2
CO_2	105	10	1.0

8.11 The solvent in Problem 8.10 is flashed at 250 psia and 60°F after leaving the absorber. The liquid stream leaving the 250-psia flash separator is then flashed at 15 psia and 60°F. The liquid leaving the 15-psia flash separator is to be stripped by using a gas that contains no H_2S or CO_2. Determine the daily volume (standard cubic feet) of stripping gas if the H_2S content of the lean solvent leaving the stripper is to be 0.005 mole fraction. The stripper contains six theoretical stages.

8.12 A typical limitation on the hydrogen sulfide concentration in "sour" crude oil is 50 ppm by weight. A crude oil and its off-gas from a separator are shown in the table below. **(a)** Determine whether the gas can be used to strip the crude oil to a concentration of 50 ppm H_2S by weight. **(b)** What is the maximum allowable concentration of H_2S in the gas if it is to be used to strip the crude oil? **(c)** For a gas containing 0.3 mol % H_2S, calculate the required gas stripping rate and the compositions of the gas and liquid leaving a stripper that contains three theoretical plates and operates at a total pressure of 20 psia and 120°F.

	mol/h		*lb_m/h*	
Component	*Gas*	*Liquid*	*Gas*	*Liquid*
H_2S	99.0	5.2	3,366	177
CO_2	256.0	1.2	11,264	53
N_2	60.0	—	1,680	—
C_1	2,328.0	1.4	37,248	22
C_2	768.0	15.4	23,040	462
C_3	503.0	115.0	22,132	5,060
iC_4	63.0	59.0	3,654	3,422
nC_4	154.0	211.0	8,932	12,238
iC_5	39.0	157.0	2,808	11,304
nC_5	70.0	359.0	5,040	25,848
Heavy	42.0	5,000.0	3,990	1,300,000
	4,382.0	5,924.2	123,154	1,358,586

Binary Distillation 9

9.1 Introduction

In industrial practice the separation of two volatile components is usually carried out on a continuous basis. This is accomplished through the use of a distillation or fractionation column as shown schematically in Figure 9-1. The feed is introduced continuously at some midpoint in the column. Heat that is introduced to the reboiler vaporizes a portion of the liquid. This vapor rises through the column because its density is less than that of the descending liquid. Stages are provided in the column to allow intimate contact of the vapor and liquid. These plates are made in a number of configurations but from the standpoint of this discussion are important only in that they allow the liquid to flow downward through the column and the vapor to rise upward through the column with periodic mixing and separation. The vapor entering a plate from the tray below is at a higher temperature than the liquid descending to that plate from the plate above. The vapor will be cooled slightly, with some condensation of heavier materials occurring; the liquid will be heated with some corresponding vaporization of lighter components. The vapor leaving the top plate of the column enters the condenser, where heat is removed by cooling water or some other cooling medium. A portion of the condensed liquid is returned to the column as reflux liquid and the remainder becomes the distillate product.

The combination of vapor generation in the reboiler and liquid condensation in the condenser, with both streams being returned to the tower, differentiates continuous distillation from the equilibrium or differential flash separations discussed in Chapter 7. The reboiler vapor and reflux liquid allow for much-

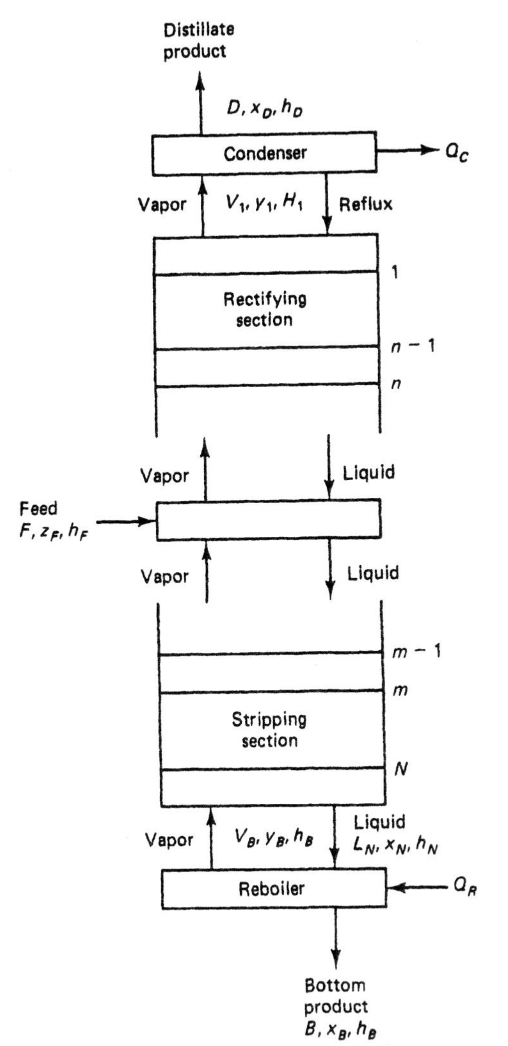

Figure 9-1 Schematic of a continuous distillation column

higher-purity products and at the same time provide a much greater recovery of the usable materials fed to the column.

The condenser may operate as either a total condenser or a partial condenser. In a total condenser all vapor entering the condenser is condensed to liquid and the reflux that is returned to the column has the same composition as the distillate or overhead product. In the case of a partial condenser only a portion of the vapor entering the condenser is condensed to liquid. In most partial condensers only sufficient liquid will be condensed to provide reflux for the tower. In some cases, however, more liquid will be condensed than is required for reflux and there will actually be two distillate products, one a liquid having the same composition as the reflux and the other a vapor product. In either type of partial condenser the vapor and liquid leaving the condenser are very close to being in equilibrium with each other.

In most cases the reboiler will operate as a partial reboiler vaporizing only a portion of the liquid entering from the bottom plate of the column. In such an operation the liquid and vapor leaving the reboiler are very nearly in equilibrium. In certain cases, such as the control of heat input, the reboiler may operate as a total reboiler with complete vaporization of the liquid that enters.

The section of the column between the feed plate and the condenser is usually called the *rectifying* or *enriching* section of the column. In the rectifying section the top plate is normally denoted as tray 1 and the plate immediately above the feed is the Nth plate. The section of plates between the feed plate and the reboiler is usually referred to as the *stripping* or *exhausting* section. The bottom plate in the stripping section is numbered stage 1 and the top plate is the Mth stage.

The nomenclature used for identifying streams and compositions in a distillation column is simple once understood. In nearly all cases stream flows and compositions are on a molar basis. Table 9-1 gives a summary of the nomenclature. For the distillate flow rate, for example, the total molar flow would be D; the composition of the liquid in the distillate product would be x_D; the composition of the vapor in the distillate product would be y_D; the moles of one component in the distillate product would be l_D; the moles of one component in the vapor phase of the distillate product would be v_D.

The design of a distillation column for continuous operation requires information about the interrelationship of three variables: the number of plates required, the reflux rate required, and the heat input required in the reboiler. These three variables determine both the physical size of the column in terms of diameter and height, and the operating costs in terms of condenser cooling and reboiler heat requirements.

9.2 Principles of Column Design

The approach used here in carrying out distillation calculations will be to consider a column composed of theoretical or equilibrium stages. The calculations carried out will provide the number of theoretical stages required to produce a specified distillate and bottom product from a feed of specified composition. After the number of theoretical plates has been determined, the number of actual plates can be calculated through the use of appropriate tray efficiencies.

A sketch of a theoretical or equilibrium plate is shown in Figure 9-2. The plate shown is an arbitrary plate in the rectifying section of the column and is designated as plate n. If the plate had been an arbitrary plate in the stripping section of the column, it would have been designated as plate m. Since complete mixing is assumed on a theoretical plate, the liquid entering the plate from the

TABLE 9-1. DISTILLATION COLUMN NOMENCLATURE

	Feed	Distillate Product	Bottom Product	Liquid Stream	Vapor Stream	Reflux
Molar flow rate	F	D	B	L	V	L_0
Liquid-phase composition	x_F	x_D	x_B	x	—	x_0
Vapor-phase composition	y_F	y_D	—	—	y	—
Moles of one component in liquid	l_F	l_D	b	l	—	l_0
Moles of one component in vapor	v_F	v_D	—	—	v	—

Liquid phase enthalpy [Btu/lb mol (energy/kgmol)], h (subscript identifies stream and/or plate)
Vapor phase enthalpy [Btu/lb mol (energy/kgmol)], H (subscript identifies stream and/or plate)
Any component in a mixture of two or more components, i
Total number of components in a mixture, C
Condenser duty (energy/unit of time or energy/unit of feed), Q_c
Reboiler duty (energy/unit of time or energy/unit of feed), Q_r
Number of plates in rectifying section, N
Any plate in rectifying section, n
Number of plates in stripping section, M
Any plate in stripping section, m

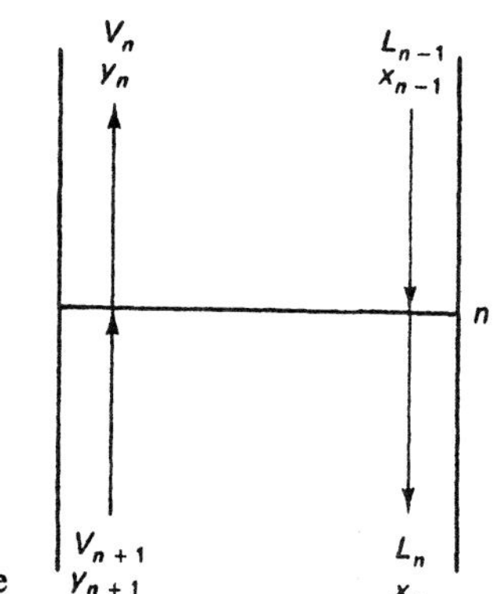

Figure 9-2 Equilibrium plate

tray above (L_{n-1}) and the vapor entering the plate from the tray below (V_{n+1}) are assumed to mix in such a way that thermodynamic equilibrium is obtained in the resulting mixture. The mixture is then separated into a vapor stream (V_n) and a liquid stream (L_n) that are in thermodynamic equilibrium. If the compositions of L_{n-1} and V_{n+1} were known and their states specified, the compositions and states of V_n and L_n could be calculated through thermodynamic relationships. This procedure gives good results in estimating the performance of actual trays in a distillation column even though the actual trays operate at steady state but not at equilibrium.

The first step in making calculations for a given distillation problem is to define the separation that is to be made. This must be done in such a way that a unique problem exists with only one solution. If more than one solution exists, there could be a great deal of confusion as to which is the better solution. In the case of a two-component, or binary, system the separation may be specified in any one of several ways.

1. The percentage composition of one component in the distillate product and the other component in the bottom product may be specified. In this case an overall material balance combined with component balances will show that a unique separation has been specified.

Example 9.1

Suppose that a feed containing 60% A and 40% B is to be fractionated. The overhead product is to contain 95% of component A and the bottom product is to contain 87% of component B. If 100 mol/h of feed enters the column, determine the quantities of the distillate and bottom products.

Solution:

The overall material balance for the column is

$$F = B + D$$

A balance around the column based on component A will be

$$0.6F = 0.13B + 0.95D$$

A balance around the column based on component B is

$$0.4F = 0.87B + 0.05D$$

These equations can be solved simultaneously for the total molar flow rates of distillate and bottom products. On the basis of 100 mol of feed to the column, 57.3 mol of distillate and 42.7 mol of bottom product will result.

2. The percentage recovery of one component in the overhead stream and the other component in the bottom stream may be specified.

Example 9.2

Using the same feed composition as that given in Example 9.1 and specifying the recovery of 90% of component A in the overhead product and 96% of component B in the bottom product, determine the quantities of the distillate and bottom products.

Solution:

The distillate is

$$D = (0.9 \times 60) + (0.04 \times 40)$$
$$= 54.0 + 1.6 = 55.6 \text{ mol}$$

For the same case the bottom product will be

$$B = (0.1 \times 60) + (0.96 \times 40)$$
$$= 6.0 + 38.4 = 44.4 \text{ mol}$$

3. The percentage composition and percentage recovery of one component in one stream can be specified.

Example 9.3

Using the same feed as before, a separation is to produce an overhead product that contains 95% component A. In addition, 90% of the A entering in the feed is to be recovered in the overhead product. Calculate the distillate flow rate.

Solution:

The amount of A in the overhead product will be

$$d_A = 0.90 \times 60 = 54 \text{ mol}$$

The total moles of distillate flow is

$$D = \frac{54}{0.95} = 56.8 \text{ mol}$$

4. The percentage of one component in the overhead and the total distillate rate may be specified.

Example 9.4

For the feed composition given in Example 9.1, a separation is to be made such that the overhead product is to contain 98% component A and 2% component B. For every 100 mol of feed entering the column, 50 mol is to be distillate product. Calculate the amount of A and B in the overhead.

Solution:

$$d_A = 50.0 \times 0.98 = 49.0 \text{ mol of } A$$

$$d_B = 1.0 \text{ mol of } B$$

Each of these procedures establishes a distillation separation in such a way that only one solution to the problem exists. A unique combination of the number of plates and reflux rate required for each of the separations will exist. Once the number of plates has been established, the required reflux rate is uniquely specified. Conversely, once a reflux rate is established, the number of plates is uniquely specified. The reboiler heat input is actually a dependent variable which is determined by other variables. Of course, the number of plates and the reboiler heat input could be considered unique, thus making the reflux rate or condenser cooling load dependent. This is not the usual approach, however.

In a binary system, once the separation to be made has been specified, the distillate composition and rate are fixed. Calculations to determine the number of plates required for the separation would necessitate going through the following steps.

1. Specify the tower operating pressure. Both temperature and pressure influence the distribution of a given component between the vapor and liquid at equilibrium. The operating pressure of the column is normally fixed by the condensing temperature of the material going overhead from the column. If the column has a total condenser, the condenser will be assumed to operate at the bubble-point temperature of the overhead product. For a partial condenser, the condenser will be assumed to operate at the dew-point temperature of the overhead product. The tower pressure is determined by making the appropriate calculation (bubble point or dew point) at the temperature that can be achieved in the condenser by the cooling medium to be used. In some cases, tower pressure cannot be fixed by the condenser temperature. Since some materials are heat sensitive, the use of cooling water in the condenser might cause an excessively high temperature to be reached in the reboiler. In this case, the tower bottom temperature should be fixed at the maximum safe value and the tower pressure fixed by a bubble-point calculation on the bottom product.

The condenser temperature would then be calculated. A satisfactory cooling medium must then be found to reach the necessary condensing temperature.

2. Specify the reflux rate for the tower and carry out heat and material balances around the condenser to determine the amount of cooling required and the composition of the vapor leaving the top tray.

3. Assume that the top tray operates as a theoretical plate and perform a dew-point calculation on the vapor leaving the top tray to find the liquid composition. This calculation also determines the temperature of the top tray.

4. Determine the amount and composition of the vapor entering the top tray from the second tray in the column. This is accomplished by assuming a flow rate for the liquid leaving the top tray and calculating the amount and composition of the vapor entering the tray by a material balance around the tray. A heat balance is then made around the tray. If the heat balance closes, the quantity assumed for the flow rate of liquid is correct; if the balance does not close, a new value for the liquid flow rate must be assumed and the material and heat balances recalculated.

5. Repeat the procedure above for the second tray of the column and for succeeding trays in the rectifying section, until the liquid leaving the tray has the same composition as the liquid in the feed.

6. Determine the heat input to the reboiler by a heat balance around the column.

7. Calculate the amount and composition of the vapor leaving the reboiler. The composition is calculated as the vapor in equilibrium with the bottom product. The vapor flow rate is calculated by assuming an amount of vapor and then making a heat balance around the reboiler. If the heat balance closes, the assumed vapor rate is correct; if the heat balance does not close, a new vapor rate must be assumed and the balances recalculated.

8. Calculate the temperature and compositions for the bottom plate of the column and each succeeding plate, until the composition of the vapor in the feed is reached.

9.3 McCabe–Thiele Method

The procedure described above for calculating the number of plates required in a tower is obviously long and tedious. One way of shortening the calculations would be through the use of an x–y diagram, as discussed in Chapter 7. Using the x–y diagram, compositions of streams can be read graphically. This still leaves the laborious material and heat balances to be calculated. However, procedures have been developed that can simplify these also.

Material balances for a single tray

The material balance for one component around a plate in the rectifying section of a distillation column, such as shown in Figure 9-2, can be written as

$$y_{n+1}V_{n+1} + x_{n-1}L_{n-1} = y_nV_n + x_nL_n \tag{9-1}$$

The component balance relating passing streams in the column is found from a component material balance around the rectifying section of a distillation column as shown in Figure 9-3. Thus

$$y_{n+1}V_{n+1} - x_n L_n = x_D D \tag{9-2}$$

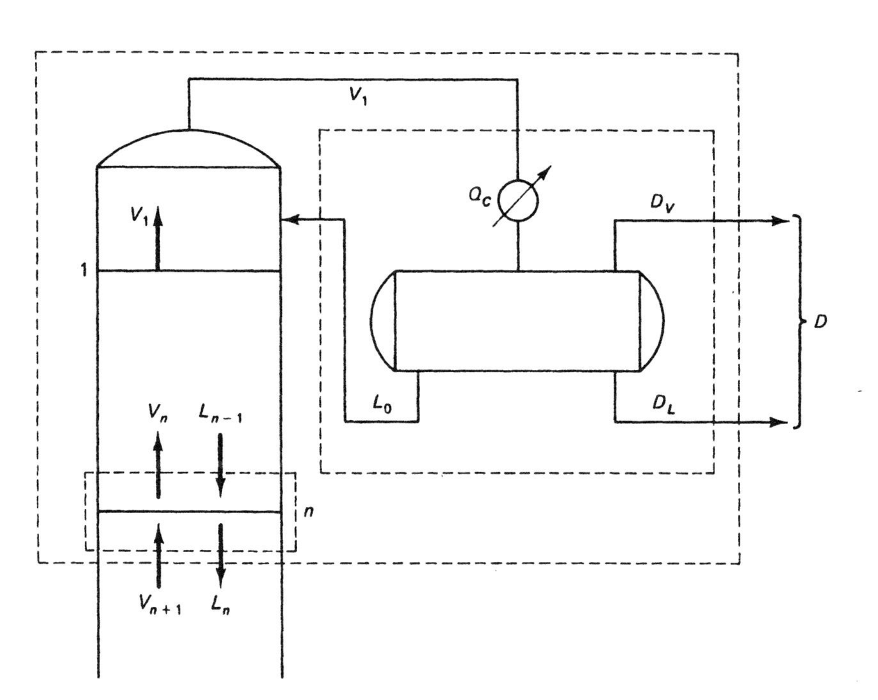

Figure 9-3 Material and energy balance envelopes for rectifying section

This is an important relationship since it points out that the difference between passing streams at any point in the rectifying section of the column is the amount of that component in the distillate product. Equation (9-2) can be rearranged as

$$y_{n+1} = \frac{x_n L_n}{V_{n+1}} + \frac{x_D D}{V_{n+1}} \tag{9-3}$$

This equation relates the composition of the vapor rising from any tray in the rectifying section to the composition of the liquid leaving the plate above. Equation (9-3) is defined as the operating line for the upper section of the column.

Equation (9-3) can be simplified for use by making the following assumptions.

1. The heat of mixing is zero or negligible for the two components in the mixture.
2. The molar latent heats of vaporization for the two components are equal or differ by a negligible amount.

When these assumptions apply

$$L_0 = L_1 = L_2 = \cdots = L_{n-1} = L_n \tag{9-4}$$

and

$$V_1 = V_2 = V_3 = \cdots = V_n = V_{n+1} \tag{9-5}$$

Equations (9-4) and (9-5) define a column operation known as *constant molar overflow*. In this type of operation the liquid overflowing succeeding plates in a given section of the column is constant in terms of total moles flowing per unit of time. The same is true for the vapor flow. Compositions of the vapor and liquid streams will change from plate to plate, but the total molar flow will remain constant. However, the vapor and liquid rates in the rectifying and stripping sections of the column will not usually be the same.

With these assumptions the material balance equation simplifies to

$$y_{n+1} = \frac{x_n L}{V} + \frac{x_D D}{V} \quad \text{or} \quad y = \frac{xL}{V} + \frac{x_D D}{V} \tag{9-6}$$

The material balance equation is now in a form that can be easily used for tray-to-tray calculations. The value for x_n is known from the dew-point calculation on the vapor leaving the tray. A material balance calculation will yield the composition of the vapor entering the tray. Once this calculation has been made, the composition of the liquid leaving the tray may be determined by a dew-point calculation on the vapor composition. Calculations proceed down the column by using the x–y plot to determine the equilibrium compositions, and by using the material balance equation to calculate the compositions of passing liquid and vapor streams.

Upon considering the material balance equation in terms of the variables on the x–y graph, we note that it has the form

$$y = mx + b \tag{9-7}$$

where

$$m = \frac{L}{V}$$

and

$$b = \frac{x_D D}{V}$$

This is the equation of a straight line on x–y coordinates. The slope of the line is L/V and the y intercept (when $x_n = 0$) is $x_D D/V$.

Following the procedure outlined above, a component material balance written around the stripping section of the column shown in Figure 9-4 will yield the operating-line equation for the bottom section of the column.

$$y_m = \frac{x_{m-1}\bar{L}}{\bar{V}} - \frac{x_B B}{\bar{V}} \quad \text{or} \quad y = \frac{x\bar{L}}{\bar{V}} - \frac{x_B B}{\bar{V}} \tag{9-8}$$

This can also be plotted on the x–y diagram as a straight line with a slope of $\bar{L}/\bar{V}$ and a y intercept of $-x_B B/\bar{V}$. The bars above the flow rates will be used for the stripping section of the column.

These two material balance equations are referred to as the *operating lines* for the column. Their relative positions on the x–y composition plot depend on

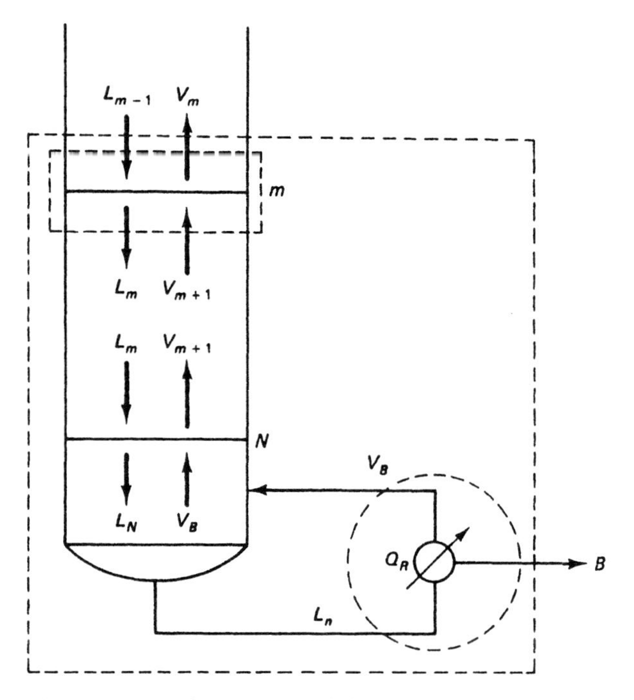

Figure 9-4 Material and energy balances for stripping section

the reflux rate and boil-up rate, which are determined by the manner in which the column operates. Equation (9-6) can be plotted as shown in Figure 9-5, and used graphically to determine the composition for each tray and the number of trays required for the separation.

The use of the two material balance or operating-line equations and the equilibrium curve to carry out plate calculations is outlined below.

1. Locate the rectifying-section operating line. This can be done in a number of ways, but perhaps the simplest is to evaluate the value for y when $x_n = x_D$. Under these conditions $y_{n+1} = x_D$. This means that the rectifying-section operating line goes through the point $y_{n+1} = x_n = x_D$. The value of the y intercept for the equation is also known. The operating line can be drawn as a straight line through the point x_D on the 45° line and the y intercept as shown in Figure 9-5.

2. Determine the composition of the liquid leaving the top tray in the column. If the column has a total condenser, $y_1 = x_D$. To determine the liquid in equilibrium with y_1, construct a horizontal line from the y_1 composition to the equilibrium curve along dashed line 1 and then draw a vertical line to the x-composition axis along dashed line 2. This gives x_1.

3. Determine the composition of the vapor entering tray 1 (y_2) by a material balance. This is accomplished graphically by drawing a vertical line from

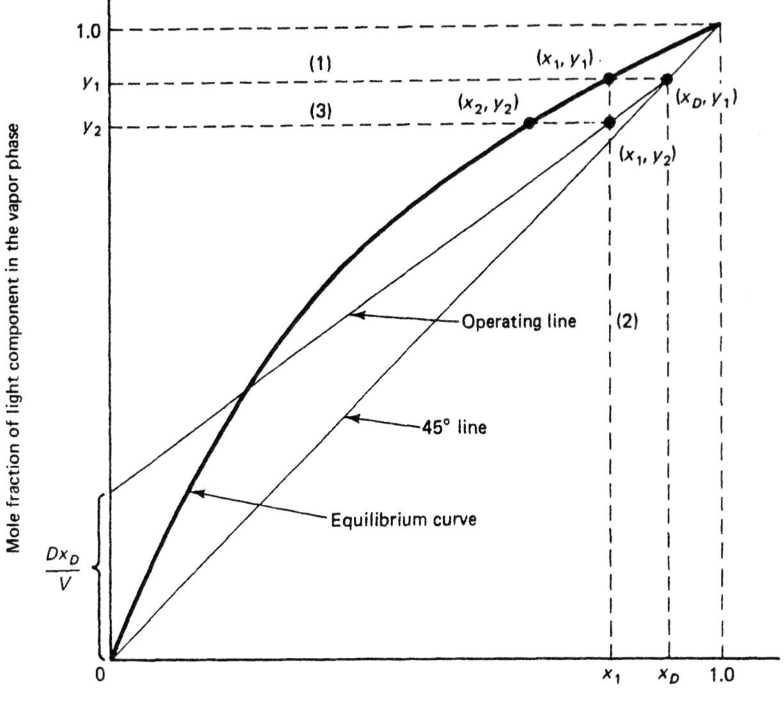

Figure 9-5 Graphical construction of equilibrium curve and operating line in rectifying section

the composition of x_1 along dashed line 2 until the operating, or material balance, line is reached and then drawing a horizontal line along dashed line 3 to the y-composition axis.

4. Once y_2 is known, the composition of the liquid in equilibrium with y_2 can be determined from the equilibrium curve. This process may be repeated as many times as needed to reach the feed plate composition.

In this way, calculations for each tray in the column can be carried out graphically. This is much more quickly and easily done than using the same procedures analytically. The number of theoretical plates required for the rectifying section of the column can be determined by applying this graphical procedure.

Location of the feed plate

The manner of operation of the feed plate in a distillation column is shown in Figure 9-6. Under the restricting assumptions applied, the liquid that leaves the bottom tray of the rectifying section adds directly with the liquid portion of

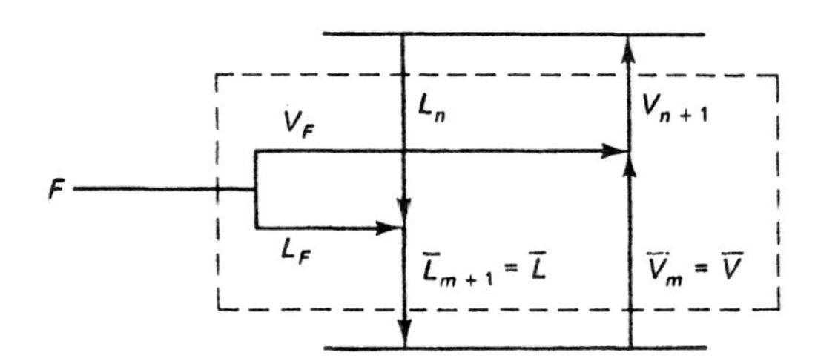

Figure 9-6 Feed plate model

the feed to form the liquid that enters the stripping section of the column. Similarly, the vapor that rises from the top tray of the stripping section adds to the vapor portion of the feed to form the vapor that enters the rectifying section of the column. Analytically, these relationships are expressed as

$$L_{m+1} = L_n + L_F \tag{9-9}$$

and

$$V_{n+1} = V_m + V_F \tag{9-10}$$

In the stripping section of the column the following notation will be used:

$$L_{m+1} = \bar{L} \quad \text{and} \quad V_m = \bar{V} \tag{9-11}$$

An overall material balance may be written around the feed plate as

$$F + L - \bar{L} = V - \bar{V} \tag{9-12}$$

If a term q is defined such that q is the moles of the saturated liquid in the feed divided by the total amount of feed, the flow rates above and below the feed plate are related as shown:

$$\bar{L} = L + qF \tag{9-13}$$

and

$$V = \bar{V} + (1 - q)F \tag{9-14}$$

By subtracting Eq. (9-8) from Eq. (9-6), we obtain

$$y(V - \bar{V}) = x(L - \bar{L}) + x_D D + x_B B \tag{9-15}$$

A component material balance around the tower shown in Figure 9-1 is

$$Fz_F = Dx_D + Bx_B \tag{9-16}$$

By combining Eqs. (9-12) through (9-16), we obtain the feed-line equation

$$y = \frac{q}{q-1}x - \frac{z_F}{q-1} \tag{9-17}$$

This line represents the locus of the intersections of the rectifying- and stripping-section operating lines. It is frequently referred to as the *q line*. Letting $x = x_F$ results in $y = x_F$ so that the q line passes through the point $(x = x_F, y = x_F)$ and has a slope of $q/(q - 1)$.

Five different possible feed conditions exist. The location of the q line for each of these feed conditions is shown in Figure 9-7 and is summarized as follows:

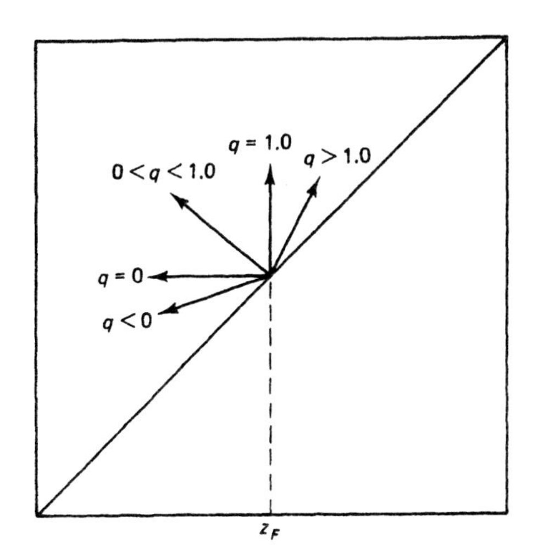

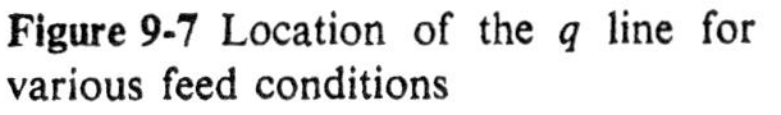
Figure 9-7 Location of the q line for various feed conditions

1. When the feed is a saturated liquid, $q = 1$. For this condition the q line is vertical.
2. If the feed is a saturated vapor, $q = 0$ and the q line is horizontal.
3. For a feed that is a mixture of saturated vapor and liquid, q has a value between 0 and 1 and the q line falls with an angle somewhere between 180 and 270° clockwise from the positive x axis.
4. When the feed is a superheated vapor, q will be less than 0 and the q line will fall 90 to 180° clockwise from the positive x axis.
5. If the feed is a subcooled liquid, the q line falls between 270 and 360° clockwise from the positive x axis.

In most cases the feed will be either a saturated liquid or a mixture of vapor and liquid. On some occasions, however, the feed will be a saturated vapor. Only rarely will subcooled liquid or superheated vapor feeds be encountered.

Equilibrium stages

The steps to be followed when constructing a McCabe–Thiele diagram are as follows:

1. Evaluate the necessary number of points and construct an equilibrium curve.
2. Draw the rectifying-section operating line through the point ($x = x_D$, $y = y_D$) to intersect with the y axis at $x_D D/V$.
3. Draw the q line through the point ($x = x_F$, $y = x_F$) with the proper slope.
4. Construct the stripping-section operating line through the point ($x = x_B$, $y = x_B$), and through the intersection of the rectifying-section operating line with the q line.

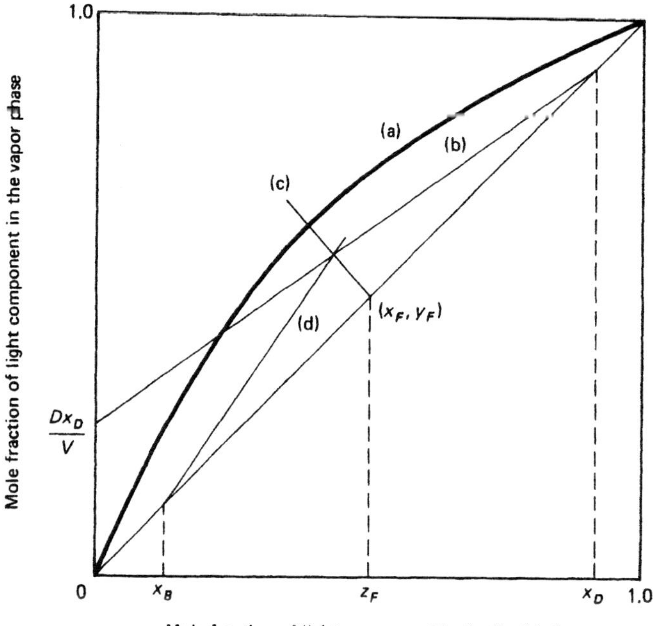

Figure 9-8 Construction for graphical solution

Figure 9-8 shows the order in which these steps are carried out and the completed construction.

The construction shown in Figure 9-8 can be used for graphical determination of the number of theoretical plates required for the rectifying and stripping sections of the column. Figure 9-9 shows the graphical construction of the plates, which is carried out as follows:

1. The diagram, including the equilibrium curve, the q line, and the rectifying- and stripping-section operating lines, is drawn as discussed previously and as shown in Figure 9-9.

2. If the column has a total condenser, the vapor leaving the top tray, the reflux, and the distillate will have the same composition, x_D. The horizontal line from the point x_D to the y axis indicates that y_1 and x_D are of the same composition. Our next step is to locate the liquid composition in equilibrium with y_1. This is found by constructing a horizontal line from y_1 to intersect the equilibrium curve and then drawing a vertical line from the equilibrium curve to the x-composition axis. This is shown on Figure 9-9 as point 1, with the composition x_1. By this construction one theoretical plate on the graphical solution has been stepped off.

3. This procedure is repeated for the second plate. The composition y_2 is found by extending a horizontal line from the operating line at composition x_1.

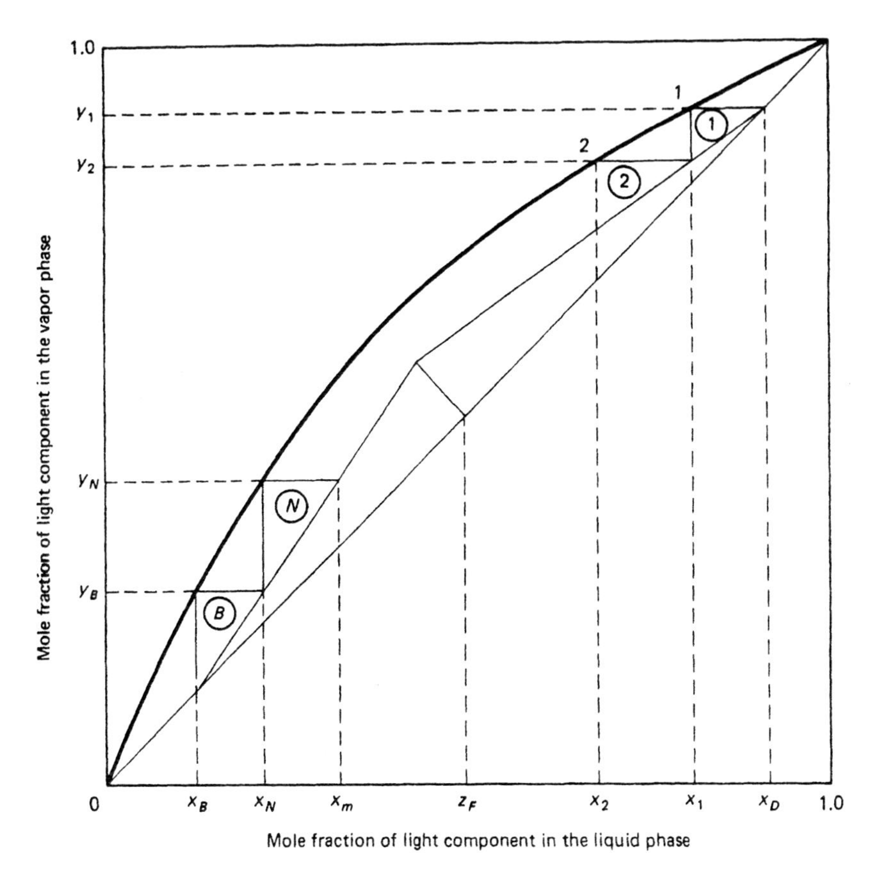

Figure 9-9 Graphical construction of theoretical plates

The liquid composition x_2, which is in equilibrium with y_2, is found by extending a vertical line from point 2 on the figure.

4. This stepwise procedure is continued until the feed plate is reached. The two theoretical plates stepped off in the rectifying section of the column are indicated by the circled numbers on the figure.

5. Plates in the stripping section are stepped off in a similar manner by starting at the point x_B. The vertical line gives the composition of the vapor leaving the reboiler and the first horizontal line gives the composition of the liquid leaving the bottom plate of the stripping section. The construction continues in this way until the feed plate is reached. The theoretical plates in the stripping section are also shown on Figure 9-9 as circled numbers. The first plate represents the equilibrium attained in the reboiler, which is assumed to operate as a theoretical plate.

If the condenser had been operating as a partial condenser instead of a total condenser, the first step on the rectifying section construction would have been slightly different from that shown in Figure 9-9. In the case of a partial condenser, a vapor overhead product with a composition of y_D is produced. The reflux is in equilibrium with y_D, so the first step would be constructed by drawing a horizontal line from the distillate composition on the y axis to intersect the equilibrium curve. A vertical line would then be drawn to the x-composition axis. This locates the composition of the reflux. A more detailed discussion of partial condensers is presented in Section 9.5.

This method of graphical solution for binary systems is known as the *McCabe–Thiele procedure* (1925). It is named for the two men who originally presented this technique of solution. The two assumptions used in developing the graphical procedure do not, in most cases, prove to be serious limitations. In most industrial applications the two components to be separated will be adjacent members of an homologous series. This means that in most cases the heat of mixing of the two components is negligible and their enthalpies are sufficiently close so that the assumption of constant molar overflow in the column is reasonably accurate.

Proper feed plate location has been an inherent assumption of the methods of graphical construction discussed so far. In plant operations the feed may or may not be introduced at the proper point in the column. The graphical construction of an improperly located feed plate is shown in Figure 9-10. If we assume that the feed is introduced at a point too low in the column, equilibrium stages would continue to be stepped off on the rectifying-section operating line past the point of its intersection with the q line. As shown, each plate stepped off in this manner will make a smaller separation than it would have if the construc-

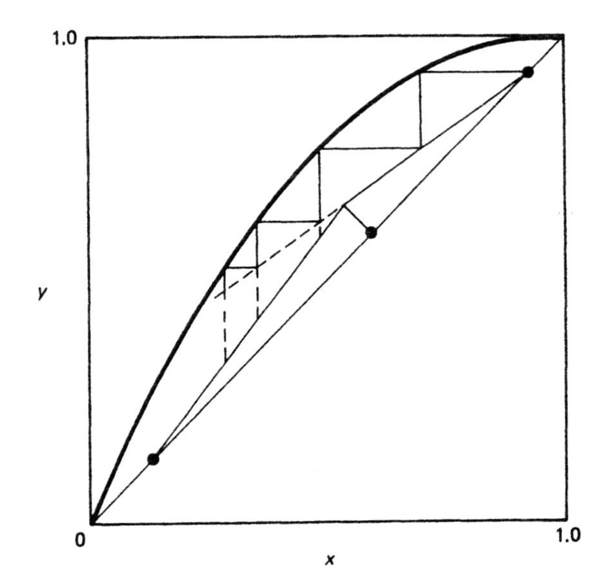

Figure 9-10 Construction showing effect of incorrect feed plate location

tion had shifted and the stripping-section operating line was used. This will be true regardless of the value for q. Changing the feed plate location does not change the column heat and material balances. The stripping section operating line will have the same slope regardless of the feed plate location. In the case of a feed that is introduced at a point too high in the column, the results will be the same except that the shift will be made to the stripping-section operating line before it intersects the q line.

Determination of minimum trays

When consideration is given to making a specified separation from a given feed, we realize that certain limits must be placed on the number of stages and on the reflux rate that can be used. For example, a simple flash separation cannot make 95% pure product from most mixtures. Consideration of the characteristics of the process suggests that some minimum number of theoretical trays is required to accomplish the specified separation; this is indeed true. This type of column operation is most commonly referred to as *total reflux*. Another and equally correct terminology is operation with *minimum trays*.

The equation for the operating line in the rectifying section of a column using the McCabe–Thiele solution is

$$y_{n+1} = \frac{Lx_n}{V} + \frac{Dx_D}{V} \tag{9-6}$$

The slope of the straight operating line, represented by Eq. (9-6), is L/V. The total material balance for the rectifying section of the column is

$$V = L + D \tag{9-18}$$

Equation (9-18) shows that L will always be less than V, at least as long as any distillate product is withdrawn. The maximum value for L/V would be unity. This can be achieved in two ways: by having D approach zero, or by having L become very large in comparison with D. The first of these choices is very easy to achieve in an operating column. The column is simply shut in and all of the overhead vapor is returned to the column in the form of reflux. The second option is more easily achieved by computer simulation. In this case the reflux is made arbitrarily large, with the distillate product retaining a value very close to that used for the operating column.

Under conditions such that L/V approaches unity, the rectifying-section operating line will coincide with a line drawn at a 45° angle through the origin. The same line of reasoning applies to the stripping-section operating line. Under conditions of total reflux or minimum trays, the two operating lines coincide with each other and with the 45° line. This construction is shown in Figure 9-11. The theoretical plates required for the separation are stepped off as before, with the only difference being that the 45° line is used as the operating line.

Study of the construction shown in Figure 9-11 shows that this corresponds to a maximum driving force for mass transfer, which will result in the minimum

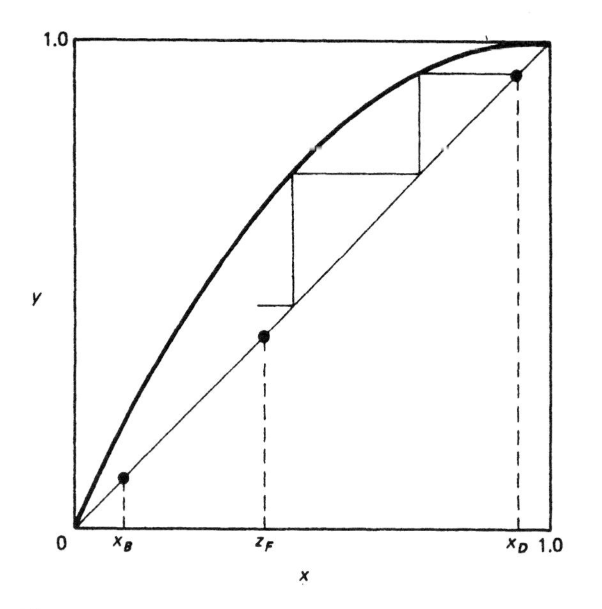

Figure 9-11 McCabe–Thiele construction for minimum trays

number of stages required for a given separation. The operating line cannot be removed farther from the equilibrium line than it is in Figure 9-11. This means that the separation obtained for each plate in the column is the maximum possible.

Minimum reflux determination

Two extremes of column operation are possible. The case of minimum trays or total reflux was just discussed. The other extreme, described as minimum reflux, is seldom encountered in an operating column. The concept of minimum reflux remains a mathematical interpretation rather than a physical reality. However, the concept of a minimum reflux is very important and will usually be calculated for any distillation separation being contemplated.

In Figure 9-12 three different rectifying-section operating lines are shown on a McCabe–Thiele plot. At the point where any one of these lines intersects the equilibrium curve, an infinite number of plates will be required to reach the point of intersection. As the increment of composition change accomplished by each tray becomes smaller, a region sometimes called a *zone of infinitude* is approached. In this zone, compositions, temperatures, and vapor and liquid flow rates are constant from tray to tray. This region of the tower may also be referred to as a *pinch zone*.

Although a minimum reflux condition occurs any time an operating line intersects the equilibrium line, only one of these represents the true minimum reflux rate for the separation. The minimum reflux of specific interest is that

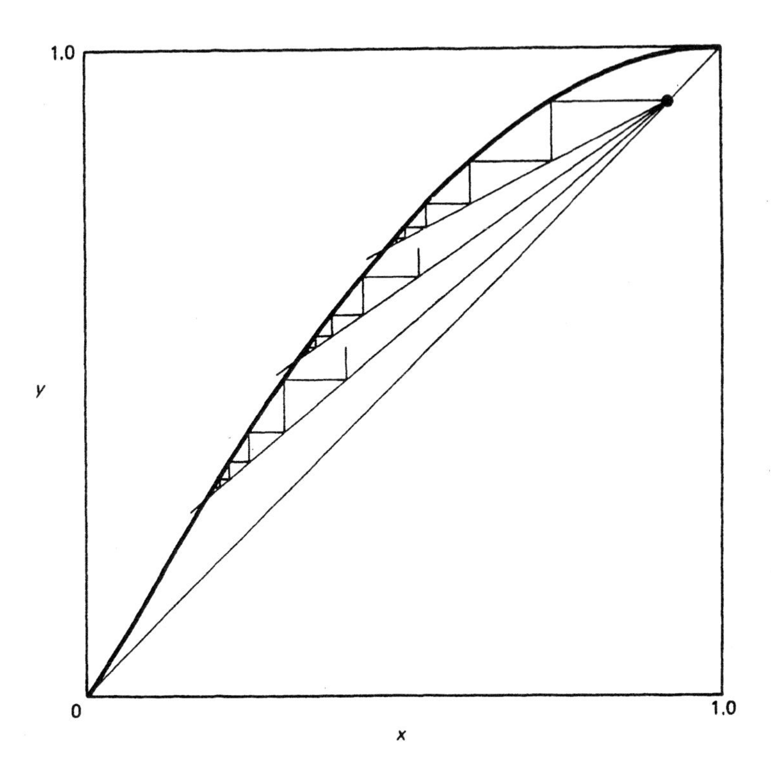

Figure 9-12 McCabe–Thiele construction showing infinite plates

rate which will permit making the required separation from the specified feed. This point will occur on the McCabe–Thiele diagram where the rectifying-section operating line, the stripping-section operating line, the feed q line, and the equilibrium curve all intersect at a common point. This construction is shown in Figure 9-13. That this is indeed the minimum reflux rate will become obvious after study of the illustration. If the reflux rate is lowered still further, the rectifying-section operating line will intersect the equilibrium curve at a value of x for the more volatile component higher than the feed composition. This means that even with an infinite number of plates, the distillate composition cannot be fractionated from the feed composition. On the other hand, if the reflux rate is increased, the intersection with the equilibrium curve will not occur because the rectifying- and stripping-section operating lines will intersect the q line away from, and inside, the equilibrium curve.

The minimum reflux rate can be calculated from the construction shown in Figure 9-13 in a number of ways. Perhaps the easiest way is to extend the rectifying-section operating line to its point of intersection with the y axis. This intersection is

$$\frac{Dx_D}{V} \tag{9-19}$$

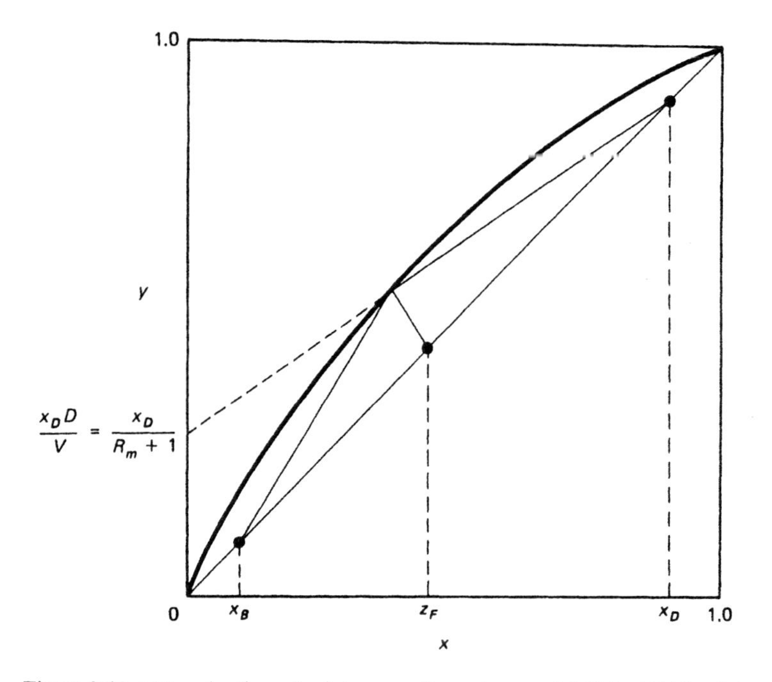

Figure 9-13 Determination of minimum reflux rate on a McCabe–Thiele plot

From an overall material balance around the rectifying section of the column, $V = L + D$. Substituting this into Eq. (9-19) directly determines the value for L, the reflux flow rate.

$$\frac{Dx_D}{L + D} = \frac{x_D}{L/D + 1} = \frac{x_D}{R + 1} \tag{9-20}$$

Since this value for the intercept was obtained under conditions of minimum reflux in the column, the value of L calculated will be the minimum reflux rate for the separation being considered. The minimum reflux rate can also be determined from the slope of the operating line at the minimum reflux condition. However, this procedure is generally not as accurate.

Example 9.5

A mixture containing 30% *n*-pentane and 70% *n*-heptane is to be continuously distilled to produce an overhead product that contains 98% *n*-pentane. Ninety-six percent of the *n*-pentane that enters the column in the feed is to be recovered in the overhead product. (a) Using the T–x diagram given in Figure 9-14, prepare an x–y diagram to use in carrying out distillation calculations by the McCabe–Thiele method. (b) Determine the minimum number of trays required to make this separation. (c) Calculate the minimum reflux ratio if the feed enters the column as a saturated liquid at its bubble point.

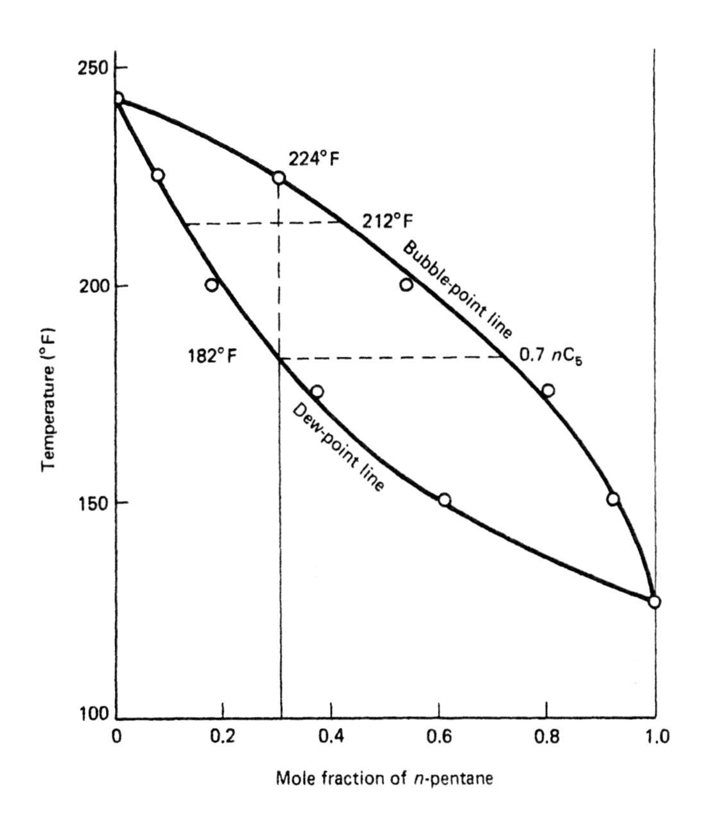

Figure 9-14 Temperature–composition diagram for the n-pentane/n-heptane binary at 25 psia

Solution:

(a) The T–x diagram presented in Figure 9-14 provides equilibrium vapor and liquid compositions. These are shown in Figure 9-15 as the equilibrium curve.

(b) Before we can determine the minimum number of trays, we must calculate the composition of the bottom stream. We will assume a basis of 1 mol of feed.

Distillate stream:

$$d_A = 0.3 \times 0.96 = 0.288 \text{ mol } n\text{-pentane in overhead}$$

$$D = \frac{d_A}{x_D} = \frac{0.288}{0.98} = 0.294 \text{ mol total overhead}$$

$$d_B = D - d_A = 0.294 - 0.288 = 0.006 \text{ mol } n\text{-heptane in overhead}$$

Bottom stream:

$$b_A = f_A - d_A = 0.3 - 0.288 = 0.012 \text{ mol } n\text{-pentane in bottom stream}$$

$$b_B = f_B - d_B = 0.7 - 0.006 = 0.694 \text{ mol } n\text{-heptane in bottom stream}$$

$$x_B = \frac{0.012}{0.012 + 0.694} = 0.017$$

The minimum number of theoretical trays is determined by stepping off plates,

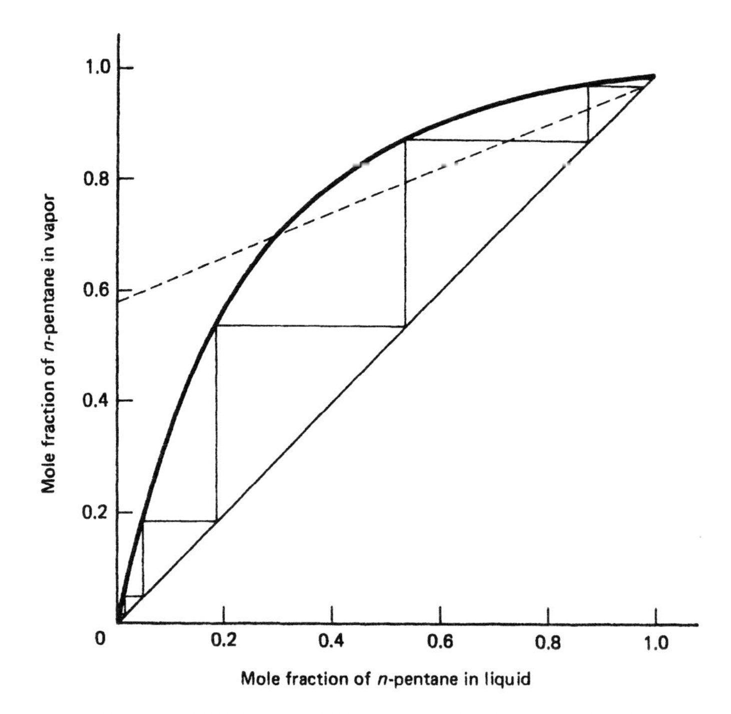

Figure 9-15 McCabe–Thiele construction for Example 9.5

starting at $x_D = 0.98$ and continuing until the bottom composition $x_B = 0.017$ is reached. As shown in Figure 9-15, the minimum number of trays is 5.0.

(c) The minimum reflux rate is determined by extending a line from the condenser composition $x_D = 0.98$ through the intersection of the equilibrium curve and q line. The q line is a vertical line that is extended from the feed composition $z_F = 0.3$. This line intersects the y axis at 0.58. Therefore,

$$\frac{x_D}{R_m + 1} = 0.58$$

$$R_m = \frac{x_D}{0.58} - 1.0$$

$$= \frac{0.98}{0.58} - 1.0$$

$$\left(\frac{L_0}{D}\right)_m = 0.69$$

The total moles of reflux is

$$L_{0,m} = D(0.69) = 0.294(0.69) = 0.203 \text{ mol}$$

The internal reflux ratio, L/V, is equal to the slope of the operating line in the rectifying section of the column. This can be found by simply measuring the slope with a ruler or calculating the value as follows:

$$\frac{L_0}{D} = \frac{L_0}{V - L_0} = \frac{L_0/V}{1 - L_0/V} = 0.69$$

Therefore, $L_0/V = 0.408$.

Example 9.5 demonstrates the method for determining the minimum number of stages and the minimum reflux ratio. The actual number of stages required to carry out a given separation depends on the actual reflux ratio. The calculation of the actual number of theoretical stages is shown in the following example.

Example 9.6

Determine the number of theoretical trays necessary to produce the overhead and bottom products in Example 9.5 if the actual reflux rate is 1.5 times the minimum value.

Solution:

$$\left(\frac{L_0}{D}\right)_{\text{actual}} = 1.5\left(\frac{L_0}{D}\right)_{\text{min}} = 1.5(0.69) = 1.035$$

A new y intercept must be calculated for the rectifying-section operating line.

$$\frac{x_D}{R+1} = \frac{0.98}{1.035 + 1.0} = 0.482$$

As shown in Figure 9-16, this line is drawn from the composition $x_D = 0.98$ to 0.482 on the y axis, but extending only to the point where it intersects the feed line. The stripping-section operating line is then extended from the reboiler composition $x_B = 0.017$ to the intersection of the rectifying-section operating line with the feed line. Stepping off trays we find that 9.5 theoretical stages are

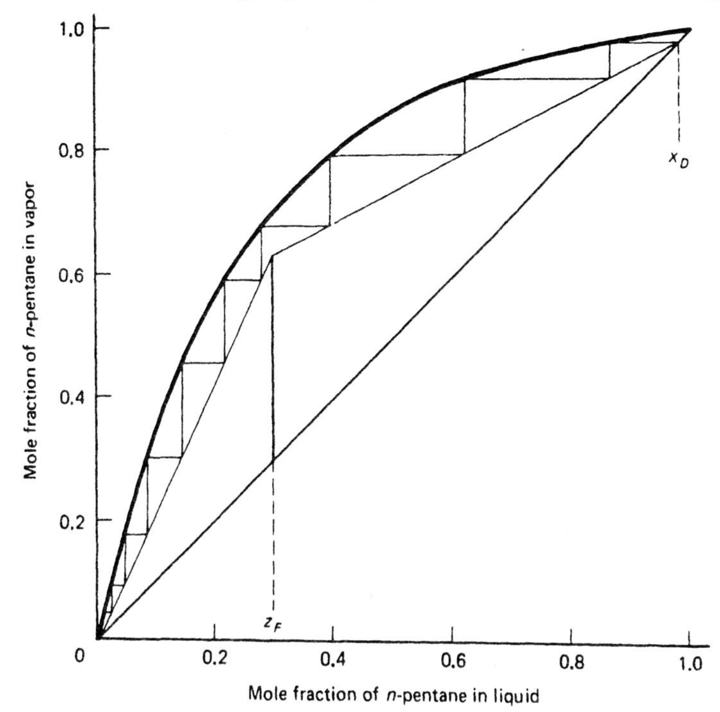

Figure 9-16 McCabe–Thiele construction for actual reflux rate for Example 9.6

required to make the separation. Since the reboiler functions essentially as a theoretical stage, the separation will require 8.5 theoretical trays plus a reboiler and a total condenser.

Multiple feeds and sidestreams

Not all distillation columns operate with a single feed and two products. Multiple product towers are used in many situations. They are particularly useful for cases in which a small amount of a volatile contaminant is present in the feed. In this case the overhead product will be the contaminant, and a side product somewhere in the rectifying section will be the desired product from the column. Multiple feed columns also occur in industrial practice. In many instances two feeds of different compositions can be fractionated more efficiently by introducing the feeds at different points in the column rather than by combining the feeds and introducing them at a single point. Although multiple feeds and side streams are not common in binary systems, use of the McCabe–Thiele diagram helps greatly in understanding the principles of operation for these columns.

Figure 9-17a shows a distillation column from which a product is withdrawn in the rectifying section of the column. In practice the side product may

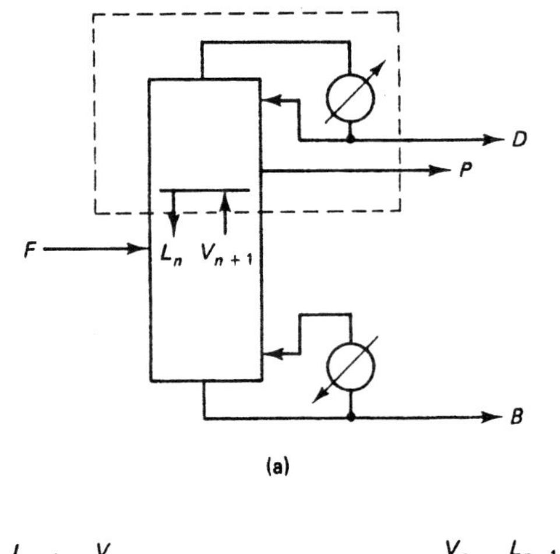

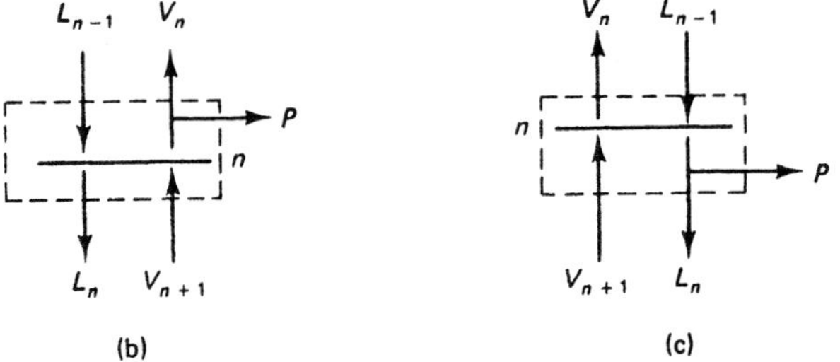

Figure 9-17 Distillation column with side product

be in either the rectifying or stripping section. The principle of operation and method of analysis are the same in either case. Figure 9-17b shows a vapor product being withdrawn from a tray in the rectifying section. For this case the material balance relationship around the plate is

$$V_{n+1}y_{n+1} + L_{n-1}x_{n-1} = (V_n + P)y_n + L_nx_n \tag{9-21}$$

where

$$V_{n+1} = V_n + P \tag{9-22}$$

Figure 9-17c shows the case for a liquid side-draw product. The plate material balance for a liquid side draw is

$$V_{n+1}y_{n+1} + L_{n-1}x_{n-1} = V_ny_n + (L_n + P)x_n \tag{9-23}$$

where

$$L_{n-1} = L_n + P \tag{9-24}$$

The side product has the composition of either the vapor or the liquid leaving the tray. Also, the withdrawal of a side product causes an upset of either the vapor or liquid flow from the tray. A material balance around the rectifying section of the column shown in Figure 9-17a for one component is

$$V_{n+1}y_{n+1} = L_nx_n + Px_P + Dx_D \tag{9-25}$$

Upon rearranging the equation above and replacing V_{n+1} by V and L_n by L, we obtain

$$y_{n+1} = \frac{Lx_n}{V} + \frac{Px_P + Dx_D}{V} \tag{9-26}$$

This represents the operating-line equation for the section of the column between the side draw plate and the feed. This line will intersect the q line for the side product and the operating line for the section of the column above the side draw at a common point. It will extend from this intersection to its y intercept, which is $(Px_P + Dx_D)/V$. Inspection of the operating-line equation for the section of column above the feed reveals that the operating line for the section between the side draw and the feed always has a slope that is less than the slope of the operating line for the section above the side draw. Construction for stepping off plates with the side-draw column is given in Figure 9-18. The construction shown is for a liquid side draw. The principles involved for a vapor side draw would be the same except that the q line for the vapor side product would be horizontal rather than vertical.

Figure 9-19 shows a schematic diagram of a distillation column with two feeds and two products. The overall material balance for this column is

$$F_1 + F_2 = D + B \tag{9-27}$$

For any component the material balance is

$$F_1z_{F,1} + F_2z_{F,2} = Dx_D + Bx_B \tag{9-28}$$

Between the top feed and the distillate product, and the bottom feed and the bottom product, the operating lines will have the same equations and form as the corresponding operating line equations for a simple column. By writing a

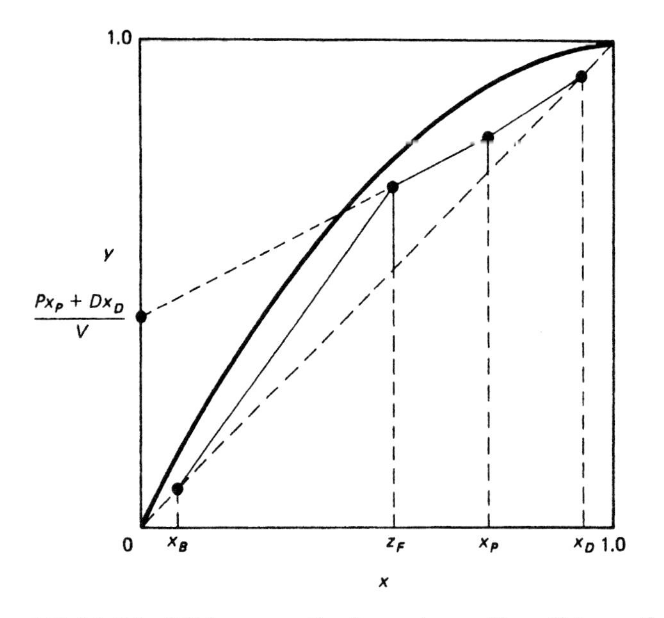

Figure 9-18 McCabe–Thiele construction for a column with rectifying-section side draw

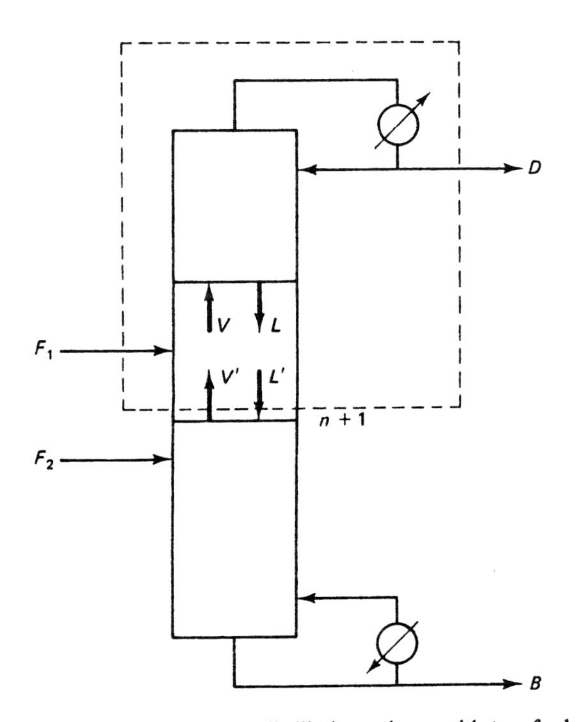

Figure 9-19 Schematic of a distillation column with two feeds

component material balance between the two feeds and around the top of the column, we obtain

$$V'y_{n+1} = L'x_n - F_1 z_{F,1} + Dx_D \tag{9-29}$$

or

$$y_{n+1} = \frac{L'x_n}{V'} + \frac{Dx_D - F_1 z_{F,1}}{V'} \tag{9-30}$$

In this equation L' and V' represent the molar flow rates of liquid and vapor, respectively, in the zone of the column between the two feeds. They are related to the liquid and vapor flows in the rectifying section of the column by the following equations:

$$L' = L + L_{F,1} \tag{9-31}$$

$$V' = V - V_{F,1} \tag{9-32}$$

If the top feed is a saturated liquid, L' will be greater than L and V' will be equal to V. In this case the slope of the operating line in the zone between the feed plates will be greater than the operating line for the section of the column above the top feed. In the case of a top feed that is a saturated vapor, L' will be equal to L and V' will be less than V. For this case also, the slope of the operating line for the section of column between the feeds will be greater than the slope in the section of column above the top feed. Thus for a saturated feed, the slope of the line between the two feed points will always be greater than the slope of the line for that section of the column above the feed. Figure 9-20 shows a McCabe–Thiele diagram for a two-feed/two-product distillation column. The two feeds are partially vaporized.

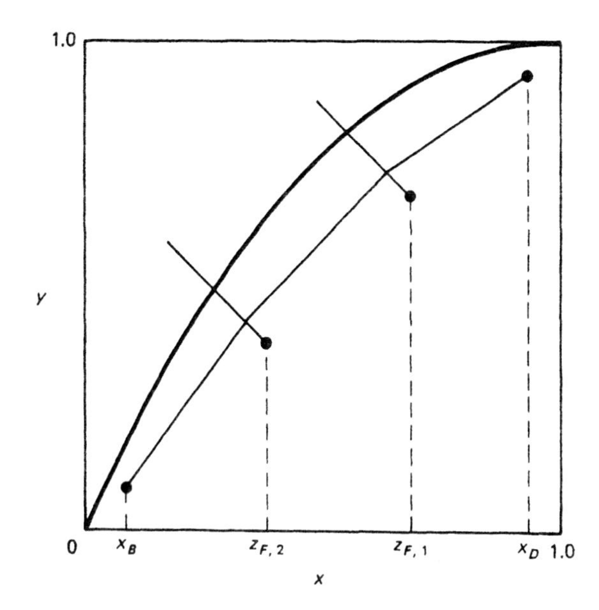

Figure 9-20 McCabe–Thiele construction of a column with two feeds

9.4 Enthalpy–Composition Method

Cases exist in which the assumptions of negligible heat of mixing and/or constant molar latent heats of vaporization do not apply. In these cases the McCabe–Thiele method will not give a good description of the separation being made. Ponchon (1921) and Savarit (1922) presented a graphical procedure which incorporates enthalpy balances as an integral part of the calculation. This procedure combines the material balance calculations with enthalpy balance calculations in such a way as to give not only vapor and liquid compositions in the column but also to provide information on the condenser and reboiler duties.

The overall material balance for the distillation column shown in Figure 9-1 is

$$F = D + B \tag{9-33}$$

For any component we can write the material balance around the column as

$$Fz_F = Dx_D + Bx_B \tag{9-34}$$

Upon writing an overall enthalpy balance for the column, we obtain

$$Fh_F + Q_R = Dh_D + Bh_B + Q_C \tag{9-35}$$

where h = enthalpy of the liquid stream, energy/mol,
Q_R = heat input to reboiler, energy/time,
Q_C = heat removed from condenser, energy/time.

The heat balance can be rearranged as

$$Fh_F = D\left(h_D + \frac{Q_C}{D}\right) + B\left(h_B - \frac{Q_R}{B}\right) \tag{9-36}$$

The enthalpy balance is rearranged in this way for convenience in plotting on the enthalpy–concentration diagram. The enthalpy ordinate on this diagram is expressed in typical units of Btu per unit of mass or moles. The total condenser and reboiler duties must be converted to these units before they can be plotted properly on the enthalpy–composition diagram. The points represented by the feed, distillate, and bottom streams can be plotted on the enthalpy–composition diagram as shown in Figure 9-21.

Substituting the overall material balance into the component material balance, we obtain

$$(D + B)z_F = Dx_D + Bx_B \tag{9-37}$$

Making the same substitution in the overall heat balance gives

$$(D + B)h_F = D\left(h_D + \frac{Q_C}{D}\right) + B\left(h_B - \frac{Q_R}{B}\right) \tag{9-38}$$

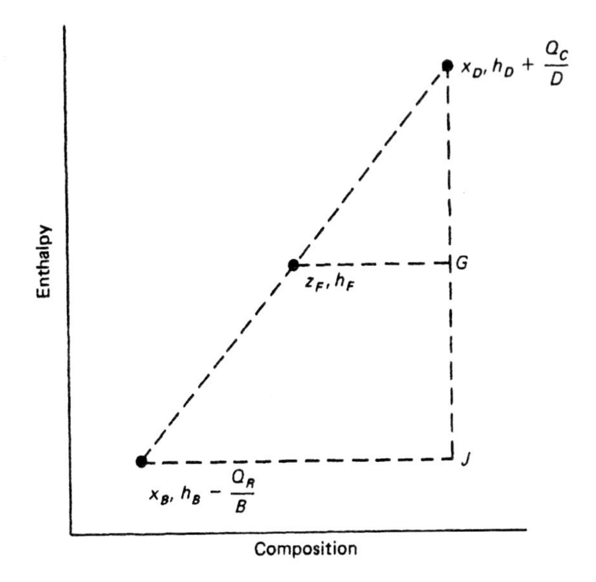

Figure 9-21 Enthalpy–concentration diagram showing enthalpy and material balances

Solving each of these two equations for D/B and equating the two results, we obtain

$$\frac{D}{B} = \frac{z_F - x_B}{x_D - z_F} = \frac{h_F - (h_B - Q_R/B)}{(h_D + Q_C/D) - h_F} \tag{9-39}$$

This can be rearranged as

$$\frac{z_F - x_B}{h_F - (h_B - Q_R/B)} = \frac{x_D - z_F}{(h_D + Q_C/D) - h_F} \tag{9-40}$$

Comparing Eq. (9-40) with the points plotted on Figure 9-21, we find that the left-hand side of Eq. (9-40) represents the slope of a straight line between the points $(x_B, h_B - Q_R/B)$ and (z_F, h_F). The right-hand side of Eq. (9-40) represents the slope of a straight line connecting the points (z_F, h_F) and $(x_D, h_D + Q_C/D)$. Since the slopes of these two lines are equal, a straight line connecting the points $(x_B, h_B - Q_R/B)$ and $(x_D, h_D + Q_C/D)$ must pass through the point (z_F, h_F). This means that all three points fall on the same straight line.

The amount of distillate has been shown in Eq. (9-39) to be proportional to the horizontal distance $\overline{z_F x_B}$, and the amount of bottoms has been shown to be proportional to the horizontal distance $\overline{x_D z_F}$. Consideration of the similar right triangles shown by the dashed lines in Figure 9-21 and the overall material balance for the column, shows that the amount of feed is proportional to the horizontal distance $\overline{x_D x_B}$. This leads to the inverse lever rule, which applies to stream quantities and enthalpies on the enthalpy–composition diagram in the same way as it applies to material balances on the temperature–composition diagram.

Consider the theoretical plate shown in Figure 9-2 and the manner in which it operates, as described on the enthalpy–composition diagram shown in Figure 9-22. The vapor feed to the plate, V_{n+1}, is a saturated vapor with composition y_{n+1}. The liquid feed from the plate above is L_{n-1} with composition x_{n-1}.

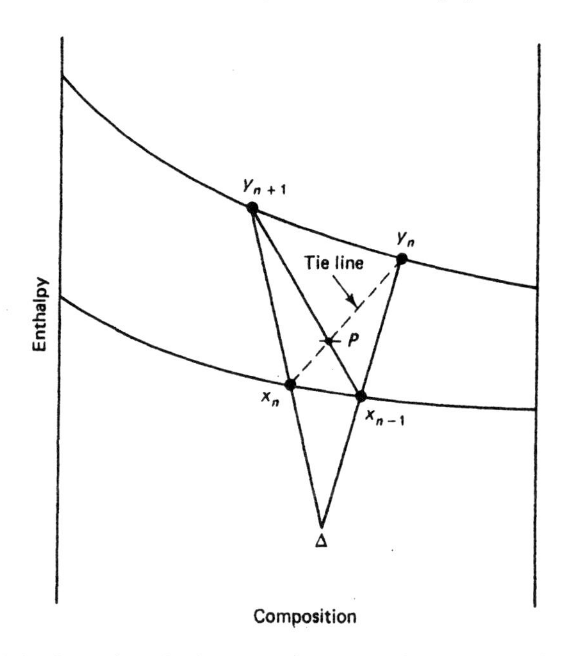

Figure 9-22 Operation of a single plate on an enthalpy–concentration diagram

The total flow to the plate is represented by point P, which falls on a straight line connecting x_{n-1} and y_{n+1}. On the basis of the inverse lever rule, point P is located such that the distance $\overline{y_{n+1}P}$ is proportional to the quantity of L_{n-1} and the distance $\overline{x_{n-1}P}$ is proportional to the quantity of V_{n+1}.

$$\frac{L_{n-1}}{V_{n+1}} = \frac{\overline{y_{n+1}P}}{\overline{x_{n-1}P}} \tag{9-41}$$

Since the vapor and liquid leaving the tray are in equilibrium with each other, they will lie on an equilibrium tie line. In addition, the sum of the liquid and vapor leaving the plate must equal the total flow to the plate. Thus the tie line must pass through point P, which represents the addition point of the vapor V_n and liquid L_n leaving the tray. The composition of these streams is found by the intersection of the tie line with the enthalpy–composition curves.

The material balances around plate n can be arranged as

$$V_{n+1} - L_n = V_n - L_{n-1} \tag{9-42}$$

and

$$V_{n+1}y_{n+1} - L_n x_n = V_n y_n - L_{n-1}x_{n-1} \tag{9-43}$$

The enthalpy balance around the plate can be written as

$$V_{n+1}H_{n+1} - L_n h_n = V_n H_n - L_{n-1}h_{n-1} \tag{9-44}$$

where H is the enthalpy of the vapor and h is the enthalpy of the liquid. These equations show that the difference in stream flows on both a material and enthalpy basis is the same above and below the plate. This means that the stream that must be added to V_{n+1} to generate L_n and the stream that must be added to V_n to generate L_{n-1} are the same. This is shown in Figure 9-22, with point Δ representing the difference point above and below the plate. This concept of a common difference point can be extended to a section of a column that contains any number of theoretical plates as long as an external feed is neither added to nor withdrawn from the column.

To set up the enthalpy–concentration diagram for determining the number of theoretical plates required to make a given separation, we must use the following procedure.

1. Develop the equilibrium curve and the enthalpy–concentration diagram for the mixture to be separated. In many cases the equilibrium curve is plotted on the same sheet of paper with the enthalpy–concentration diagram, as a matter of convenience.

2. Develop the split for the column and calculate the compositions of the distillate and bottom products. Locate these two compositions on the enthalpy–concentration diagram as shown in Figure 9-23. Also locate the feed composition and enthalpy.

3. Establish the reflux rate for the separation and locate the rectifying-section difference point, Δ_R. The point Δ_R will have a composition coordinate equal to the distillate composition and an enthalpy coordinate of $h_D + Q_C/D$. Assuming a total condenser for the column, the difference point will be located a distance Q_C/D above the saturated distillate product point x_D. The construction in Figure 9-23 shows that point y_1 represents the addition point of x_D and Δ_R. Line $\overline{x_D\Delta_R}$ is a material balance/heat balance equation around the condenser of the column.

4. Write a heat and material balance around the column and locate the stripping-section difference point Δ_S. The point Δ_S will have a composition coordinate represented by x_B, the bottom product. A straight line from Δ_R through the feed has already been shown to locate Δ_S. A straight line is drawn from Δ_R through x_F and extended to its intersection with the x_B composition coordinate. This intersection locates the stripping-section difference point Δ_S as shown in Figure 9-23.

5. Step off the plates graphically for the rectifying section of the column. The composition and enthalpy of y_1 have been determined by the balances around the condenser. The composition x_1, which represents the liquid leaving the top plate, is determined by the equilibrium curve since it is in equilibrium with y_1, the vapor leaving the plate. This is indicated by the dashed line on Figure 9-23. The composition y_2, which is then determined by a heat and material balance between x_1 and Δ_R, will fall on the saturated vapor line. This is located by drawing a line from x_1 to Δ_R. This procedure is continued until the feed plate is reached.

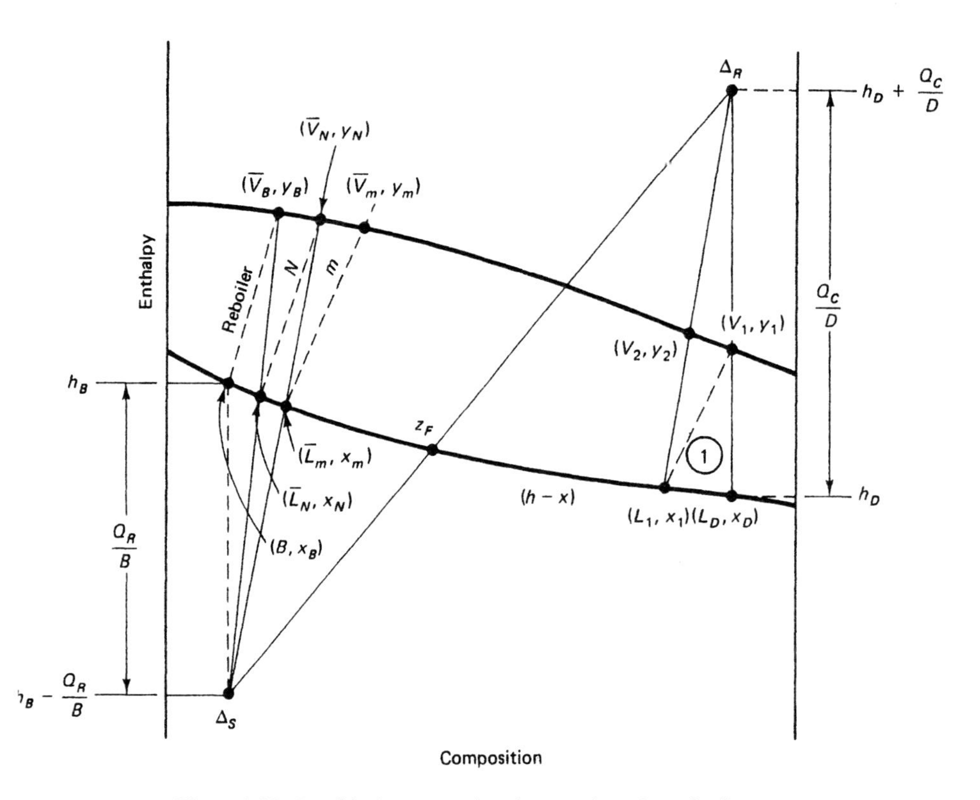

Figure 9-23 Graphical construction for Ponchon–Savarit diagram

6. Step off the theoretical plates for the stripping section. In the stripping section the first step is the equilibrium step to determine the vapor composition leaving the reboiler. A material balance and heat balance is then written to determine x_N for the liquid leaving the first plate in the stripping section. This is determined by the heat and material balance line connecting y_B and Δ_S. The vapor composition in equilibrium with x_N is then determined by extending a tie line from x_N to intersect the saturated vapor curve. This is shown as y_N. The procedure is repeated until the feed plate is reached.

The reflux rate can be found on an enthalpy–composition diagram by making an energy balance around the condenser. The energy balance is

$$V_1 H_1 = L_D h_D + D h_D + Q_C \tag{9-45}$$

or

$$\frac{V_1}{D} H_1 = \frac{L_D}{D} h_D + h_D + \frac{Q_C}{D} \tag{9-46}$$

By substituting $V_1 = L_D + D$ into Eq. (9-46) and rearranging, we obtain

$$\frac{L_D}{D} = \frac{(h_D + Q_C/D) - H_1}{H_1 - h_D} = \frac{\overline{\Delta_R H_1}}{\overline{H_1 h_D}} \tag{9-47}$$

where $\overline{\Delta_R H_1}$ and $\overline{H_1 h_D}$ are the lengths of the lines between points Δ_R and H_1, and H_1 and h_D, respectively. The internal reflux ratio above the first stage is

$$\frac{L_D}{V_1} = \frac{(h_D + Q_C/D) - H_1}{(h_D + Q_C/D) - h_D} = \frac{\overline{\Delta_R H_1}}{\overline{\Delta_R h_D}} \tag{9-48}$$

The internal reflux ratio between any two stages in the rectifying section can be written as

$$\frac{L_n}{V_{n+1}} = \frac{(h_D + Q_C/D) - H_{n+1}}{(h_D + Q_C/D) - h_n} = \frac{\overline{\Delta_R H_{n+1}}}{\overline{\Delta_R h_n}} \tag{9-49}$$

The internal reflux ratio in the stripping section of the column is

$$\frac{\bar{L}_{m-1}}{\bar{V}_m} = \frac{H_m - (h_B - Q_R/B)}{h_{m-1} - (h_B - Q_R/B)} \tag{9-50}$$

For the bottom tray N the internal reflux ratio becomes

$$\frac{\bar{L}_N}{\bar{V}_B} = \frac{H_B - (h_B - Q_R/B)}{h_N - (h_B - Q_R/B)} = \frac{\overline{H_B \Delta_S}}{\overline{h_N \Delta_S}} \tag{9-51}$$

Balances around the overall column provide relationships between the distillate and bottom products in terms of compositions and enthalpies on the enthalpy–composition diagram. The overall energy balance around the column is

$$Fh_F + Q_R = Dh_D + Bh_B + Q_C \tag{9-52}$$

Introducing the overall material balance, $F = D + B$, into Eq. (9-52) gives

$$\frac{D}{B} = \frac{h_F - (h_B - Q_R/B)}{(h_D + Q_C/D) - h_F} = \frac{\overline{z_F \Delta_S}}{\overline{z_F \Delta_R}} \tag{9-53}$$

Minimum trays

In the Ponchon–Savarit solution, the coordinates of the difference points in the rectifying section of the column are $(x_D, h_D + Q_C/D)$. If D approaches zero, the enthalpy coordinate of the difference point approaches infinity, regardless of the individual values of h_D and Q_C. On the other hand, Q_C becomes large if L becomes very large with respect to D. The condenser loading is related directly to the amount of liquid condensed in the condenser. Also for this case the enthalpy coordinate of the difference point will become very large. In the limit, the difference point for the rectifying section of the column will approach infinity.

The same line of reasoning applies to the difference point for the stripping section. The coordinates for this difference point are $(x_B, h_B - Q_R/B)$. As B approaches zero, the enthalpy coordinate will approach negative infinity. If the liquid loading in the column becomes very large with respect to B, the reboiler heat duty also becomes very large and the difference point approaches negative infinity.

Construction of the Ponchon–Savarit diagram under conditions of minimum trays is shown in Figure 9-24. The solid lines in Figure 9-24 represent the

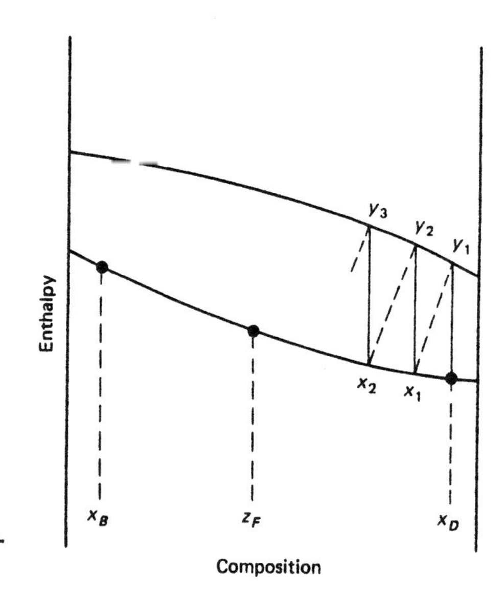

Figure 9-24 Ponchon–Savarit construction for minimum trays

material and heat balance equations around the trays, and the dashed lines represent equilibrium compositions leaving the trays. The operating lines are vertical lines throughout the column because both difference points are located at infinity. The reader should note that, at minimum trays, the reboiler operates as a theoretical plate. In case the column has a partial condenser rather than a total condenser, the partial condenser will function simply as another theoretical plate in the column. No basic changes are needed in the construction with either the McCabe–Thiele solution or the Ponchon–Savarit solution to account for a partial condenser or partial reboiler.

The thermal state of the feed has no effect on the minimum number of trays required for a separation. Since the operating lines coincide with the 45° line on the McCabe–Thiele construction, the q line becomes a point on the 45° line. On the Ponchon–Savarit diagram, the two difference points are located at infinity and the thermal state of the feed again has no effect on the minimum number of trays required.

Minimum reflux

A zone of infinite plates occurs on the Ponchon–Savarit diagram any time an operating line (material and heat balance equations) and an equilibrium tie line coincide. Unless abnormalities occur in the vapor–liquid equilibrium curve of the system, the minimum reflux for the process will normally occur at the feed plate. The minimum reflux rate for a specified separation is found by extending the tie line through the feed composition to intersect a vertical line drawn through x_D. Both the minimum reflux rate and the condenser loading at minimum reflux can be determined from the location of the rectifying-section difference point. The line can be extended to intersect the x_B composition line. This determines the boil-up rate and the reboiler heat duty at minimum reflux. The procedure for determining minimum reflux is demonstrated in Figure 9-25.

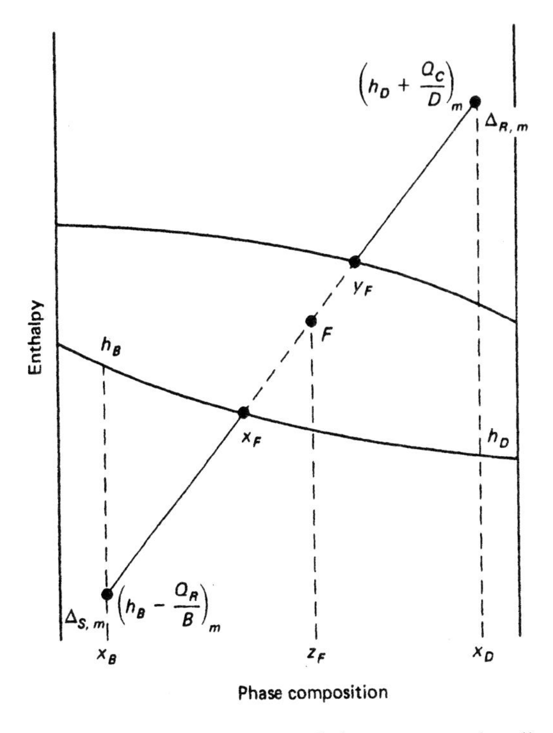

Figure 9-25 Minimum reflux on an enthalpy–concentration diagram

Example 9.7

A total of 100 lb-mol per hour of a 40 mol% methanol and 60 mol% water mixture is to be separated at 1 atm to give a distillate that contains 92 mol% methanol and a bottom product that contains 4 mol% methanol. A total condenser is to be used and the reflux will be returned to the column as a saturated liquid at its bubble point. An external reflux ratio of 1.5 times the minimum will be used. The feed is introduced into the column as a saturated liquid at its bubble point. Use the Ponchon–Savarit method and determine the following: (a) the minimum number of theoretical stages, (b) the minimum reflux ratio, (c) the heat loads of the condenser and reboiler for the condition of minimum reflux, (d) the quantities of the distillate and bottom streams using the actual reflux ratio, (e) the actual number of theoretical stages, (f) the heat load of the condenser for the actual reflux ratio, and (g) the internal reflux ratio between the second and third stages from the top of tower.

Solution:

(a) The minimum number of stages is found by drawing vertical operating lines that intersect as $\Delta_R \longrightarrow \infty$. As shown in Figure 9-26, four stages are required.

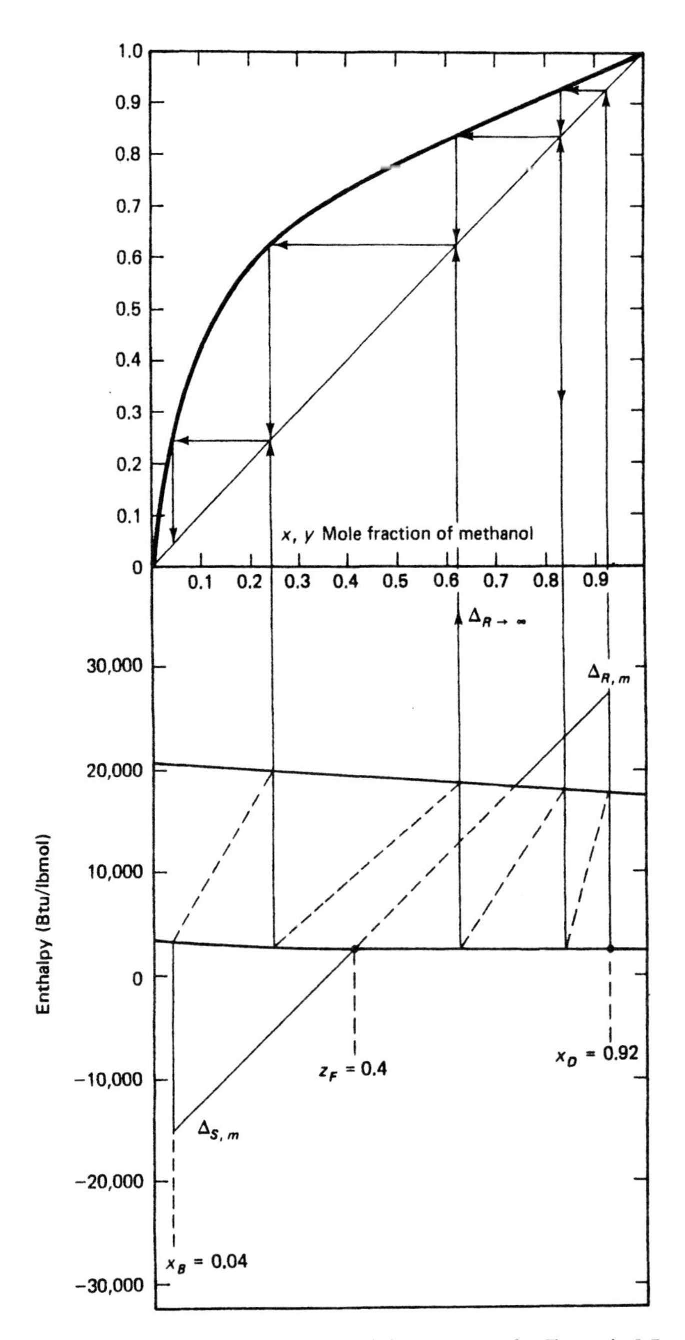

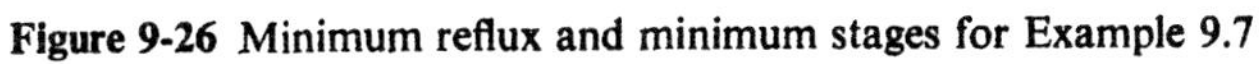
Figure 9-26 Minimum reflux and minimum stages for Example 9.7

(b) Using Eq. (9-47), the minimum reflux is

$$\frac{L_D}{D} = \frac{27{,}000 - 17{,}500}{17{,}500 - 2300} = 0.625$$

(c) From the figure,

$$h_B - \frac{Q_R}{B} = -15{,}000 \text{ Btu/lbmol}$$

$$\frac{Q_R}{B} = 15{,}000 + 3100 = 18{,}100 \text{ Btu/lbmol}$$

Also
$$h_D + \frac{Q_C}{D} = 27{,}000 \text{ Btu/lbmol}$$

$$\frac{Q_C}{D} = 27{,}000 - 17{,}500 = 9500 \text{ Btu/lbmol}$$

(d) The distillate and bottom flow rates are found by solving the overall and component material balances simultaneously.

$$100 = D + B$$

$$0.4(100) = 0.92D + 0.04B$$

Thus
$$D = 40.9 \text{ lbmol/h}$$

$$B = 100 - 40.9 = 59.1 \text{ lbmol/h}$$

(e)
$$\left(\frac{L}{D}\right)_{\text{act}} = 1.5\left(\frac{L}{D}\right)_{\text{m}} = 1.5(0.625) = 0.9375$$

Using Eq. (9-47) gives a new value for Δ_R:

$$0.9375 = \frac{(h_D + Q_C/D) - 17{,}500}{17{,}500 - 2300}$$

$$h_D + \frac{Q_C}{D} = 31{,}750 \text{ Btu/lbmol}$$

A new Δ_R is located in Figure 9-27 and the actual number of theoretical stages is found. As shown, slightly less than eight stages are required.

(f) From the above,

$$h_D + \frac{Q_C}{D} = 31{,}750 \text{ Btu/lbmol}$$

$$\frac{Q_C}{D} = 31{,}750 - 17{,}500 = 14{,}250 \text{ Btu/lbmol}$$

(g) Reading enthalpies for the points (a) and (b) and using Eq. (9-49) gives

$$\frac{L}{V} = \frac{31{,}750 - 18{,}000}{31{,}750 - 2500} = 0.47$$

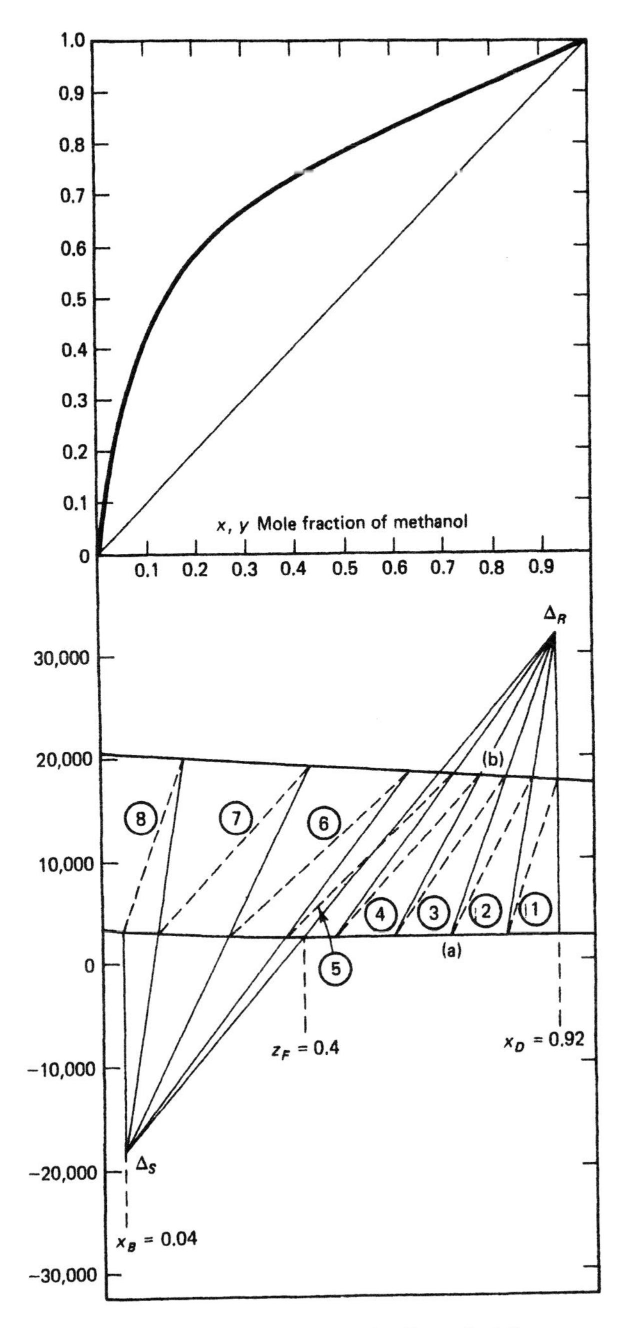

Figure 9-27 Actual stages for Example 9.7

9.5 Partial Condensers

Up to this point in the discussion, we have considered columns that operate with a total condenser, a bubble-point liquid feed, and a partial reboiler. In many situations one or more of these conditions will be changed in an operating column. For this reason the graphical construction for these different conditions of column operation must be considered.

As discussed earlier, a partial condenser operates with a vapor distillate product, which leaves the condenser in equilibrium with the condensed liquid. In most cases the condensed liquid will be returned to the column as reflux. In some cases, however, a portion of the condensed liquid may also be withdrawn as a product from the column.

Figure 9-28a shows the construction of a McCabe–Thiele diagram for a column operating with a partial condenser. Inspection of the equation for the

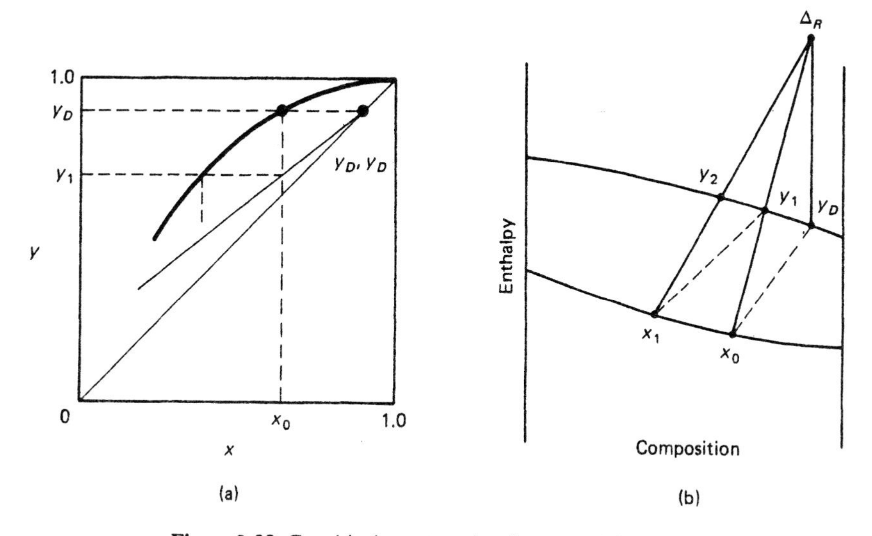

Figure 9-28 Graphical construction for a partial condenser

rectifying-section operating line, Eq. (9-3), will show that the line will pass through point ($x = y_D, y = y_D$). The liquid reflux to the tower will have the composition x_0, which is in equilibrium with y_D. This composition is determined from the equilibrium curve. The remaining plates in the rectifying section are stepped off exactly as would be done for a total condenser.

Figure 9-28b shows the construction on the Ponchon–Savarit diagram for a partial condenser. The distillate product is y_D and the rectifying-section difference point, Δ_R, is located with this composition and the appropriate enthalpy coordinate. The reflux composition, x_0, is determined by extending a tie line from y_D to intersect the saturated liquid curve. The remaining plates are stepped off in the same manner as for a total condenser. The reader should note that the line $\overline{x_0\Delta_R}$ constitutes the heat and material balance line around the condenser. The molar flow of the distillate is proportional to the distance $\overline{x_0y_1}$, and the molar flow of reflux is proportional to the distance $\overline{y_1\Delta_R}$.

$$\frac{L}{D} = \frac{\overline{y_1\Delta_R}}{\overline{x_0y_1}} \tag{9-54}$$

If the distillate product is a mixture of vapor and liquid, the overall composition of the distillate, z_D, must be determined from the expression

$$z_D = \frac{V_Dy_D + L_Dx_D}{D} \tag{9-55}$$

where $D = V_D + L_D$. The rectifying-section operating line will pass through the point ($x = z_D$, $y = z_D$) as shown in Figure 9-29a. The vapor portion of the distillate product and the reflux to the tower will be in equilibrium with each other, and the trays will be stepped off as shown previously.

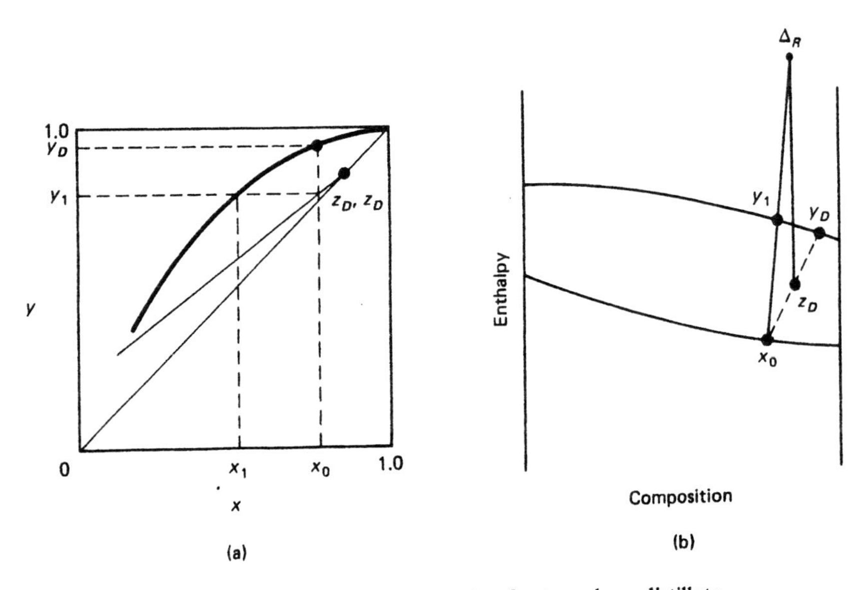

Figure 9-29 Graphical construction for two-phase distillate

Figure 9-29b shows the same construction on the enthalpy–concentration plot. The composition x_0 represents the liquid in equilibrium with the vapor portion of the product y_D. The overall distillate composition z_D is located on the equilibrium tie line and positioned such that the distance $\overline{x_0z_D}$ is proportional to the amount of vapor in the product, and the distance $\overline{z_Dy_D}$ is proportional to the amount of liquid in the distillate product. The rectifying-section difference point has the composition of the overall distillate product z_D. It is positioned such that the distance $\overline{x_0y_1}$ is proportional to the amount of distillate product D and the distance $\overline{y_1\Delta_R}$ is proportional to the amount of liquid reflux as shown by Eq. (9-54). The proportion of liquid and/or vapor in the distillate product must be determined from the relationships on the line $\overline{x_0z_Dy_D}$.

The q line and its location on the McCabe–Thiele diagram have been discussed. The construction for a partially vaporized feed on the enthalpy-concentration diagram is shown in Figure 9-30. The overall composition of the feed is

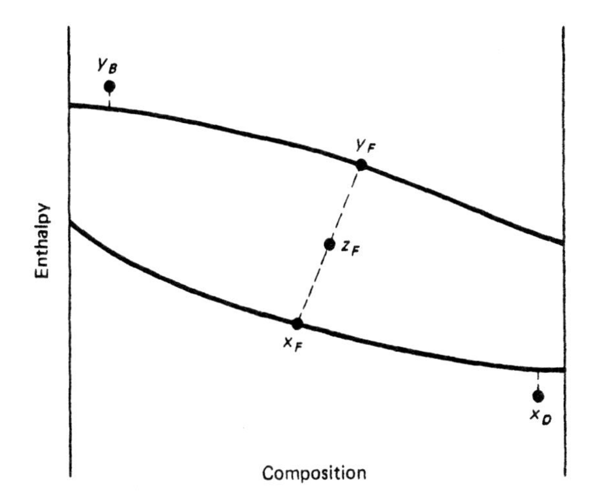

Figure 9-30 Graphical construction showing different enthalpy conditions

represented by the point z_F. Since the vapor and liquid portions of the feed are in equilibrium with each other, z_F must fall on an equilibrium tie line such that y_F and x_F are in equilibrium with each other. In addition, the distance $\overline{z_Fx_F}$ will be proportional to the amount of vapor in the feed and the distance $\overline{z_Fy_F}$ will be proportional to the amount of liquid in the feed.

The construction for a subcooled liquid distillate product (reflux) and a superheated reboiler vapor are also shown in Figure 9-30. The subcooled distillate would simply be a liquid with composition x_D that has had enough heat removed such that it falls below the saturated liquid curve. The superheated vapor y_B has been heated above its saturation point with a corresponding higher enthalpy coordinate than the saturated vapor would have.

REFERENCES

McCabe, W. L., and E. W. Thiele, *Ind. Eng. Chem.*, *17*, 605 (1925).

Ponchon, M., *Tech. Mod.*, *13*, 20 (1921).

Savarit, R., *Arts et métiers*, pp. 65, 142, 178, 241, 266, 307 (1922).

NOTATIONS

b = flow rate of any component in the bottom product, mol/t; intercept of a straight line
B = total flow rate of the bottom product, mol/t
d = flow rate of any component in the distillate product, mol/t
D = total flow rate of the distillate, mol/t
F = total flow rate of the feed, mol/t
h = enthalpy of the liquid, E/mol
h_B = enthalpy of the bottom product, E/mol
h_D = enthalpy of the liquid distillate, E/mol
h_m = enthalpy of the liquid leaving tray m, E/mol
h_n = enthalpy of the liquid leaving tray n, E/mol
H = enthalpy of the vapor, E/mol
H_B = enthalpy of the vapor leaving the reboiler, E/mol
H_m = enthalpy of the vapor leaving tray m, E/mol
l = flow rate of any component in the liquid stream, mol/t
l_D = flow rate of any component in the liquid distillate, mol/t
l_F = flow rate of any component in the liquid feed, mol/t
l_0 = flow rate of any component in the liquid reflux, mol/t
L = total flow rate of the liquid stream in the rectifying section, mol/t
L_F = total flow rate of the liquid in the feed, mol/t
L_m = total flow rate of the liquid leaving tray m, mol/t
L_n = total flow rate of the liquid leaving tray n, mol/t
L_0 = total flow rate of the liquid reflux stream, mol/t
$\bar{L}$ = total flow rate of the liquid in the stripping section, mol/t
L' = total flow rate of the liquid in the zone between the two feeds, mol/t
m = any tray in the stripping section
M = total number of plates in the stripping section
n = any tray in the rectifying section
N = total number of plates in the rectifying section
P = flow rate of any component in the side stream, mol/t
q = mole fraction of saturated liquid in the feed stream
Q_C = condenser duty, E/t
Q_R = reboiler duty, E/t
R = external reflux ratio, L_0/D
R_m = minimum external reflux ratio
v = flow rate of any component in the vapor, mol/t
v_D = flow rate of any component in the distillate vapor, mol/t
v_F = flow rate of any component in the vapor feed, mol/t
V = total flow rate of the vapor in the rectifying section, mol/t
V_B = total flow rate of the vapor leaving the reboiler, mol/t
V_F = total flow rate of the vapor in the feed, mol/t

V_m = total flow rate of the vapor leaving tray m, mol/t
V_n = total flow rate of the vapor leaving tray n, mol/t
$\bar{V}$ = total flow rate of the vapor in the stripping section, mol/t
V' = total flow rate of the vapor in the zone between the two feeds, mol/t
x = mole fraction of any component in the liquid
x_B = mole fraction of any component in the bottom product
x_D = mole fraction of any component in the liquid distillate
x_F = mole fraction of any liquid component in the feed
x_n = mole fraction of any component leaving tray n
x_0 = mole fraction of any liquid component in the reflux stream
y = mole fraction of any component in the vapor
y_B = mole fraction of any vapor component leaving the reboiler
y_D = mole fraction of any vapor component in the distillate
y_F = mole fraction of any component in the vapor feed
z_F = mole fraction of any component in the feed

Greek Letters

Δ_R = rectifying-section difference point defined in Section 9.4
$\Delta_{R,m}$ = rectifying-section difference point for the determination of minimum reflux
Δ_S = stripping-section difference point defined in Section 9.4
$\Delta_{S,m}$ = stripping-section difference point for the determination of minimum reflux

---

PROBLEMS

9.1 A fractionating column is to be used to separate 100 mol/h of a liquid mixture that contains 57 mol% *n*-pentane and 43 mol% *n*-hexane. Calculate the flow rates and compositions of the distillate and bottom products for the following cases. **(a)** The overhead product is to contain 93% *n*-pentane and the bottom product is to contain 92% *n*-hexane. **(b)** For the feed above, 91% of the *n*-pentane is to be recovered in the distillate and 89% of the *n*-hexane is to be recovered in the bottom product. **(c)** The distillate is to contain 97% *n*-pentane and 90% of the *n*-pentane initially present in the feed is to be recovered as distillate. **(d)** The bottom flow rate is to be 60 mol/h, and 97% of the *n*-hexane in the feed is to be recovered in the bottom product.

9.2 A distillation column produces an overhead product that contains 8% ethane, 90% propane, and 2% isobutane. Determine the operating pressure of the column if the temperature in the condenser is 120°F. Assume that the overhead product is a saturated liquid.

9.3 The operating pressure of the column in Problem 9.2 is set at 325 psia and the distillate product is to be produced as a saturated liquid. Determine the temperature of the condenser for this case.

9.4 In Problem 9.2 the operating pressure of the column is to be 325 psia and the overhead product is produced as a saturated vapor. Calculate the temperature in the top of the column.

9.5 A distillation column produces an overhead vapor that contains 8% ethane, 90% propane, and 2% isobutane. The vapor enters a partial condenser where 30% of it is condensed and returned to the column as reflux. The remaining vapor is to be the overhead product. **(a)** Determine the operating pressure of the partial condenser if the condenser temperature is 120°F, and **(b)** calculate the composition of the vapor that leaves the partial condenser. **(c)** The vapor in part (b) is to enter a total condenser where it is totally condensed. Determine the pressure if the temperature is 120°F.

9.6 A binary mixture that contains 30 mol of *n*-pentane and 70 mol of *n*-hexane is to be separated in a distillation column that operates at a pressure of 10 atm. The liquid-to-vapor ratio in the rectifying section just above the feed plate is 0.5 and the vapor flow rate is 120 mol/h. Determine the liquid-to-vapor ratio in the bottom of the column for the following feed conditions: **(a)** the feed is 10% vaporized, **(b)** the feed is 50% vaporized, and **(c)** the feed enters the column as a saturated vapor. **(d)** If a saturated liquid feed contains 30% propane and 70% *n*-hexane, and the pressure is 10 atm, is the assumption of equal molar heats of vaporization valid? Explain.

9.7 A saturated liquid mixture containing 40 mol% *n*-pentane and 60 mol% *n*-hexane is to be separated by continuous distillation. The overhead product is to contain 95% of the *n*-pentane in the feed and the bottom product is to contain 97% of the *n*-hexane in the feed. The distillation column is to operate at a total pressure of 25 psia. A total condenser and partial reboiler are to be used for this process. **(a)** Determine the minimum number of trays required for the separation, and **(b)** find the minimum reflux ratio. **(c)** If the operating reflux ratio is 1.2 times the minimum reflux ratio, how many trays will be required to make the separation? Use the McCabe–Thiele method.

9.8 Repeat Problem 9.7 using a total column pressure of 100 psia.

9.9 A mixture containing 50 mol% *n*-pentane and 50 mol% *n*-hexane is to be separated by continuous distillation at 25 psia into an overhead that contains 94% of the *n*-pentane in the feed and a bottom product that contains 95% of the *n*-hexane in the feed. The feed that enters the column will be 30% vaporized. The boil-up ratio, $\bar{V}/B$, from the reboiler is to be 2.0. **(a)** Determine the minimum reflux ratio, **(b)** calculate the internal reflux ratio in the top and bottom of the column, and **(c)** determine the number of theoretical stages in the rectifying and stripping sections of the column.

9.10 Solve Problem 9.7 using an enthalpy–concentration diagram. Also determine the condenser and the reboiler duties.

9.11 A total condenser was specified in Problem 9.7. Repeat Problem 9.7 using a partial condenser. Use an enthalpy–concentration diagram instead of the McCabe–Thiele method.

9.12 Solve Problem 9.9 by using an enthalpy–concentration diagram. Determine the condenser and reboiler duties.

9.13 A mixture containing 40 mol% *n*-pentane and 60 mol% *n*-hexane is to be separated into a distillate that contains 95% *n*-pentane and a bottom product that contains 95% *n*-hexane. Determine the following: **(a)** the minimum number of stages, **(b)** the minimum reflux ratio, and **(c)** the actual number of stages at 1.2 times the minimum reflux ratio if the feed is introduced as (i) a saturated vapor, (ii) a saturated liquid, and (iii) as a mixture that is 50% vaporized. The column is to be operated at a pressure of 25 psia.

9.14 Two saturated liquid mixtures of *n*-pentane and *n*-hexane are to be separated by continuous distillation into an overhead that contains 95% *n*-pentane and a bottom product that contains 95% *n*-hexane. One feed contains 65% *n*-pentane and 35% *n*-hexane, and the other feed contains 40% *n*-pentane and 60% *n*-hexane. A total of 100 mol/h of each feed is to be introduced at the optimum point in the column and processed at 25 psia. Determine **(a)** the minimum reflux ratio, and **(b)** the number of stages if a reflux ratio equal to 1.2 times the minimum is used.

9.15 Repeat Problem 9.14 for the case in which the two feed streams are mixed prior to entering the column.

10 Multicomponent Distillation

10.1 Introduction

Multicomponent separations are carried out by using the same type of distillation columns, reboilers, condensers, heat exchangers, and so on, as those used for binary separations. However, some fundamental differences exist between the two operations that must be thoroughly understood by the designer if he or she is to properly design a multicomponent system. These differences involve the use of the phase rule to specify the thermodynamic conditions of a stream at equilibrium. Since most fractionations are part of an overall system, the total system analysis also plays a part.

When specifying a binary separation, the designer is free to specify the amount of the more volatile component in the distillate. In a two-component system the remainder of the distillate must be the second component. In multicomponent systems the same degree of specificity is not achieved because of the presence of other components in the overhead product, which cannot be specified independently. In general, the overhead from a multicomponent distillation column will appear as shown in Figure 10-1 for any one component. As that component begins to appear in the distillate it will be at a low concentration. As more distillate is withdrawn, the composition of that component will increase to a maximum and then begin to decrease. Typically, two distillate rates occur at which the composition of any component will be the same. An infinite number of distillate rates will exist in between these values at which the composition will meet or exceed some specified value for the overhead product.

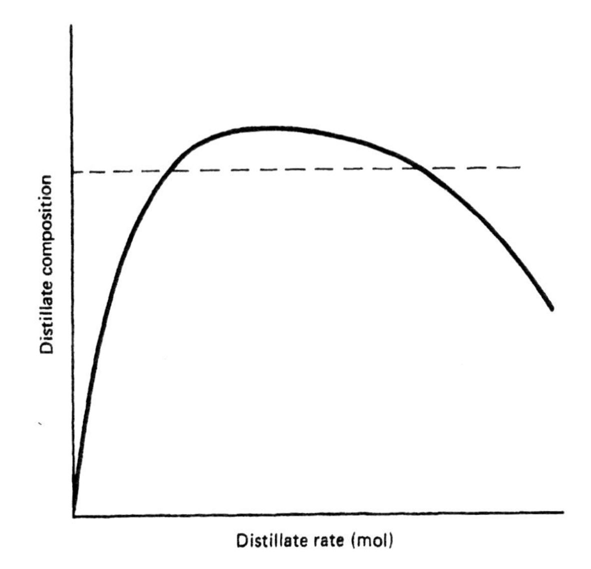

Figure 10-1 Variation of distillate composition with distillate rate

For this reason a different method of specification must be used for multicomponent systems.

The primary goal of the designer is to specify the multicomponent separation in such a way that only one unique solution exists, and any procedure using those specifications will arrive at the same unique solution. To accomplish this for multicomponent systems, two components known as the *key components* must be selected. The *light key* component is that component about which enough information is available to specify its recovery in the distillate product. The *heavy key* component is that component about which enough information is available to specify its recovery in the bottom product. In most cases the light key will be the lightest component that will appear in the bottom stream in any appreciable quantity and the heavy key will be the heaviest component appearing in the distillate in any appreciable quantity. The latter condition is not a restriction on selecting the key components.

Several shortcut methods are presently in use for carrying out calculations in multicomponent systems. Generally, these involve an estimation of the minimum number of trays, the estimation of the minimum reflux rate, and the use of some correlation to predict the relationship between the actual number of theoretical plates and the actual reflux rate necessary to produce the specified product.

Proper selection of key components is important if a multicomponent separation is to be adequately specified. Improper choice of the keys can lead to an improperly designed column.

10.2 Minimum Number of Trays

Simple fractionators (Fenske equation)

For the case of separation of a multicomponent mixture, a minimum number of trays is required for a given separation. Fenske (1932) was the first to present an equation that applied to a multicomponent mixture. The development that follows is based on his original derivation. Figure 10-2 shows the nomenclature that will be used in developing the Fenske equation; subscripts on all variables apply to tray numbers only.

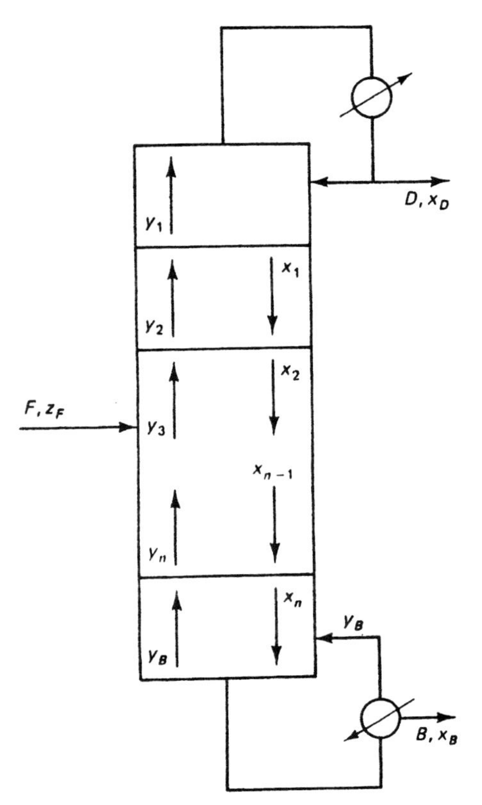

Figure 10-2 Multicomponent column at minimum trays

For the light key component on the top tray

$$y_1 = K_1 x_1 \tag{10-1}$$

Since $y_1 = x_D$ for a total condenser,

$$x_D = K_1 x_1 \tag{10-2}$$

We can write an overall material balance below the top tray and around the top of the column as

$$V_2 = L_1 + D \tag{10-3}$$

Under conditions of minimum trays or total reflux, D is equal to zero. Thus

$$V_2 = L_1 \tag{10-4}$$

A component material balance around the first plate and the top of the column for the light key component is

$$V_2 y_2 = L_1 x_1 + D x_D \tag{10-5}$$

Under the restrictions of minimum trays, this equation simply becomes $y_2 = x_1$. The equilibrium relationship on the second plate is

$$y_2 = K_2 x_2$$

Now since $y_2 = x_1$ this becomes

$$x_1 = K_2 x_2$$

By substituting the expression above into Eq. (10-2), we get

$$x_D = K_1 K_2 x_2 \tag{10-6}$$

Continuing this for the entire column, we obtain

$$x_D = K_1 K_2 \cdots K_n K_B x_B \tag{10-7}$$

Following the same development for the heavy key component, we obtain

$$x'_D = K'_1 K'_2 \cdots K'_n K'_B x'_B \tag{10-8}$$

Dividing Eq. (10-7) by Eq. (10-8) gives

$$\frac{x_D}{x'_D} = \frac{K_1 K_2 \cdots K_n K_B x_B}{K'_1 K'_2 \cdots K'_n K'_B x'_B} \tag{10-9}$$

However, the ratio of the K values is equal to the relative volatility.

$$\frac{K_1}{K'_1} = \frac{y_1 x'_1}{y'_1 x_1} = \alpha_1 \tag{10-10}$$

Equation (10-9) thus becomes

$$\frac{x_D}{x'_D} = \alpha_1 \alpha_2 \cdots \alpha_n \alpha_B \frac{x_B}{x'_B} \tag{10-11}$$

Assuming that an average value of the relative volatility applies for all trays in the column and remembering that the column is operating under conditions of minimum trays, we can simplify Eq. (10-11) to

$$\left(\frac{x_D}{x'_D}\right) = \alpha_{\text{avg}}^{S_m} \frac{x_B}{x'_B} \tag{10-12}$$

where S_m is the minimum number of theoretical trays required.

Equation (10-12) is a form of the original Fenske equation. However, mole fractions are not normally specified in determining a multicomponent separation. Instead, the amount of recovery of a selected component is specified.

Thus we express Eq. (10-12) as

$$\alpha_{avg}^{S_m} = \frac{d}{b}\frac{b'}{d'} \tag{10-13}$$

where b and d represent the molar flow rates of the light key in the bottom and distillate streams, respectively. The terms b' and d' are used to represent the molar flow rates of the heavy key component. This form of the Fenske equation gives results identical with Eq. (10-12). However, Eq. (10-13) is a form more suited for multicomponent usage.

By treating each component in the mixture in turn as the light key component and leaving the heavy key component fixed, we can use Eq. (10-13) to calculate the composition of the distillate product for a multicomponent case. For estimating purposes, the composition thus calculated will be the best estimate of product compositions that can be made. Of course, if a tray-by-tray solution is available, the composition determined from that solution will be a better estimate than that provided by Eq. (10-13).

At one time considerable controversy existed among designers as to the best value of α_{avg} to use. Consideration of the separation occurring in the column will help eliminate some problems in the selection of the average value. Since most columns operate with a partial reboiler, the last separation between vapor and liquid occurring at the bottom of the column will be at the reboiler temperature. This is the bubble point of the bottom product. In the case of a total condenser, the last vapor–liquid separation contact will be on the top tray of the column, which operates at the dew-point temperature of y_1 or the dew point of the distillate product. In the case of a partial condenser, the last vapor–liquid contact will be in the condenser, which operates at the dew-point temperature of the distillate product. If the average tower temperature is assumed to be the average of the dew point of the distillate product and the bubble point of the bottom product, we can evaluate the average relative volatility as the relative volatility of the key components at the average tower temperature. When the average relative volatility is evaluated in this way, the results of the minimum tray calculations found from Eq. (10-13) will be within 5% of the values obtained from tray-by-tray calculations unless a large temperature difference exists between the reboiler and the overhead condenser.

Example 10.1

A liquid has the following composition: propane 1.36%, *i*-butane 14.33%, *n*-butane 16.37%, *i*-pentane 15.66%, *n*-pentane 17.88%, and *n*-hexane 34.40%. This liquid is to be separated into a distillate product which contains 95% of the *n*-butane originally contained in the feed and a bottom product that contains 98% of the *i*-pentane contained in the feed. (a) If the column operates at 25 psia, what will be the minimum number of trays required for the separation? (b) Determine the most probable composition of distillate and bottom products that would result from making this separation in an actual column.

Solution:

(a) The first step in the solution requires the determination of the average tower temperature in order to determine the average relative volatility for these components. The average tower temperature may be taken as the average of the dew-point temperature of the overhead and the bubble-point temperature of the bottom. To estimate these values, the quantities of the key components in the overhead and the bottom streams must be determined.

For the light key component (n-butane):

$$16.37 \times 0.95 = 15.55 \text{ mol } nC_4 \text{ to distillate}$$

$$16.37 - 15.55 = 0.82 \text{ mol } nC_4 \text{ to bottom}$$

For the heavy key component (i-pentane):

$$15.66 \times 0.98 = 15.35 \text{ mol } iC_5 \text{ to bottom product}$$

$$15.66 - 15.35 = 0.31 \text{ mol } iC_5 \text{ to distillate product}$$

For purposes of estimating the temperatures, all of the other components will be assumed to appear in the distillate or bottom streams. For components lighter than the light key, the assumption will be that all of these components go to the distillate. For components heavier than the heavy key, the assumption will be that all of those components go to the bottom stream. This results in the estimated distillate and bottom stream compositions shown in the table below. Temperatures of 60°F and 40°F were assumed for the dew point of the overhead and are seen to bracket the dew-point temperature. Temperatures of 140°F and 160°F were assumed for the reboiler temperature and are seen to bracket the bubble point of the bottom stream.

Component	z_F	d	K^{60}	d/K^{60}	K^{40}	d/K^{40}	b	K^{140}	$K^{140}b$	K^{160}	$K^{160}b$
C_3	1.36	1.36	3.9	0.35	3.0	0.45					
iC_4	14.33	14.33	1.45	9.88	1.0	14.33					
nC_4	16.37	15.55	1.01	15.40	0.7	22.21	0.82	3.3	2.71	4.1	3.36
iC_5	15.66	0.31	0.38	0.82	0.24	1.29	15.35	1.5	23.03	2.0	30.70
nC_5	17.88						17.88	1.2	21.46	1.57	28.07
C_6	34.40						34.40	0.43	14.79	0.62	21.33
		31.55		26.45		38.28	68.45		61.99		83.46

The bubble-point and dew-point temperatures can be estimated by linear interpolation. Because of the structure of the figures from which the K values are read, this will be as accurate as trying to interpolate equilibrium constants at intervals closer than 20°.

$$T_{DP} = 40 + \left(\frac{38.28 - 31.55}{38.28 - 26.45}\right)20 = 40 + 11.4 = 51.4°\text{F}$$

$$T_{BP} = 140 + \left(\frac{68.45 - 61.99}{83.46 - 61.99}\right)20 = 140 + 6 = 146°\text{F}$$

The average tower temperature will be the average of these two temperatures or

$$T_{avg} = \frac{146 + 51.4}{2} = 98.7°F; \qquad \text{therefore, use } 100°F$$

(b) The distillate and bottom compositions can be estimated by using the Fenske equation, Eq. (10-13). However, the calculation of the minimum number of trays is made first.

$$\alpha_{avg}^{S_m} = \frac{d}{b}\frac{b'}{d'}$$

Using the specified values for the light and heavy key component splits and the K values at 100°F and 25 psia, we obtain

$$\left(\frac{1.95}{0.82}\right)_{avg}^{S_m} = \frac{15.55}{0.82}\left(\frac{15.35}{0.31}\right)$$

$$S_m = 7.89 \text{ minimum theoretical trays}$$

Calculations for the other components can be completed by considering each in turn to be the light key component. This is necessary because all relative volatility values are referred to the heavy key component (i-pentane). These calculations are shown below.

For iC_4, we obtain

$$(3.29)^{7.89} = \frac{d}{b}\left(\frac{15.35}{0.31}\right)$$

or

$$b = 0.059$$

For nC_5, we find

$$(0.77)^{7.89} = \frac{d}{b}\left(\frac{15.35}{0.31}\right)$$

or

$$d = 0.046$$

The remaining components can be determined in a similar manner and are summarized in the following table.

Component	z_F	K_{25}^{100}	α	α^{S_m}	d	b
C_3	1.36	6.5	7.93	1.25×10^7	1.36	5.39×10^{-6}
iC_4	14.33	2.7	3.29	1.20×10^4	14.27	0.059
nC_4	16.37	1.95	2.38		15.55	0.82
iC_5	15.66	0.82	1.0		0.31	15.35
nC_5	17.88	0.63	0.77	0.1272	0.046	17.83
C_6	34.40	0.21	0.26	2.42×10^{-5}	1.68×10^{-5}	34.40

The last two columns in the table immediately above represent the most probable compositions of the distillate and bottom products that will result from the specified separation in an actual column.

For cases in which a wide temperature difference exists between the reboiler and condenser of the column, the method presented by Winn (1958) should be used. He found that by plotting K values of similar components at a given pressure on a log-log graph, the K values could be represented as straight-line functions of each other. For this case the K values are related by

$$K = \beta(K')^{\theta}$$

where $K = K$ value of the light key component,
$K' = K$ value of the heavy key component,
$\beta =$ constant,
$\theta =$ slope of line relating K and K' on log-log coordinates.

Making this substitution in Eq. (10-9), we obtain

$$\beta^{S_m} = \frac{x_D}{x_B}\left(\frac{x'_B}{x'_D}\right)^{\theta} \tag{10-14}$$

Equation (10-14) is written in terms of product rates as

$$\beta^{S_m} = \frac{d}{b}\left(\frac{b'}{d'}\right)^{\theta}\left(\frac{B}{D}\right)^{1-\theta} \tag{10-15}$$

Solution of Eq. (10-15) is obviously trial and error since the total rates of flow for the distillate and bottom products are not known. However, Eq. (10-15) is relatively insensitive to the initial B/D assumption and, as a result, it converges rapidly even from a poor first guess.

Equation (10-15) can also be used to calculate the distribution of nonkey components. The heavy key is retained and each component is in turn treated as the light key in order to develop individual component values for β and θ. As a matter of fact, this solution must be completed for all components in order to arrive at values of total flow (D and B) to check the initial assumption.

Complex fractionators (Joyner–Erbar–Maddox equation)

Calculations can also be made on complex fractionators to determine the minimum number of trays. Joyner et al. (1962) presented equations for a three-product fractionator. A sketch of the fractionator they considered is shown in Figure 10-3. The section of the column between the distillate and the side draw is denoted as section I and the section of the column between the side draw and the bottom product is section II. The procedure used for deriving the expression for the minimum number of stages in each section of the column is exactly the same as that described in the original derivation of the Fenske equation. The equation for section I is

$$\alpha_{\text{avg}}^{S_{m\text{I}}} = \frac{x_D}{x'_D}\frac{x'_P}{x_P} \tag{10-16}$$

and for section II the equation is

$$\alpha_{\text{avg}}^{S_{m\text{II}}} = \frac{x_P}{x'_P}\frac{x'_B}{x_B} \tag{10-17}$$

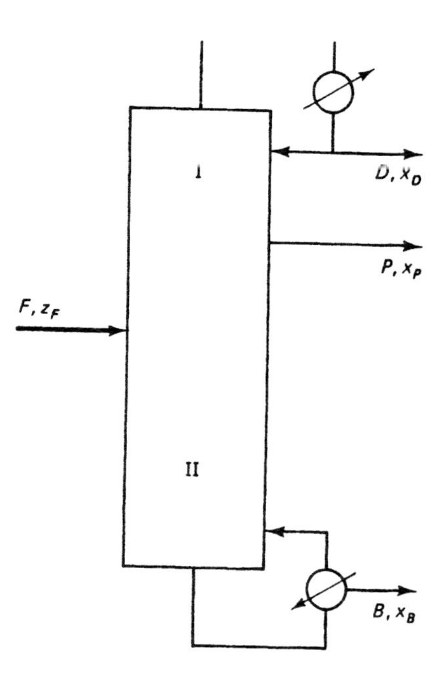

Figure 10-3 Schematic of a complex fractionator for minimum trays

Both of these equations can be converted to the molar flow form as follows:

$$\alpha_{\text{avg}}^{S_{m\text{I}}} = \frac{d}{p}\frac{p'}{d'} \tag{10-18}$$

and

$$\alpha_{\text{avg}}^{S_{m\text{II}}} = \frac{p}{b}\frac{b'}{p'} \tag{10-19}$$

The total minimum number of stages required for this separation is

$$S_m = S_{m\text{I}} + S_{m\text{II}} \tag{10-20}$$

Using the approach of Winn for the definition of volatility, expressions similar to Eq. (10-15) can be derived. Thus

$$\beta_{\text{I}}^{S_{m\text{I}}} = \frac{d}{p}\left(\frac{p'}{d'}\right)^{\theta_1}\left(\frac{P}{D}\right)^{1-\theta_1} \tag{10-21}$$

and

$$\beta_{\text{II}}^{S_{m\text{II}}} = \frac{p}{b}\left(\frac{b'}{p'}\right)^{\theta_2}\left(\frac{B}{P}\right)^{1-\theta_2} \tag{10-22}$$

Equations (10-18) and (10-19), or (10-21) and (10-22) can be used to determine the total minimum number of theoretical stages required in a complex column. Obviously, they can be extended to columns that contain additional side streams. They also can be used to predict product compositions in a tower. Joyner et al. compared the results of calculations from these equations with those obtained from tray-by-tray solutions. The results gave good agreement between the shortcut and regular procedures.

When using Eqs. (10-18) and (10-19), or (10-21) and (10-22), the reader should note that the average relative volatilities for the two sections need not be the same. In addition, different key components may be used in the different sections of the column. This should cause no difficulty if one remembers that when key components are changed, the relative volatility used must be based on the heavy key component shown.

The procedure outlined above can be applied to any number of side streams by simply dividing the column such that the number of sections equals the number of side streams plus one. Also, the side streams may be in either the rectifying or stripping sections of the column. Equations (10-18) and (10-19), and (10-21) and (10-22) can also be applied to determine the minimum number of trays for a multiple feed fractionator. In the material balance equations, the feeds are simply lumped together to provide a single stream representing the total feed to the column. After this, the procedure is exactly the same as outlined above.

10.3 Minimum Reflux

Simple fractionators

A component material balance around the rectifying-section pinch zone and the top of the column shown in Figure 10-4 is

$$V_{\mathrm{RP}} y_{\mathrm{RP}} = L_{\mathrm{RP}} x_{\mathrm{RP}} + D x_D \tag{10-23}$$

where the subscript RP represents the pinch zone in the rectifying section. We can write the material balance in terms of molar flows for one component as

$$v_{\mathrm{RP}} = l_{\mathrm{RP}} + d \tag{10-24}$$

In the pinch zone the compositions and temperatures are constant from plate to plate. This means that the liquid entering a plate and the liquid leaving a plate have the same composition. Since the passing streams in the pinch zone are in equilibrium, we obtain

$$y_{\mathrm{RP}} = K_{\mathrm{RP}} x_{\mathrm{RP}} \tag{10-25}$$

or

$$\frac{v_{\mathrm{RP}}}{V_{\mathrm{RP}}} = K_{\mathrm{RP}} \frac{l_{\mathrm{RP}}}{L_{\mathrm{RP}}} \tag{10-26}$$

Substituting Eq. (10-26) into Eq. (10-24) gives

$$v_{\mathrm{RP}} = v_{\mathrm{RP}} \left(\frac{L}{KV}\right)_{\mathrm{RP}} + d \tag{10-27}$$

or

$$v_{\mathrm{RP}} = \frac{d}{1 - (L/KV)_{\mathrm{RP}}} \tag{10-28}$$

Following the procedure described above in the stripping section of the column, we obtain

$$v_{\mathrm{SP}} = \frac{b}{(L/KV)_{\mathrm{SP}} - 1} \tag{10-29}$$

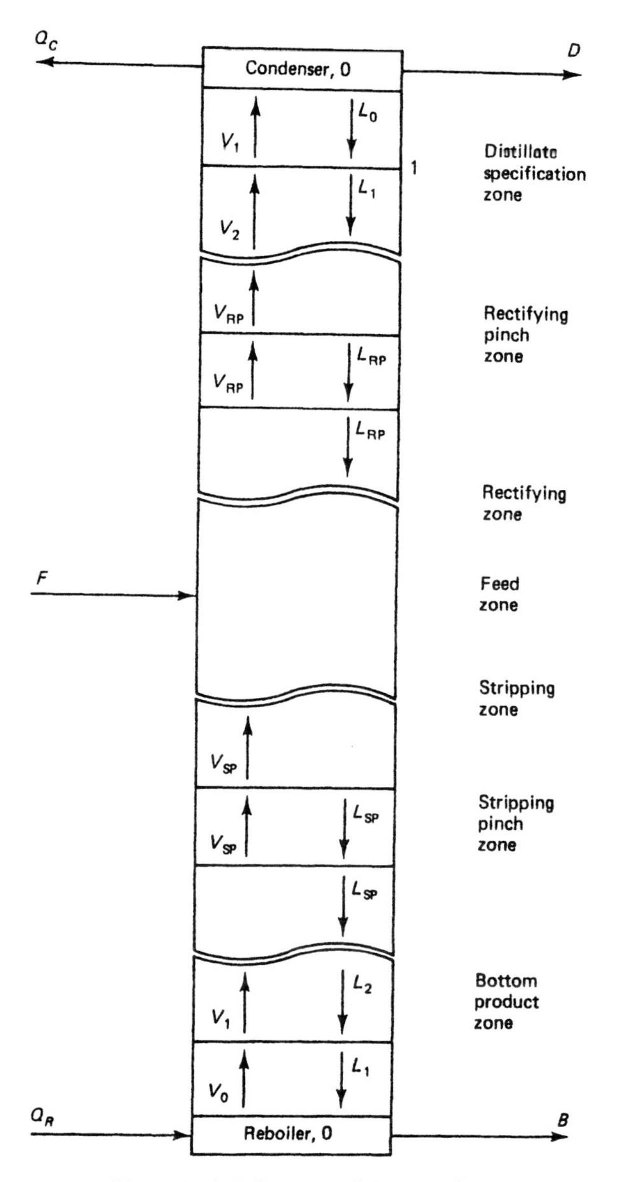

Figure 10-4 Column at minimum reflux

Equations (10-28) and (10-29) apply equally well to binary or multicomponent systems. The difficulty in using them quantitatively in a multicomponent system arises from the difficulty of estimating conditions in the pinch zone. Temperatures and flow rates in the pinch zone must be known to evaluate the L/KV term in the denominator of both equations. For binary systems Eqs. (10-28) and (10-29) are not needed since pinch-zone conditions can be determined from either the McCabe–Thiele or Ponchon–Savarit diagrams.

Consideration of Eqs. (10-28) and (10-29) as applied to multicomponent systems leads to some interesting conclusions. Consider a component in the rectifying section of the column whose K value is small enough that the denominator in Eq. (10-28) becomes negative. If this component appears in the distillate, the molar flow rate for that component in the vapor in the rectifying pinch zone will be negative. Since in any real system a negative flow rate is not possible, the interpretation applied to this result is that the component cannot have a finite value in the distillate; the distillate flow rate for that component must be zero. The same line of reasoning applies to Eq. (10-29) for the case in which the K value for the component becomes large enough to make the denominator of Eq. (10-29) negative. The composition of that component in the bottom product must be zero.

The foregoing line of reasoning gives rise to the concept of *nondistributed components*. Any one component in a multicomponent system must appear in either the distillate or the bottom stream. However, in contrast to behavior for binary systems, any single component does not necessarily have to appear in both product streams. This raises intriguing possibilities in the calculation of the minimum reflux rate for multicomponent systems.

In a detailed analysis of these phenomena, Erbar and Maddox (1962) arrived at the following conclusions:

1. For a completely distributed multicomponent system (all components in the feed appear in both products of the column), the zone of infinite plates occurs at the feed plate. The rectifying pinch zone and the stripping pinch zone coincide at the feed. Vapor and liquid compositions in the pinch zone have the same compositions as the equilibrium vapor and liquid in the feed.

2. For the case of one or more components not appearing in the distillate product, the rectifying pinch zone will be removed from the feed plate. Stream compositions in the rectifying pinch zone will be different from those of the equilibrium feed and some components will not be present in the pinch-zone streams. The stripping pinch zone will coincide with the location of the feed, and the vapor and liquid compositions in the stripping pinch zone will be the same as the equilibrium vapor and liquid compositions of the feed.

3. For the case of one or more components not appearing in the bottom product, the rectifying pinch zone will coincide with the location of the feed. Vapor and liquid compositions in the rectifying pinch zone will be the same as those of the vapor and liquid feed. The stripping pinch zone will be removed from the feed and the nondistributed components will not appear in the liquid and vapor streams in the pinch zone. Stripping pinch-zone compositions will be different from those of the feed.

4. For the case of one or more components not appearing in the distillate and one or more components not appearing in the bottom product, both pinch zones are removed from the location of the feed. Stream compositions of both zones will be different from those of the vapor and liquid feed.

Based on the analysis above, Erbar and Maddox (1962) described seven possible zones in a distillation column which operate at minimum reflux with a multicomponent feed. These zones and their relative locations in a column are shown in Figure 10-4.

The feed zone is between the two plates on either side of the point at which the feed is introduced. In the feed zone the feed is flashed and combined with the vapor and liquid streams entering from adjacent sections. The resulting liquid and vapor streams are the feeds to the stripping and rectifying sections of the column.

The rectifying zone begins above the upper feed plate and extends to the plate at which all components not appearing in the distillate have been fractionated to zero. An infinite number of plates is required to fractionate any component to zero composition. However, temperatures and compositions do change from plate to plate in the rectifying zone. In this way its behavior differs from that normally associated with a zone of infinite plates.

The rectifying pinch zone extends from the plate on which temperature and compositions have become constant. All components not appearing in the distillate (nondistributed components) have been fractionated to zero. Since there is no composition or temperature change from plate to plate, the rectifying pinch zone extends for an infinite number of plates. Because of the constant temperatures and compositions, passing streams are in equilibrium with each other. The rectifying pinch zone extends outward in the column until a change occurs in the temperature and composition.

The distillate specification zone extends from the plate above the rectifying pinch zone to the top of the tower. The first plate in the distillate specification zone is the plate on which temperatures and compositions differ from those of the rectifying pinch zone. The purpose of the distillate specification zone is to fractionate from the pinch-zone composition to the composition of the desired overhead product.

Conditions in the stripping section of the column are similar to those in the rectifying section. The functions of the stripping zone, the stripping pinch zone, and the bottom product zone are analogous to those of their individual counterparts in the rectifying section of the column. The total vapor rate in the rectifying and stripping pinch zones in the column can be obtained by summing Eqs. (10-28) and (10-29) over all components as follows:

$$V_{\mathrm{RP}} = \sum_{i=1}^{c} \frac{d_i}{1 - (L/K_i V)_{\mathrm{RP}}} \tag{10-30}$$

and

$$V_{\mathrm{SP}} = \sum_{i=1}^{c} \frac{-b_i}{1 - (L/K_i V)_{\mathrm{SP}}} \tag{10-31}$$

Equations (10-30) and (10-31) apply to the rectifying and stripping pinch zones, respectively. Dividing both Eqs. (10-30) and (10-31) by α_i and utilizing the definition $\alpha_i = K_i/K'$, we obtain

$$V_{\mathrm{RP}} = \sum_{i=1}^{c} \frac{\alpha_i d_i}{\alpha_i - (L/K'V)_{\mathrm{RP}}} \tag{10-32}$$

and

$$V_{\mathrm{SP}} = \sum_{i=1}^{c} \frac{-\alpha_i b_i}{\alpha_i - (L/K'V)_{\mathrm{SP}}} \tag{10-33}$$

For a totally distributed system the two pinch zones coincide with the feed plate and

$$V_{\mathrm{RP}} = V_{\mathrm{SP}} + V_F \tag{10-34}$$

Upon substituting Eqs. (10-32) and (10-33) into Eq. (10-34), we obtain

$$\sum_{i=1}^{c} \frac{\alpha_i d_i}{\alpha_i - (L/K'V)_{\mathrm{RP}}} = \sum_{i=1}^{c} \frac{-\alpha_i b_i}{\alpha_i - (L/K'V)_{\mathrm{SP}}} + V_F \tag{10-35}$$

or

$$\sum_{i=1}^{c} \frac{\alpha_i f_i}{\alpha_i - (L/K'V)_{\mathrm{RP}}} = V_F \tag{10-36}$$

These equations were first developed and presented by Underwood (1948) who wrote his result in the form

$$\sum_{i=1}^{c} \frac{\alpha_i z_{i,F}}{\alpha_i - \phi_j} = 1 - q = \frac{V_{\mathrm{RP}} - V_{\mathrm{SP}}}{F} = \frac{V_F}{F} \tag{10-37}$$

The two forms of the equations are interchangeable and give identical results for the minimum reflux rate. However, in the calculation of the rectifying pinch zone vapor rate, compositions for the distillate must be at minimum reflux or some other operating reflux rate that might be available from tests on an operating fractionator.

The two equations utilized in calculating the minimum reflux rate for a multicomponent system are Eqs. (10-32) and (10-37). We can rewrite these equations in conventional form as

$$\sum_{i=1}^{c} \frac{\alpha_i d_i}{\alpha_i - \phi_j} = V_{\mathrm{RP}} = \sum_{i=1}^{c} v_{\mathrm{RP}} \tag{10-38}$$

and

$$\sum_{i=1}^{c} \frac{\alpha_i f_i}{\alpha_i - \phi_j} = V_F = \sum_{i=1}^{c} v_F \tag{10-39}$$

We can also express Eq. (10-38) as

$$\sum_{i=1}^{c} \frac{\alpha_i x_{i,D}}{\alpha_i - \phi_j} = 1 + R_m \tag{10-40}$$

A multicomponent mixture containing C components will represent a C-order equation in ϕ_j in Eqs. (10-38) and (10-39). There will be C values of ϕ_j which satisfy these equations, with $C - 1$ of these values positive and bounded by adjacent values of relative volatility such that

$$\alpha_i > \phi_j > \alpha_{i+1} \tag{10-41}$$

The last value of ϕ_j to satisfy the equations will vary with feed conditions but will always be negative for a saturated feed. Since the groups in parentheses in the denominators of Eqs. (10-35) and (10-36) could never realistically take on negative values, only the positive real values of ϕ_j that satisfy Eqs. (10-38) and (10-39) are of concern.

Proper use of the Underwood procedure involves determining the $C - 1$ values of ϕ_j that will satisfy Eq. (10-39). Equation (10-39) is chosen because both the feed composition and the feed conditions are known at the start of a distillation problem. After the $C - 1$ values of ϕ_j that satisfy Eq. (10-39) have been determined, the $C - 1$ equations of the form of Eq. (10-38) are written. Consideration of these equations shows that $C + 1$ unknowns must be determined from the $C - 1$ equations, which is an impossibility. This problem is resolved by specifying the distillate flow rate for two components in the mixture (the two key components). This gives a set of $C - 1$ equations of the form of Eq. (10-38) with $C - 1$ unknowns.

Equations (10-38) and (10-39) were developed by assuming a distributed system. In general, this will not be the case because of the nondistributed components in the mixture. These components may be recognized by their unrealistic distillate flow rates calculated from the system of equations. In the case of a distillate rate for a component which is calculated to be larger than the flow rate of that component found in the entering feed, the component will be nondistributed in the stripping section of the column. Thus all of that component will appear in the distillate product. Conversely, a negative value for the distillate rate means a nondistributed component in the rectifying section, and all of that component that enters in the feed will appear in the bottom product. Should either of these possible alternatives occur, the distillate flow rate for that component should be fixed. The matrix of equations then will be reduced by one and must be re-solved.

The procedure outlined above for finding the Underwood value of the minimum reflux is lengthy and tedious. It is well suited for computer calculations but is less well adapted to hand calculations because of its length. Shiras et al. (1950) presented a method for estimating whether or not a component in a mixture would be nondistributed. Their equation is

$$\frac{d_i}{f_i} = \frac{\alpha_i - 1.0}{\alpha - 1.0}\frac{d}{f} + \frac{\alpha - \alpha_i}{\alpha - 1.0}\frac{d'}{f'} \tag{10-42}$$

If the value of d_i/f_i calculated from Eq. (10-42) is greater than 1, the component will be nondistributed in the stripping section and all of that component entering with the feed will appear in the distillate. If the calculated value of d_i/f_i is less than zero, the component will be nondistributed in the rectifying section and all of that component entering with the feed will appear in the bottom product. The authors have solved literally hundreds of minimum reflux cases using the estimating procedure of Shiras et al. and have yet to find a case in which the predictions of Eq. (10-42) are not correct.

For the case of a sharp separation in which only the two key components are distributed, Van Winkle and Todd (1971) presented a graphical solution for the Underwood equation. Figure 10-5 shows the chart used by Van Winkle and

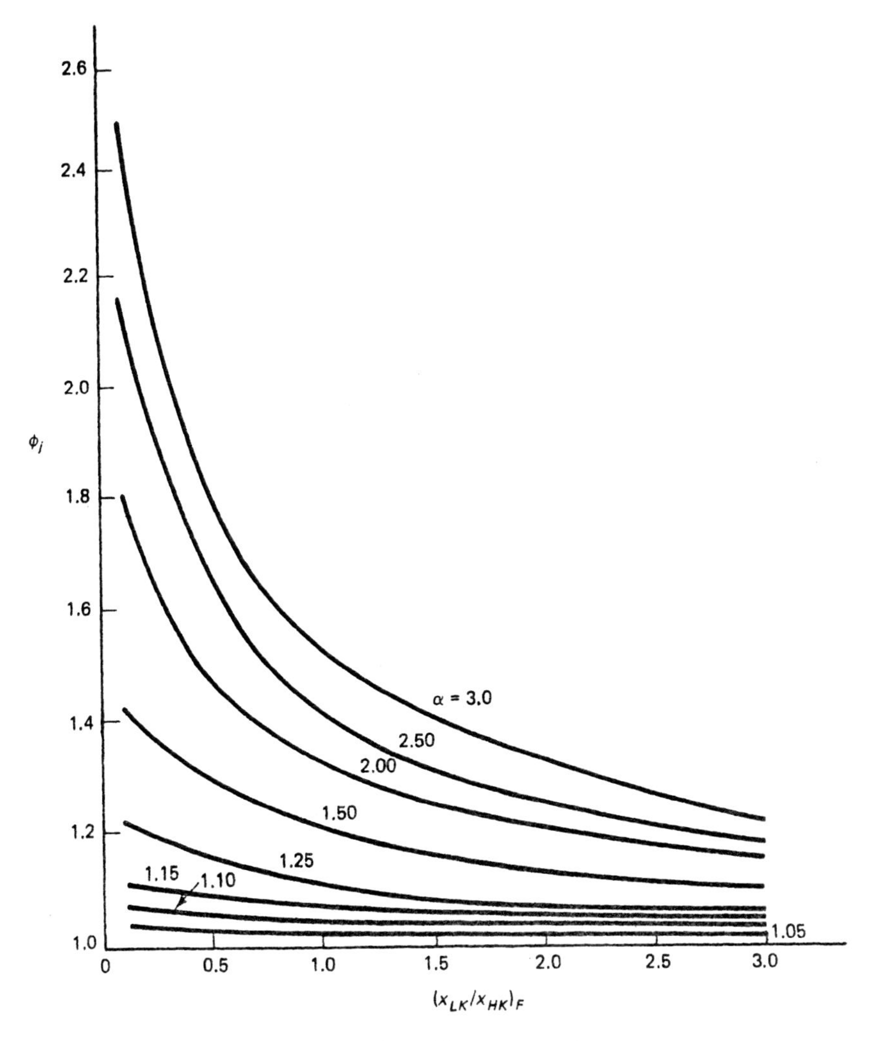

Figure 10-5 ϕ_j and key ratios in feed

Todd in their solution. The value of ϕ_j is determined from Figure 10-5 by using the value of α and the ratio of compositions in the feed for the key components. We can determine the multiplication factors for the individual distillate compositions in the Underwood calculation. The reader should note that the value plotted in this figure is the reciprocal of the actual multiplier. This means that the value read from the chart must be inverted before it can be used properly in the calculation. The inverse is plotted because it becomes a straight line on the graph and provides for easier reading and interpretation. Use of this figure and Eq. (10-42) greatly simplifies the solution for the minimum reflux rate obtained from the Underwood equations. The user should always check the results of the minimum reflux calculation when using Figure 10-5 to make certain that the proper value for the minimum reflux rate has been determined.

Example 10.2

Calculate the minimum reflux rate for the separation of Example 10.1. Assume that the relative volatility values determined in Example 10.1 also apply to the calculation of minimum reflux.

Solution:

The first step in the solution is to determine the distributed and nondistributed components. For this determination, Eq. (10-42) is used.

For the first component lighter than the light key component (*i*-butane):

$$\left(\frac{d}{f}\right)_{iC_4} = \frac{3.29 - 1.0}{2.38 - 1.0}\left(\frac{15.55}{16.37}\right) + \frac{2.38 - 3.29}{2.38 - 1.0}\left(\frac{0.31}{15.66}\right)$$
$$= 1.58 - 0.013 = 1.57$$

Since the calculated value of $(d/f)_{iC_4}$ is greater than 1, *i*-butane and all components in the feed with volatilities higher than *i*-butane will be nondistributed in the stripping section of the column. The distillate rates for these components will be set equal to the amount of that component entering with the feed.

For the first component heavier than the heavy key component (*n*-pentane):

$$\left(\frac{d}{f}\right)_{nC_6} = \frac{0.77 - 1.0}{2.38 - 1.0}\left(\frac{15.55}{16.37}\right) + \frac{2.38 - 0.77}{2.38 - 1.0}\left(\frac{0.31}{15.66}\right)$$
$$= -0.16 + 0.02 = -0.14$$

Since the calculated $(d/f)_{nC_6}$ is less than 0, *n*-pentane and all components heavier than *n*-pentane will be nondistributed in the rectifying section of the column. Their distillate rates will be set to zero and all of that component entering with the feed stream will be assumed to appear in the bottom product.

The calculations above indicate that the only distributed components in the system are the key components. This means that the only value of ϕ required for Eq. (10-39) must satisfy the conditions $1.0 < \phi < 2.38$. An initial estimate of the proper value of ϕ can be obtained from Figure 10-5. The ratio of the light key to heavy key concentration in the feed is

$$\left(\frac{x_{LK}}{x_{HK}}\right)_F = \frac{16.37}{15.66} = 1.05$$

For a value of $(x_{LK}/x_{HK})_F = 1.05$ and for $\alpha = 2.38/1.0$ a value of $\phi = 1.39$ is estimated from Figure 10-5. This value must be checked by calculation since Figure 10-5 gives only an estimated value for ϕ. This calculation is shown in the first five columns of the table below. Equation (10-37) or (10-39) must sum to zero since $q = 1$ for a saturated liquid feed. Instead of summing to zero, a value of $\phi = 1.39$ gives a value of -4.45. Thus another value of ϕ must be selected.

Component	f_i	α_i	$\alpha_i f_i$	(v_F) $\phi = 1.39$	(v_F) $\phi = 1.413$
C_3	1.36	7.93	10.78	1.65	1.65
iC_4	14.33	3.29	47.15	24.82	25.12
nC_4	16.37	2.38	38.96	39.35	40.29
iC_5	15.66	1.0	15.66	−40.15	−37.92
nC_5	17.88	0.77	13.77	−22.21	−21.42
C_6	34.40	0.26	8.94	−7.91	−7.75
				−4.45	−0.02

The last column in the table shows the calculations of the terms in Eq. (10-39) for the value of $\phi = 1.413$. Since the summation to -0.02 is sufficiently close to 0, the value of ϕ to use in the calculation of minimum reflux for the separation is 1.413.

The distillate composition at minimum reflux can be estimated from the results of the calculations by using Eq. (10-42). Column 2 of the table below shows this distillate composition and the remainder of the table shows the calculation of the vapor rate in the rectifying pinch zone at minimum reflux. The last column is obtained by applying Eq. (10-38).

Component	d_i	α_i	$\alpha_i d_i$	v_{RP}
C_3	1.36	7.93	10.78	1.65
iC_4	14.33	3.29	47.15	25.12
nC_4	15.55	2.38	37.01	38.27
iC_5	0.31	1.0	0.31	−0.75
nC_5	—			—
C_6	—			—
	31.55			64.29

The quantity of liquid to be returned to the column as reflux can be calculated from a material balance. The basis of this material balance is the assumption of constant molar overflow between the rectifying pinch zone and the condenser. Under these conditions the minimum reflux rate for the specified separation is

$$L_{0_m} = 64.29 - 31.55$$
$$= 32.74 \text{ mol of reflux per 100 mol of feed}$$

The value of ϕ read from Figure 10-5 should be exact for a binary mixture which consists only of the light and heavy key components. For mixtures that contain more than two components, the value read from the figure is an estimate only. The greater the extent to which the key components dominate the tower feed composition, the closer the value of α will be. Never, however, for a multicomponent mixture should a value of ϕ be read from Figure 10-5 and then used directly to calculate the minimum reflux rate. The ϕ value read from Figure 10-5 must always be checked by calculation.

Complex fractionators

The minimum reflux rate for a side-draw fractionator can also be determined. The slope of the operating line between the side draw and feed will always be such as to intercept the equilibrium line. For this reason, whether the side draw is in the rectifying or stripping section of the column, the pinch zone will always occur at the feed plate on the McCabe–Thiele diagram. This construction is shown in Figure 10-6.

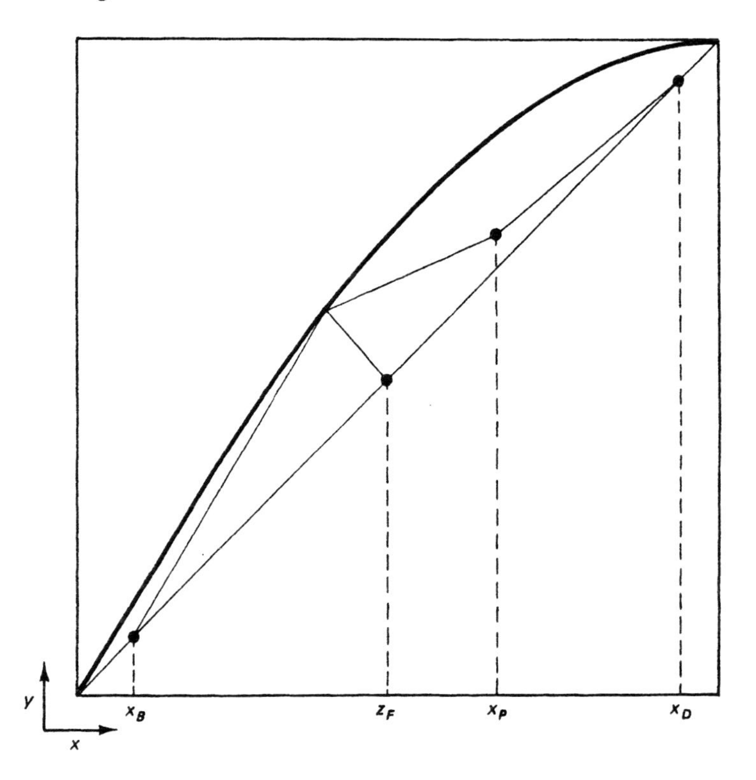

Figure 10-6 Minimum reflux for complex fractionator

Writing a material balance for one component around the top of a fractionator that contains a side draw in the rectifying pinch zone, we obtain

$$v_{RP} = l_{RP} + p + d \tag{10-43}$$

But

$$\frac{l_{RP}}{v_{RP}} = \left(\frac{L}{KV}\right)_{RP} \tag{10-44}$$

Therefore,

$$v_{RP} = \frac{d + p}{1 - (L/KV)_{RP}} \tag{10-45}$$

A pseudodistillate rate in the rectifying section of the column is defined such that $\bar{d}$ is equal to $d + p$. The Underwood equations for a simple column, Eqs. (10-38) and (10-39), can then be used to determine the minimum reflux rate in the complex fractionator. The only substitution that must be made is $\bar{d}$ for d. When calculating the minimum reflux rate, we follow the same procedure as that used for a simple column, including the prediction of the nondistributed components.

10.4 Operating Reflux and Plates

A number of authors over the years have presented correlations that relate operating to minimum reflux rates and plates. Perhaps the most popular of these is the one proposed by Gilliland (1940) shown in Figure 10-7, and that presented

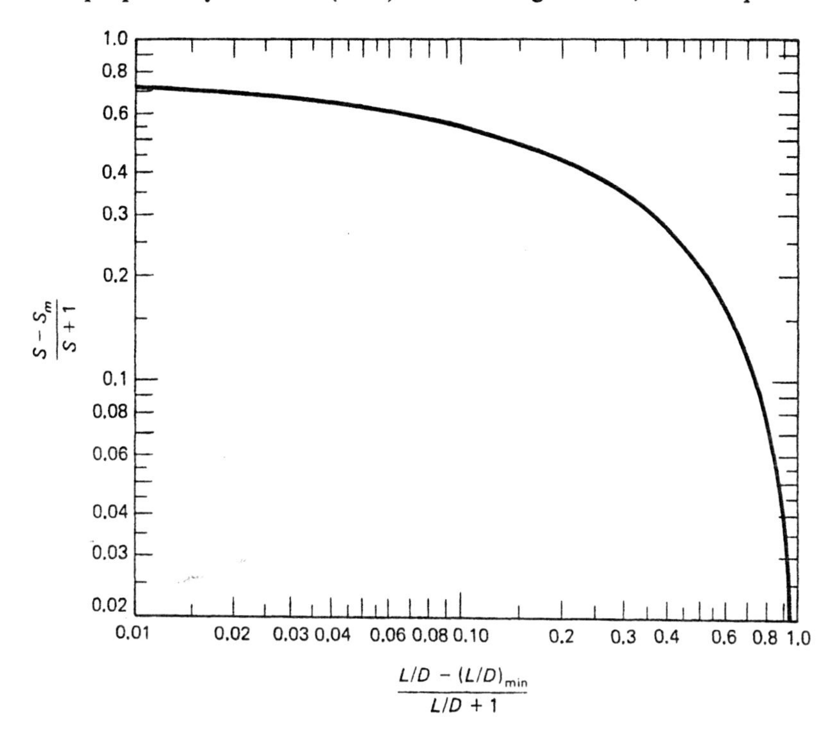

Figure 10-7 Gilliland graphical correlation of reflux and plates (from Gilliland, 1940)

by Brown and Martin (1939) shown in Figure 10-8. Although these correlations are useful for making preliminary economic analyses, a somewhat better correlation has been developed by Gray (1960) and by Erbar and Maddox (1961). Their correlation is shown in Figure 10-9. A given curve on Figure 10-9 repre-

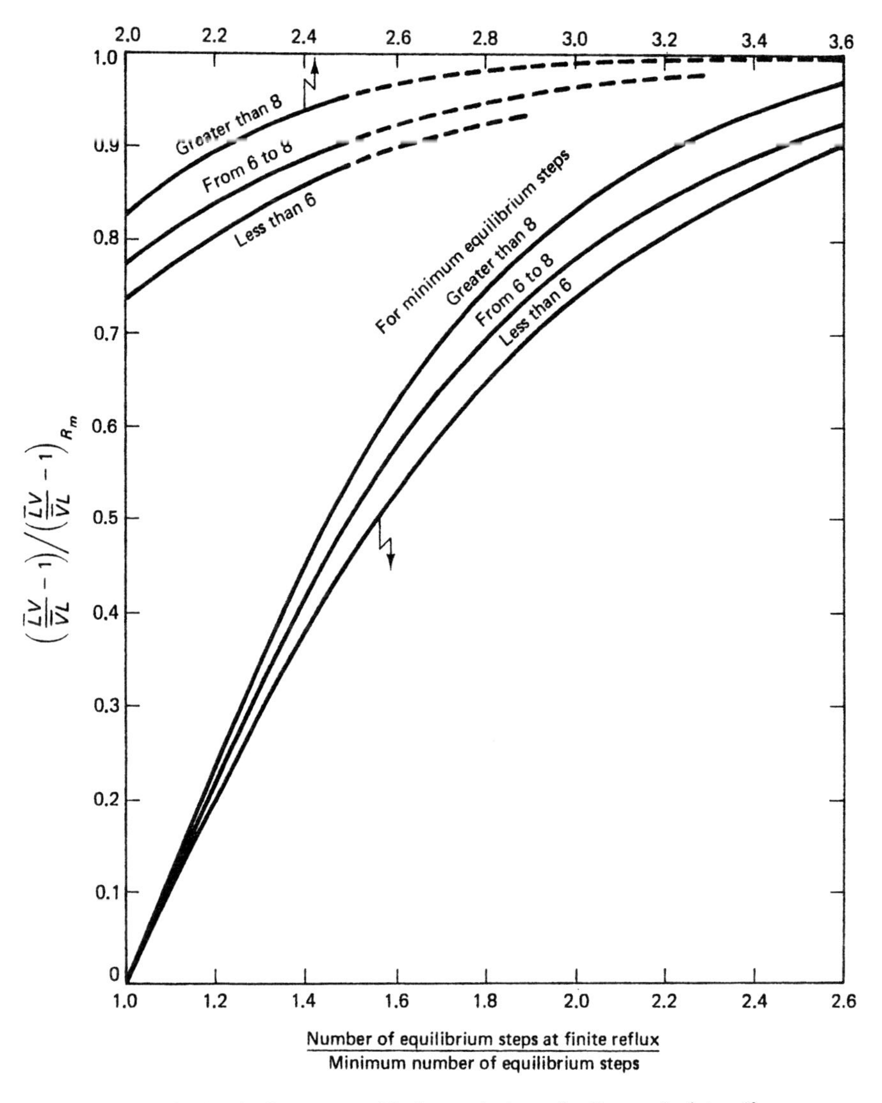

Figure 10-8 Martin–Brown graphical correlation of reflux and plates (from Martin and Brown, 1939)

sents the minimum reflux ratio $(L_0/V_1)_m$ for a specified separation. By reading vertically from the X axis (S_m/S) to the minimum reflux ratio, the operating value of L_0/V_1 can be determined. A material balance written around the condenser will yield the actual reflux rate that must be employed for that number of stages. The use of this procedure will be better understood by studying the following example.

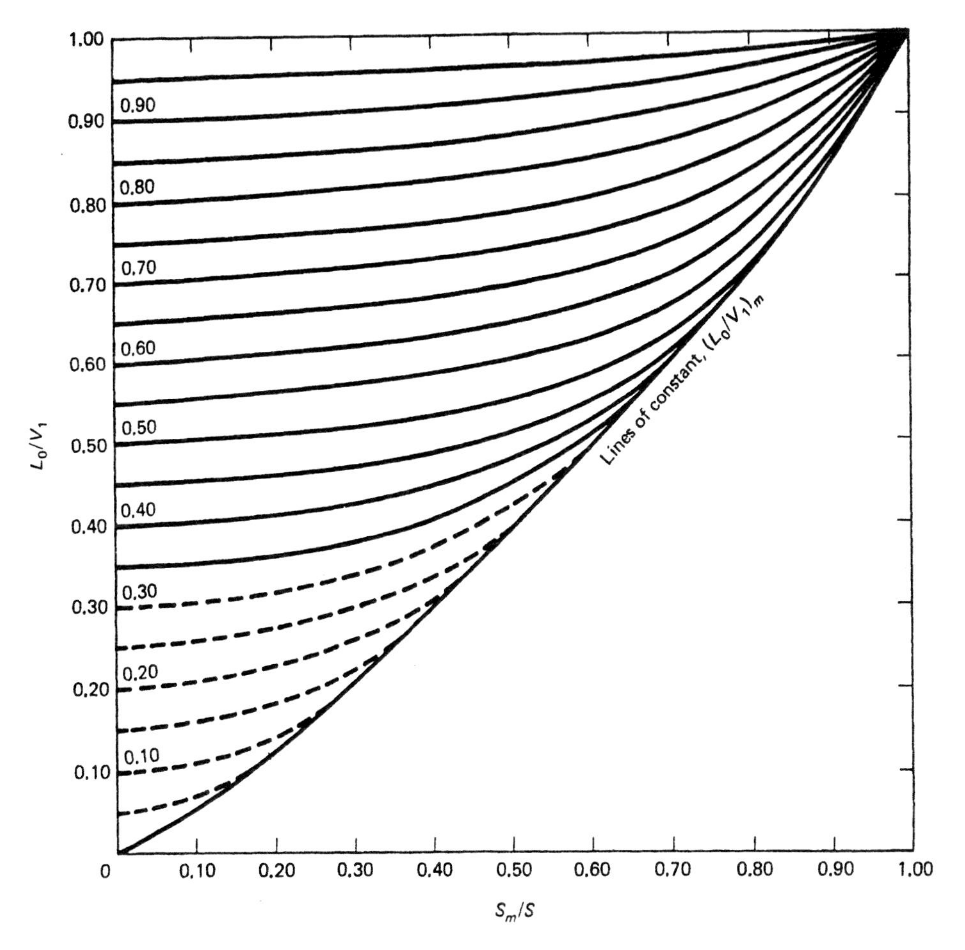

Figure 10-9 Erbar–Gray–Maddox correlation. Dashed lines: extrapolated. (from Erbar and Maddox, 1961)

Example 10.3

Using the minimum tray results calculated from Example 10.1 and the minimum reflux obtained from Example 10.2, determine the operating reflux and the actual number of theoretical plates for $S_m/S = 0.2$ to 0.7 for the specified depropanizer separation.

Solution:

From Example 10.1 the minimum number of trays is $S_m = 7.89$. From Example 10.2 the minimum external reflux rate is $L_0 = 32.74$ mol/100 mol of feed. The minimum vapor rate is $V_1 = 32.74 + 31.55 = 64.29$. Therefore, the minimum reflux ratio is

$$\left(\frac{L_0}{V_1}\right)_m = \frac{32.74}{64.29} = 0.51$$

We are now able to complete the following table by using Figure 10-9.

$\frac{S_m}{S}$	$\frac{L_0}{V_1}$	$1 - \frac{L_0}{V_1}$	L_0	S
0.2	0.518	0.482	33.91	39.5
0.4	0.554	0.446	39.19	19.7
0.7	0.70	0.30	73.62	11.3

The relationship between the reflux rate and the actual number of stages is shown in Figure 10-10. As the number of plates increases, the actual reflux rate approaches asymptotically the minimum reflux rate of 32.7 mol per 100 mol of feed. As the reflux rate increases, the number of required theoretical trays approaches the minimum number of theoretical trays, $S_m = 7.89$. The general hyperbolic form of the reflux–trays relationship is typical of hydrocarbon separations.

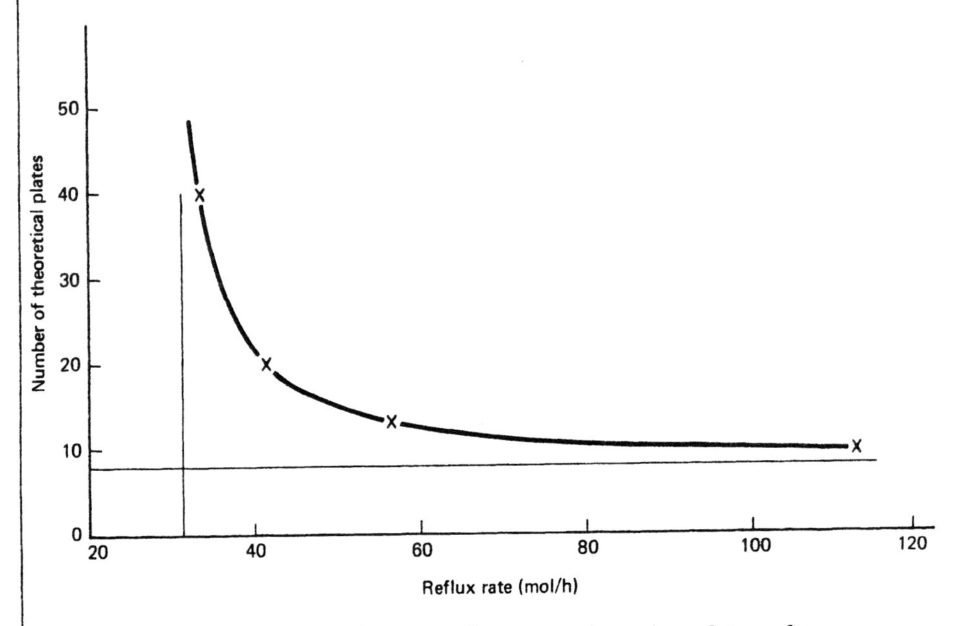

Figure 10-10 Relationships between reflux rate and number of trays for depropanizer

Vaporized feeds

The discussion thus far has centered on feeds that enter the column at the bubble point, although the feed may frequently be partially vaporized. Although Eqs. (10-38) and (10-39) are often used for partially vaporized feeds, Erbar and Maddox (1961) found that these equations do not provide reliable estimates for the reflux. Instead, the recommended procedure is to estimate the operating reflux rate as outlined for the feed at the bubble point and then adjust the operating reflux rate to account for the feed vaporization. The adjustment is made by using the following equation:

$$V_{1u} = V_{1k} + \frac{(H_{Fu} - H_{Fk})(1 - D/F)}{Q_C/L_{CT}} \tag{10-46}$$

In the equation above L_{CT} represents the total number of moles of liquid formed in the condenser per unit time. The subscripts k and u represent the conditions for a bubble-point feed and partially vaporized feed, respectively.

10.5 Feed Plate Location

All of the work done thus far has been based on the implicit assumption that the feed will be introduced into the column at the "optimum" location. The optimum, which is based on fractionation considerations only, is difficult to clearly state but can be said in one of two ways: For a given reflux rate the feed plate is that which will require the smallest number of theoretical contacts to achieve the desired separation, or for a given number of trays the feed plate will be that which will require the smallest reflux rate to achieve the specified separation.

The Fenske (1932) relationship provides a basis for estimating the feed plate location. Basically, the assumption is made that the proportion of trays in the rectifying section of the column will be constant and independent of the reflux rate. Using this as a basis, the equations developed for the rectifying and stripping sections of the column are

$$\alpha^N = \left(\frac{d}{f}\right)_{\text{LK}}\left(\frac{f}{d}\right)_{\text{HK}} = \frac{d}{f}\frac{f'}{d'} \tag{10-47}$$

and

$$\alpha^M = \left(\frac{f}{b}\right)_{\text{LK}}\left(\frac{b}{f}\right)_{\text{HK}} = \frac{f}{b}\frac{b'}{f'} \tag{10-48}$$

where N represents the number of trays above the feed plate and M is the number of trays in the stripping section. When applying Eqs. (10-47) and (10-48), the relative volatility is normally assumed constant across the column. The relative volatility for the average temperature and pressure of the column should be used.

An empirical method for estimating feed tray location has been presented by Kirkbride (1944).

$$\log \frac{N}{M} = 0.206 \log \left[\frac{B}{D} \frac{z_{F,HK}}{z_{F,LK}} \left(\frac{x_{B,LK}}{x_{D,HK}} \right)^2 \right] \tag{10-49}$$

Akashah et al. (1979) made an extensive study of feed plate location by making tray-to-tray calculations. They concluded that the feed plate could best be estimated by using a modified form of Eq. (10-49). The number of stages above the feed plate is given as

$$N = N_{\text{Eq. (10-49)}} - 0.5 \log(N_T) \tag{10-50}$$

where N_T is the total number of theoretical stages in the column. Equations (10-49) and (10-50) require that the compositions of the feed, distillate, and bottom streams be known. If not otherwise available (such as from plant test data), suitable compositions can be estimated by using the Fenske equation as outlined in Example 10.1.

Example 10.4

Using the results in Example 10.1 as a basis, estimate the feed plate location.

Solution:

Using the Fenske relationship for the rectifying section, we obtain

$$\alpha^N = \frac{d}{f} \frac{f'}{d'} = \frac{15.55}{16.37} \left(\frac{15.66}{0.31} \right) = 47.99$$

$$N = 4.47 \text{ rectifying trays}$$

In an analogous manner for the stripping section, we get

$$\alpha^M = \frac{16.37}{0.82} \left(\frac{15.35}{15.66} \right) = 19.57$$

$$M = 3.43$$

$$\frac{N}{N+M} = \frac{4.47}{7.90} = 0.57$$

The feed plate location is found by using the Kirkbride relationship as shown:

$$\log \frac{N}{M} = 0.206 \log \left[\frac{68.45}{31.55} \left(\frac{0.1566}{0.1637} \right) \left(\frac{0.012}{0.0098} \right)^2 \right]$$

$$\frac{N}{M} = 1.26$$

$$\frac{N}{N+M} = 0.56$$

In this case the two estimating procedures give essentially the same results. This frequently will be true.

10.6 Tray-by-Tray Calculations

Tray-by-tray calculations represent, in one sense, the ultimate tool to be used in the design and/or evaluation of the performance of stagewise processes. The results of the tray calculations include the composition, flow rate, and temperature of the overflow and underflow streams leaving each tray or plate in the column. An analysis of the detailed results of the tray calculations may indicate potential design or operating problems and suggest their solution.

Tray-by-tray calculations are usually based on the assumption that a column which contains theoretical plates can be used to model or simulate the performance of a column which contains actual plates. Tray efficiencies can be introduced into the tray-by-tray calculation. However, in the opinion of the authors, current methods for predicting tray efficiencies are not sufficiently accurate to justify the additional effort involved in incorporating in the tray calculations a separate tray efficiency for each tray in the column.

In a computational sense, tray-by-tray calculations simply represent the simultaneous incorporation of material balances, energy balances, and equilibrium relationships for each tray in the column. Some of the earlier tray-by-tray procedures did not incorporate heat balances. This was because the calculations were carried out by hand and incorporation of the heat balances greatly extended the calculational procedure. Tray-by-tray calculations are highly repetitious and require a considerable amount of time to carry out by hand. For this reason they are ideally suited for digital computer solution. The real difficulty in computer solution of tray-by-tray calculations is achieving a suitable material balance closure for all components for the particular set of column variables specified. When making hand calculations, the judgment of the engineer reduces the difficulty and complexity of the closure problem. Programming this judgment for a computer solution is difficult, if not impossible, since a special set of problems results. These problems have not yet been completely resolved, although tremendous strides have been made toward achieving a computer program that will provide a closed solution for all stagewise problems.

The Lewis–Matheson (1932) procedure represents perhaps the best method for presenting the concepts involved in tray-by-tray calculations. The procedure is not frequently used as a contemporary computational tool because the computational techniques developed specifically for computer solution are far more powerful. The Thiele–Geddes (1933) calculational procedure is a second method that was developed primarily for hand use. This procedure is still in contemporary use but, in contrast to the earlier versions, heat balances are currently incorporated into the calculations.

Lewis–Matheson method

The conventional method of initiating the Lewis–Matheson calculations is to assume the composition for the distillate and bottom product streams and begin the calculations at the column terminals. The calculation is continued toward the feed zone using a specified reflux rate. If the composition of the stream entering the feed zone is not identical to the compositions of streams leaving the feed zone as determined by tray-by-tray calculations, the terminal compositions are adjusted and the calculations repeated. For hand calculation purposes the distillate and bottom product compositions and rates are based on the *key-component splits* and assumed distributions for the remaining components. When the ratio of key-component concentrations calculated from the terminals essentially matches the ratio of those components in the feed, the feed zone is assumed to have been reached. The composition of the non key components is adjusted to force a material balance agreement at the feed zone. When the Lewis–Matheson calculational procedure is applied to computers, the number of plates and feed plate location are normally specified rather than the key-component splits. The complete product distribution for all components is then calculated. Regardless of which variables are initially specified, the calculational procedure proceeds in the same manner.

The first calculation in any stagewise calculational procedure normally involves feed conditions. The composition and rate of the liquid and vapor portions of the feed, together with the total feed enthalpy, are calculated. Usually, the Lewis–Matheson calculations begin at the condenser. Since the distillate composition is known (or has been assumed), the temperature of the reflux accumulator and the composition of the reflux can be determined. Writing a component material balance around the condenser–reflux accumulator, as shown in Figure 10-11, we obtain

$$V_1 = L_0 + D \tag{10-51}$$

Since the reflux rate is known, Eq. (10-51) can be used to determine the composition and flow rate of the vapor leaving the top tray of the column. The top tray is assumed to be an equilibrium tray, so the vapor leaves at its dew point. This fixes the temperature of the vapor leaving the top tray. The condenser duty can be calculated from a heat balance around the condenser.

$$Q_C = H_V V_1 - (h_D D_L + h_L L_0 + H_D D_V) \tag{10-52}$$

The dew-point calculation for the vapor leaving the top tray will also yield the composition of the liquid leaving the tray.

For the top tray, the following information is now available: the composition, flow rate, and temperature of the liquid (the reflux) entering the tray; the composition, flow rate, and temperature of the vapor leaving the tray; and the composition and temperature of the liquid leaving the tray. The composition, flow rate, and temperature of the vapor entering the top tray are not known and neither is the flow rate of the liquid leaving the tray. These variables, however, can be determined by assuming a flow rate for either the vapor entering the tray or the liquid leaving the tray. The flow rate of the other can then be determined by a material balance. The assumed flow rate can then be checked by a heat balance around the top tray. If the flow rate of the liquid leaving the tray is

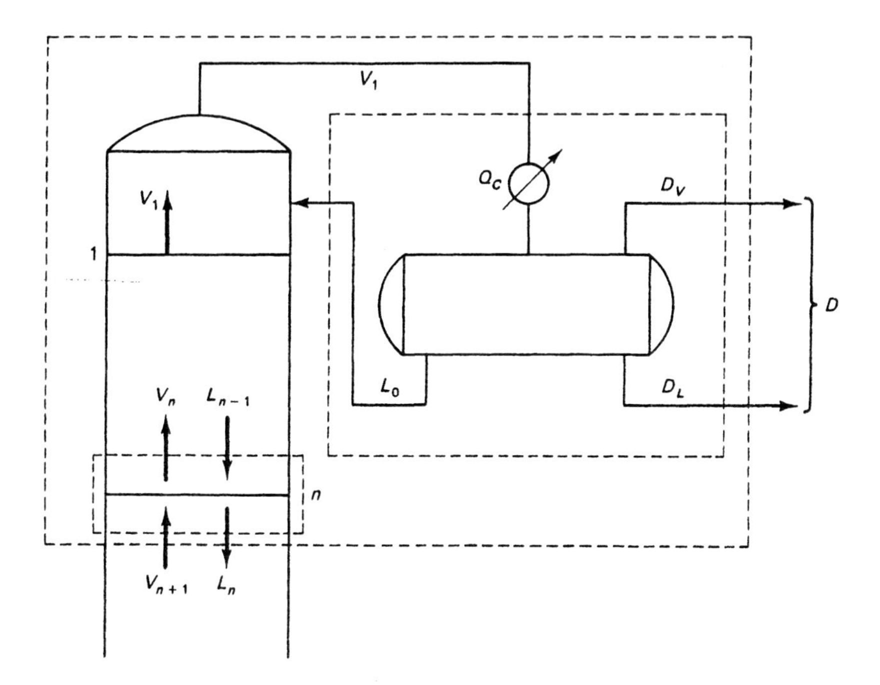

Figure 10-11 Material and energy balance envelopes for rectifying section

estimated, the flow rate and composition of the vapor entering the tray can be calculated by one of two material balances.

$$V_2 = L_1 + D \tag{10-53a}$$

or

$$V_2 = V_1 + L_1 - L_0 \tag{10-53b}$$

Equation (10-53a) is the simpler to use but either is correct. Once Eq. (10-53a) has been applied for all components, the flow rate and composition of V_2 (vapor entering the top tray) are known. The temperature of V_2 can now be determined by a dew-point calculation. Once the temperature is known, we can write a heat balance around tray n as

$$H_{n+1}V_{n+1} = Q_C + h_nL_n + (H_DD_V + h_DD_L) \tag{10-54a}$$

or

$$H_{n+1}V_{n+1} = H_nV_n + h_nL_n - h_{n-1}L_{n-1} \tag{10-54b}$$

If the heat balance equality is not satisfied, a new liquid rate must be assumed and the calculations repeated until a satisfactory heat balance is obtained.

A guide for reestimating the next assumption for the liquid rate would be desirable. If the overall material balance

$$V_{n+1} = L_n + (D_V + D_L) \tag{10-55}$$

is substituted into Eq. (10-54a), the following equation will result:

$$H_{n+1}(D_L + D_V + L_n) = h_DD_L + H_DD_V + Q_C + h_nL_n \tag{10-56}$$

We can rearrange the equation above as

$$L_n = \frac{h_D D_L + H_D D_V + Q_C - H_{n+1}(D_L + D_V)}{H_{n+1} - h_n} \tag{10-57}$$

Equation (10-57) can be used to estimate the next value of L_n. Notice that Eq. (10-57) utilizes the vapor enthalpy which was based on the vapor rate calculated from the assumed value for L_n. If a poor value of L_n is assumed, this value of vapor enthalpy can be seriously in error and will lead to the prediction of unreasonable values for L_n.

Once the heat balance on a particular tray has been closed, calculations proceed to the next tray in the column. The dew-point calculation on the vapor entering the tray yields the composition of the liquid leaving the tray below so that the next tray calculation starts with the assumption of a value for the new liquid rate. These calculations are repeated until the feed zone is reached.

A special problem exists in hand calculations of the Lewis–Matheson type of tray-by-tray solution. Normally, "nondistributed" components will occur in the product streams. These components appear in such small quantities that they are assumed to be zero. To achieve a feed zone match, the heavy components missing from the distillate must be introduced somewhere in the rectifying section so that they will appear in the feed zone compositions. Proper introduction of the heavy components is very much an art. They must be introduced on some arbitrary tray in the rectifying section in small enough concentration so that they can build to their proper value at the feed plate. Most computer applications of the Lewis–Matheson procedure do not encounter this problem. Rather, the problem is avoided by introducing the heavier components in the distillate (and the lighter components in the bottom) at a suitably small concentration. These concentrations, together with all others, are adjusted as required to achieve matching feed-zone compositions for all components. The technique of introducing the heavy components in the distillate (and introducing the light component in the stripping section) can also be used when making hand calculations.

The procedure outlined above is somewhat lengthy and complex. However, study of the logic diagram shown in Figure 10-12, which outlines the computational procedure, will enhance understanding. Figure 10-12 shows the procedure as it might be programmed for computer solution, with the introduction of the nondistributed components in the terminal streams at very low concentrations and subsequent adjustment of these compositions to achieve feed-zone composition matching.

Stripping-section calculations are very similar to those outlined for the rectifying section. The difference is that they are initiated at the reboiler and then proceed upward until the feed plate is reached. The necessary equations will be presented with an abbreviated discussion.

Figure 10-13 shows the column variables used for determining the bottom composition and flow rate from an overall material balance. The component

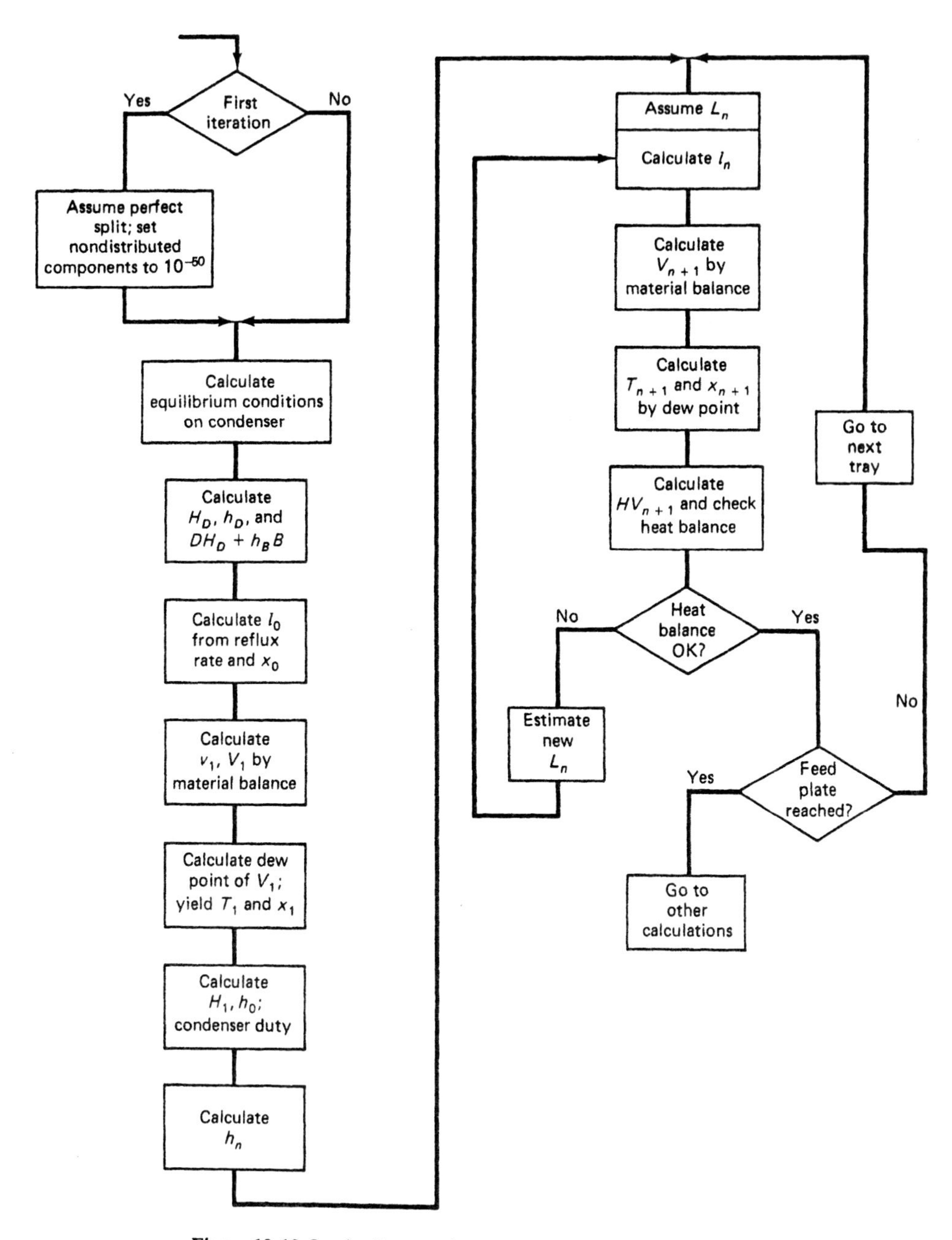

Figure 10-12 Logic diagram for rectifying section calculations

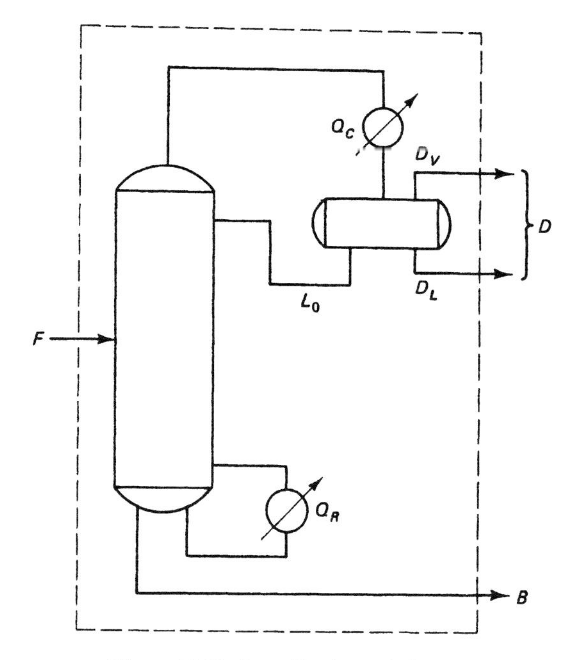

Figure 10-13 Overall column balances

and overall material balances are

$$b_i = f_i - d_i$$

and

$$B = F - D$$

The temperature of the liquid leaving the bottom of the column can be determined by a bubble-point calculation. This also determines the composition of the vapor leaving the reboiler. The reboiler duty can be determined by an overall column heat balance.

$$Q_R = Q_C + H_D D_V + h_D D_L - (H_F F_V + h_F F_L) + h_B B \tag{10-58}$$

Tray calculations are now ready to be initiated. Figure 10-14 shows the material and energy balance envelopes for the stripping section of the column. The first step is to assume a vapor rate V_m. A material balance is

$$L_{m+1} = V_m + B \tag{10-59a}$$

or

$$L_{m+1} = L_m + V_m - V_{m-1} \tag{10-59b}$$

The term V_{m-1} will be zero for the reboiler. A bubble-point calculation yields the temperature of stream L_{m+1} and also the composition of stream V_{m+1}. A heat balance around the bottom of the column is

$$h_{m+1} L_{m+1} = H_m V_m - Q_R + h_B B \tag{10-60a}$$

or

$$h_{m+1} L_{m+1} = h_m L_m + H_m V_m - H_{m-1} V_{m-1} \tag{10-60b}$$

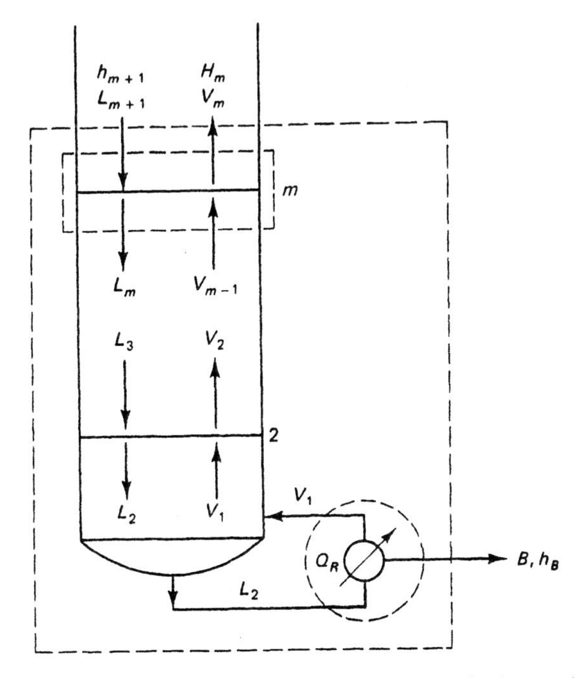

Figure 10-14 Material and energy balances for stripping section

The heat balance determines the correctness of the assumed vapor rate. If the heat balance is not satisfied, the vapor rate must be adjusted and the calculations must be repeated. An expression to reestimate the vapor rate can be developed from Eq. (10-60a). By substituting the overall material balance into Eq. (10-60a), we obtain

$$h_{m+1}(V_m + B) = H_m V_m - Q_R + h_B B \tag{10-61}$$

We can rearrange Eq. (10-61) as shown:

$$V_m = \frac{h_B B - Q_R - h_{m+1} B}{h_{m+1} - H_m} \tag{10-62}$$

Equation (10-62) is exactly like Eq. (10-57). Poor initial estimates of V_m will result in inaccurate values of H_{m+1}, which lead to unreasonable new estimates for V_m.

The foregoing procedure is repeated until a satisfactory heat balance closure is achieved. When that occurs calculations are initiated for tray $m + 1$ using the equilibrium vapor composition determined from the bubble-point calculation of L_{m+1}. These calculations are repeated until the feed zone is reached. A logic diagram for a computer program for the stripping-section calculations is shown in Figure 10-15.

The next step is to complete the feed-zone calculations. In order to do this, a feed-zone model must be selected. Of the models available, the "simple" feed

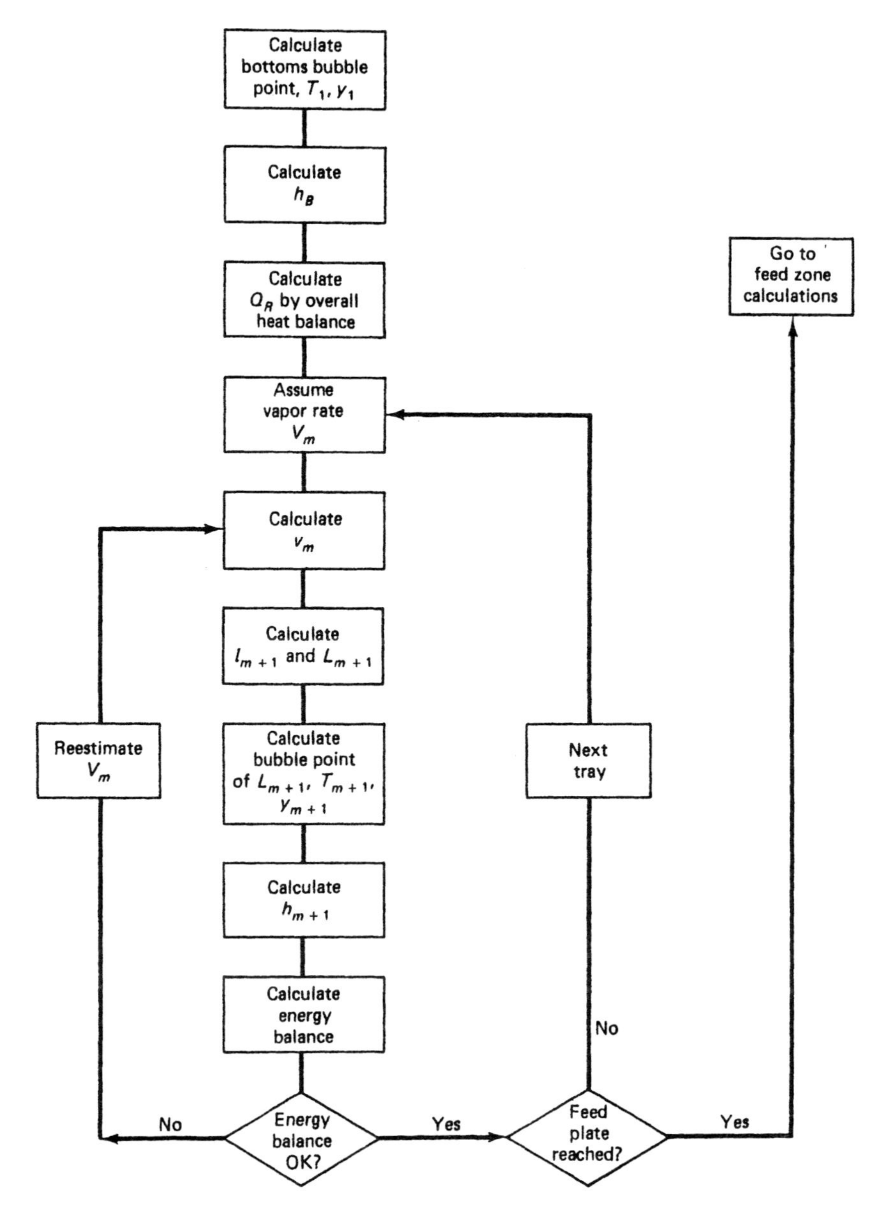

Figure 10-15 Logic diagram for stripping-section calculations

zone used in the McCabe–Thiele solution will be selected. In this the liquid leaving the rectifying section mixes with the liquid portion of the feed to form the liquid entering the stripping section; the vapor leaving the top tray of the stripping section mixes with the vapor portion of the feed to form the vapor feed to the rectifying section. These mixing processes are assumed to be adiabatic. An additional assumption is that no vaporization or condensation will occur as the result of mixing the two saturated streams. Calculations on a limited number of systems indicate that this assumption is correct. A sketch of this feed zone is shown in Figure 10-16.

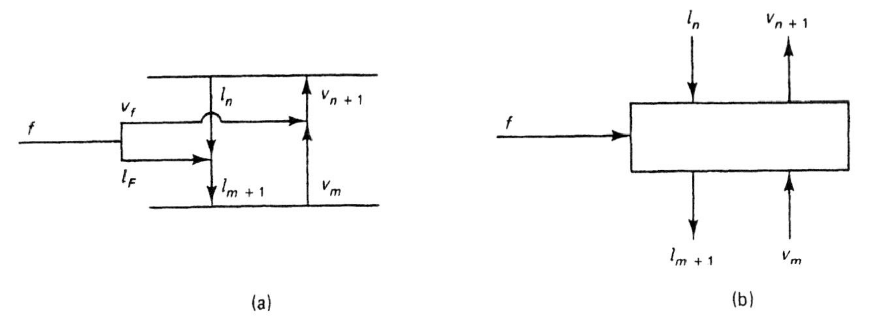

Figure 10-16 Models of (a) a simple feed zone, and (b) an equilibrium feed zone

Using the simple feed zone, the following relationships, which will be written for component i, determine whether a satisfactory feed-zone match between streams calculated from opposite ends of the column has been reached.

$$|v_m + v_F - v_{n+1}| \leq \epsilon \quad (10\text{-}63a)$$

$$|l_m + l_F - l_{m-1}| \leq \epsilon \quad (10\text{-}63b)$$

The subscript i has been omitted for convenience. Equations (10-63a) and (10-63b) must close within the specified tolerance, ϵ. If they do not, the assumed product compositions must be adjusted and the calculations repeated. The following procedure is recommended for making the adjustments in the product distributions.

If $d \leq b$,

$$d^{k+1} = d^k \frac{v_m + v_F}{v_{n+1}} \quad (10\text{-}64a)$$

and

$$b^{k+1} = f - d^{k+1} \quad (10\text{-}64b)$$

In some cases the value of d calculated for use in the next trial will be "unrealistic" or larger than the amount of that component in the feed.

If $d^{k+1} \geq f$,

$$b^{k+1} = b^k \frac{v_{n+1}}{v_m + v_F} \quad (10\text{-}65a)$$

and

$$d^{k+1} = f - d^{k+1} \quad (10\text{-}65b)$$

If $d > b$,

$$b^{k+1} = b^k \frac{v_{n+1}}{v_m + v_F} \quad (10\text{-}66a)$$

and

$$d^{k+1} = f - b^{k+1} \quad (10\text{-}66b)$$

If the calculated value for b is unrealistic, Eq. (10-64b) is used. A logic diagram used to illustrate this adjusting and checking procedure is shown in Figure 10-17.

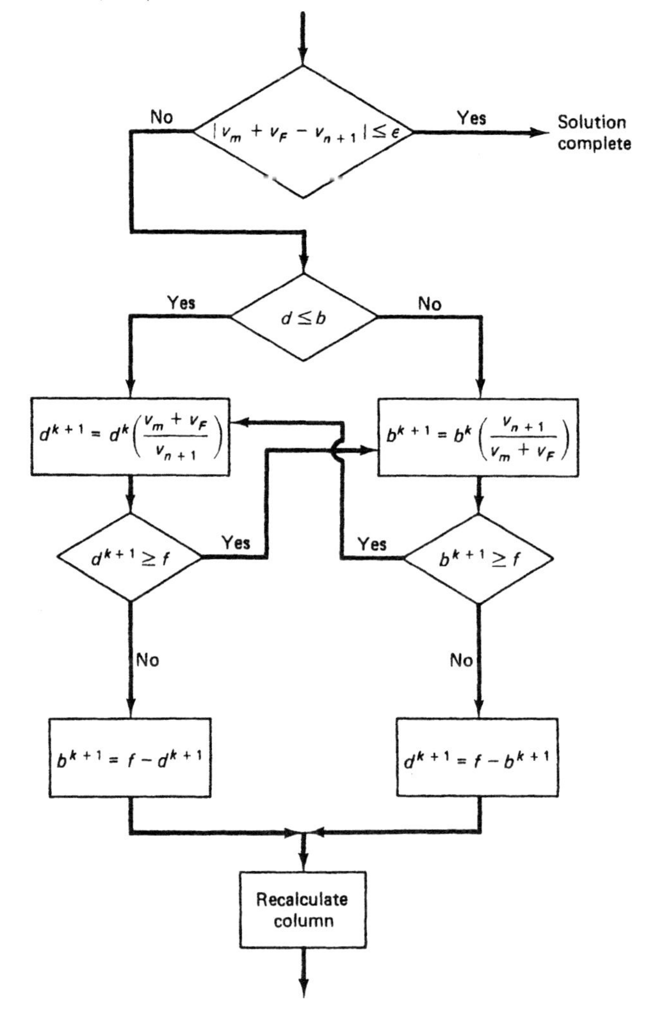

Figure 10-17 Logic diagram for feed zone diagrams

After the assumed product distributions have been adjusted, the entire calculational procedure must be repeated until feed-zone closure according to Eqs. (10-63a) and (10-63b) is satisfactory. This obviously is a long and involved procedure and makes clear the reason why computer solution is encouraged.

Thiele–Geddes method

The Thiele–Geddes method is another of the classical procedures used for carrying out tray-by-tray calculations. Basic assumptions of the calculation are that the liquid and vapor rates leaving each tray and the temperature of each tray are known. Usually, this requires that the flow rates on the tray and temperature be assumed with subsequent checking and adjusting as necessary.

The usual column specifications are:

1. Column pressure
2. Reflux and distillate rates
3. Number of trays and feed plate location
4. Type of condenser

Equations for use in the Thiele–Geddes calculations are developed from material balance considerations for the column. The development of the Thiele–Geddes equations follows the outline shown below. For the rectifying section a component material balance is

$$V_{n+1}y_{n+1} = L_n x_n + D x_D \tag{10-67}$$

We can rearrange Eq. (10-67) as

$$\frac{y_{n+1}}{x_D} = \frac{L_n}{V_{n+1}}\frac{x_n}{x_D} + \frac{D}{V_{n+1}} \tag{10-68}$$

We obtain the following from an overall material balance around the rectifying section:

$$\frac{D}{V_{n+1}} = 1 - \frac{L_n}{V_{n+1}} \tag{10-69}$$

Upon substituting Eq. (10-58) into Eq. (10-57) and rearranging, we obtain

$$\frac{y_{n+1}}{x_D} = \frac{L_n}{V_{n+1}}\left(\frac{x_n}{x_D} - 1\right) + 1 \tag{10-70}$$

The relationship between y_n/x_D and x_n/x_D can be developed from the condition of equilibrium

$$\frac{y_n}{x_D} = K_n \frac{x_n}{x_D} \tag{10-71}$$

Equation (10-71) is valid for all trays in the column, including the reboiler and a partial condenser. If the column has a total condenser, the equilibrium constant K_n must be replaced by unity.

Equations for use in stripping-section calculations are developed in a similar manner. We can write a component material balance for the stripping section as

$$L_{m+1}x_{m+1} = V_m y_m + B x_B \tag{10-72}$$

Rearranging as before yields

$$\frac{x_{m+1}}{x_B} = \frac{V_m}{L_{m+1}}\left(\frac{y_m}{x_B} - 1\right) + 1 \tag{10-73}$$

The relationship between x_m/x_B and y_m/x_B is similar to Eq. (10-71). Thus

$$\frac{y_m}{x_B} = K_m \frac{x_m}{x_B} \tag{10-74}$$

Application of Eqs. (10-68) through (10-74) to a given column for a specific separation results in internal tray compositions which are expressed as the ratio of the tray composition to the appropriate terminal stream composition. These ratios are made usable through knowledge of the terminal stream compositions. These compositions are developed from the feed composition and material balance considerations. A material balance around the column for any component is

$$Fz_F = Dx_D + Bx_B \tag{10-75}$$

Thus

$$x_D = \frac{z_F}{D/F(1 - x_B/x_D) + x_B/x_D} \tag{10-76a}$$

or

$$x_B = \frac{z_F}{B/F(1 - x_D/x_B) + x_D/x_B} \tag{10-76b}$$

By definition $\sum x_D$ and $\sum x_B$ must equal unity. The proper value of D/F (or B/F) is found by trial-and-error solution of the appropriate equation, either

$$\sum x_D = \sum \frac{z_F}{D/F(1 - x_B/x_D) + x_B/x_D} \tag{10-77a}$$

or

$$\sum x_B = \sum \frac{z_F}{B/F(1 - x_D/x_B) + x_D/x_B} \tag{10-77b}$$

The ratio x_D/x_B or x_B/x_D for either bubble-point or dew-point feeds can be determined from the internal ratios. For a saturated liquid feed, $y_{n+1} = y_m$ and

$$\frac{x_B}{x_D} = \frac{y_{n+1}/x_D}{y_m/x_B} \tag{10-78a}$$

For a saturated vapor feed, $x_n = x_{m+1}$ and

$$\frac{x_B}{x_D} = \frac{x_n/x_D}{x_{m+1}/x_B} \tag{10-78b}$$

For partially vaporized feeds the material balances for the combined streams $(V_m + V_F)$ or $(L_n + L_F)$ must be solved simultaneously with either Eq. (10-77a) or (10-77b) to determine the correct value for the ratio y_m/x_B or x_n/x_D.

The Thiele–Geddes calculation is initiated by assuming liquid and vapor flows and temperature profile for the entire column. The iterative calculations for the Thiele–Geddes method are summarized below.

1. Solve Eq. (10-70) for the condenser using the appropriate value of K.
2. Solve Eqs. (10-70) and (10-71) for each component on each tray in the rectifying section. The assumed temperatures are used to determine the K values, and the assumed vapor and liquid flows are used to compute L_n and V_{n+1} for each tray.
3. Solve Eqs. (10-73) and (10-74) for each component in the reboiler and on each plate in the stripping section.
4. Solve either Eq. (10-77a) or (10-77b) simultaneously with the appropriate material balance if a partially vaporized feed is present to determine x_D, x_B, D, and B.
5. Convert the ratios x_n/x_D (or x_m/x_B) to mole fractions for each tray in both the rectifying and stripping sections of the column.
6. Determine the condenser duty and reboiler duty by heat balances.

7. Calculate the bubble point for each tray and compare these temperatures with the estimated tray temperatures.

8. Perform a heat balance calculation on each tray using the temperature computed in step 7, and adjust the liquid and vapor rates as necessary. This heat-balancing step is not a part of the original Thiele–Geddes calculational procedure.

9. Repeat steps 1 through 8 until satisfactory agreement between the assumed and calculated liquid, vapor, and temperature profiles is obtained.

A logic diagram for this procedure is given in Figure 10-18.

The Thiele–Geddes calculational procedure offers distinct advantages over the Lewis–Matheson procedure. Problems of estimating the distillate compostion for nondistributed components are avoided and the Thiele–Geddes procedure can be adapted to multifeed and multiproduct fractionators more easily. The major disadvantage of the Thiele–Geddes method in comparison with the Lewis–Matheson procedure is the additional computational effort required.

Holland–Thiele–Geddes method

Holland and co-workers (1963) have applied a form of the Thiele–Geddes method to a variety of column configurations ranging from a single-feed two-product column to multifeed, three-or-more-product columns. Using the Holland approach, we can write Eq. (10-68) for any component as

$$\frac{v_{n+1}}{d} = A_n\left(\frac{v_n}{d}\right) + 1 \tag{10-79}$$

where

$$A_n = \frac{L_n}{K_n V_n}$$

We can write Eq. (10-73) as

$$\frac{l_{m+1}}{b} = S_m\left(\frac{l_m}{b}\right) + 1 \tag{10-80}$$

where

$$S_m = \frac{K_m V_m}{L_m}$$

The feed-zone equation is found from a balance around the top of the column, including the feed plate shown in Figure 10-16a. Thus

$$v_F + v_m = l_n + d \tag{10-81}$$

Since $f = v_F + l_F$ and $f = d + b$, we may rearrange Eq. (10-81) as

$$\frac{b}{d} = \frac{l_n/d + l_F/f}{v_m/b + v_F/f} \tag{10-82}$$

The basic calculational procedure and assumptions outlined for the original Thiele–Geddes procedure were also used by Holland. However, some problems were encountered in the computer solution. The most severe problem was caused by the fact that the calculated total distillate from the overall

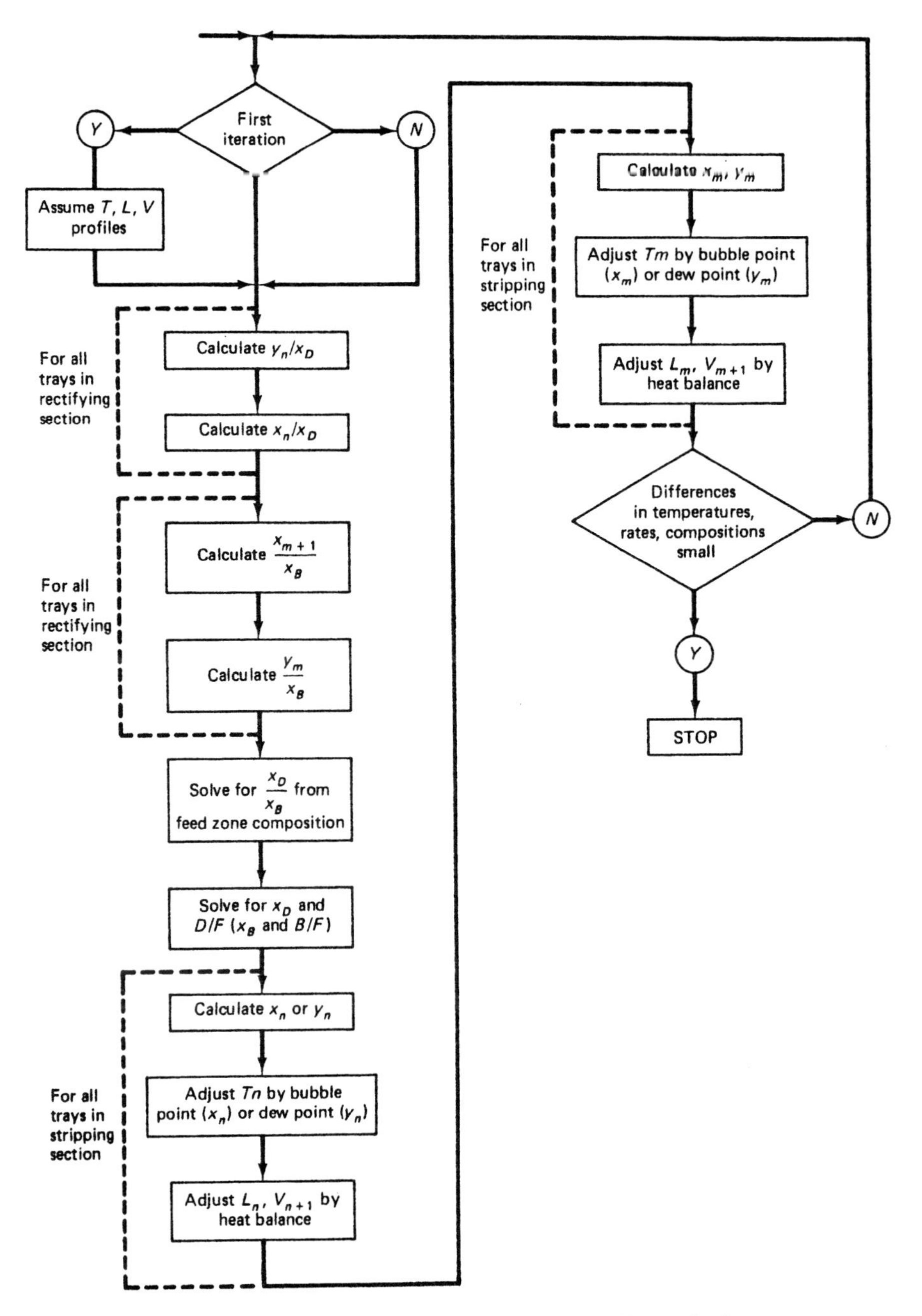

Figure 10-18 Logic diagram for Thiele–Geddes method

balance $d = f/(1 + b/d)$ and the specified total distillate flow rates usually did not agree. An arbitrary correction procedure was developed to adjust the calculated b/d ratios to force agreement between the calculated and specified total distillate rates. Holland referred to this procedure as the *θ method* and has described it in considerable detail. The θ method is based on the following. If $\sum d = \sum f/(1 + b/d) \neq D_{spec}$, find a value of θ such that $\sum d = \sum f/(1 + \theta b/d) = D_{spec}$. The value of θ used in the calculation is the same for all components. It must be found by a trial-and-error procedure; Holland recommends the Newton–Raphson method as follows:

$$g(\theta) = \sum \frac{f}{1 + \theta b/d} - D_{spec} \tag{10-83}$$

$$g'(\theta) = \frac{\partial[\sum f/(1 + \theta b/d)]}{\partial \theta} = -\sum \frac{fb/d}{(1 + \theta b/d)^2} \tag{10-84}$$

$$\theta_{n+1} = \theta_n - \frac{g(\theta_n)}{g'(\theta_n)} \tag{10-85}$$

The Holland adaptation of the Thiele–Geddes calculational procedure has been applied to many different column configurations and is described in considerable detail in the original presentation. The θ method does provide a convenient method for adjusting column material balance calculations. This procedure can be used satisfactorily in most instances.

10.7 Comparison of Shortcut and Tray-by-Tray Calculations

By using the results of Examples 10.1 through 10.4, a tray-by-tray calculation was carried out in order to compare the manual calculations with those made on a computer. The programs used in this calculation are: "MINISIM," "DIST," and "MINITXT" (Maddox and Erbar, 1981). The output from a microcomputer program is shown in Table 10-1. The program used to obtain this table incorporated Eq. (10-13) to predict the minimum number of trays, Eqs. (10-38) and (10-39) to predict the minimum reflux rate, and the Erbar–Gray–Maddox reflux-plates correlation. The hand calculation gave a total of 7.89 minimum theoretical trays, whereas the computer calculation gave 7.76. The hand-calculated minimum reflux rate is 32.74 mol per 100 mol of feed and the computer-calculated value is 31.53. These values are in very good agreement. Differences in the hand and computer calculated values stem from several sources; an error of 2 to 3% in the equilibrium constants is introduced by reading graphs such as those shown in Appendix C. The microcomputer program uses the Soave–Redlich–Kwong equation of state (Redlich and Kwong, 1949; Soave, 1972) to determine the equilibrium constants. The calculated values will differ slightly from those shown in Appendix C.

TABLE 10-1. RESULTS OF COMPUTER CALCULATIONS FOR MINIMUM TRAYS AND MINIMUM REFLUX FOR A BUBBLE-POINT FEED

Component	*Feed (lbmol)*	*Distillate (lbmol)*	*Bottom (lbmol)*
C_3H_8	1.36	1.36	0.00
iC_4H_{10}	14.33	14.27	0.06
nC_4H_{10}	16.37	15.55	0.82
iC_5H_{12}	15.66	0.31	15.35
nC_5H_{12}	17.88	0.05	17.83
nC_6H_{14}	34.40	0.00	34.40
	100.00	31.54	68.46
Temperature (°F)	92.50	41.92	155.46
Pressure (psia)	25.00	25.00	30.00
Enthalpy (Btu)	−81,520	−67,940	132,200

Minimum number of stages = 7.76
Minimum reflux rate = 31.53 lbmol

Number of Stages in Column, (Including Reboiler)	*Reflux Rate (lbmol)*	*Condenser Duty (Btu)*	*Reboiler Duty (Btu)*	*Feed Tray (Reboiler = Tray 1)*
38.80	32.11	594×10^3	740×10^3	17.9
25.87	33.52	607	753	11.9
19.40	36.41	634	780	8.9
15.52	41.72	684	830	7.1
12.93	51.22	773	919	6.0

The minimum reflux rate and minimum number of stages are typically used as input for a tray-by-tray calculation. The results of a tray-by-tray calculation using 25 theoretical plates and an external reflux rate of 33.5 mol per 100 mol of feed are shown in Tables 10-2 and 10-3. The separation is to be made such that the distillate product contains 95% of the *n*-butane originally contained in the feed and the bottom product contains 98% of the *i*-pentane that enters with the feed. The product compositions shown in Table 10-2 indicate that the column is not performing up to the required specifications. The distillate contains only 87.9% of the entering *n*-butane and the bottom product contains 91.2% of the entering *i*-pentane. The profiles shown in Table 10-3 do not indicate that the feed plate is seriously misplaced. While some improvement in fractionation might be gained by raising the feed tray one or two plates, there seems to be little possibility that this would improve the separation to meet specifications.

One available alternative that we may use to improve the separation is to increase the reflux rate. The component distribution for a 25-tray column that operates with a reflux pump-back rate of 40.5 mol per 100 mol of feed is shown in Tables 10-4 and 10-5. This summary represents about as good a match between

TABLE 10-2. TRAY-BY-TRAY CALCULATION RESULTS FOR A BUBBLE-POINT FEED AND A REFLUX OF 33.5 MOL/100 MOL FEED

Component	*Feed (lbmol)*	*Distillate (lbmol)*	*Bottom (lbmol)*
C_3H_8	1.36	1.36	0.00
iC_4H_{10}	14.33	14.26	0.07
nC_4H_{10}	16.37	14.39	1.98
iC_5H_{12}	15.66	1.38	14.28
nC_5H_{12}	17.88	0.11	17.77
nC_6H_{14}	34.40	0.00	34.40
	100.00	31.50	68.50
Temperature (°F)	92.50	43.67	152.98

Condenser duty = -603.5×10^3 Btu
Reboiler duty = 718.4×10^3 Btu

TABLE 10-3. TRAY TEMPERATURE, LIQUID, AND VAPOR RATE PROFILES FOR TABLE 10-2

Tray Number	*Temperature (°F)*	*Pressure (psia)*	*Liquid Leaving the Tray (lbmol)*	*Vapor Leaving the Tray (lbmol)*
26	43.7	25.0	33.50	0.00
25	53.6	25.2	32.47	65.00
24	60.6	25.4	31.59	63.97
23	66.7	25.6	30.86	63.09
22	71.6	25.8	30.36	62.36
21	75.3	26.0	30.03	61.86
20	77.8	26.2	29.80	61.53
19	79.4	26.3	29.64	61.30
18	80.6	26.5	29.49	61.14
17	81.5	26.7	29.34	60.99
16	82.4	26.9	29.13	60.84
15	83.4	27.1	28.67	60.63
14	85.4	27.3	27.55	60.17
13	90.0	27.5	25.23	59.05
12	99.4	27.7	126.02	56.73
11	101.6	27.9	126.49	57.52
10	103.4	28.1	126.81	57.99
9	105.1	28.3	127.04	58.31
8	107.0	28.5	127.23	58.54
7	109.3	28.7	127.40	58.73
6	112.2	28.8	127.59	58.90
5	115.8	29.0	127.86	59.09
4	120.4	29.2	128.28	59.36
3	125.8	29.4	128.82	59.78
2	132.1	29.6	129.14	60.32
1	140.1	29.8	128.41	60.64
0 (reboiler)	153.0	30.0	68.50	59.91

TABLE 10-4. TRAY-BY-TRAY CALCULATION RESULTS FOR A BUBBLE-POINT FEED AND A REFLUX OF 40.5 MOL/100 MOL FEED

Component	*Feed (lbmol)*	*Distillate (lbmol)*	*Bottom (lbmol)*
C_3H_8	1.36	1.36	0.00
iC_4H_{10}	14.33	14.31	0.02
nC_4H_{10}	16.37	15.54	0.83
iC_5H_{12}	15.66	0.28	15.38
nC_5H_{12}	17.88	0.01	17.87
nC_6H_{14}	34.40	0.00	34.40
	100.00	31.50	68.50
Temperature (°F)	92.50	42.56	155.49

Condenser duty = -662.3×10^3 Btu
Reboiler duty = 785.2×10^3 Btu

TABLE 10-5. TRAY TEMPERATURE, LIQUID, AND VAPOR RATE PROFILES FOR TABLE 10-4

Tray Number	*Temperature (°F)*	*Pressure (psia)*	*Liquid Leaving the Tray (lbmol)*	*Vapor Leaving the Tray (lbmol)*
26	42.6	25.0	40.50	0.00
25	49.1	25.2	40.32	72.00
24	52.3	25.4	39.93	71.82
23	55.2	25.6	39.35	71.43
22	58.6	25.8	38.63	70.85
21	62.8	26.0	37.85	70.13
20	67.3	26.2	37.15	69.35
19	71.5	26.3	36.59	68.65
18	75.0	26.5	36.17	68.09
17	77.8	26.7	35.84	67.67
16	79.9	26.9	35.50	67.34
15	81.8	27.1	34.96	67.00
14	84.3	27.3	33.71	66.46
13	89.1	27.5	30.95	65.21
12	99.5	27.7	131.77	62.45
11	102.4	27.9	132.28	63.27
10	105.0	28.1	132.60	63.78
9	107.9	28.3	132.87	64.10
8	111.2	28.5	133.13	64.37
7	115.1	28.7	133.49	64.63
6	119.6	28.8	134.02	64.99
5	124.5	29.0	134.74	65.52
4	129.4	29.2	135.56	66.24
3	134.0	29.4	136.27	67.06
2	138.6	29.6	136.36	67.77
1	144.6	29.8	135.04	67.86
0 (reboiler)	155.5	30.0	68.50	66.54

the specified separation and calculated values as can be expected. As shown, 94.9% of the entering *n*-butane will appear in the distillate and 98.2% of the *i*-pentane will appear in the bottom product.

In industrial practice, distillation columns seldom operate with a saturated liquid feed. For reasons of energy conservation as well as equalizing the loading between the rectifying and stripping sections of the column, the feed is normally partially vaporized. With this in mind, shortcut computations were carried out with 25 mol% of the feed to the distillation column being vaporized. We show the microcomputer results for the shortcut calculation in Table 10-6. The minimum number of stages remains fixed at 7.76. However, the minimum reflux rate has increased by some 20% to 41 mol of external reflux pump-back per 100 mol of feed. Also note that the feed tray location predicted by the shortcut method has remained the same. This is because Eqs. (10-47) through (10-50) do not in any way take into consideration the impact of feed vaporization.

TABLE 10-6. RESULTS OF COMPUTER CALCULATIONS FOR MINIMUM TRAYS AND MINIMUM REFLUX FOR A 25% VAPORIZED FEED

Component	*Feed (lbmol)*	*Distillate (lbmol)*	*Bottom (lbmol)*
C_3H_8	1.36	1.36	0.00
iC_4H_{10}	14.33	14.27	0.06
nC_4H_{10}	16.37	15.55	0.82
iC_5H_{12}	15.66	0.31	15.35
nC_5H_{12}	17.88	0.05	17.83
nC_6H_{14}	34.40	0.00	34.40
	100.00	31.54	68.46
Temperature (°F)	112.38	41.92	155.46
Pressure (psia)	27.00	25.00	30.00
Enthalpy (Btu)	240,700	−67,940	132,200

Minimum number of stages = 7.76
Minimum reflux rate = 41.02 lbmol

Number of Stages in Column (Including Reboiler)	*Reflux Rate (lbmol)*	*Condenser Duty (Btu)*	*Reboiler Duty (Btu)*	*Feed Tray (Reboiler = Tray 1)*
38.80	41.69	684×10^3	507×10^3	17.9
25.87	43.31	699	522	11.9
19.40	46.64	730	553	8.9
15.52	52.74	787	610	7.1
12.93	63.68	889	713	6.0

Generally speaking, vaporizing a portion of the feed will increase the temperature. As a result of this temperature increase the feed normally will be introduced at a slightly lower point in the column.

To estimate more correctly the impact of feed vaporization on the reflux requirement, the feed enthalpies from Tables 10-1 and 10-4 will be used in combination with Eq. (10-46) as follows:

$$V_{1u} = V_{1k} + \frac{(H_{Fu} + H_{Fk})(1 - D/F)}{Q_C/L_{CT}} \tag{10-46}$$

$$V_{1k} = L_0 + D = 33.5 + 31.54 = 65.0$$

$$L_{CT} = L_0 + D = 65.0$$

$$V_{1u} = 65 + \frac{[240{,}700 - (-81{,}520)](1 - 0.315)}{607{,}000/65}$$

$$= 88.6$$

$$L_0 = 88.6 - 31.5 = 57.1 \text{ mol per 100 mol of feed}$$

Using an external reflux pump-back of 0.55 mol per mole of feed, another tray-by-tray calculation was performed. The results are shown in Tables 10-7 and 10-8. The reflux rate does not appear to be quite large enough. However, if we compare the condenser and reboiler duties from Tables 10-2 and 10-7, the effects of feed vaporization are obvious; the condenser duty increases but the reboiler duty decreases.

TABLE 10-7. TRAY-BY-TRAY CALCULATION RESULTS FOR A 25% VAPORIZED FEED AND A REFLUX OF 55.0 MOL/100 MOL FEED

Component	*Feed (lbmol)*	*Distillate (lbmol)*	*Bottom (lbmol)*
C_3H_8	1.36	1.36	0.00
iC_4H_{10}	14.33	14.22	0.11
nC_4H_{10}	16.37	15.09	1.28
iC_5H_{12}	15.66	0.83	14.83
nC_5H_{12}	17.88	0.01	17.87
nC_6H_{14}	34.40	0.00	34.40
	100.00	31.50	68.50
Temperature (°F)	112.38	43.10	154.27
Condenser duty =	-798.6×10^3 Btu		
Reboiler duty =	622.5×10^3 Btu		

TABLE 10-8. TRAY TEMPERATURE, LIQUID, AND VAPOR RATE PROFILES FOR TABLE 10-7

Tray Number	*Temperature (°F)*	*Pressure (psia)*	*Liquid Leaving the Tray (lbmol)*	*Vapor Leaving the Tray (lbmol)*
26	43.1	25.0	55.00	0.00
25	51.1	25.2	54.08	86.50
24	57.1	25.4	52.89	85.58
23	63.5	25.6	51.55	84.39
22	70.4	25.8	50.43	83.05
21	76.5	26.0	49.68	81.93
20	81.2	26.2	49.25	81.18
19	84.3	26.3	49.00	80.75
18	86.3	26.5	48.83	80.50
17	87.6	26.7	48.70	80.33
16	88.5	26.9	48.57	80.20
15	89.2	27.1	48.43	80.07
14	89.9	27.3	48.27	79.93
13	90.6	27.5	48.06	79.77
12	91.4	27.7	47.71	79.56
11	92.6	27.9	46.91	79.21
10	95.2	28.1	44.97	78.41
9	101.0	28.3	41.10	76.47
8	112.5	28.5	117.74	48.77
7	115.7	28.7	118.25	49.24
6	119.3	28.8	118.77	49.75
5	123.3	29.0	119.38	50.27
4	127.6	29.2	120.09	50.88
3	132.2	29.4	120.82	51.59
2	137.1	29.6	121.25	52.32
1	143.4	29.8	120.71	52.75
0 (reboiler)	154.3	30.0	68.50	52.21

REFERENCES

AKASHAH, S. A., J. H. ERBAR, and R. N. MADDOX, *Chem. Eng. Commun.*, *3*, 461 (1979).

BROWN, G. G., and H. Z. MARTIN, *Trans. AIChE*, *35*, 679 (1939).

ERBAR, J. H., and R. N. MADDOX, *Hydrocarbon Process. Petrol. Refiner*, *40*, No. 5, 183 (1961).

ERBAR, R. C., and R. N. MADDOX, *Can. J. Chem. Eng.*, *40*, No. 1, 25 (1962).

FENSKE, M. R., *Ind. Eng. Chem.*, *24*, 482 (1932).

GILLILAND, E. R., *Ind. Eng. Chem.*, *32*, 1220 (1940).

GRAY, J. R., "Reflux-to-Trays Correlation for Multicomponent Distillation Systems," M.S. thesis, Oklahoma State University, 1960.

HOLLAND, C. D., *Multicomponent Distillation*, Prentice-Hall, Englewood Cliffs, N.J., 1963.

JOYNER, R. S., J. H. ERBAR, and R. N. MADDOX, *Petro/Chem Eng.*, *34*, No. 2, 169 (1962).

KIRKBRIDE, C. G., *Petrol. Refiner*, *23*, No. 9, 321 (1944).

LEWIS, W. K., and G. L. MATHESON, *Ind. Eng. Chem.*, *24*, 494 (1932).

MADDOX, R. N., and J. H. ERBAR, *Gas Conditioning and Processing*, Vol. 3: *Advanced Techniques and Applications*, Campbell Petroleum Series, J. M. Campbell, Norman, Okla., 1981.

REDLICH, O., and J. N. S. KWONG, *Chem. Rev.*, *44*, 233 (1949).

SHIRAS, R. N., D. N. HANSON, and C. W. GIBSON, *Ind. Eng. Chem.*, *42*, 871 (1950).

SOAVE, G., *Chem. Eng. Sci.*, *27*, 1197 (1972).

THIELE, E. W., and R. L. GEDDES, *Ind. Eng. Chem.*, *25*, 289 (1933).

UNDERWOOD, A. J. V., *Chem. Eng. Prog.*, *44*, No. 8, 603 (1948).

VAN WINKLE, M., and W. G. TODD, *Chem. Eng.*, *78*, 136 (Sept. 1971).

WINN, F. W., *Petrol. Refiner*, *37*, No. 5, 216 (1958).

NOTATIONS

---

b = flow rate of the light key component in the bottom product, mol/t
b' = flow rate of the heavy key component in the bottom product, mol/t
B = total flow rate of the bottom product, mol/t
d = flow rate of the light key component in the distillate, mol/t
d_i = flow rate of any component in the distillate, mol/t
d' = flow rate of the heavy key component in the distillate, mol/t
D = total flow rate of the distillate, mol/t
D_L = total flow rate of liquid in the distillate, mol/t
D_V = total flow rate of vapor in the distillate, mol/t
f = flow rate of the light key component in the feed, mol/t
f_i = flow rate of any component in the feed, mol/t
f' = flow rate of the heavy key component in the feed, mol/t
F = total flow rate of the feed, mol/t
h_B = enthalpy of the bottom product, E/mol
h_F = enthalpy of the feed, E/mol
h_{FK} = enthalpy of a bubble-point feed, E/mol
h_L = enthalpy of the liquid reflux, E/mol
h_m = enthalpy of the liquid leaving tray m, E/mol
h_n = enthalpy of the liquid leaving tray n, E/mol
H_D = enthalpy of the vapor distillate, E/mol
H_F = enthalpy of the vapor feed, E/mol

H_{Fu} = vapor enthalpy for a partially vaporized feed, E/mol
K_n = distribution coefficient for the light key
K'_n = distribution coefficient for the heavy key
l_F = flow rate of any component in the liquid feed, mol/t
l_m = flow rate of any component in the liquid leaving tray m, mol/t
L = total liquid flow rate, mol/t
L_{CT} = total number of moles of liquid formed in the condenser, mol/t
L_n = total flow rate of the liquid leaving tray n, mol/t
L_{RP} = liquid flow rate in the rectifying pinch zone, mol/t
L_0 = flow rate of liquid reflux, mol/t
L_{0_m} = minimum flow rate of liquid reflux, mol/t
M = number of plates in the stripping section
N = number of plates in the rectifying section
N_T = total number of plates in the column
p = flow rate of the light key in the side product, mol/t
p' = flow rate of the heavy key in the side product, mol/t
Q_C = condenser duty, E/t
Q_R = reboiler duty, E/t
S = actual number of plates in the column
S_m = minimum number of plates in the column
T_{avg} = average tower temperature, T
T_{BP} = bubble-point temperature, T
T_{DP} = dew-point temperature, T
v_F = flow rate of any component in the vapor feed, mol/t
v_n = flow rate of any component in the vapor leaving tray n, mol/t
v_{RP} = vapor flow rate of any component in the rectifying-section pinch zone, mol/t
v_{SP} = vapor flow rate of any component in the stripping-section pinch zone, mol/t
V = total vapor flow rate, mol/t
V_1 = total vapor flow rate leaving the first plate, mol/t
V_F = total vapor flow rate in the feed stream, mol/t
V_n = total vapor flow rate leaving tray n, mol/t
V_{RP} = total vapor flow rate in the rectifying-section pinch zone, mol/t
$x_{B,LK}$, x_B = mole fraction of the light key in the bottom product
x_D = mole fraction of the light key in the distillate product
$x_{D,HK}$, x'_D = mole fraction of the heavy key in the distillate product
x_{HK} = mole fraction of the heavy key in the feed product
x_{LK} = mole fraction of the light key in the feed product
x_P = mole fraction of the light key in the side-stream product
x_{RP} = mole fraction of any component in the rectifying-section pinch zone
x'_B = mole fraction of the heavy key in the bottom product
x'_P = mole fraction of the heavy key in the side-stream product
y_B = mole fraction of any component in the vapor leaving the reboiler
y_{RP} = mole fraction of any component in the vapor of the rectifying-section pinch zone
$z_{i,F}$ = mole fraction of i in the feed

Greek Letters

α_{avg} = average relative volatility in the column

α_n = relative volatility for plate n

β = constant value defined in Winn's equation

ϵ = specified tolerance as defined in Eq. (10-63)

θ = slope of the K–K' plot on log-log coordinate as defined in Winn's equation

ϕ_j = specified constant value bounded by adjacent values of relative volatility as defined in Eq. (10-41)

PROBLEMS

10.1 A saturated liquid mixture containing 40 mol% n-pentane and 60 mol% n-hexane is to be separated by continuous distillation. The overhead product is to contain 95% of the n-pentane in the feed and the bottom product is to contain 97% of the n-hexane in the feed. The distillation column is to operate at a total pressure of 25 psia. A total condenser and partial reboiler are to be used for this process. Determine the minimum number of stages by applying the equation of Fenske, and compare the result to that obtained by using the McCabe–Thiele method.

10.2 A saturated liquid mixture consisting of 25% propane, 25% n-butane, 25% n-pentane, and 25% n-hexane is to be separated by continuous distillation. The overhead product is to contain 95% propane and the bottom product is to contain no more than 1% propane. Determine the minimum number of stages for column operating pressures of **(a)** 100 psia, and **(b)** 250 psia. **(c)** Calculate the compositions of the distillate and bottom products. The Fenske equation should be used for this problem.

10.3 Solve Problem 10.2 by using the procedure proposed by Winn.

10.4 The operating pressure of a condenser is determined by the temperature of the available cooling medium. Determine the operating temperature of the total condenser described in Problem 10.2 for tower pressures of **(a)** 100 psia, and **(b)** 250 psia. **(c)** If water is to be used as the cooling medium for the condenser, estimate the maximum temperature of the water for each column pressure.

10.5 **(a)** Calculate the minimum reflux ratio for Problem 10.1 by using the equation of Underwood. **(b)** Compare your result with the value obtained by using the McCabe–Thiele method.

10.6 A saturated liquid mixture consisting of 60% propane, 30% n-butane, and 10% n-pentane is to be separated into an overhead product that contains 98% of the propane that enters with the feed. The bottom product is to contain 95% of the n-butane that enters with the feed. Determine the minimum reflux ratio for tower pressures of **(a)** 100 psia, and **(b)** 300 psia.

10.7 The mixture in Problem 10.6 is to be separated at a pressure of 300 psia into the distillate and bottom product specified above. Calculate the minimum reflux ratio if the feed is **(a)** 50% vaporized, and **(b)** 100% vaporized.

10.8 The separation described in Problem 10.6 is to be carried out at a tower pressure of 100 psia and a reflux ratio equal to 1.2 times the minimum value. Determine the number of theoretical stages required for the separation by using **(a)** the Gilliland correlation, and **(b)** the Erbar–Gray–Maddox correlation. **(c)** Determine the feed plate location by using the Kirkbride equation.

10.9 Determine the number of trays above and below the feed plate for Problem 10.2 at a tower pressure of 250 psia.

10.10 Use the McCabe–Thiele method to estimate the minimum number of stages and the minimum reflux ratio for the separation described in Problem 10.2. Assume that the separation of the light key and heavy key components can be used to represent the overall mixture. The operating pressure is 250 psia.

Liquid-Liquid Extraction 11

11.1 Introduction

Although distillation is the most often used separation process because of its simplicity, a number of separations cannot be carried out by this method, for various reasons. Following are some situations in which distillation is not feasible:

1. The components have low volatility
2. The components in the solution have essentially the same volatility
3. The components are sensitive to the high temperatures that would be required for separation by distillation
4. The solute in the feed solution is present in a relatively small amount

These situations are found in essentially all of the process industries, including the pharmaceutical, chemical, and petroleum industries.

An alternative separation process frequently encountered is liquid–liquid extraction. This process consists of the recovery of a solute from a solution by mixing with a solvent. The extracting solvent used must be insoluble or at least soluble to only a limited extent in the solution to be extracted. In addition, the solute to be extracted should have a high affinity for the extraction solvent. Liquid–liquid extraction consists of two basic steps: (1) intimate mixing of the extracting solvent with the solution to be extracted followed by (2) separation of the mixed solution into two immiscible liquid phases. Other operations are generally involved in the total extraction process. One of these is the separation of solute from the extracting solvent and the subsequent recovery of the solvent for further extractions.

In liquid–liquid extraction the factor that determines the amount of solute in the two liquid phases is the distribution coefficient. The liquid phase that contains the greater concentration of solvent and the smaller concentration of feed is referred to as the *extract*. The other liquid phase, which contains a greater concentration of the feed liquid and a smaller concentration of solvent, is referred to as the *raffinate*. The actual extraction process can be carried out in much the same way as distillation and absorption. Towers containing sieve trays, bubble cap trays, and various packings are frequently employed. Other types of extractors, however, are also encountered. Several extraction methods will be discussed in this chapter, including: (1) single-stage contact, (2) simple multiple or crosscurrent contact, and (3) multiple countercurrent contact.

11.2 Application of Ternary Data

The extraction of a solute from a binary feed can be readily carried out by using a ternary diagram of the solution and the appropriate solvent. For multiple counter-current contact in which several stages may be required to achieve the desired separation, the ternary diagram can be easily converted to either the McCabe–Thiele or Ponchon–Savarit diagram to simplify the determination of the number of stages.

Type I diagrams: one pair of partially miscible liquids

A ternary system in which only one pair of partially miscible liquids exists is known as a *type I system*. A typical right-angle ternary diagram for this type is shown in Figure 11-1a. The compositions of the extract and raffinate phases, which are represented by the contact of the ends of the tie lines with the saturation line, are found from the equilibrium distribution curve shown in Figure 11-1b. The *plait point* represents a point on the phase diagram at which the two conjugate phases have the same concentration of A. This is indicated by its position on the 45° line, as shown in Figure 11-1b. In this figure the concentration of A in the extract is greater than in the raffinate phase. This results in a distribution coefficient, y_E^*/x_R, greater than 1.0, since the equilibrium curve lies above the diagonal. For systems in which the tie line slopes in the opposite direction, the distribution curve will fall below the 45° line. Solute A will then show a stronger affinity for the carrier solvent B than for the extracting solvent S.

In Figure 11-1, A refers to the solute to be recovered, S is the extracting solvent, and B is the dilutent or carrier solvent. Although mass fractions were used in the figure, ternary data can be expressed conveniently by using solvent-free coordinates. The solute concentration in the raffinate phase is written on a

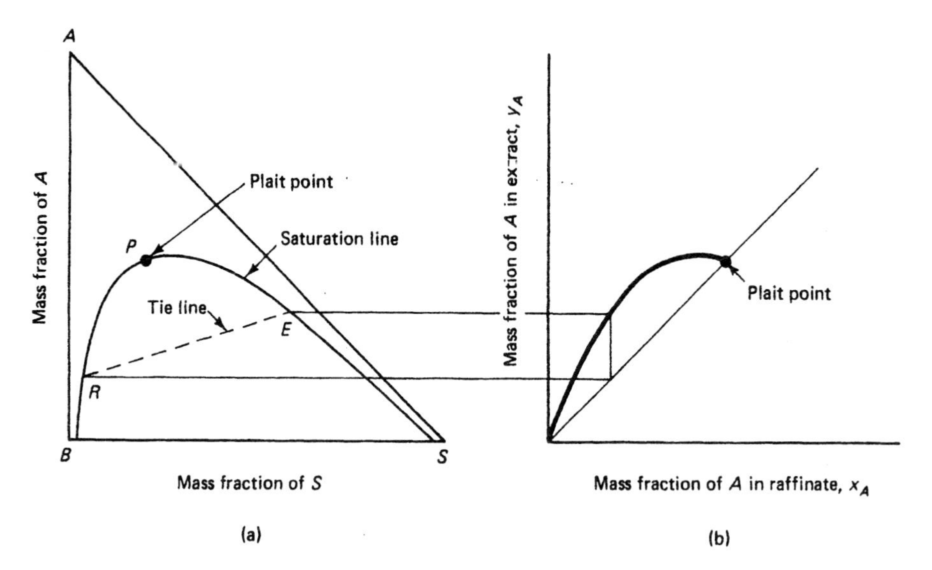

Figure 11-1 (a) Ternary diagram, and (b) equilibrium distribution curve

solvent-free basis as

$$\bar{X}_A = \frac{\text{mass of } A}{\text{mass of } A + \text{mass of } B} = \frac{x_A}{x_A + x_B} \tag{11-1}$$

For the extract phase the concentration is

$$\bar{Y}_A = \frac{\text{mass of } A}{\text{mass of } A + \text{mass of } B} = \frac{y_A}{y_A + y_B} \tag{11-2}$$

We can express the concentration of the solvent S on a solvent-free basis for the raffinate phase as

$$\bar{Z}_R = \frac{\text{mass of } S}{\text{mass of } A + \text{mass of } B} = \frac{x_S}{x_A + x_B} \tag{11-3}$$

For the extract phase we have

$$\bar{Z}_E = \frac{\text{mass of } S}{\text{mass of } A + \text{mass of } B} = \frac{y_S}{y_A + y_B} \tag{11-4}$$

The phase is indicated by the subscripts E for extract and R for raffinate. Weight fractions are given by x and y.

By selecting points along the saturation line shown in Figure 11-1a on either side of the plait point and applying Eqs. (11-1) through (11-4), the ternary data can be converted to a Ponchon–Savarit type diagram. Representation of data in this manner for liquid–liquid systems is known as the *Maloney–Schubert method* (1940). Tie lines for this type of diagram can be obtained by transforming the equilibrium distribution curve to a solvent-free basis. This is shown in Figure 11-2. Typical tie-line and mutual solubility data are presented in Table 11-1.

TABLE 11-1. MUTUAL SOLUBILITY AND TIE-LINE DATA FOR THE ETHYLENE GLYCOL–WATER–ETHYL METHYL KETONE SYSTEM AT 30°C (RAO AND RAO, 1957)

Mutual Solubility Data (wt %)

Ethyl Methyl Ketone	*Water*	*Ethylene Glycol*
89.0	11.0	—
78.5	11.7	9.8
74.3	12.2	13.5
67.6	13.8	18.6
63.4	15.2	21.2
57.0	17.9	25.1
52.7	19.8	27.5
45.4	22.6	32.0
40.8	25.6	33.6
37.7	28.2	34.1
35.2	30.7	34.1
32.4	33.8	33.8
30.1	37.7	32.2
28.6	41.4	30.0
25.2	47.5	27.3
24.0	50.0	26.0
22.5	56.0	21.5
21.5	60.0	18.5
20.8	64.7	14.5
20.6	79.4	—

Tie-Line Data (wt %)

Solvent Layer			*Water Layer*		
Ethyl Methyl Ketone	*Water*	*Ethylene Glycol*	*Ethyl Methyl Ketone*	*Water*	*Ethylene Glycol*
88.4	11.1	0.5	20.8	69.7	9.5
87.1	11.2	1.7	21.0	65.6	13.4
84.9	11.3	3.8	22.1	58.3	19.6
82.7	11.6	5.7	23.6	52.4	24.0
80.6	11.8	7.6	26.1	46.1	27.8
50.0	20.5	29.5		Plait point	

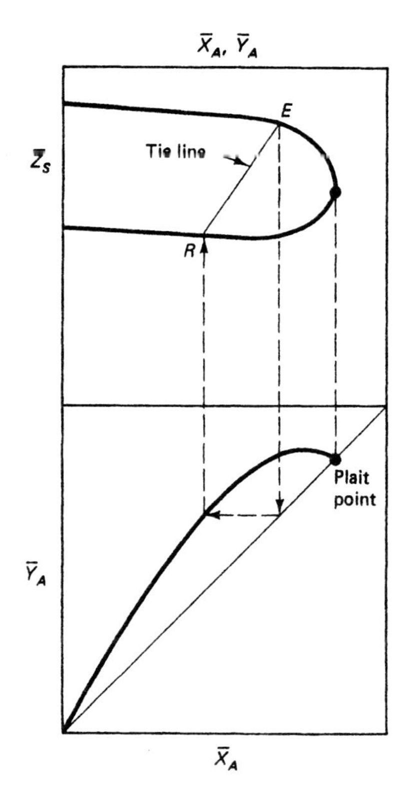

Figure 11-2 Solvent-free coordinates

Example 11.1

Using the tie-line data given in Table 11-1, obtain a Maloney–Schubert diagram such as that shown in Figure 11-2.

Solution:

The tie-line data must first be written on a solvent-free basis, as follows:

Extract (Solvent Layer)				*Raffinate (Water Layer)*			
y_A	$\bar{Y}_A$	y_S	$\bar{Z}_E$	x_A	$\bar{X}_A$	x_S	$\bar{Z}_R$
0.005	0.043	0.884	7.62	0.095	0.120	0.208	0.263
0.017	0.132	0.871	6.75	0.134	0.170	0.210	0.266
0.038	0.251	0.849	5.62	0.196	0.252	0.221	0.284
0.057	0.329	0.827	4.78	0.240	0.314	0.236	0.309
0.076	0.392	0.806	4.15	0.278	0.376	0.261	0.353
0.295	0.590	0.500	1.0		Plait point		

Figure 11-3 was obtained using the data above. In the figure we see that the equilibrium distribution curve crosses the diagonal at a value of $\bar{Y}_A \approx 0.25$. On the Maloney–Schubert diagram this corresponds to a vertical tie line as shown. It is important to note that mutual solubility data must generally be used to obtain the complete saturation line.

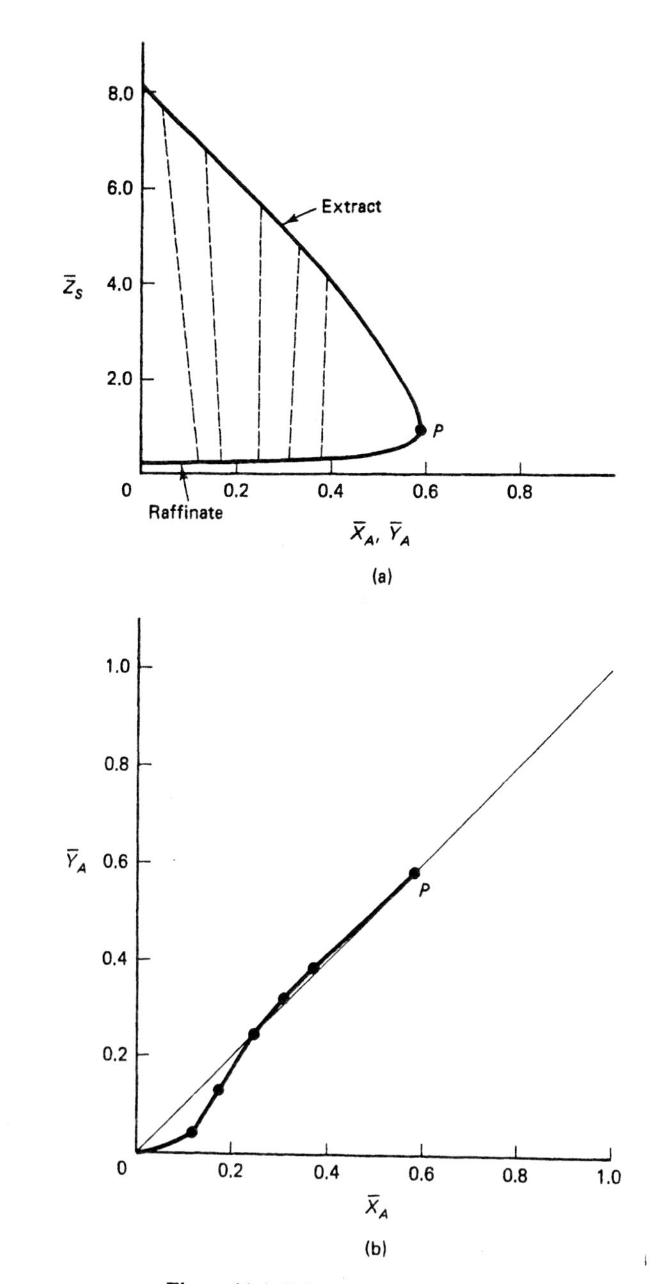

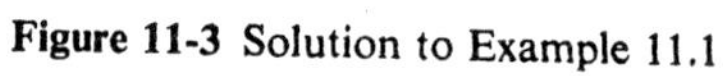
Figure 11-3 Solution to Example 11.1

The amount of solute extracted in any type of extraction process depends on the slope of the tie line. A large equilibrium distribution coefficient, as indicated by a large tie-line slope, provides for a better separation than does a small one. The size of the two-phase region also has an influence on the degree of separation that can be achieved in an extraction process. If the two-phase separation zone is small, mixtures containing only small amounts of solute can be extracted. An increase in temperature tends to decrease the size of the separation zone. At sufficiently high temperatures the two-phase region will disappear entirely. Liquid–liquid extraction is then not possible.

Type II diagrams: systems consisting of two pairs of immiscible liquids

When the temperature of a system is lowered, the separation zone becomes larger. If the temperature is low enough and the individual components remain in the liquid phase, a second immiscible liquid pair will form. This is described as a type II system and is shown in Figure 11-4, where $T_4 < T_3 < T_2 < T_1$.

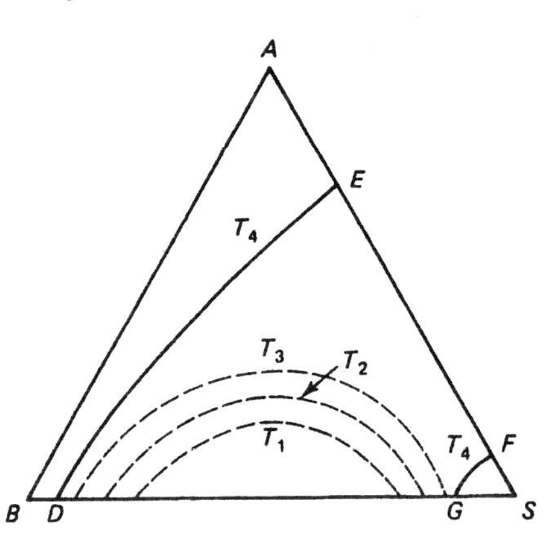

Figure 11-4 Type II ternary diagram

A type II diagram and its associated distribution curve are shown in Figure 11-5. Points along the curve *DF* represent compositions of the raffinate phase and points along *GH* represent extract compositions. Equilibrium raffinate and extract compositions are indicated by the tie lines between the *DF* and *GH* lines shown in Figure 11-5a. Since the equilibrium curve shown in Figure 11-5b lies below the diagonal, the distribution coefficient is less than 1. Thus solute *A* has a stronger affinity for *B* than for the extracting solvent *S*. This does not mean that solvent *S* cannot be used to extract *A* from *B*, but that more solvent will be required to make the desired separation.

A type II ternary diagram can be represented on a solvent-free basis in a manner similar to that used for a type I system. In contrast to the Maloney–Schubert diagram shown in Figure 11-3, a type II system does not have a plait point. Thus the extract and raffinate phases have compositions ranging from 0 to 1 as shown in Figure 11-6. Because of the similarity between this type of

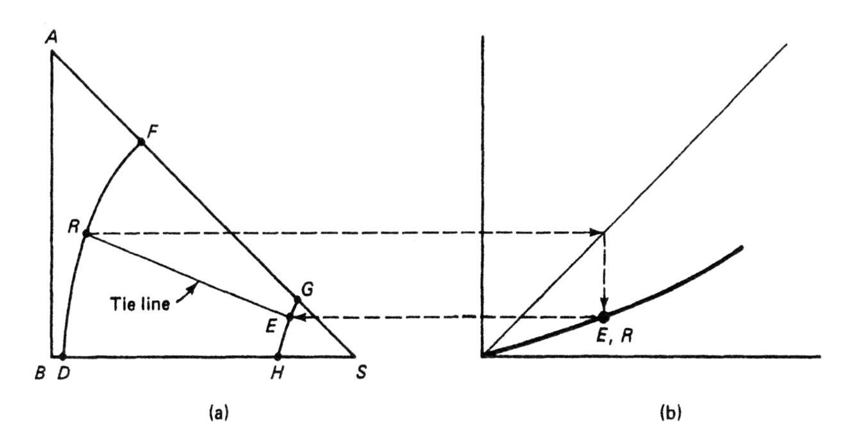

Figure 11-5 (a) Type II ternary diagram, and (b) equilibrium distribution curve

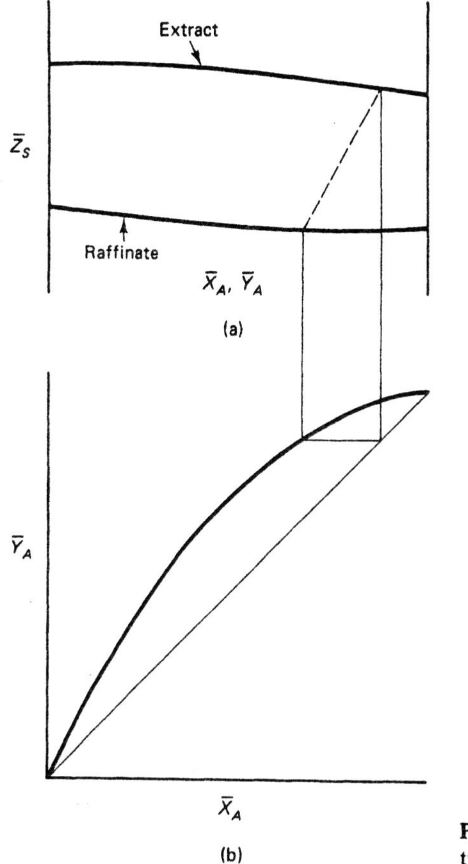

Figure 11-6 Solvent-free diagram for type II systems

diagram and the enthalpy–composition diagram, procedures used in making distillation calculations can be applied to liquid–liquid extraction.

11.3 Single-Stage Contact

Single-stage contact is easy to carry out physically since it consists of contacting the solution to be treated with the extracting solvent in a single stage, followed by separation of the phases for further processing. A single-stage process is shown in Figure 11-7. The separator is included in the single-stage process since extraction of the solute from the feed solution continues as long as the two phases are in contact or until equilibrium is reached.

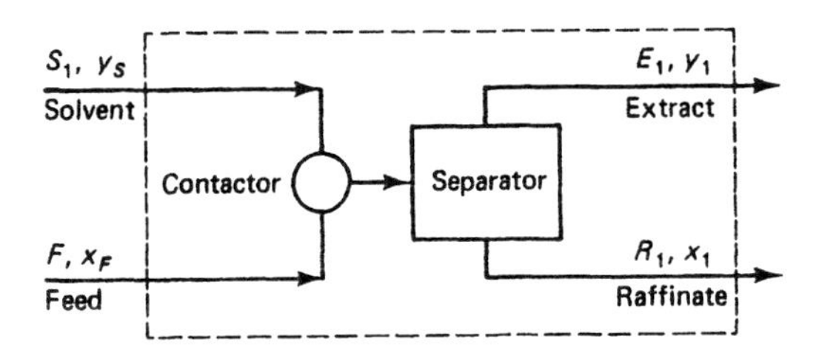

Figure 11-7 Single-stage process

Extraction calculations are very similar to those used in distillation and absorption, and are readily carried out when the feed solution contains only two components. The amount of solute present in the extract and raffinate phases can be found by using either the ternary diagram, as suggested by Hunter and Nash (1935), and mass balances around the stage, or by using solvent-free diagrams accompanied by solvent-free material balances.

The mass balance for the system shown in Figure 11-7 is

$$F + S_1 = E_1 + R_1 = M \tag{11-5}$$

The material balance for any component is

$$Fx_F + S_1 y_S = E_1 y_1 + R_1 x_1 = M x_M \tag{11-6}$$

where F, S_1, E_1, and R_1 are mass flow rates, and the compositions x_F, y_S, y_1, and x_1 are weight fractions of the solute to be extracted. In the expression above, M is equal to the total mass either entering or leaving the contactor. The concentration, x_M, is the mass fraction of the solute in the total mixture that enters the contactor. From Eq. (11-6) we see that M is a point with composition x_M that lies on a line joining the compositions x_F and y_S. The composition x_M also lies on a line joining the compositions x_1 and y_1. The location of M on a ternary diagram can be determined from the relative amounts of feed and solvent that enter the stage by applying the lever arm principle.

A material balance for a single stage can be written on a solvent-free basis as follows:

$$\bar{F} + \bar{S}_1 = \bar{E}_1 + \bar{R}_1 = \bar{M} \tag{11-7}$$

The bar above each term indicates that the streams are free of the extracting solvent. The component balance for solute A is

$$\bar{F}\bar{X}_F + \bar{S}_1\bar{X}_S = \bar{E}_1\bar{Y}_1 + \bar{R}_1\bar{X}_1 = \bar{M}\bar{X}_M \tag{11-8}$$

Equations (11-7) and (11-8) can be applied in much the same manner as Eqs. (11-5) and (11-6). The relationships between the flow rates for the mass fraction and solvent-free coordinates are

$$E = \bar{E}(1 + \bar{Z}_E) \quad \text{and} \quad R = \bar{R}(1 + \bar{Z}_R) \tag{11-9}$$

Example 11.2

In a continuous single-stage extraction process, 50 kg/min of water is mixed with 50 kg/min of a solution containing 35% by weight of acetic acid and 65% methyl isobutyl ketone. (a) Determine the amount of acetic acid in the total mixture and the compositions of the extract and raffinate phases. (b) Find the flow rate of each phase.

Solution:

(a) Using Figure 11-8, the contact of the two streams is carried out graphically by drawing a line between the compositions of the entering streams. The

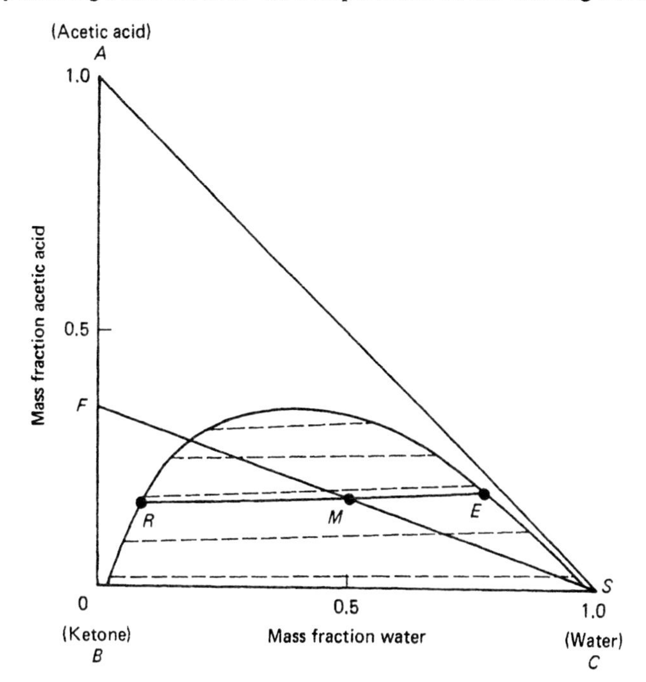

Figure 11-8 Graphical solution for a single-stage process (data from Othmer et al., 1941)

quantity of acetic acid in the total mixture is obtained by combining Eqs. (11-5) and (11-6). Thus we obtain

$$x_M = \frac{Fx_F + S_1 y_S}{M} = \frac{Fx_F + S_1 y_S}{F + S_1}$$

$$= \frac{50(0.35) + 50(0)}{100} = 0.175$$

The composition is shown as point M on the line that connects the entering streams. The compositions of the extract and raffinate phases are found by extending a tie line through point M to the phase envelope as denoted by points E and R. The raffinate phase, point R, has a composition $x_A = 0.16$, $x_C = 0.089$, and $x_B = 1 - 0.16 - 0.089 = 0.751$. Composition of the extract phase is $y_A = 0.185$, $y_C = 0.78$, and $y_B = 1 - 0.78 - 0.185 = 0.035$. (b) The flow rates of the extract and raffinate phases are also found by applying Eqs. (11-5) and (11-6).

$$R_1 x_1 + E_1 y_1 = M x_M$$

or

$$R_1 x_1 + (M - R_1) y_1 = M x_M$$

Thus

$$R_1(0.16) + (M - R_1)(0.185) = 100(0.175)$$

$$R_1 = 40 \text{ kg/min}$$

$$E_1 = M - R_1 = 100 - 40 = 60 \text{ kg/min}$$

11.4 Crosscurrent Multistage Extraction

In multistage crosscurrent extraction, the raffinate phase that results from a single-stage contact is further contacted by fresh solvent in a second stage. The process of contacting the raffinate from each stage by a fresh solvent is continued until the desired solute concentration of the raffinate stream is achieved. A simple multistage crosscurrent extraction process is shown in Figure 11-9.

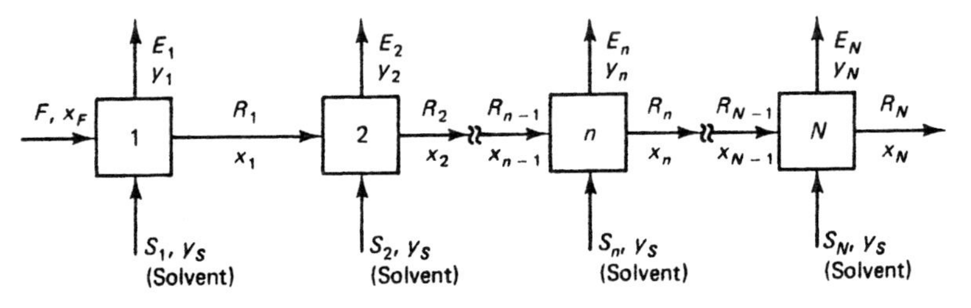

Figure 11-9 Crosscurrent process

Partially miscible systems

The compositions shown in Figure 11-9 represent the weight fractions of the solute to be extracted from the feed. The overall material balance for stage n is

$$R_{n-1} + S_n = R_n + E_n = M_n \tag{11-10}$$

where M_n is the total mass flow rate to stage n. In the notation above, the feed to the first stage is denoted as R_0 or F. The material balance for the solute is

$$R_{n-1}x_{n-1} + S_n y_S = R_n x_n + E_n y_n = M_n x_{M,n} \tag{11-11}$$

On a solvent-free basis the equation above becomes

$$\bar{R}_{n-1}\bar{X}_{n-1} + \bar{S}\bar{Y}_S = \bar{R}_n\bar{X}_n + \bar{E}_n\bar{Y}_n = \bar{M}\bar{X}_{M,n} \tag{11-12}$$

Contact for a two-stage crosscurrent process is shown graphically in Figure 11-10 for the case in which the extracting solvent contains species A and B.

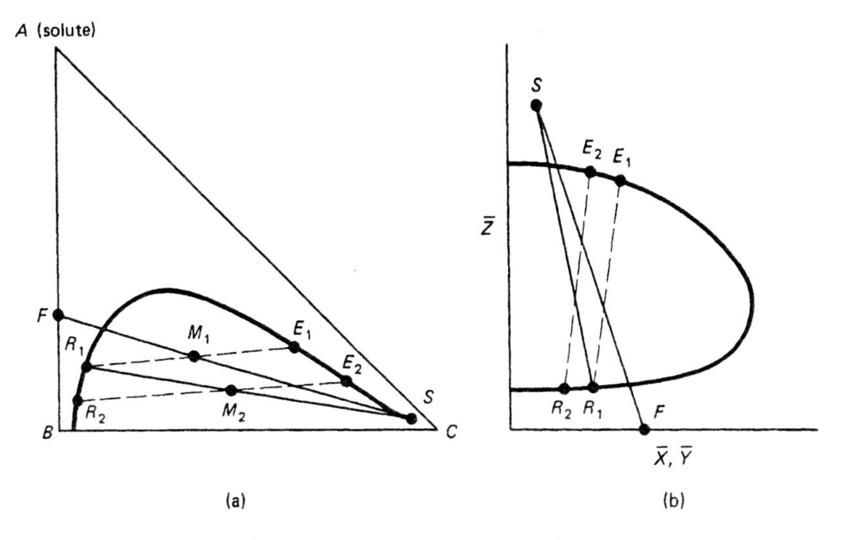

Figure 11-10 Crosscurrent extraction

Example 11.3

A feed containing 35% by weight of acetic acid and 65% methyl isobutyl ketone is contacted by pure water in a two-stage crosscurrent extraction process. The flow rate of the entering feed is 100 kg/min. Fresh water is added to each stage at a rate of 100 kg/min. Determine the compositions of the extract and raffinate phases after each stage. Find the flow rates for all streams leaving the contractors.

Solution:

The calculations are carried out by contacting the entering feed with the solvent in the first stage and then contacting the raffinate from the first stage with fresh solvent in the second stage as shown in Figure 11-11. For the first

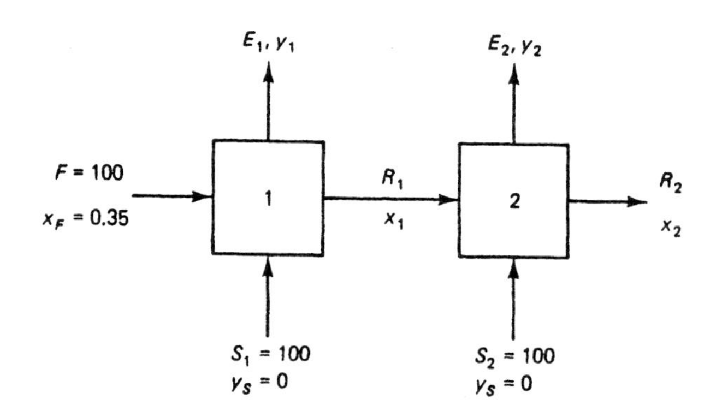

Figure 11-11 Two-stage extraction process

stage, $R_0 = F$ and $x_0 = x_F$. The mass balance for the first stage is

$$F + S_1 = R_1 + E_1 = M_1$$

Thus $$M_1 = 100 + 100 = 200 \text{ kg/min}$$

The total composition of solute for this stage is obtained from Eq. (11-11).

$$x_{M,1} = \frac{Fx_F + S_1 y_S}{M_1} = \frac{100(0.35) + 100(0)}{200} = 0.175$$

The total flow entering the system, M_1, can be found from Figure 11-12 by locating $x_{M,1}$ on the line drawn between the feed and the solvent. Point M_1 can also be found by applying the lever arm principle. Using the lever arm for line segments between F and S_1, we obtain

$$M_1(\overline{M_1S_1}) = F(\overline{FS_1})$$

or $$\overline{M_1S_1} = \frac{F}{M_1}(\overline{FS_1}) = \frac{100}{200}(\overline{FS_1}) = 0.5(\overline{FS_1})$$

where $\overline{M_1S_1}$ and $\overline{FS_1}$ are the lengths of the lines between the points M_1 and S_1, and F and S_1, respectively. The line segment $\overline{M_1S_1}$ thus locates M_1. The intersection of the tie line that passes through M_1 with the saturation curve locates the concentration of the raffinate and extract phases for stage 1. For acetic acid, $x_1 = 0.16$ and $y_1 = 0.185$. The amounts of extract and raffinate are found by writing Eqs. (11-10) and (11-11) for stage 1.

$$(M_1 - E_1)x_1 + E_1 y_1 = M_1 x_{M,1}$$

or $$E_1 = \frac{M_1(x_{M,1} - x_1)}{y_1 - x_1} = \frac{200(0.175 - 0.16)}{0.185 - 0.16} = 120 \text{ kg/min}$$

From Eq. (11-10) we find that

$$R_1 = 200 - 120 = 80 \text{ kg/min}$$

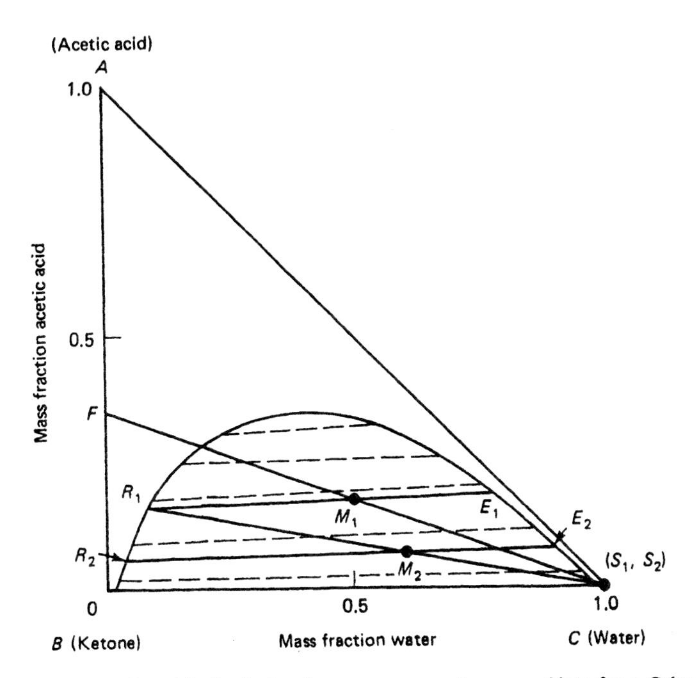

Figure 11-12 Graphical solution for a crosscurrent process (data from Othmer et al., 1941)

For the second stage Eq. (11-10) gives

$$R_1 + S_2 = R_2 + E_2 = M_2$$

$$80 + 100 = M_2 = 180$$

Applying Eq. (11-11) we obtain

$$x_{M,2} = \frac{R_1 x_1 + S_2 y_S}{M_2} = \frac{80(0.16) + 100(0)}{180} = 0.071$$

By drawing a tie line through M_2, we obtain the concentrations of the raffinate and extract phases leaving the second stage, $x_2 = 0.058$ and $y_2 = 0.077$. Equations (11-10) and (11-11) can be used to find the amounts of raffinate and extract that leave the second stage. Because the concentrations of the solute in the two phases are very close to each other, the concentration of water in the two phases should be used. Thus using the water concentrations in the two phases, we get

$$E_2 = \frac{M_2(x_{M,2} - x_2)}{y_2 - x_2} = \frac{180(0.595 - 0.04)}{0.905 - 0.04} = 113.4 \text{ kg/min}$$

and $$R_2 = 180 - 113.4 = 66.6 \text{ kg/min}$$

Crosscurrent contact of insoluble liquids

The simplest case of liquid–liquid extraction involves the use of two completely insoluble liquid solvents. Although systems of this type are not encountered frequently in practice, a number of systems exist in which the solvents are nearly insoluble over a limited concentration range. The system shown in Figure 11-13 is of this type for solute concentrations less than 30%.

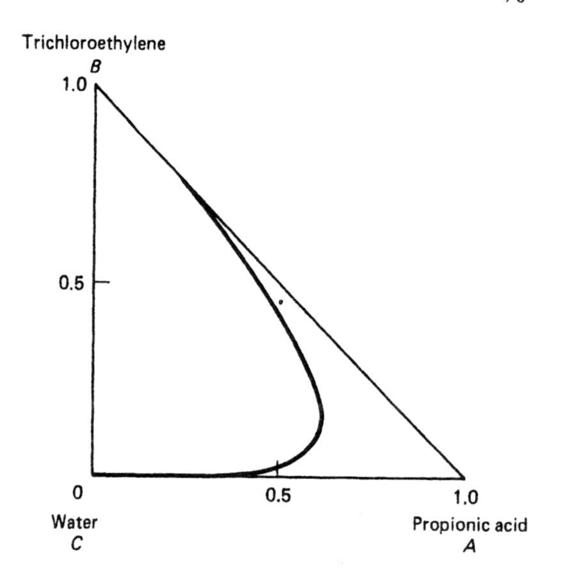

Figure 11-13 Ternary diagram for the propionic acid–water–trichloroethylene system (data from Rao and Rao, 1956)

Since the solvent B, which contains the solute A, and the extracting solvent C are totally insoluble, the concentrations y'_A and x'_A can be represented as

$$y'_A = \frac{y_A}{1 - y_A} = \frac{\text{mass of } A \text{ in extract phase}}{\text{mass of solvent } C} \tag{11-13}$$

$$x'_A = \frac{x_A}{1 - x_A} = \frac{\text{mass of } A \text{ in raffinate}}{\text{mass of solvent } B} \tag{11-14}$$

For stage n in a crosscurrent extraction process, we can write Eq. (11-11) in terms of the foregoing concentration units for species A as

$$R'_{n-1}x'_{n-1} + S'_n y'_S = R'_n x'_n + E'_n y'_n \tag{11-15}$$

where R'_{n-1} and R'_n are the mass flow rates of carrier solvent B, and S'_n and E'_n are the flow rates of the solvent C. The carrier solvent flow rate remains constant for these concentration units. Thus $R'_{n-1} = R'_n = R'$ and $S'_n = E'_n$. We can write the material balance as

$$R'(x'_{n-1} - x_n) = S'_n(y'_n - y'_S) \tag{11-16}$$

or

$$-\frac{R'}{S'_n} = \frac{y'_S - y'_n}{x'_{n-1} - x'_n} \tag{11-17}$$

Equation (11-17) is the equation of the operating line with slope $-R'/S'_n$. In a single-stage process, R' is the flow rate of the entering carrier solvent and is denoted as F'. The parallel operating lines shown in Figure 11-14 indicate that the same quantity of extracting solvent is added to each stage. This value may be changed as desired.

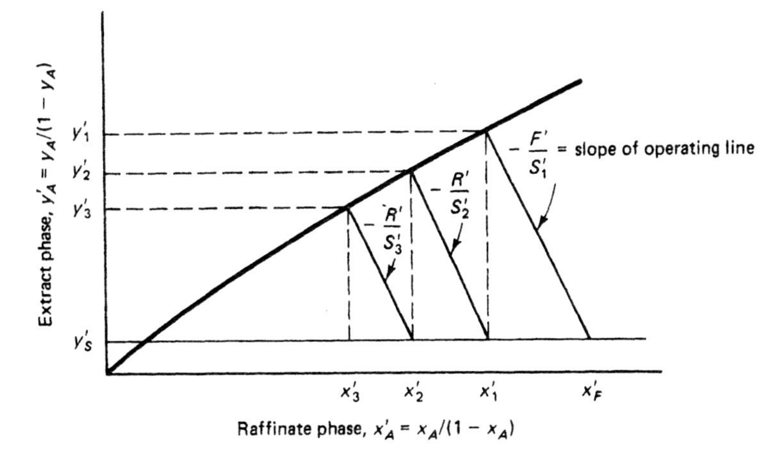

Figure 11-14 Graphical solution for insoluble liquids

Example 11.4

Propionic acid is to be extracted from trichloroethylene in a continuous three-stage crosscurrent extraction process by using water as the solvent. The entering feed stream contains 30% propionic acid by weight and 70% trichloroethylene. The feed enters at a flow rate of 100 kg/min. It is to be contacted by 100 kg/min of water in stages 1 and 2, and by 50 kg/min of water in stage 3. The distribution of propionic acid between the extract and raffinate phases can be expressed as $y_A = 0.38x_A$. Determine the concentration of propionic acid in the extract and raffinate phases leaving the third stage.

Solution:

The entering feed contains 70 kg/min of trichloroethylene. The slope of the operating line for stage 1 is $-(F'/S'_1) = -(70/100) = -0.7$. For stage 2, $-(R'/S'_2) = -(70/100) = -0.7$, and for stage 3, $-(R'/S'_3) = -(70/50) = -1.4$. The feed concentration is $x'_A = 0.3/(1 - 0.3) = 0.429$ and $y'_A = 0$. The equilibrium diagram and operating line are shown in Figure 11-15. From the figure $x'_3 = 0.155$ and $y'_3 = 0.055$.

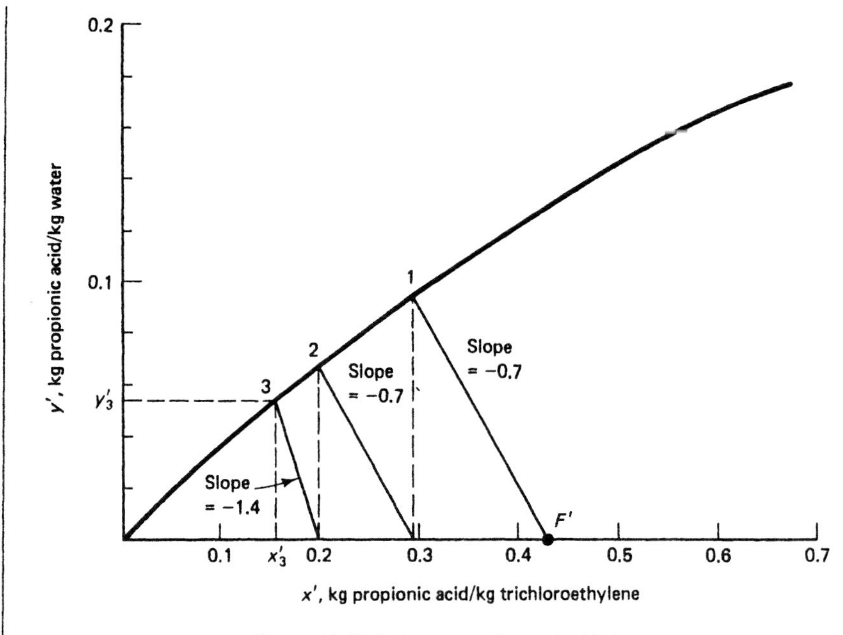

Figure 11-15 Solution to Example 11.4

For a crosscurrent extraction process in which a limited quantity of solvent is available, the greatest recovery of solute is achieved if the solvent is distributed equally to each stage. If a fixed amount of solvent is available, the quantity of solute extracted in a multistage crosscurrent process frequently is not much greater than the amount recovered in a single-stage extraction.

11.5 Countercurrent Multistage Processes

In countercurrent extraction, the feed that contains the solute to be extracted and the extracting solvent enter at opposite ends of a multistage extraction apparatus. A multistage countercurrent process is shown in Figure 11-16. For a fixed quantity of extracting solvent and a fixed number of stages, a countercurrent process is more efficient than a crosscurrent process.

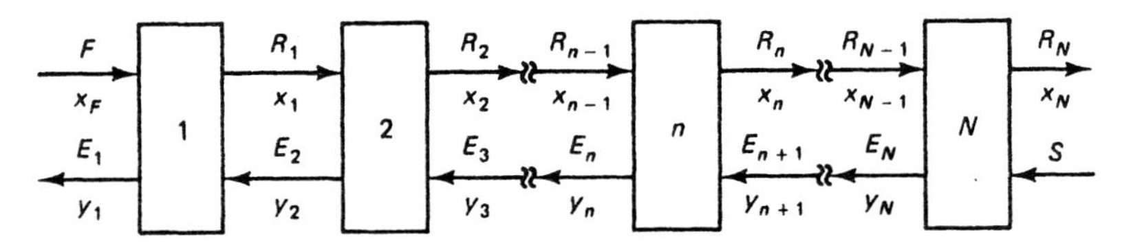

Figure 11-16 Countercurrent extraction process

Countercurrent extraction of partially miscible systems

Multistage calculations for partially miscible systems can be carried out on both ternary and Maloney–Schubert diagrams by applying material balances around the system. The overall material balance and component material balance for the extraction process shown in Figure 11-16 are

$$F + S = R_N + E_1 = M \tag{11-18}$$

and

$$Fx_F + Sy_S = R_N x_N + E_1 y_1 = M x_M \tag{11-19}$$

where M is equal to the total mass flow rate entering and leaving the process and x_M is the total composition. From Eq. (11-18) we see that points F, S, and M fall on a straight line. Similarly, points R_N, E_1, and M also fall on a straight line. Since point M represents the sum of two flow rates, it will lie between the points F and S, and between points R_N and E_1. Obviously, M represents the intersection of the lines that connect points F and S, and R_N and E_1. We can locate point M on a diagram by writing Eq. (11-19) as

$$x_M = \frac{Fx_F + Sy_S}{M} = \frac{Fx_F + Sy_S}{F + S} \tag{11-20}$$

By writing Eq. (11-18) in terms of the difference in flow rates of passing streams, we obtain

$$F - E_1 = R_N - S = \Delta \tag{11-21}$$

Points F, E_1, and Δ lie on a single straight line as do points R_N, S, and Δ. Since these points represent different lines, the point Δ must be at the intersection. Now let us consider the overall material balance for stage n. We have

$$R_{n-1} + E_{n+1} = R_n + E_n \tag{11-22}$$

or

$$R_{n-1} - E_n = R_n - E_{n+1} = \Delta \tag{11-23}$$

We see that a line which passes through the points R_{n-1} and E_n, and a line that passes through the points R_n and E_{n+1} will also pass through Δ. Thus for a multistage process, point Δ represents the focal point for all lines that are extended through all sets of points that represent passing streams. This is demonstrated in Figure 11-17 for a four-stage process.

The solvent-to-feed ratio is found by applying the lever arm principle about the point M. Thus we obtain

$$S(\overline{SM}) = F(\overline{FM}) \tag{11-24}$$

or

$$\frac{S}{F} = \frac{\overline{FM}}{\overline{SM}} \tag{11-25}$$

In the equations above, $\overline{SM}$ and $\overline{FM}$ represent the lengths of the line segments between points S and M, and F and M, respectively. The quantity of raffinate that leaves the fourth stage, R_4, and the quantity of extract that leaves the first stage, E_1, can also be found by applying the lever arm principle. The amount of extract leaving the process is calculated from the expression

$$E_1 = \frac{M(\overline{R_4 M})}{\overline{R_4 E_1}} \tag{11-26}$$

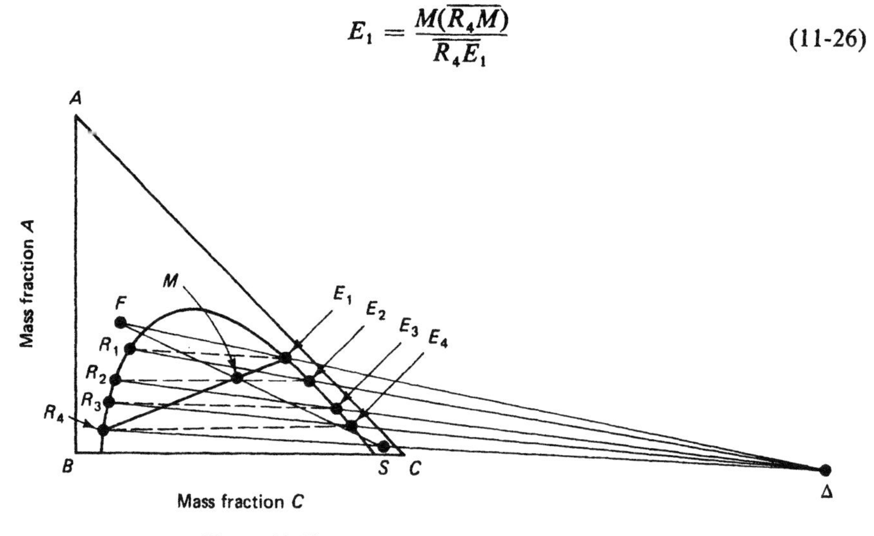

Figure 11-17 Four-stage countercurrent process

The construction of Figure 11-17 is carried out by locating the feed F and solvent S on the ternary diagram at the appropriate compositions. After drawing a line through these two points, we locate point M by applying Eq. (11-25). To complete the construction, the composition of either the final extract, E_1, or the raffinate, R_4, must be known. If R_4 is known, then a line is drawn from R_4 through M to the phase envelope; this locates E_1. The Δ point is found by extending lines from F through E_1 and from R_4 through S. The raffinate composition, R_1, is found by extending a tie line from E_1 to the other side of the phase envelope. The construction is continued by drawing a line from R_1 to Δ to locate E_2. The process is repeated until the final raffinate composition is obtained.

Minimum solvent-to-feed ratio

The solvent-to-feed ratio cannot be fixed arbitrarily. For any given separation process, a minimum solvent-to-feed ratio exists that corresponds to an infinite number of stages. This value can be determined by extending tie lines on a ternary diagram to intersect the line through R_N and S, as shown in Figure 11-18. However, we must consider the case shown in Figure 11-18a, in which the tie lines slope away from the solvent, and the case shown in Figure 11-18b, in which the tie lines slope toward the solvent.

Although the tie line that passes through the feed composition usually determines Δ_{min}, this is not always the case, as shown in 11-18a. If point Δ_{min} is used to determine the number of stages in an extraction process, the construction lines that radiate from Δ_{min} will become coincident with a tie line. This corresponds to a pinch point in the process and would result in an infinite number of stages. The tie line that passes through the feed composition in this case would

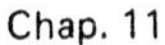

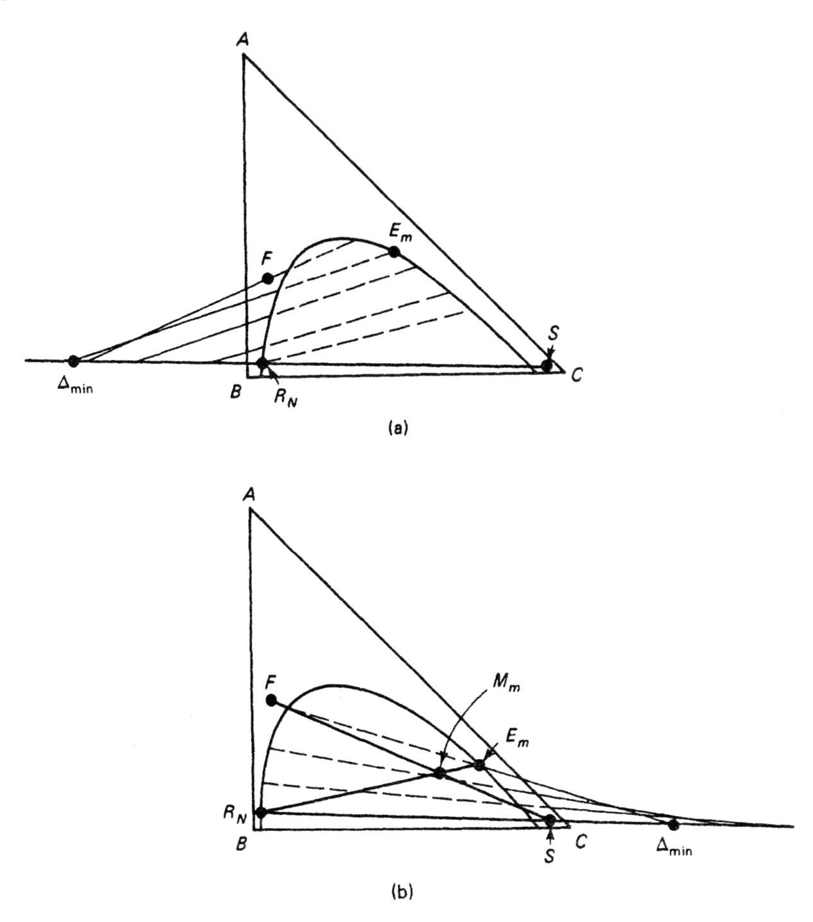

Figure 11-18 Minimum solvent-to-feed ratio

result in a solvent-to-feed ratio less than the minimum. Thus the intersection of the tie line that passes through R_N and S farthest from S gives the actual minimum value. If the tie lines slope toward the solvent S, as shown in Figure 11-18b, the tie line that intersects the line which passes through R_N and S closest to the point S determines Δ_{min}. The minimum solvent-to-feed ratio is calculated by constructing lines from R_N to E_m and from F to S. The intersection of these lines locates the point M_m. The minimum value is calculated from the lengths of the lines as shown in Figure 11-18b. Thus

$$\frac{S}{F} = \frac{\overline{FM_m}}{\overline{SM_m}} \tag{11-27}$$

Example 11.5

A feed containing 35% by weight of solute A and 65% B is to be contacted by pure solvent C in a multistage countercurrent extraction process. The flow rate of the entering feed is 100 kg/min. Fresh solvent enters the extraction battery at a rate equal to 1.5 times the minimum value. (a) Determine the minimum solvent-to-feed ratio if the final raffinate composition is to be 0.04. (b) Find the number of stages for this separation process. The ternary diagram is shown in Figure 11-19.

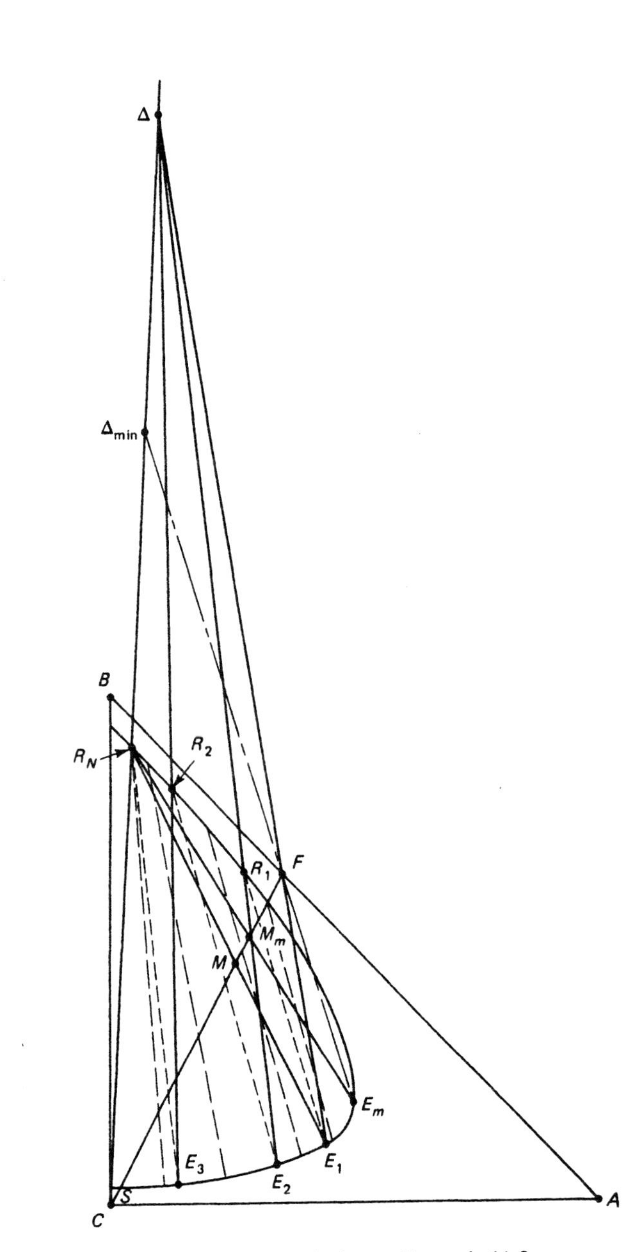

Figure 11-19 Solution to Example 11.5

Solution:

(a) The feed and raffinate compositions are shown on the diagram given in Figure 11-19. Using Eq. (11-27), we obtain

$$\left(\frac{S}{F}\right)_{\min} = \frac{0.65 - 0.525}{0.525 - 0} = 0.238$$

and

$$\left(\frac{S}{F}\right)_{\text{act}} = 1.5(0.238) = 0.357$$

Thus

$$S = 0.357(100) = 35.7 \text{ kg/min}$$

(b) The actual value for M must be located on the diagram. Applying Eq. (11-20) for solute A, we find that

$$x_M = \frac{100(0.35) + 35.7(0)}{135.7} = 0.258$$

After locating M, a line is drawn from R_N through M to locate E_1. The point is located by extending a line from E_1 through F to intersect the line that passes through S and R_N. Using the Δ point and counting stages, we find that three stages are necessary to achieve the final raffinate composition.

As noted in the example, the use of ternary diagrams to carry out equilibrium stage calculations becomes quite tedious when several stages are required. The Maloney–Schubert diagram frequently is more convenient to use, particularly if the extraction process involves a type II system. The calculation procedure for this type of diagram and system is similar to that used for a ternary diagram since the material balance equations are the same.

Countercurrent contact of insoluble liquids

Equilibrium stage calculations become somewhat easier if the solvents involved are nearly insoluble in the concentration range over which the calculations are to be made. Solute-free concentration units y'_A and x'_A defined by Eqs. (11-13) and (11-14) are conveniently used for a countercurrent process. The overall material and component balances for the extraction battery are

$$F' + S' = R'_N + E'_1 \tag{11-28}$$

and

$$F'x'_F + S'y'_S = R'_N x'_N + E'_1 y'_1 \tag{11-29}$$

Since the flow rates of the carrier solvent and extracting solvent will not change, $F' = R'_N$ and $S' = E'_1$. Thus we can write Eq. (11-28) as

$$\frac{F'}{S'} = \frac{y'_1 - y'_S}{x'_F - x'_N} \tag{11-30}$$

The expression above is the equation of the operating line that passes through the points (y'_1, x'_F) and (y'_S, x'_N), and has the slope F'/S'. This is shown in Figure 11-20 for a four-stage process. Equilibrium stages are counted by following the same procedure used for counting stages in stripping and absorption processes.

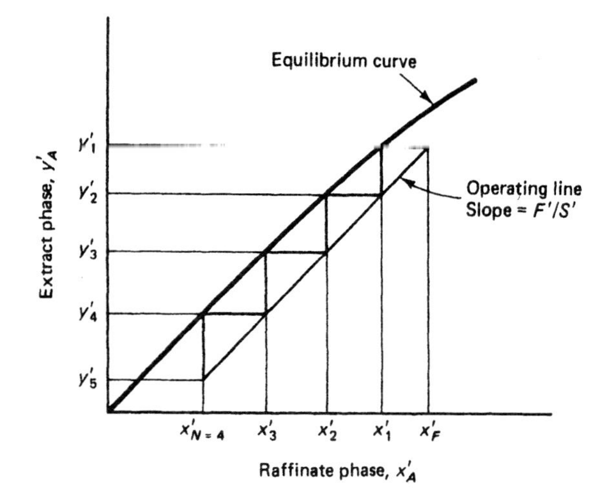

Figure 11-20 Countercurrent extraction using insoluble solvents

11.6 Extraction Equipment

Liquid–liquid extraction processes can be carried out by using many of the same types of equipment that are employed for distillation and absorption, since the primary goal of a contactor is to provide intimate contact of the liquid phases. A wide variety of continuous multistage contactors have been discussed in the literature. However, these can be divided into two broad categories depending on whether they operate as stagewise or differential contactors. Several contactors are listed by Oberg and Jones (1963) and Hanson (1968) and are summarized in Table 11-2 according to type of operation.

TABLE 11-2. INDUSTRIAL CONTACTORS

Stagewise Contactors	*Differential Contact Extractors*
1. Plate column	1. Spray column
2. Pulsed sieve plates	2. Baffle-plate column
3. Mixer-settlers	3. Packed column
4. Pulsed mixer-settlers	4. Rotary disk contactor
5. Scheibel column	5. Pulsed packed column
6. Treybal contactor	6. Oldshue–Rushton column
7. Asymmetric rotating disk contactor	7. Graesser raining bucket contactor
	8. Podbielniak extractor

Unagitated and pulsed column contactors

Spray columns, such as those shown in Figure 11-21, are the simplest of the differential contactors. In this type of contactor the heavy liquid is introduced at the top of the column and the lighter phase is introduced into the bottom.

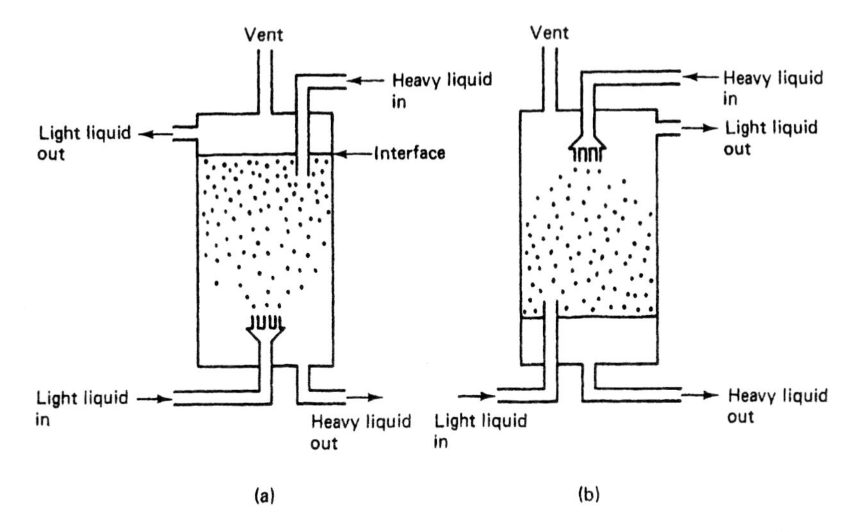

Figure 11-21 Schematic of spray columns: (a) light phase dispersed, and (b) heavy phase dispersed

Either the heavy or lighter liquid may be introduced into the column as the dispersed phase. Spray columns typically exhibit low efficiency because of poor mixing and as a result are not widely used in industry.

The efficiency of packed extraction columns is somewhat better than spray towers because of the presence of the packing, which promotes mixing and distribution of the two phases. Packed columns are used on a large scale by the petroleum industry for the separation of aromatics from aliphatics. Since one theoretical stage may be the equivalent of several feet of packing, packed towers are usually limited to separations that require only a few theoretical stages. The efficiency of the packed column can be increased significantly by applying an oscillating pulse to the fluids contained in the column. The pulse, which may be generated by either a valveless pump or a diaphragm, can be a sine wave, a square wave, or a sawtooth wave.

Sieve trays provide a reliable and efficient method for carrying out liquid–liquid separation processes. Although the heavy phase will be introduced at the top of the column, either the light or heavy phase may be dispersed. Examples of perforated plates are shown in Figure 11-22. As with packed towers, the efficiency of a perforated plate column can be increased by pulsing. The pulsing action promotes coalescence and redispersion of the dispersed phase. A pulsed column is shown in Figure 11-23. For design purposes, an orifice diameter of $\frac{1}{8}$ to $\frac{1}{4}$ in is typically used. The total free area of the plate should be as large as possible, but is limited by coalescence of the droplets as they emerge from the perforations. Typically, the free area ranges from 15 to 25%.

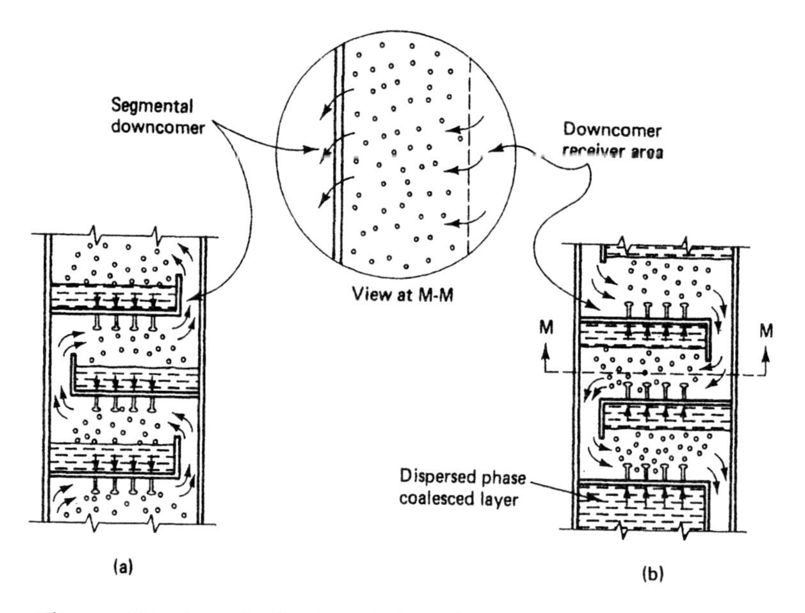

Figure 11-22 Schematic of perforated plates: (a) heavy phase dispersed, and (b) light phase dispersed

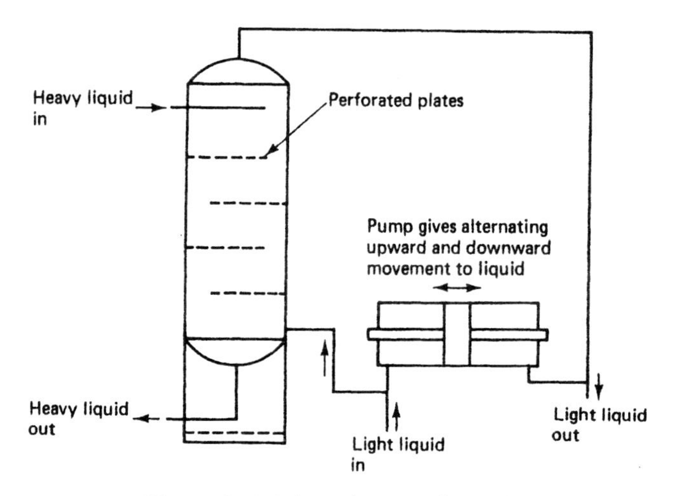

Figure 11-23 Schematic of a pulsed column

Mechanically agitated column contactors

The mass transfer performance of a column can be greatly enhanced by introducing mechanical agitation. Since the mechanical agitation increases the interfacial contact area of the two fluids per unit volume, the column efficiency is greatly increased. Typical mechanically agitated columns are: the Scheibel

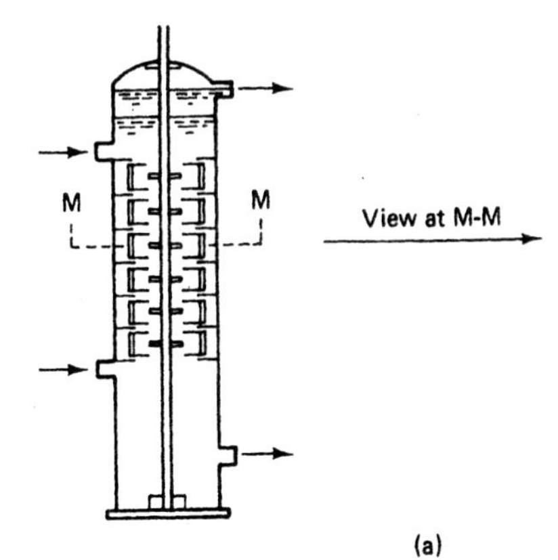

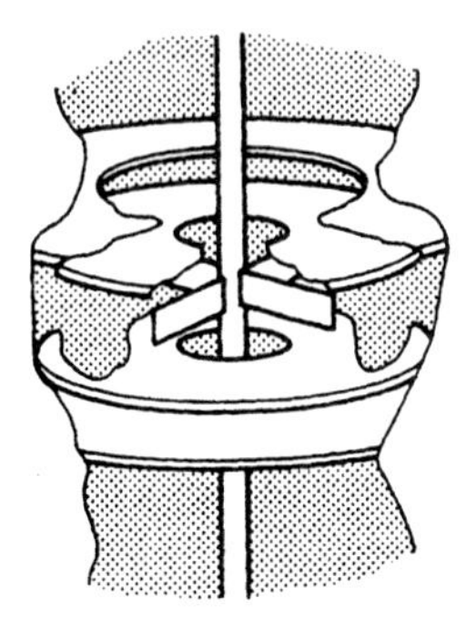

(a)

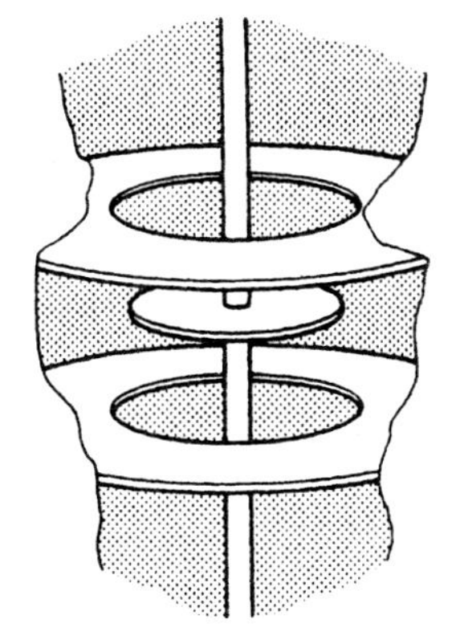

(b)

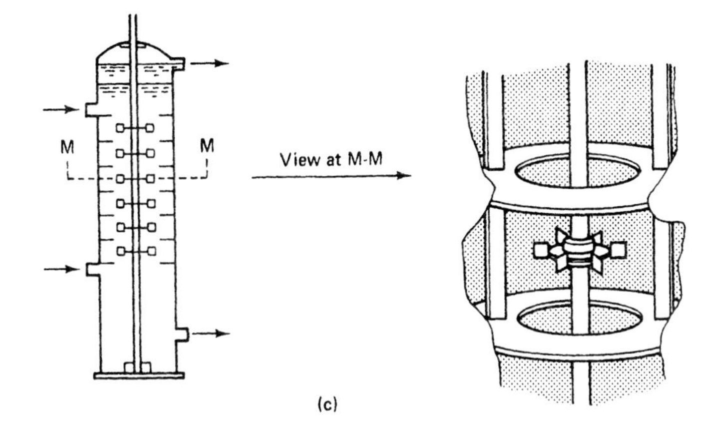

(c)

Figure 11-24 Three types of agitated columns: (a) Scheibel column; (b) rotating disk column; and (c) Oldshue–Rushton column

column (Scheibel, 1948); the rotating disk contactor (Reman and Olney, 1955); and the Oldshue–Rushton column (Oldshue and Rushton, 1952). These agitated columns are shown in Figure 11-24.

Contactor selection

The criteria to be considered when selecting a contactor are: settling characteristics, number of stages required, space availability, volume throughput, and contact time. Contactor selection charts have been developed by Hanson (1968) and by Laddha and Degaleesan (1978). The selection chart given in Figure 11-25 summarizes the important factors to be considered when selecting

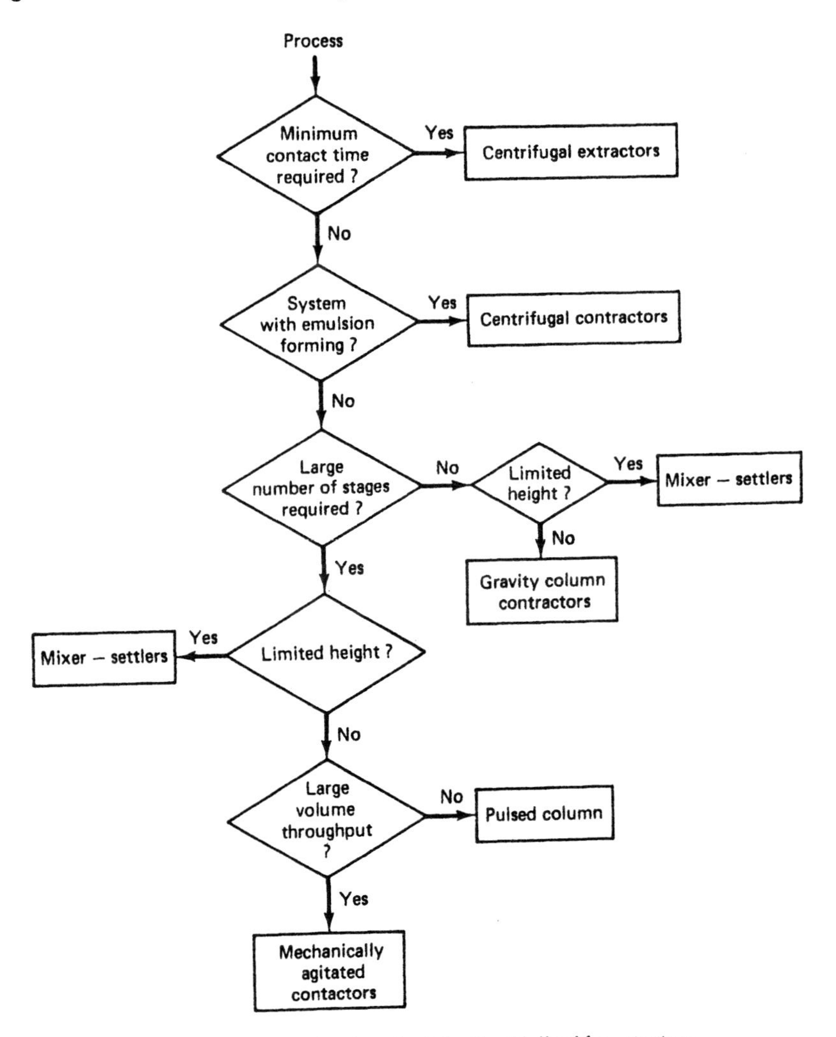

Figure 11-25 Selection chart for liquid–liquid contactors

an extractor. For a more detailed discussion of liquid–liquid extraction, the discussions given by Laddha and Degaleesan (1978), Treybal (1963), and Oliver (1966) are recommended. Laddha and Degaleesan (1978) give an extensive discussion of contactors and their design methods.

REFERENCES

BRANCKER, A. V., T. G. HUNTER, and A. W. NASH, *J. Phys. Chem.*, *44*, 683 (1940).

HANSON, C., *Chem. Eng.*, *75*, No. 18, 76 (1968).

HUNTER, T. G., and A. W. NASH, *Ind. Eng. Chem.*, *27*, 836 (1935).

LADDHA, G. S., and T. E. DEGALEESAN, *Transport Phenomena in Liquid Extraction*, McGraw-Hill, New York, 1978.

MALONEY, J. O., and A. E. SCHUBERT, *Trans. AIChE*, *36*, 741 (1940).

OBERG, A. G., and S. C. JONES, *Chem. Eng.*, *70*, No. 15, 119 (1963).

OLDSHUE, J. Y., and J. H. RUSHTON, *Chem. Eng. Prog.*, *48*, No. 6, 297 (1952).

OLIVER, E. D., *Diffusional Separation Processes*, Wiley, New York, 1966.

OTHMER, D. F., R. E. WHITE, and E. TRUEGER, *Ind. Eng. Chem.*, *33*, 1240 (1941).

RAO, M. R., and C. V. RAO, *J. Appl. Chem.*, *6*, 269 (1956).

RAO, M. R., and C. V. RAO, *J. Appl. Chem.*, *7*, 659 (1957).

REMAN, G. H., and R. B. OLNEY, *Chem. Eng. Prog.*, *51*, No. 3, 141 (1955).

SCHEIBEL, E. G., *Chem. Eng. Prog.*, *44*, 681 (1948).

TREYBAL, R. E., *Liquid Extraction*, 2nd ed., McGraw-Hill, New York, 1963.

NOTATIONS

E = mass flow rate of extract phase, M/t
E_n = mass flow rate of extract leaving contactor n, M/t
$\bar{E}_n$ = mass flow rate of extract leaving contactor n written on a solvent-free basis, M/t
E_n' = mass flow rate of the solvent used for insoluble systems, M/t
F = mass flow rate of the feed, M/t
$\bar{F}$ = mass flow rate of the feed written on a solvent-free basis, M/t
F' = mass flow rate of the feed used for insoluble systems, M/t
M = total mass entering or leaving the contactor, M/t
$\bar{M}_n$ = total mass entering or leaving the contactor n written on a solvent-free basis, M/t
S_n = mass flow rate of the solvent entering contactor n, M/t
$\bar{S}_n$ = mass flow rate of the solvent entering contactor n written on a solvent-free basis, M/t
S_n' = mass flow rate of the solvent entering contactor n for insoluble systems, M/t
T = temperature
x_A = mass fraction of A in raffinate phase
x_A' = mass of A in raffinate per unit mass of solvent B, defined as $x_A/(1 - x_A)$
x_B = mass fraction of B in the raffinate phase

x_F = mass fraction of solute in the feed
x'_F = feed composition written as $x_F/(1 - x_F)$
x_M = mass fraction of the solute in the total mixture that enters the extractor
x_n = mass fraction of the solute in the raffinate phase that leaves extractor n
x'_n = raffinate composition that leaves extractor n written as $x_n/(1 - x_n)$
x_S = mass fraction of solvent S in the raffinate phase
$\bar{X}_A$ = concentration of A in the raffinate phase written on a solvent-free basis
$\bar{X}_F$ = concentration of A in the feed written on a solvent-free basis
$\bar{X}_{M,n}$ = concentration of A in the total mixture that enters the contactor written on a solvent-free basis
$\bar{X}_n$ = concentration of A in the raffinate phase written on a solvent-free basis
y_A = mass fraction of A in the extract phase
y'_A = mass of A in extract phase per unit mass of solvent C, defined as $y_A/(1 - y_A)$
y_B = mass fraction of B in the extract phase
y_n = mass fraction of the solute in the extract phase that leaves extractor n
y'_n = extract composition that leaves extractor n written as $y_n/(1 - y_n)$
y_S = mass fraction of solvent S in the extract phase
$\bar{Y}_A$ = concentration of A in the extract phase written on a solvent-free basis
$\bar{Y}_n$ = solute concentration in the extract phase leaving contactor n written on a solvent-free basis
$\bar{Z}_E$ = concentration of solvent S in the extract phase expressed on a solvent-free basis
$\bar{Z}_R$ = concentration of solvent S in the raffinate phase written on a solvent-free basis

Greek Letters

Δ = difference point representing the difference in flow rates of passing streams as defined by Eq. (11-21)
Δ_m = difference point corresponding to minimum solvent rate

PROBLEMS

11.1 An ethylene glycol–methyl ethyl ketone (MEK) solution that contains 40% by weight of ethylene glycol and 60% MEK is to be contacted with pure water in a continuous single-stage extraction process. The flow rates of the glycol solution and the water are 40 kg/min and 60 kg/min, respectively. **(a)** Determine the compositions of the extract and raffinate phases, and **(b)** find the flow rate of each phase.

11.2 A solution that contains 65% by weight of methyl ethyl ketone and 35% ethylene glycol is to be contacted with pure water in a continuous single-stage extraction process. The flow rate of the ethylene glycol–MEK solution is 50 kg/min. **(a)** Determine the flow rate of water necessary to produce an extract phase that contains 28% by weight of ethylene glycol, and **(b)** determine the composition of the raffinate phase.

11.3 Pure water is used to separate a chloroform–acetone mixture in a three-stage crosscurrent extraction process. The feed contains 45% by weight of chloroform and 55% acetone, and flows at a rate of 100 kg/min. Pure water is added to each stage at a rate of 50 kg/min. **(a)** Determine the compositions of the extract and raffinate phases for each stage, and **(b)** find the flow rates for all streams leaving the contactors. The extraction process is carried out at 25°C. The equilibrium data for this system are given below.

Mutual solubility data (wt % at 25°C), obtained by Brancker et al. (1940), are as follows:

Acetone	*Water*	*Chloroform*
57.3	7.3	35.4
60.5	11.0	28.5
60.0	18.0	22.0
59.2	23.0	17.8
58.5	27.0	14.5
56.6	32.4	11.0
55.5	34.4	10.0
54.0	37.4	8.6
53.2	38.8	8.0
51.6	41.4	7.0
49.0	45.4	5.6

Tie-line data at 25°C are:

Solvent Layer (wt %)			*Raffinate Layer (wt %)*		
Acetone	*Water*	*Chloroform*	*Acetone*	*Water*	*Chloroform*
3.0	96.0	1.0	9.0	1.0	90.0
8.3	90.5	1.2	23.7	1.3	75.0
13.5	85.0	1.5	32.0	1.6	66.4
17.4	81.0	1.6	38.0	2.0	60.0
22.1	76.1	1.8	42.5	2.5	55.0
31.9	66.0	2.1	50.5	4.5	45.0
44.5	51.0	4.5	57.0	8.0	35.0

11.4 A solution containing acetic acid and methyl isobutyl ketone is to be separated by contacting it with pure water in a three-stage crosscurrent extraction process. The feed contains 40 % by weight of acetic acid and enters the first contactor at a flow rate of 100 kg/min. Equal amounts of water are to enter each stage. Determine the quantity of water for each stage necessary to give a final raffinate composition of 3 % by weight acetic acid.

11.5 Styrene is to be extracted from ethylbenzene in a three-stage crosscurrent extraction process by using diethylene glycol as the solvent. Assume that the feed contains 28 % styrene by weight and 72 % ethylbenzene, and enters at a rate of 100 kg/min. It is to be contacted by 70 kg/min of diethylene glycol in each stage. Determine the concentration of styrene in the extract and raffinate phases leaving each stage. The extraction process is carried out at 25°C.

11.6 A feed containing 35% by weight of acetic acid and 65% methyl isobutyl ketone is to be separated by using pure water in a multistage countercurrent extraction process. The flow rates of the entering feed and the water are 100 kg/min each. **(a)** Determine the number of stages necessary to make the separation if the final raffinate composition is to be 5% by weight of acetic acid. **(b)** How many stages are needed if the water flow rate is increased to 200 kg/min? **(c)** Compare the result obtained from part (b) with that obtained in Example 11.3.

11.7 Acetic acid is to be separated from methyl isobutyl ketone by contacting the solution with pure water in a multistage countercurrent extraction process. The feed, which contains 30% by weight of acetic acid and 70% methyl isobutyl ketone, enters the extractor at a flow rate of 50 kg/min. It is desired to obtain a final raffinate that contains no more than 5% by weight of acetic acid. Determine **(a)** the minimum allowable water flow rate, and **(b)** the minimum number of stages for this separation.

11.8 Styrene is to be extracted from ethylbenzene in a continuous countercurrent process by using diethylene glycol as the solvent. The feed contains 28% styrene and 72% ethylbenzene by weight, and it enters the contactor at a flow rate of 100 kg/min. Determine **(a)** the minimum flow rate of diethylene glycol if the final raffinate is to contain 13.4% by weight of styrene, and **(b)** the final raffinate composition if a two-stage contactor is used and the diethylene glycol enters at a flow rate of 140 kg/min.

Mass Transfer in Continuous Differential Contactors

12

12.1 Introduction

Absorption, stripping, distillation, and liquid–liquid extraction, although different, involve the transfer of one or more components from one phase to another. Since all of these processes can often be better carried out in a continuous differential contactor, such as a packed column or spray tower, instead of a tray tower, a discussion of the calculation methods for mass transfer in a differential contactor is presented.

The obvious question asked by one involved in tower design is: When should a packed or spray tower be used instead of a tray tower? Packed towers are preferred for separations in which corrosive liquids and gases are being processed. The pressure drop in packed towers is considerably less than in tray towers. In addition, packed towers are often less expensive to construct. However, packed towers frequently cannot be used over a wide capacity range because of channeling at low flow rates. When fouling is a problem, tray towers are preferred because they can be cleaned more easily. Spray towers can also be used for processing corrosive gases and liquids but typically give poorer performance than packed towers because of coalescence of the droplets. Coalescence results in a lower surface area and subsequently a lower rate of mass transfer.

The rate of mass transfer in a separation process depends on the area of contact between the phases. Intimate contact of passing streams is provided in packed towers as a result of the distribution of the phases over the packing surface. A variety of packings are available for accomplishing this goal, but because of their differences, the rate of mass transfer is a function of the packing type and packing size. Intuitively, we assume that if two different types of packing with the same surface area are used in a separation process, the same degree of separation should be achieved. This, however, is not necessarily true. The distribution of the phases over the surface and the wettability of the packing are a function of the type of material from which the packing is constructed and the type of fluid being processed. Thus for different packing materials, the difference in wettability causes a change in the mass transfer coefficient and subsequently the degree of fluid separation.

12.2 Use of Local Mass Transfer Coefficients to Determine Tower Heights

The height of a differential contactor, such as a packed tower, depends on the properties and flow rates of the contacting streams as well as on the type of packing being used. In this section we develop equations that utilize local mass transfer coefficients for determining contactor heights.

Since differential contactors can be used for absorption, stripping, liquid–liquid extraction, and distillation processes, a general equation will be derived and modified as necessary for the type of process being considered. Let us consider Figure 12-1, in which an isothermal separation process is carried out in a differential contactor. Since the change in the number of moles in one phase must

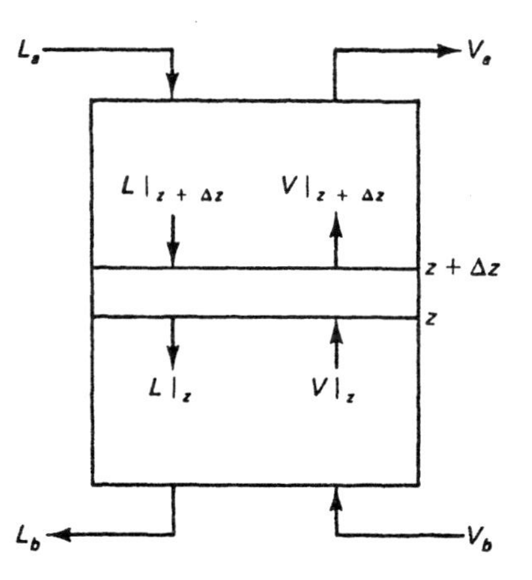

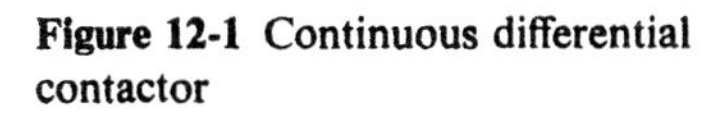
Figure 12-1 Continuous differential contactor

equal the change in the number of moles in the other, we can write an overall material balance over a differential section of the column as

$$L|_{Z+\Delta Z} + V|_Z = V|_{Z+\Delta Z} + L|_Z \tag{12-1}$$

Upon rearranging and dividing both sides by ΔZ, we can express Eq. (12-1) in differential form as

$$\frac{dL}{dZ} = \frac{dV}{dZ} \tag{12-2}$$

In the same manner, we can write the balance over the differential element for component A as

$$\frac{d(Lx_A)}{dZ} = \frac{d(Vy_A)}{dZ} \tag{12-3}$$

where we note that the equation above has typical units kgmol/h·m³. For a column with cross-sectional area S, the change of component A in the differential volume, $S\,dZ$, can be related to the flux N_A if the area available for the mass transfer for a particular type of packing can be determined. Since the surface area is difficult to obtain and it changes with the size and type of packing, a factor a is introduced to represent the mass transfer area per unit volume of packing. Thus we can write the molar flux as shown:

$$N_A a = \frac{\text{kgmol}}{\text{h}\cdot\text{m}^2} \cdot \frac{\text{m}^2}{\text{m}^3} = \frac{d(Vy_A)}{dZ} = \frac{d(Lx_A)}{dZ} \tag{12-4}$$

Upon combining Eq. (5-64) with the expression above for the gas phase, we have

$$N_A = \frac{d(Vy_A)}{a\,dZ} = \frac{k'_y}{\beta_{V-i}}(y_A - y_{Ai}) \tag{12-5}$$

The height of tower packing is thus obtained by separating the variables in Eq. (12-5) and integrating over the concentration range.

$$Z = \int_0^Z dZ = \int_{y_{A,a}}^{y_{A,b}} \frac{\beta_{V-i}}{k'_y a} \frac{d(Vy_A)}{y_A - y_{Ai}} \tag{12-6}$$

For the liquid phase, we can express the equation above as

$$Z = \int_0^Z dZ = \int_{x_{A,a}}^{x_{A,b}} \frac{\beta_{i-L}}{k'_x a} \frac{d(Lx_A)}{x_{Ai} - x_A} \tag{12-7}$$

For an absorber, integration is carried out from the top of the column, a, to the bottom, b, to give a positive value for the column height.

Transfer through a stagnant film

Although the bulk flow correction terms are cumbersome to use, they can be reduced for the limiting cases of transfer through a stagnant film and equimolar counter transfer. For the case of transfer of a single component A through a stagnant film, $N_B = 0$, and $\beta_{V-i} = (1 - y_A)_{iM}$ and $\beta_{i-L} = (1 - x_A)_{iM}$. Thus we can write Eqs. (12-6) and (12-7) as

$$Z = \int_{y_{A,a}}^{y_{A,b}} \frac{(1 - y_A)_{iM}}{k'_y a} \frac{d(Vy_A)}{y_A - y_{Ai}} \tag{12-8}$$

and

$$Z = \int_{x_{A,a}}^{x_{A,b}} \frac{(1 - x_A)_{iM}}{k'_x a} \frac{d(Lx_A)}{x_{Ai} - x_A} \tag{12-9}$$

Although the total vapor flow rate and composition vary over the length of the tower, the flow rate of the carrier gas remains constant. Thus to simplify Eq.

(12-8) further, we can write the flow rates and compositions on a solute-free basis. Thus

$$d(Vy_A) = d(\bar{V}Y_A) = d\left(\frac{\bar{V}y_A}{1 - y_A}\right) = \bar{V}\frac{dy_A}{(1 - y_A)^2} \tag{12-10}$$

But $\bar{V} = (1 - y_A)V$ and

$$d(Vy_A) = \frac{V\,dy_A}{1 - y_A} \tag{12-11}$$

Similarly for the liquid phase, we obtain

$$d(Lx_A) = \frac{L\,dx_A}{1 - x_A} \tag{12-12}$$

We may write Eqs. (12-8) and (12-9) as follows:

$$Z = \int_{y_{A,a}}^{y_{A,b}} \frac{V}{k'_y a} \frac{(1 - y_A)_{iM}}{1 - y_A} \frac{dy_A}{y_A - y_{Ai}} \tag{12-13}$$

and

$$Z = \int_{x_{A,a}}^{x_{A,b}} \frac{L}{k'_x a} \frac{(1 - x_A)_{iM}}{1 - x_A} \frac{dx_A}{x_{Ai} - x_A} \tag{12-14}$$

The left-hand terms under the integrals in the equations above have the dimensions of length and are defined as the height of a transfer unit. The heights of the gas-phase and liquid-phase transfer units are

$$H_V = \frac{V}{k'_y a} \tag{12-15}$$

and

$$H_L = \frac{L}{k'_x a} \tag{12-16}$$

The height of a transfer unit is assumed to be nearly independent of concentration and can be removed from the integral. The height of the transfer unit does vary, however, and to minimize any calculational errors the average value over the length of the column should be used. The integral of the remaining terms is defined as the number of transfer units as shown:

$$N_V = \int_{y_{A,a}}^{y_{A,b}} \frac{(1 - y_A)_{iM}}{1 - y_A} \frac{dy_A}{y_A - y_{Ai}} \tag{12-17}$$

and

$$N_L = \int_{x_{A,a}}^{x_{A,b}} \frac{(1 - x_A)_{iM}}{1 - x_A} \frac{dx_A}{x_{Ai} - x_A} \tag{12-18}$$

Therefore we have

$$Z = \left(\frac{V}{k'_y a}\right)_{\text{avg}} \int_{y_{A,a}}^{y_{A,b}} \frac{(1 - y_A)_{iM}}{1 - y_A} \frac{dy_A}{y_A - y_{Ai}} = (H_V)_{\text{avg}} N_V \tag{12-19}$$

and

$$Z = \left(\frac{L}{k'_x a}\right)_{\text{avg}} \int_{x_{A,a}}^{x_{A,b}} \frac{(1 - x_A)_{iM}}{1 - x_A} \frac{dx_A}{x_{Ai} - x_A} = (H_L)_{\text{avg}} N_L \tag{12-20}$$

Equations (12-19) and (12-20) can be simplified by replacing the logarithmic mean concentration difference by the arithmetic mean as follows:

$$(1 - y_A)_{iM} = \frac{(1 - y_{Ai}) + (1 - y_A)}{2} \tag{12-21}$$

Thus we can calculate the column heights from the expressions

$$Z = \left(\frac{V}{k'_y a}\right)_{\text{avg}} \int_{y_{A,a}}^{y_{A,b}} \frac{dy_A}{y_A - y_{Ai}} + \frac{1}{2} \ln \frac{1 - y_{A,a}}{1 - y_{A,b}} \tag{12-22}$$

and

$$Z = \left(\frac{L}{k'_x a}\right)_{\text{avg}} \int_{x_{A,a}}^{x_{A,b}} \frac{dx_A}{x_{Ai} - x_A} + \frac{1}{2} \ln \frac{1 - x_{A,b}}{1 - x_{A,a}} \tag{12-23}$$

For dilute solutions (mole fractions less than about 8 to 10%), the equilibrium curve is frequently relatively straight, and since only a small amount of solute is removed in the column, the operating line is also nearly straight. Although $(1 - y_A)_{iM}$ and $1 - y_A$ will vary over the column length, the ratio $(1 - y_A)_{iM}/(1 - y_A)$ is approximately constant. For dilute solutions it can be shown that $(1 - y_A)_{iM} \approx 1 - y_A$ and $(1 - x_A)_{iM} \approx 1 - x_A$. Thus the tower height can be readily determined by calculating the height of the transfer unit and numerically evaluating the integrals given in Eqs. (12-24) and (12-25).

$$Z = \left(\frac{V}{k'_y a}\right)_{\text{avg}} \int_{y_{A,a}}^{y_{A,b}} \frac{dy_A}{y_A - y_{Ai}} = (H_V)_{\text{avg}} N_V \tag{12-24}$$

and

$$Z = \left(\frac{L}{k'_x a}\right)_{\text{avg}} \int_{x_{A,a}}^{x_{A,b}} \frac{dx_A}{x_{Ai} - x_A} = (H_L)_{\text{avg}} N_L \tag{12-25}$$

The integrals presented in Eqs. (12-24) and (12-25) are obviously easier to evaluate numerically than those given by Eqs. (12-19) and (12-20), and give acceptable results for dilute solutions.

Example 12.1

Ammonia is to be recovered from air by contact with water in a packed absorption tower. The entering gas stream contains 8 mol% NH_3 and the exiting gas is to contain 0.5 mol% NH_3. The absorber is to operate at 20°C and at a pressure of 1 atm absolute. The rate at which water enters the column is to be 1.5 times the minimum flow. The gas and liquid mass transfer coefficients are $k'_y a = 85.0$ kgmol/h·m³ and $k'_x a = 170.0$ kgmol/h·m³, respectively. The entering gas flow rate is 100 kgmol/h·m². Determine the column height.

Solution:

The minimum liquid-to-vapor ratio is found, as discussed previously, by using the equilibrium curve shown in Figure 12-2. Since the equilibrium curve is concave toward the operating line, the minimum $\bar{L}/\bar{V}$ ratio can be determined by using Eq. (8-4). Thus $(x_b)_{\min} = 0.086$ as found from the figure. Introducing

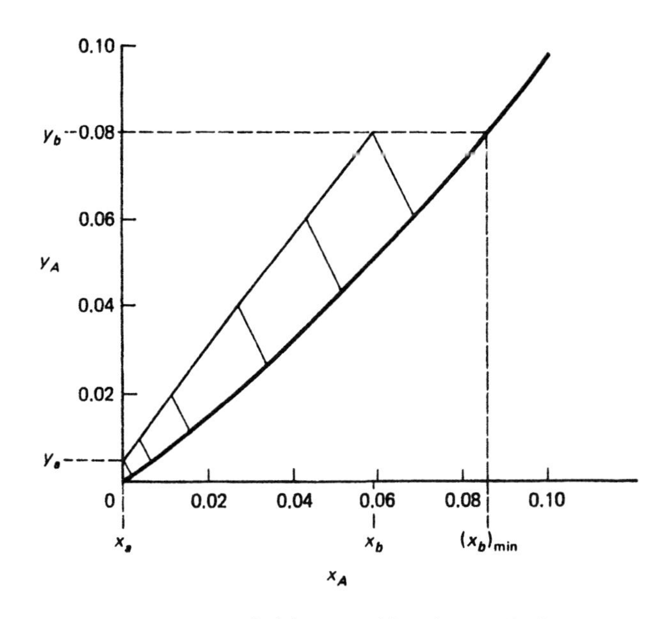

Figure 12-2 Interfacial compositions in a packed tower

Eq. (8-4), we obtain

$$\bar{L}\frac{x_a}{1-x_a}+\bar{V}\frac{y_b}{1-y_b}=\bar{L}\frac{x_b}{1-x_b}+\bar{V}\frac{y_a}{1-y_a} \tag{8-4}$$

Upon rearranging Eq. (8-4) and substituting values for the concentrations, we get

$$\left(\frac{\bar{L}}{\bar{V}}\right)_{min}=\frac{\dfrac{0.005}{1-0.005}-\dfrac{0.08}{1-0.08}}{0-\dfrac{0.086}{1-0.086}}=0.8708$$

Thus $$\left(\frac{\bar{L}}{\bar{V}}\right)_{act}=1.5\left(\frac{\bar{L}}{\bar{V}}\right)_{min}=1.5(0.8708)=1.307$$

We find the actual concentration of the exiting liquid stream by substituting the actual $\bar{L}/\bar{V}$ ratio into Eq. (8-4) as shown:

$$1.307=\frac{\dfrac{0.005}{1-0.005}-\dfrac{0.08}{1-0.08}}{0-\dfrac{x_b}{1-x_b}}$$

or $$\frac{x_b}{1-x_b}=\bar{X}_b=-\frac{\dfrac{0.005}{1-0.005}-\dfrac{0.08}{1-0.08}}{1.307}$$

$$\bar{X}_b=0.0627$$

$$x_b=\frac{\bar{X}_b}{1+\bar{X}_b}=0.059$$

Using the value above for x_b, we can construct the operating line as shown in Example 8-1. Thus

$$\bar{Y}_A = \frac{\bar{L}}{\bar{V}}(\bar{X}_A - \bar{X}_{A,b}) + \bar{Y}_{A,b}$$

$$= 1.307\left(\bar{X}_A - \frac{0.059}{1 - 0.059}\right) + \frac{0.08}{1 - 0.08}$$

$$= 1.307\bar{X}_A + 0.005$$

By selecting values for x_A, we can calculate $\bar{X}_A$, $\bar{Y}_A$, and y_A. As shown in Figure 12-2, the operating line is nearly straight.

The height of the column packing can be found by using either Eq. (12-19) or Eq. (12-20). For this problem, however, the gas-phase equation will be used since the gas-phase mass transfer coefficient is smaller than that for the liquid phase.

$$Z = \left(\frac{V}{k'_y a}\right)_{\text{avg}} \int_{y_{A,a}}^{y_{A,b}} \frac{(1 - y_A)_{iM}}{1 - y_A} \frac{dy_A}{y_A - y_{Ai}} \tag{12-19}$$

The equation above must be solved numerically. Bulk compositions and the corresponding interfacial concentrations must be known at intervals over the concentration range of interest. Since this problem deals with the transport of a single component, the interfacial compositions must be found by the trial-and-error procedure shown in Chapter 5. To obtain an initial value for interfacial compositions, we use the following equation.

$$-\frac{k'_x a}{k'_y a} = \frac{y_A - y_{Ai}}{x_A - x_{Ai}} = -\frac{170.0}{85.0} = -2.0$$

The interfacial compositions that correspond to the bulk concentrations at the bottom of the tower $y_{A,b} = 0.08$ and $x_{A,b} = 0.059$ are first determined. By constructing a line from $(x_{A,b}, y_{A,b})$ with slope -2, we obtain the interfacial compositions $x_{Ai} = 0.069$ and $y_{Ai} = 0.06$.

Therefore, $$(1 - x_A)_{iM} = \frac{(1 - x_A) - (1 - x_{Ai})}{\ln \dfrac{1 - x_A}{1 - x_{Ai}}} = \frac{(1 - 0.059) - (1 - 0.069)}{\ln \dfrac{1 - 0.059}{1 - 0.069}}$$

$$= 0.936$$

and $$(1 - y_A)_{iM} = \frac{(1 - y_{Ai}) - (1 - y_A)}{\ln \dfrac{1 - y_{Ai}}{1 - y_A}} = \frac{(1 - 0.06) - (1 - 0.08)}{\ln \dfrac{1 - 0.06}{1 - 0.08}}$$

$$= 0.930$$

Thus $$-\frac{k'_x a/(1 - x_A)_{iM}}{k'_y a/(1 - y_A)_{iM}} = -\frac{170/0.936}{85/0.93} = -1.987$$

Drawing a line from $(x_{A,b}, y_{A,b})$ with the slope above, we obtain the new values for the interfacial compositions; these are nearly the same as the assumed values. Proceeding in a similar manner, we get the following corresponding bulk and interfacial compositions:

y_A	y_{Ai}	$1 - y_A$	$(1 - y_A)_{iM}$	$y_A - y_{Ai}$	$\frac{(1 - y_A)_{iM}}{(1 - y_A)(y_A - y_{Ai})}$
0.005	0.002	0.995	0.998	0.003	334.3
0.01	0.005	0.99	0.995	0.005	201.0
0.02	0.012	0.98	0.988	0.008	126.0
0.04	0.027	0.96	0.973	0.013	78.0
0.06	0.043	0.94	0.957	0.017	59.9
0.08	0.06	0.92	0.94	0.02	51.1

The graphical integration is shown in Figure 12-3. The total area under the curve is found here by using the trapezoidal rule.

$$N_V = \text{area} = (0.08 - 0.06)\frac{51.1 + 59.9}{2} + (0.06 - 0.04)\frac{78.0 + 59.9}{2}$$
$$+ (0.04 - 0.02)\frac{126.0 + 78.0}{2} + (0.02 - 0.01)\frac{201.0 + 126.0}{2}$$
$$+ (0.01 - 0.005)\frac{334.3 + 201.0}{2}$$
$$= 7.50$$

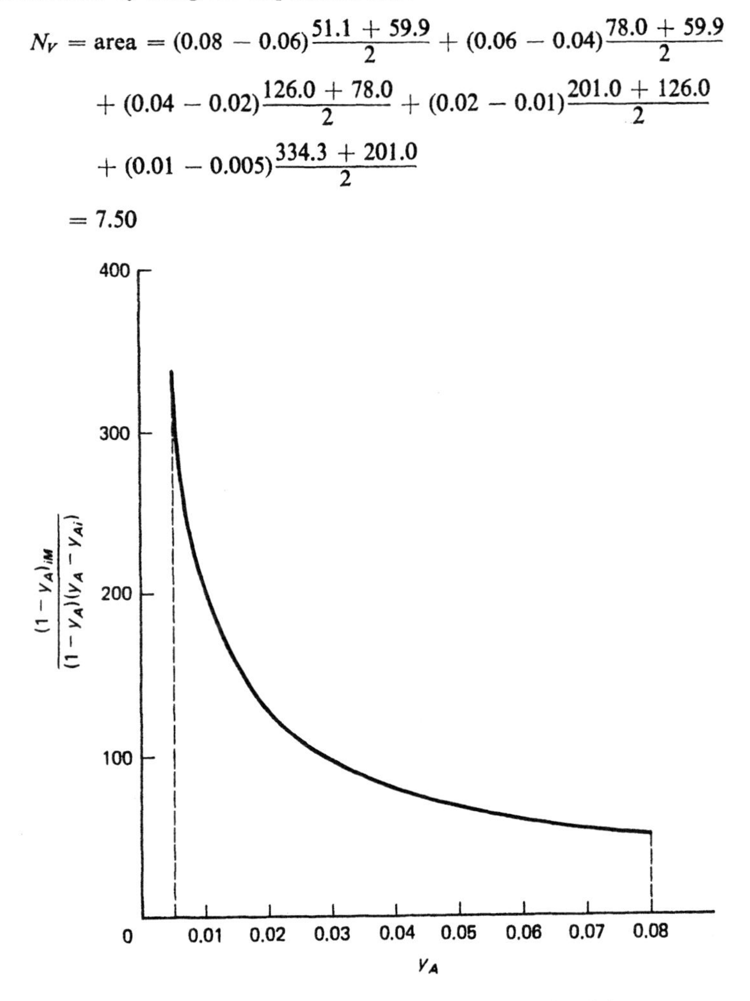

Figure 12-3 Graphical integration for Example 12.1

To calculate the height of the transfer unit, we use the average vapor flow rate.

$$V_b = 100.0\ \text{kgmol/h}\cdot\text{m}^2$$

$$\bar{V} = V_b(1 - 0.08) = 100.0(0.92) = 92.0$$

$$V_a = \frac{\bar{V}}{1 - y_a} = \frac{92.0}{1 - 0.005} = 92.5$$

$$V_{avg} = \frac{100.0 + 92.5}{2} = 96.25$$

$$(H_V)_{avg} = \frac{96.25}{85.0} = 1.132$$

$$Z = (H_V)_{avg} N_V = 1.132(7.5) = 8.49\ \text{m}$$

Equimolar counter transfer

The development presented in the preceding section finds application in absorption and for liquid–liquid extraction in which the solvents are mutually insoluble. For distillation, however, the use of the more general expressions given by Eqs. (12-6) and (12-7) are necessary. If the simplifying assumptions associated with the McCabe–Thiele method are valid, the height of the tower packing can be found by modifying these equations for the case of equimolar counter transfer. Since $N_A = -N_B$, the bulk flow correction terms are equal to 1. Using the assumption of equal molar overflow as required by the McCabe–Thiele method, we can assume that the vapor and liquid flow rates will be constant at any point in the column. Thus

$$d(Vy_A) = V\,dy_A \quad \text{and} \quad d(Lx_A) = L\,dx_A \tag{12-26}$$

Equations (12-6) and (12-7) become

$$Z = \int_{y_{A,a}}^{y_{A,b}} \frac{V}{k'_y a} \frac{dy_A}{y_A - y_{Ai}} = \left(\frac{V}{k'_y a}\right)_{avg} \int_{y_{A,a}}^{y_{A,b}} \frac{dy_A}{y_A - y_{Ai}} \tag{12-27}$$

and

$$Z = \int_{x_{A,a}}^{x_{A,b}} \frac{L}{k'_x a} \frac{dx_A}{x_{Ai} - x_A} = \left(\frac{L}{k'_x a}\right)_{avg} \int_{x_{A,a}}^{x_{A,b}} \frac{dx_A}{x_{Ai} - x_A} \tag{12-28}$$

In practice, a temperature gradient exists over the length of a distillation column and the difference in the latent heats of vaporization of the components being separated results in a change in the liquid and vapor flow rates from the top to the bottom of the tower. If the change in L and V in the column must be considered, we should use Eqs. (12-6) and (12-7).

It is important to note that all of the equations presented in Section 12.2

are for the case in which the equilibrium curve is located below the operating line, such as for absorption. The driving force for mass transfer is a positive value given by either $y_A - y_{Ai}$ or $x_{Ai} - x_A$. Thus, when the number of transfer units is calculated from the integral equation, a positive value is obtained. If these same equations are used for stripping or distillation calculations, a negative value will be obtained for N_V and N_L. This should be expected since for these cases the equilibrium curve is located above the operating line, and the concentration differences $y_A - y_{Ai}$ and $x_{Ai} - x_A$ are negative. Therefore, for stripping and distillation calculations the concentration differences should be changed from $y_A - y_{Ai}$ to $y_{Ai} - y_A$ and from $x_{Ai} - x_A$ to $x_A - x_{Ai}$.

Example 12.2

A total of 100 mol of a saturated liquid feed containing 40 mol% acetone and 60 mol% ethanol is to be separated in a packed tower into an overhead product containing 85% acetone and a bottom product containing 5% acetone. The distillation is to be carried out at 1 atm. Using a total condenser and a reflux ratio of $L/D = 3.25$, determine the number of transfer units. The local mass transfer coefficients for this process are $k'_y a = 0.2$ kgmol/m³·s (mole fraction) and $k'_x a = 1.6$ kgmol/m³·s (mole fraction). Assume equal molar overflow.

Solution:

The equilibrium diagram for this system is shown in Figure 12-4. We proceed by locating the compositions of the distillate and bottom products, $x_D = 0.85$ and $x_B = 0.05$. The feed point is then located and a vertical q line for the saturated liquid feed is drawn. The operating line is drawn from $x_D = 0.85$ to intersect the q line. The slope of the operating line is

$$\frac{L}{V} = \frac{L}{L + D} = \frac{L/D}{L/D + 1} = \frac{3.25}{3.25 + 1.0} = 0.765$$

Its intersection with the y intercept is obtained by applying Eq. (9-17).

$$\frac{Dx_D}{V} = \frac{Dx_D}{L + D} = \frac{x_D}{L/D + 1} = \frac{0.85}{4.25} = 0.2$$

The bottom operating line is located by extending a line from $x_B = 0.05$ to the intersection of the upper operating line and q line. The number of transfer units is determined by selecting points along the operating line and extending lines with the slope $-k'_x a/k'_y a = -1.6/0.2 = -8.0$ to intersect the equilibrium curve. This locates the interfacial compositions. The number of transfer units for either the vapor or liquid phase can be determined by modifying either Eq. (12-27) or (12-28) to apply to distillation. In this problem the gas phase will be evaluated by using the expression

$$N_V = \int_{y_{A,a}}^{y_{A,b}} \frac{dy_A}{y_{Ai} - y_A}$$

The numerical integration is carried out by using the information given in the following table.

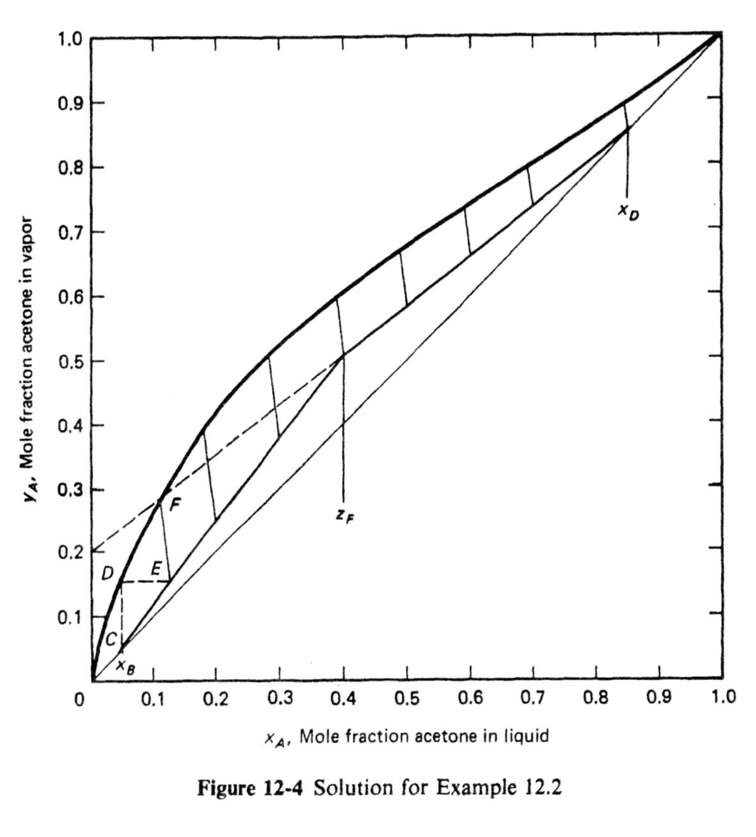

Figure 12-4 Solution for Example 12.2

x_A	y_A	y_{Ai}	$\dfrac{1}{y_{Ai} - y_A}$
0.85	0.850	0.892	23.81
0.70	0.736	0.796	16.67
0.60	0.660	0.773	13.70
0.50	0.582	0.667	11.76
0.40	0.505	0.597	10.87
0.30	0.378	0.504	7.93
0.20	0.248	0.392	6.94
0.05	0.155	0.280	8.0

In the table the vapor composition that corresponds to a bottom composition $x_B = 0.05$ is found by extending a vertical line from C to D. The interfacial composition is obtained by extending a horizontal line from D to intersect the operating line at point E, and then constructing a line with a slope of -8 to intersect the equilibrium curve at F. Graphical integration gives $N_V = 5.33$ for the rectifying section and $N_V = 2.86$ for the stripping section. Examination of the figure shows that the number of transfer units and number of equilibrium stages are approximately equal.

12.3 Application of Overall Mass Transfer Coefficients

Overall mass transfer coefficients can be used for cases in which the local coefficients are constant over the length of the tower and the equilibrium curve is nearly straight. Although the conditions above exist for dilute solutions, overall mass transfer coefficients can be used in concentrated regions provided that the conditions for their use are similar to those conditions under which they were obtained. Overall coefficients are easier to obtain experimentally since they are based on the difference between the bulk and equilibrium concentrations. Furthermore, they are simpler to use than local coefficients because a determination of the interfacial compositions is not necessary.

The derivation of equations for calculating tower heights using overall coefficients is analogous to that previously shown. The molar flux is written as the product of an overall mass transfer coefficient and the difference between the bulk and equilibrium concentrations as

$$N_A = \frac{d(Vy_A)}{a\,dZ} = \frac{K'_y}{\beta_{V-V^\bullet}}(y_A - y_A^*) \tag{12-29}$$

By rearranging the equation above, we obtain

$$Z = \int_0^Z dZ = \int_{y_{A,a}}^{y_{A,b}} \frac{\beta_{V-V^\bullet}}{K'_y a} \frac{d(Vy_A)}{y_A - y_A^*} \tag{12-30}$$

For liquids an equation similar to the one above is

$$Z = \int_0^Z dZ = \int_{x_{A,a}}^{x_{A,b}} \frac{\beta_{L^\bullet-L}}{K'_x a} \frac{d(Lx_A)}{x_A^* - x_A} \tag{12-31}$$

Transfer through a stagnant film

As for the local coefficients, equations for the overall mass transfer coefficients can be written for transfer of component A through a stagnant film and for equimolar counter transfer. For transfer through a stagnant film $\beta_{V-V^\bullet} = (1 - y_A)_{\bullet M}$ and $\beta_{L^\bullet-L} = (1 - x_A)_{\bullet M}$, where the log mean concentration differences are given by Eqs. (5-90c) and (5-90d). After simplifying, we can write Eqs. (12-30) and (12-31) as

$$Z = \int_{y_{A,a}}^{y_{A,b}} \frac{V}{K'_y a} \frac{(1 - y_A)_{\bullet M}}{1 - y_A} \frac{dy_A}{y_A - y_A^*} \tag{12-32}$$

and

$$Z = \int_{x_{A,a}}^{x_{A,b}} \frac{L}{K'_x a} \frac{(1 - x_A)_{\bullet M}}{1 - x_A} \frac{dx_A}{x_A^* - x_A} \tag{12-33}$$

On the basis of the assumptions given for the application of the overall coefficients, the terms $V/K'_y a$ and $L/K'_x a$ should be nearly constant. Using an average value for these terms, we get

$$Z = \left(\frac{V}{K'_y a}\right)_{\text{avg}} \int_{y_{A,a}}^{y_{A,b}} \frac{(1-y_A)_{\cdot M}}{1-y_A} \frac{dy_A}{y_A - y_A^*} \tag{12-34}$$

and

$$Z = \left(\frac{L}{K'_x a}\right)_{\text{avg}} \int_{x_{A,a}}^{x_{A,b}} \frac{(1-x_A)_{\cdot M}}{1-x_A} \frac{dx_A}{x_A^* - x_A} \tag{12-35}$$

The heights of the transfer units for the overall mass transfer coefficients are defined as

$$H_{OV} = \frac{V}{K'_y a} \tag{12-36}$$

and

$$H_{OL} = \frac{L}{K'_x a} \tag{12-37}$$

The number of transfer units for the gas and liquid phases are related to the overall concentration differences as follows:

$$N_{OV} = \int_{y_{A,a}}^{y_{A,b}} \frac{(1-y_A)_{\cdot M}}{1-y_A} \frac{dy_A}{y_A - y_A^*} \tag{12-38}$$

and

$$N_{OL} = \int_{x_{A,a}}^{x_{A,b}} \frac{(1-x_A)_{\cdot M}}{1-x_A} \frac{dx_A}{x_A^* - x_A} \tag{12-39}$$

Equimolar counter transfer

The bulk flow factors, β_{V-V^*} and β_{L^*-L}, are equal to 1 for the case of equimolar counter transfer. Thus we have for column height

$$Z = \int_{y_{A,a}}^{y_{A,b}} \frac{V}{K'_y a} \frac{dy_A}{y_A^* - y_A} \tag{12-40}$$

and

$$Z = \int_{x_{A,a}}^{x_{A,b}} \frac{L}{K'_x a} \frac{dx_A}{x_A - x_A^*} \tag{12-41}$$

The numbers of transfer units for this case are

$$N_{OV} = \int_{y_{A,a}}^{y_{A,b}} \frac{dy_A}{y_A - y_A^*} \tag{12-42}$$

and

$$N_{OL} = \int_{x_{A,a}}^{x_{A,b}} \frac{dx_A}{x_A^* - x_A} \tag{12-43}$$

Approximate expressions

As was shown previously, the logarithmic concentration difference can be approximated with the arithmetic mean value. Thus Eqs. (12-38) and (12-39) become

$$N_{OV} = \int_{y_{A,a}}^{y_{A,b}} \frac{dy_A}{y_A - y_A^*} + \frac{1}{2} \ln \frac{1 - y_{A,a}}{1 - y_{A,b}} \tag{12-44}$$

and

$$N_{OL} = \int_{x_{A,a}}^{x_{A,b}} \frac{dx_A}{x_A^* - x_A} + \frac{1}{2} \ln \frac{1 - x_{A,b}}{1 - x_{A,a}} \tag{12-45}$$

For dilute solutions $(1 - y_{A,a}) \approx (1 - y_{A,b})$ and $(1 - x_{A,b}) \approx (1 - x_{A,a})$. Since the second terms in Eqs. (12-44) and (12-45) are very small, we can reduce these equations to

$$N_{OV} = \int_{y_{A,a}}^{y_{A,b}} \frac{dy_A}{y_A - y_A^*} \tag{12-46}$$

and

$$N_{OL} = \int_{x_{A,a}}^{x_{A,b}} \frac{dx_A}{x_A^* - x_A} \tag{12-47}$$

Each of the equations above can be written in terms of solute-free coordinates. These are summarized below:

$$Z = \left(\frac{V}{K'_y a}\right)_{\text{avg}} \left[\int_{\bar{Y}_{A,a}}^{\bar{Y}_{A,b}} \frac{d\bar{Y}_A}{\bar{Y}_A - \bar{Y}_A^*} + \frac{1}{2} \ln \frac{1 + \bar{Y}_{A,b}}{1 + \bar{Y}_{A,a}}\right] \tag{12-48}$$

and

$$Z = \left(\frac{L}{K'_x a}\right)_{\text{avg}} \left[\int_{\bar{X}_{A,a}}^{\bar{X}_{A,b}} \frac{d\bar{X}_A}{\bar{X}_A^* - \bar{X}_A} + \frac{1}{2} \ln \frac{1 + \bar{X}_{A,a}}{1 + \bar{X}_{A,b}}\right] \tag{12-49}$$

For dilute solutions,

$$Z = \left(\frac{V}{K'_y a}\right)_{\text{avg}} \int_{\bar{Y}_{A,a}}^{\bar{Y}_{A,b}} \frac{d\bar{Y}_A}{\bar{Y}_A - \bar{Y}_A^*} = \left(\frac{V}{K'_y a}\right)_{\text{avg}} \int_{y_{A,a}}^{y_{A,b}} \frac{dy_A}{y_A - y_A^*} \tag{12-50}$$

and

$$Z = \left(\frac{L}{K'_x a}\right)_{\text{avg}} \int_{\bar{X}_{A,a}}^{\bar{X}_{A,b}} \frac{d\bar{X}_A}{\bar{X}_A^* - \bar{X}_A} = \left(\frac{L}{K'_x a}\right)_{\text{avg}} \int_{x_{A,a}}^{x_{A,b}} \frac{dx_A}{x_A^* - x_A} \tag{12-51}$$

Example 12.3

If the overall mass transfer coefficient is constant over the length of the column and the equilibrium curve is nearly straight, the number of transfer units can be determined by using the difference between the equilibrium and bulk concentrations as the driving force for mass transfer. Assuming that this is true for the absorption process shown in Example 12.1, determine the number of transfer units.

Solution:

Since ammonia absorption is characteristic of transport of a single component, the number of transfer units can be determined by using either Eq. (12-38) or (12-44). Equation (12-44) will be used here. Bulk concentrations and the corresponding equilibrium concentrations used in this calculation are shown below. For any bulk concentration, the equilibrium concentration is found by extending a vertical line to intersect the equilibrium curve as shown in Figure 12-5.

y_A	y_A^*	$\frac{1}{y_A - y_A^*}$
0.005	0.0	200.0
0.01	0.0028	138.9
0.02	0.0087	88.5
0.04	0.021	52.6
0.06	0.0345	39.2
0.08	0.05	33.3

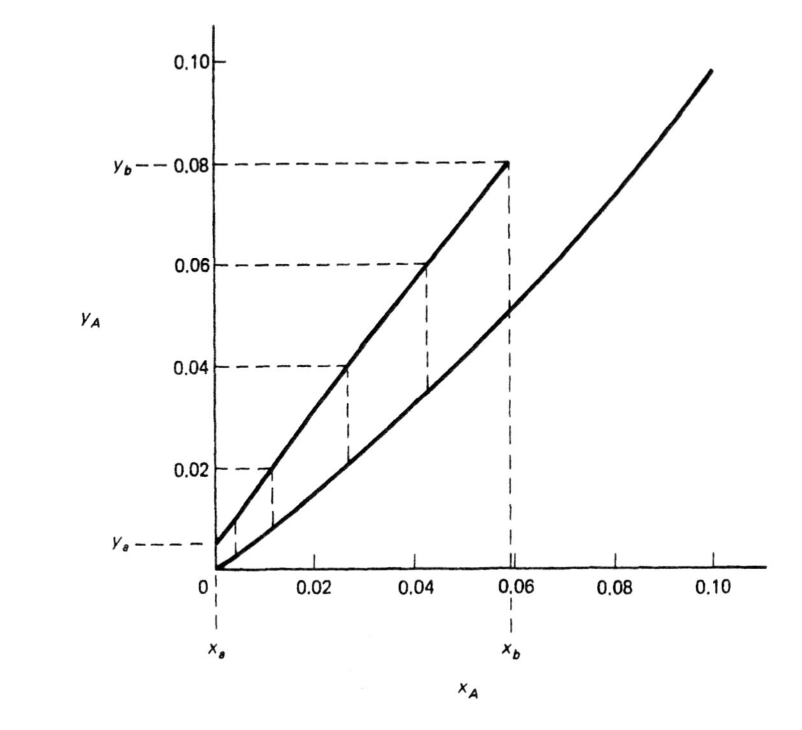

Figure 12-5 Solution to Example 12.3

Using Eq. (12-44), we obtain

$$N_{OV} = \int_{0.005}^{0.08} \frac{dy_A}{y_A - y_A^*} + \frac{1}{2} \ln \frac{1 - 0.005}{1 - 0.08}$$

$$= 5.04 + 0.039 = 5.08$$

This is appreciably smaller than N_V found in Example 12.1.

For the special case in which the equilibrium curve is a straight line over the concentration range of interest, we can integrate Eqs. (12-46) and (12-47) analytically to obtain an expression for the number of transfer units. If the equilibrium curve is a straight line,

$$y_A^* = mx_A + C \tag{12-52}$$

If the solution is dilute, the liquid and vapor flow rates will not change over the column length. Thus, a total material balance around the bottom of the column is

$$L(x_A - x_{A,b}) = V(y_A - y_{A,b}) \tag{12-53}$$

or

$$x_A = x_{A,b} + \frac{V}{L}(y_A - y_{A,b}) \tag{12-54}$$

We can determine the number of transfer units by combining Eqs. (12-46), (12-52), and (12-54) as follows:

$$\begin{aligned} N_{OV} &= \int_{y_{A,a}}^{y_{A,b}} \frac{dy_A}{y_A - y_A^*} \\ &= \int_{y_{A,a}}^{y_{A,b}} \frac{dy_A}{y_A - m[x_{A,b} + (V/L)(y_A - y_{A,b})] - C} \end{aligned} \tag{12-55}$$

Integrating gives

$$N_{OV} = \frac{1}{1 - mV/L} \ln \frac{y_{A,b} - mx_{A,b} - C}{y_{A,a} - m[x_{A,b} + (V/L)(y_{A,a} - y_{A,b})] - C} \tag{12-56}$$

Writing Eq. (12-52) for the top and bottom of the column, we obtain

$$m = \frac{y_{A,a}^* - y_{A,b}^*}{x_{A,a} - x_{A,b}} \tag{12-57}$$

Combining Eqs. (12-56) and (12-57) and simplifying gives

$$N_{OV} = \frac{y_{A,b} - y_{A,a}}{(y_A - y_A^*)_b - (y_A - y_A^*)_a} \ln \frac{(y_A - y_A^*)_b}{(y_A - y_A^*)_a} \tag{12-58}$$

$$= \frac{y_{A,b} - y_{A,a}}{(y_A - y_A^*)_M} \tag{12-59}$$

For the liquid phase

$$N_{OL} = \frac{x_{A,b} - x_{A,a}}{(x_A^* - x_A)_b - (x_A^* - x_A)_a} \ln \frac{(x_A^* - x_A)_b}{(x_A^* - x_A)_a} \tag{12-60}$$

$$= \frac{x_{A,b} - x_{A,a}}{(x_A^* - x_A)_M} \tag{12-61}$$

For the case of dilute solutions, in which the equilibrium curve and operating line are both straight, Colburn (1939) developed analytical expressions for the number of transfer units. For the gas phase

$$N_{OV} = \frac{1}{1 - mV/L} \ln \left[\left(1 - \frac{mV}{L}\right) \frac{y_{A,b} - mx_{A,a}}{y_{A,a} - mx_{A,a}} + \frac{mV}{L} \right] \tag{12-62}$$

The analogous equation for the liquid phase is

$$N_{OL} = \frac{1}{1 - L/mV} \ln \left[\left(1 - \frac{L}{mV}\right) \frac{x_{A,b} - y_{A,a}/m}{x_{A,a} - y_{A,a}/m} + \frac{L}{mV} \right] \tag{12-63}$$

A graphical solution for Eqs. (12-62) and (12-63) is shown in Figure 12-6. These approximate solutions find application in absorption, stripping, and extraction.

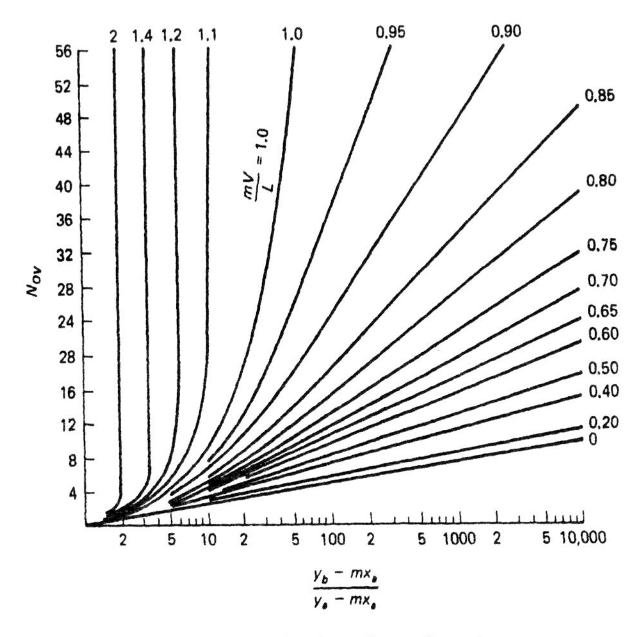

Figure 12-6 Number of transfer units

Relationships between overall and local transfer units

The relationship between local and overall transfer unit heights is obtained by using the relationship between the local and overall mass transfer coefficients developed in Chapter 5. The relationship is

$$\frac{1}{K'_y/\beta_{V-V^*}} = \frac{1}{k'_y/\beta_{V-i}} + \frac{m'}{k'_x/\beta_{i-L}}$$

By dividing each term by a and multiplying by V, we obtain for the gas phase

$$\frac{V}{K'_ya/\beta_{V-V^*}} = \frac{V}{k'_ya/\beta_{V-i}} + \frac{m'V}{L}\frac{L}{k'_xa/\beta_{i-L}} \tag{12-64}$$

or

$$\frac{H_{OV}}{\beta_{V-V^*}} = \frac{H_V}{\beta_{V-i}} + \frac{m'V}{L}\frac{H_L}{\beta_{i-L}} \tag{12-65}$$

Similarly, for liquids

$$\frac{H_{OL}}{\beta_{L^*-L}} = \frac{H_L}{\beta_{i-L}} + \frac{L}{m''V}\frac{H_V}{\beta_{V-i}} \tag{12-66}$$

where m' and m'' are the slopes of the equilibrium curve in different concentration regions as shown in Figure 5-7. We can write Eqs. (12-65) and (12-66) for transfer of component A through a stagnant film as

$$\frac{H_{OV}}{(1-y_A)_{*M}} = \frac{H_V}{(1-y_A)_{iM}} + \frac{m'V}{L}\frac{H_L}{(1-x_A)_{iM}} \tag{12-67}$$

and
$$\frac{H_{OL}}{(1-x_A)_{\cdot M}} = \frac{H_L}{(1-x_A)_{IM}} + \frac{L}{m''V}\frac{H_V}{(1-y_A)_{IM}} \tag{12-68}$$

Equations (12-65) and (12-66) can be written for equimolar counter transfer as

$$H_{OV} = H_V + \frac{m'V}{L}H_L \tag{12-69}$$

and
$$H_{OL} = H_L + \frac{L}{m''V}H_V \tag{12-70}$$

12.4 Mass Transfer Correlations for Packed Columns

Transfer unit heights

Mass transfer data are frequently correlated in terms of the height of a transfer unit, since it changes less over the length of a column than does the mass transfer coefficient. The first extensive investigation of transfer unit heights was conducted by Sherwood and Holloway (1940) for the desorption of hydrogen, oxygen, and carbon dioxide from water. They found that the height of the liquid phase transfer unit, H_L, was independent of the gas flow rate. Sherwood and Holloway proposed an empirical relationship to correlate the height of a transfer unit for the liquid phase as a function of liquid flow rate and the Schmidt number. The Sherwood and Holloway correlation is

$$H_L = \beta\left(\frac{L}{\mu_L}\right)^n (\mathrm{Sc})^{0.5} \tag{12-71}$$

where H_L = height of transfer unit, ft,
L = liquid flow rate, lb/h·ft²,
μ_L = viscosity of the liquid, lb/ft·h,
Sc = Schmidt number for the liquid phase,
β, n = constants for different packings, Table 12-1.

TABLE 12-1. CONSTANTS FOR DETERMINING H_L

Packing	β	n
Raschig rings		
3/8 in	0.00182	0.46
1/2 in	0.00357	0.35
1 in	0.0100	0.22
1 1/2 in	0.0111	0.22
2 in	0.0125	0.22
Berl saddles		
1/2 in	0.00666	0.28
1 in	0.00588	0.28
1 1/2 in	0.00625	0.28

The liquid film correlation applies for liquid flow rates that range from 400 to 15,000 lb/h·ft². A correlation was proposed by van Krevelen and Hoftijzer (1948) in which H_L was related to the Schmidt number raised to the $\frac{2}{3}$ power and the Reynolds number raised to the $\frac{1}{3}$ power. The Reynolds number was based on the effective interfacial area. Onda et al. (1959) used a similar relationship to show that the Schmidt number varied with the $\frac{1}{2}$ power and the Reynolds number varied with the 0.51 power.

Literature data were correlated for absorption, stripping, and distillation by Cornell et al. (1960 a, b), and extended by Bolles and Fair (1979). They proposed an empirical relationship for the height of a transfer unit for rings, saddles, and spiral tile. Their correlation is related to the Schmidt number for the liquid phase by the expression

$$H_L = \phi(\mathrm{Sc}_L)^{0.5}(C)\left(\frac{Z}{10}\right)^{0.15} \tag{12-72}$$

where H_L = height of liquid phase transfer unit, ft,
ϕ = parameter from Figures 12-7 through 12-10,
Sc_L = liquid phase Schmidt number, $\mu_L/\rho_L D_L$,
C = flooding correction factor from Figure 12-11,
Z = height of column packing, ft.

A number of relationships have been proposed for correlating the number of gas phase transfer units in terms of the gas and liquid flow rates. Some of these, however, do not take into account the liquid holdup in the voids of the packing and subsequently are of limited use. Another difficulty in obtaining

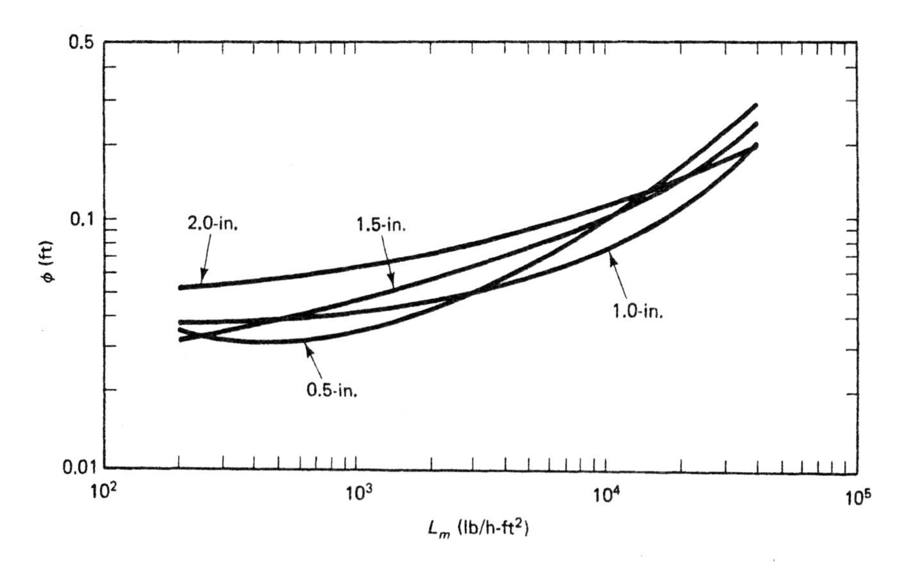

Figure 12-7 Packing parameter for metal Pall rings (from Bolles and Fair, 1979)

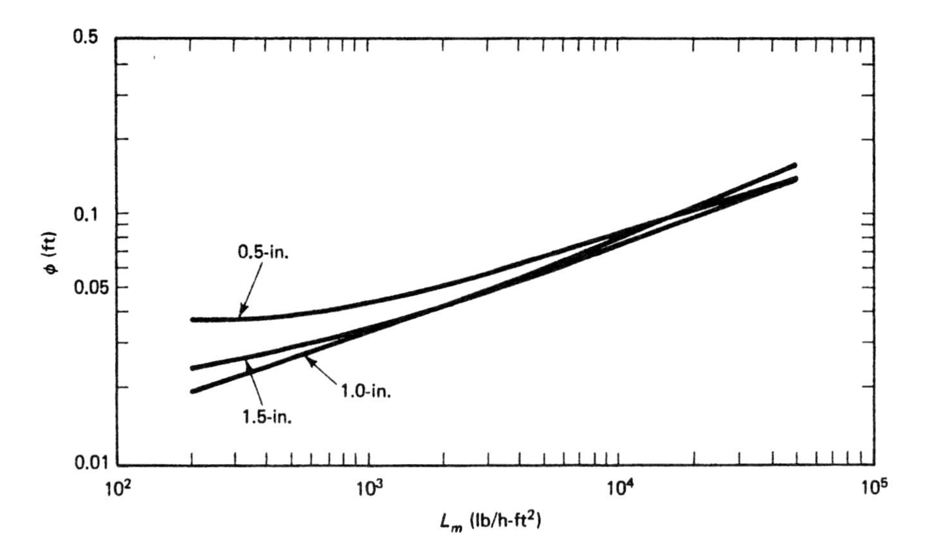

Figure 12-8 Liquid-phase packing parameter for ceramic Berl saddles (from Bolles and Fair, 1979)

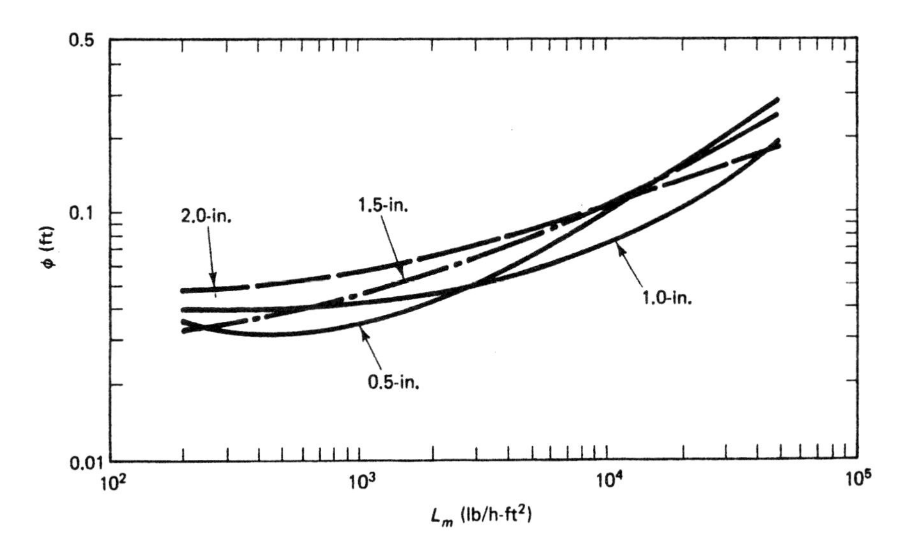

Figure 12-9 Liquid-phase packing parameter for metal Raschig rings (from Bolles and Fair, 1979)

accurate gas phase correlations results from mass transfer data taken on gas–liquid systems in which the resistance to mass transfer is attributed to both phases. The absorption of NH_3 in water provides a system in which nearly all of the resistance to mass transfer lies in the gas phase. Fellinger's (1941) absorption study of NH_3 in water provides the primary source of data for gas film correlations.

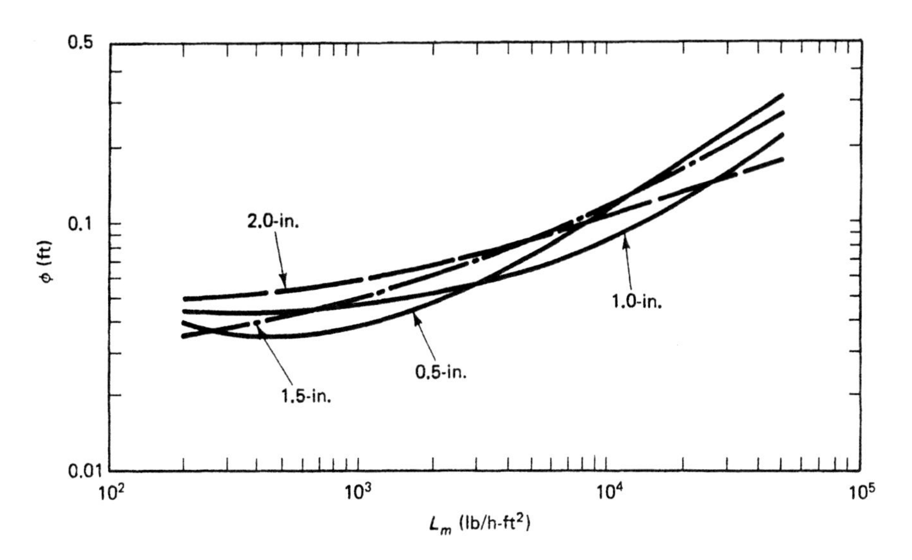

Figure 12-10 Liquid-phase packing parameter for ceramic Raschig rings (from Bolles and Fair, 1979)

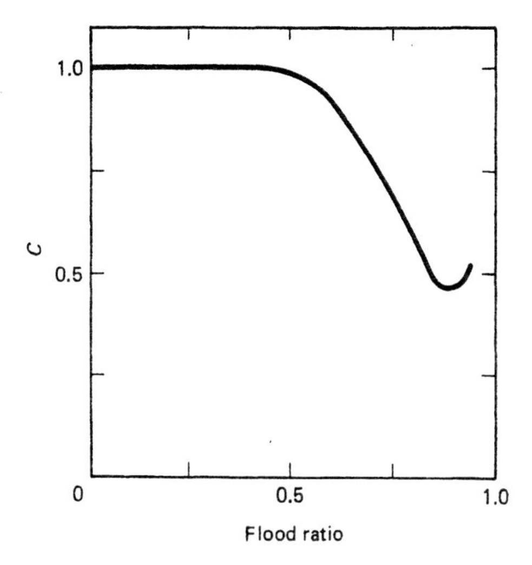

Figure 12-11 Liquid-phase vapor load factor (from Bolles and Fair, 1979)

Cornell et al. (1960 a, b), and Bolles and Fair (1979) correlated a large body of experimental data and proposed an equation for saddles and rings. For saddles

$$H_V = \frac{\psi(\mathrm{Sc}_V)^{0.5}}{(L_m f_1 f_2 f_3)^{0.5}}\left(\frac{d_c}{12}\right)^{1.11}\left(\frac{Z}{10}\right)^{1/3} \tag{12-73}$$

For rings

$$H_V = \frac{\psi(\mathrm{Sc}_V)^{0.5}}{(L_m f_1 f_2 f_3)^{0.6}}\left(\frac{d_c}{12}\right)^{1.24}\left(\frac{Z}{10}\right)^{1/3} \tag{12-74}$$

where H_V = height of gas film transfer unit, ft,
Sc_V = vapor-phase Schmidt number,
d_c = column diameter, in,
Z = packing height, ft,
ψ = parameter from Figures 12-12 through 12-15,
L_m = liquid flow rate, lb/h·ft²,

$$f_1 = \left(\frac{\mu_L}{1.005}\right)^{0.16}; \quad \mu_L(\text{cP}) \tag{12-75}$$

$$f_2 = \left(\frac{1}{\rho_L}\right)^{1.25}; \quad \rho_L(\text{g/cm}^3), \tag{12-76}$$

$$f_3 = \left(\frac{72.8}{\sigma}\right)^{0.8}; \quad \sigma(\text{dyn/cm}). \tag{12-77}$$

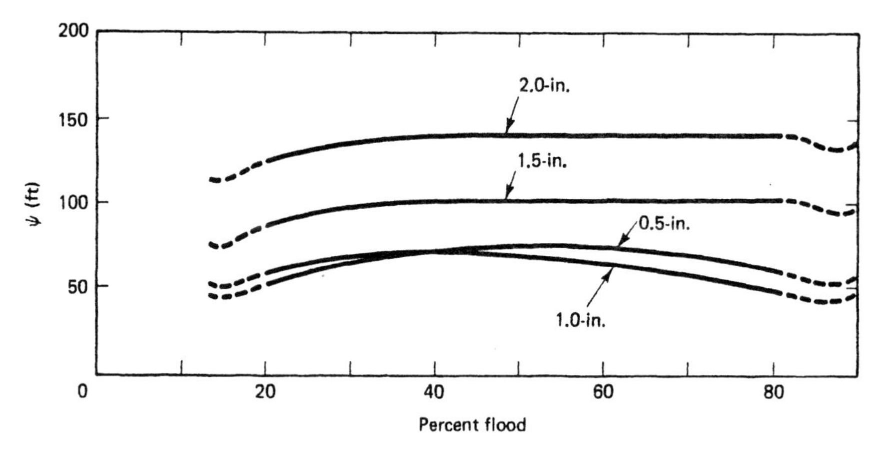

Figure 12-12 Vapor-phase packing parameter for metal Pall rings (from Bolles and Fair, 1979)

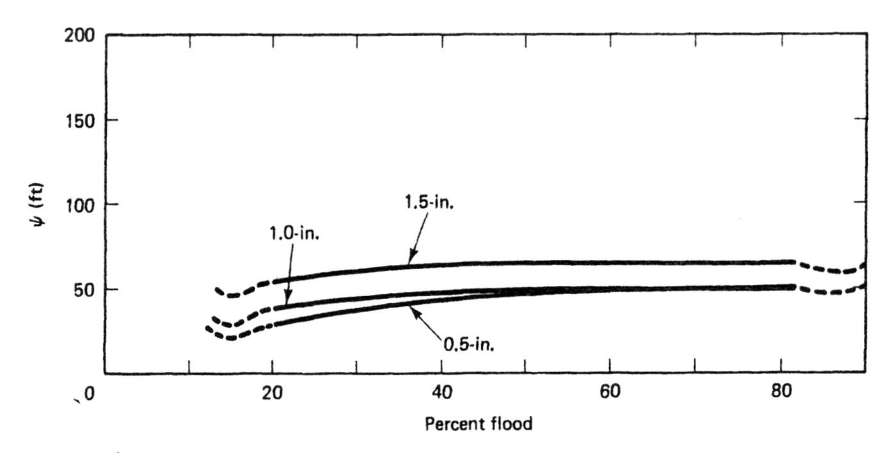

Figure 12-13 Vapor-phase packing parameter for ceramic Berl saddles (from Bolles and Fair, 1979)

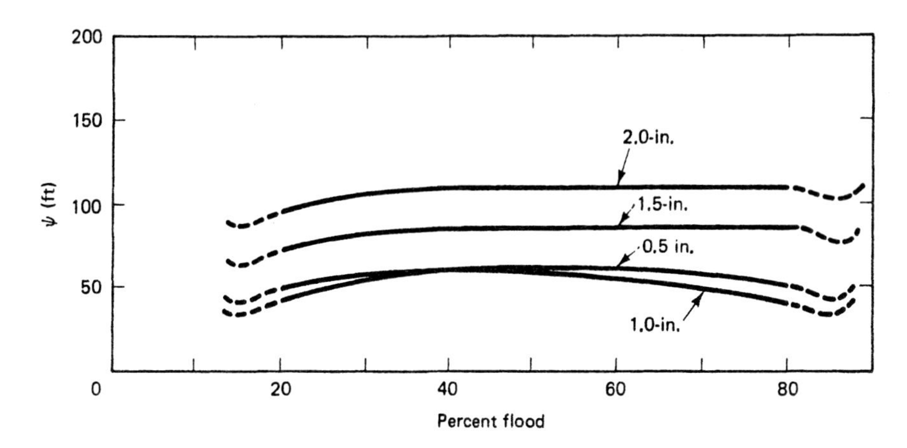

Figure 12-14 Vapor-phase packing parameter for metal Raschig rings (from Bolles and Fair, 1979)

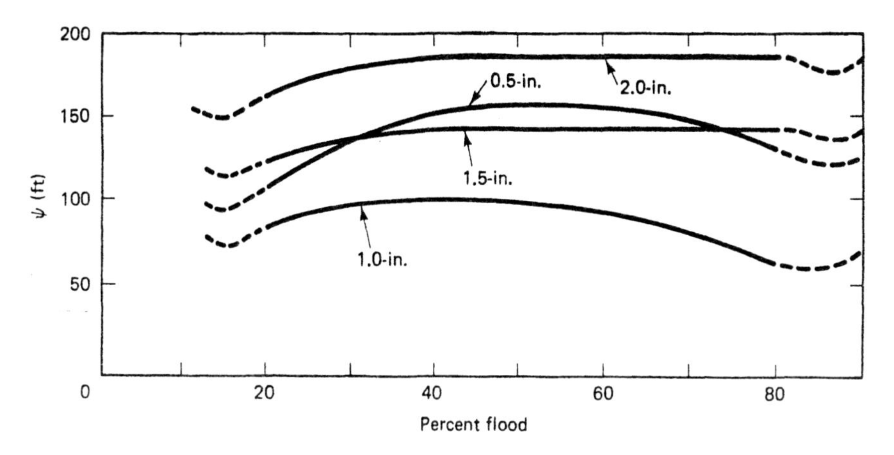

Figure 12-15 Vapor-phase packing parameter for ceramic Raschig rings (from Bolles and Fair, 1979)

The obvious difficulty that we encounter when using Eqs. (12-72) through (12-74) is the requirement that an estimate of the packing height be known. The correlations developed by Bolles and Fair (1979) are valid for both absorption and distillation. Their work represents the most extensive and probably the best correlation of mass transfer data in packed columns at this time. Although these equations provide reasonable results for some systems, data for the system of interest should be used when possible.

Frequently, the height of packing that is equivalent to a theoretical plate (HETP) is needed and as a result a great deal of effort has been expended to relate packed and tray towers. If a reliable value for the HETP is available, the total height of a packed tower can be determined by multiplying the number of equilibrium trays by the HETP. We can relate the height that is equivalent to a theoretical plate to the overall height of a transfer unit by the expression

$$\text{HETP} = H_{OV} \frac{\ln(mV/L)}{mV/L - 1} \tag{12-78}$$

The equation above was developed for straight operating and equilibrium lines. If $mV/L = 1$, HETP $= H_{OV}$. The variation of HETP with vapor velocity is shown in Figure 12-16. In the region B to C the HETP is relatively constant.

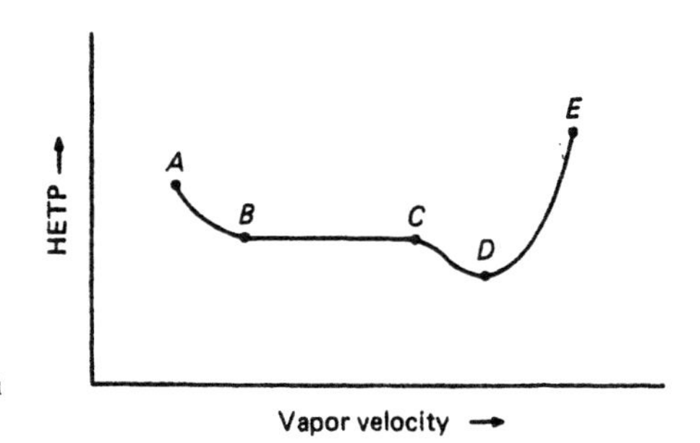

Figure 12-16 Effect of vapor velocity on HETP

Most designers use values obtained from this region. In the region A to B the liquid rate is too low to wet the packing completely, and channeling of the liquid and vapor occurs. This results in the loss of efficiency and gives higher values for HETP. From D to E the packing begins to flood and the efficiency decreases rapidly. Accurate HETP values are published by packing manufacturers and should be used for design purposes. Since most multicomponent distillation and absorption calculations are made on the basis of equilibrium plates, reliable HETP correlations are very important.

12.5 Column Design

A typical absorption tower is shown in Figure 12-17. The top of the tower contains a liquid distributor to distribute the liquid evenly over the packing. A demister, which may consist of extra packing or several inches of wire mesh above the liquid inlet, is installed to remove entrained liquid from the exiting gas stream. Quite often, additional liquid distributors are installed in a column to prevent channeling and the accompanying decrease in packing effectiveness. Redistributors are also used at points in the column where either additional feed streams are introduced or side streams are withdrawn.

Packing

A wide variety of packing types are presently in use; several of these are shown in Figure 12-18. Although Raschig rings and Berl saddles were the most popular packings for many years, these have been largely replaced by higher-

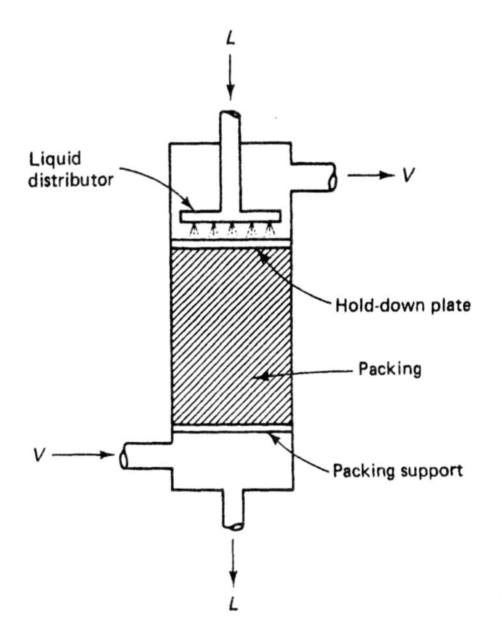

Figure 12-17 Packed tower

capacity and more efficient packings, such as Pall rings, Intalox and Super Intalox saddles, and Flexipak. The type of packing selected for a process depends on several factors. Desirable properties of the packing are:

1. Large void volume to decrease pressure drop
2. Chemically inert to the fluids being processed
3. Large surface area per unit volume of packing
4. Lightweight but strong
5. Good distribution of fluids
6. Good wettability

Tower packings are usually available in a variety of materials, including ceramic, metal, plastic, and carbon. In addition to the desirable properties of a packing, one limitation on the packing is that its size be no greater than one-eighth of the tower diameter. If the size of the packing for a particular tower is too large, a decrease in operating performance will result because of channeling along the column wall.

Flow rate and pressure drop in packed columns

Packing can be loaded into a tower by stacking or by filling the tower with water and dumping the packing into the water. The pressure drop through the random packing is several times greater than through stacked packing. As indicated by Leva (1953), the primary factors that influence the pressure drop are: percentage of void space in the packed tower, packing size and shape, and

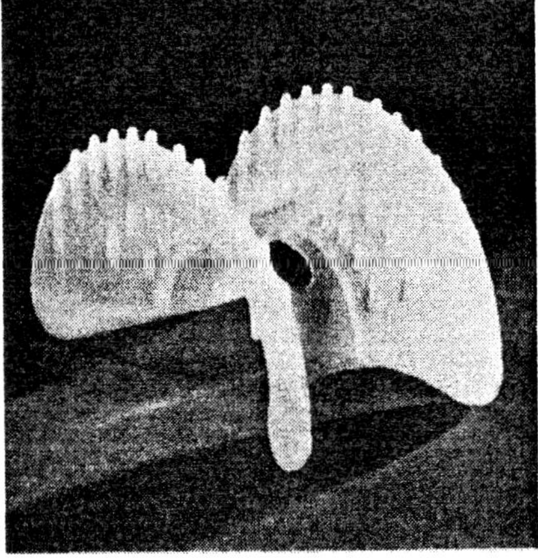
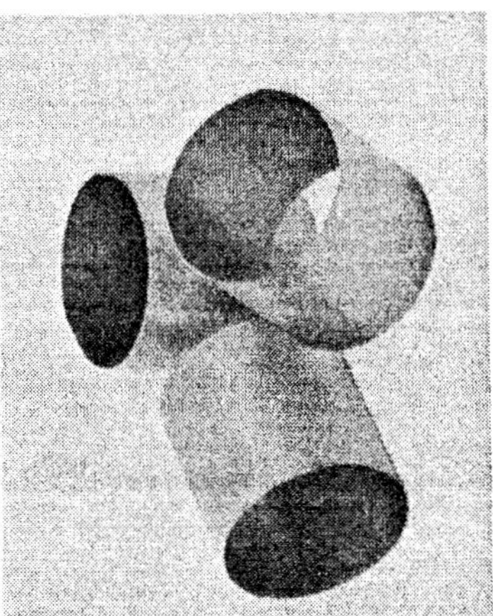

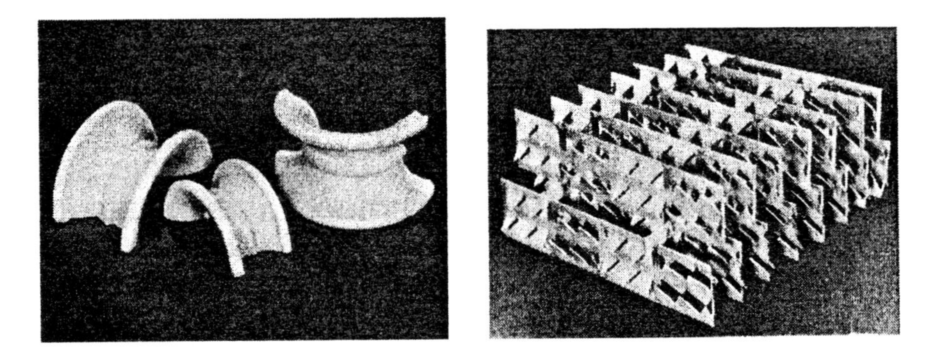

Figure 12-18 Tower packing (courtesy of Koch Engineering Company, Inc.)

the densities and mass velocities of the gas and liquid streams. The primary effect of the liquid rate is to fill the void space and thus decrease the space available for gas flow. The liquid retained in the voids of the packing is called the *dynamic holdup*. For a constant liquid rate, the pressure drop increases with increasing gas flow rate. If the gas rate becomes sufficiently large, a region is reached in which a significant holdup of liquid on the packing occurs. This is defined as the region of loading. From a log-log plot of pressure drop versus gas mass velocity, as shown in Figure 12-19, the loading point can be defined as a point on the curve where the slope is greater than 2.0. After reaching the

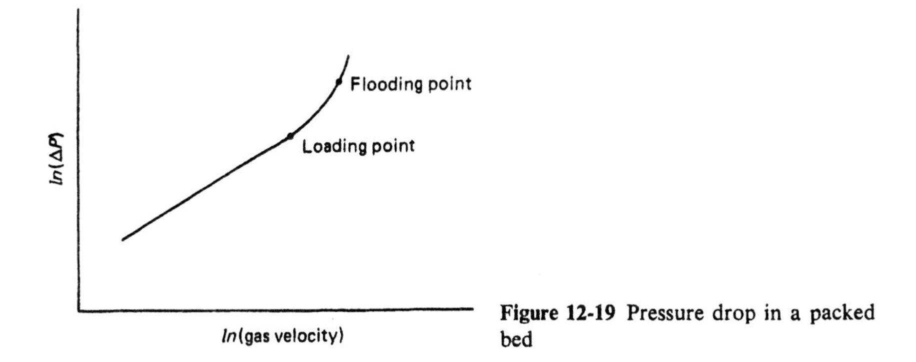

Figure 12-19 Pressure drop in a packed bed

loading point in the column, the pressure drop increases more rapidly with increasing gas flow until the flooding point is reached. Flooding is defined as the condition at which all of the void space in the packing fills with liquid and the liquid will not flow through the column. This gives a pressure drop of 2 to 3 in of water per foot (1.63 to 2.45 kPa/m) of packing depth. A column cannot be operated under these conditions. Packed towers are typically operated at a gas velocity that corresponds to about 50 to 80% of flooding. This condition is usually near the loading point and will result in a pressure drop that ranges from 0.5 to 1.0 in of water/ft (408.2 to 816.4 Pa/m) of packing. A large decrease in flow rate, typically referred to as the *turn-down ratio*, will cause channeling through the packed bed and will frequently result in poor column performance. This, however, depends on the operating range.

Two different approaches are presently being used to determine the tower diameter and flow rates through the tower. One is to select an allowable pressure drop in the bed and the other is to select some fraction of the flooding capacity. The preferred method, however, is to design the column to operate at a specified flooding capacity. This is due in part to the uncertainty in the generalized pressure correlations for the tower packings. The pressure drop and gas flow rates can be calculated by applying the correlations shown in Figure 12-20. The calculation of pressure drop in a packed tower and the determination of tower diameter are demonstrated in Example 12.4.

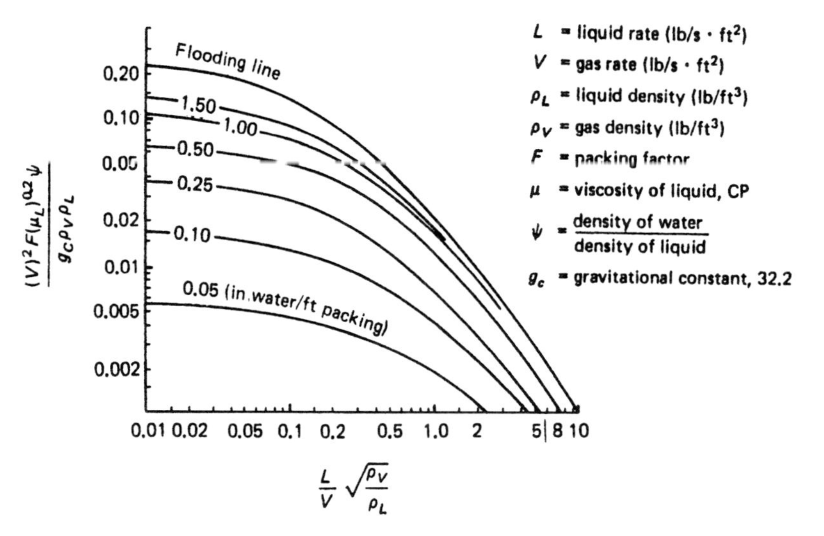

Figure 12-20 Flooding and pressure drop correlation for random packings (from Eckert, 1970)

Example 12.4

A tower packed with 1-in Raschig rings is to be used to absorb NH_3 from air by contacting the gas stream with water. The entering gas flow rate is 40 lbmol/h and contains 5.0% NH_3. Ninety percent of the NH_3 is to be removed from the air. The entering water flow rate is 3200 lbm/h Absorption is to be carried out at a pressure of 1 atm absolute and 20°C. (a) Calculate the tower diameter for 70% of flooding. (b) Determine the pressure drop per foot of packing.

Solution:

(a) When calculating the tower diameter, conditions at the bottom of the tower should be used since this corresponds to the maximum flow rates. At the bottom of the tower

$$MW_{gas} = 0.05(17) + 0.95(29) = 28.4 \text{ lb/lbmol}$$

$$NH_3 \text{ removed from the air} = 0.9(0.05)(40) = 1.8 \text{ lbmol/h}$$
$$= 30.6 \text{ lb/h}$$

$$\text{gas flow rate at the bottom} = 40 \text{ lbmol/h } (28.4 \text{ lb/lbmol})$$
$$= 1136 \text{ lb/h} = 0.316 \text{ lb/s}$$

$$\text{liquid flow rate at the bottom} = 3200 + 30.6 = 3230.6 \text{ lb/h}$$

$$\rho_V = \frac{1}{359}\left(\frac{273}{293}\right)(28.4) = 0.0737 \text{ lb/ft}^3$$

$$\rho_L = 62.3 \text{ lb/ft}^3$$

$$\mu_L = 1.0 \text{ cP}$$

TABLE 12-2. PACKING FACTORS (FT^{-1}) (ECKERT, 1975)

Type of Packing	Material	Nominal Packing Size (in)										
		$\frac{1}{4}$	$\frac{3}{8}$	$\frac{1}{2}$	$\frac{5}{8}$	$\frac{3}{4}$	1	$1\frac{1}{4}$	$1\frac{1}{2}$	2	3	$3\frac{1}{2}$
Super Intalox	Ceramic	—	—	—	—	—	60	—	—	30	—	—
Super Intalox	Plastic	—	—	—	—	—	33	—	—	21	16	—
Intalox saddles	Ceramic	725	330	200	—	145	98	—	52	40	22	—
Hy-Pak rings	Metal	—	—	—	—	—	42	—	—	18	15	—
Pall rings	Plastic	—	—	—	97	—	52	—	40	25	—	16
Pall rings	Metal	—	—	—	70	—	48	—	28	20	—	16
Berl saddles	Ceramic	900*	—	240*	—	170†	110†	—	65†	45*	—	—
Raschig rings	Ceramic	1600[a],*	1000[b],*	580[c]	380[c]	255[c]	155[d]	125[e],*	95[e]	65[f]	37[g],*	—
Raschig rings, $\frac{1}{32}$-in wall	Metal	700*	390*	300*	170	155	115*	—	—	—	—	—
Raschig rings, $\frac{1}{16}$-in wall	Metal	—	—	410	290	220	137	110*	83	57	32*	—
Tellerettes	Plastic	—	—	—	—	—	40	—	—	20	—	—
Maspak	Plastic	—	—	—	—	—	—	—	—	32	20	—
Lessing exp.	Metal	—	—	—	—	—	—	—	30	—	—	—
Cross-partition	Ceramic	—	—	—	—	—	—	—	—	—	70	—

[a] $\frac{1}{32}$ wall. [b] $\frac{1}{16}$ wall. [c] $\frac{3}{32}$ wall. [d] $\frac{1}{8}$ wall. [e] $\frac{3}{16}$ wall. [f] $\frac{1}{4}$ wall. [g] $\frac{3}{8}$ wall.
*Extrapolated.
†Packing factors obtained in 16- and 30-in-I.D. towers.

The abscissa for Figure 12-20 is

$$\frac{L}{V}\sqrt{\frac{\rho_V}{\rho_L}} = \frac{3230.6}{1136}\sqrt{\frac{0.0737}{62.3}} = 0.0978$$

Thus from Figure 12-20,

$$\frac{(V)^2F(\mu_L)^{0.2}\psi}{g_c\rho_V\rho_L} = 0.14$$

From Table 12-2, $F = 155$. Since the absorbing liquid is water, $\psi = 1.0$.

$$V = \left[\frac{0.14(g_c)(\rho_V)(\rho_L)}{F(\mu_L)^{0.2}}\right]^{1/2}$$

$$= \left[\frac{(0.14)(32.2)(0.0737)(62.3)}{155(1.0)^{0.2}}\right]^{1/2}$$

$$= 0.365 \text{ lb/s}\cdot\text{ft}^2$$

For 70% of flooding,

$$V = 0.7(0.365) = 0.256 \text{ lb/s}\cdot\text{ft}^2$$

The tower cross-sectional area is

$$S = \frac{0.316}{0.256} = 1.23 \text{ ft}^2$$

Therefore

$$d = \left(\frac{4S}{\pi}\right)^{1/2} = \left[\frac{4(1.23)}{\pi}\right]^{1/2}$$

$$= 1.25 \text{ ft} = 15.0 \text{ in}$$

A tower with a diameter of 15 in should be selected.

(b) To determine the pressure drop, we must first calculate a value for the ordinate using the actual gas flow rate and the cross-sectional area. Thus

$$V = \frac{0.316}{1.23} = 0.257 \text{ lb/s}\cdot\text{ft}^2$$

and

$$\frac{V^2F(\mu_L)^{0.2}\psi}{g_c\rho_V\rho_L} = \frac{(0.257)^2(155)(1)^{0.2}}{(32.2)(0.0737)(62.3)}\frac{62.3}{62.3} = 0.069$$

Using an abscissa of 0.0978 and a value of 0.069 for the ordinate in Figure 12-20, the pressure drop is 0.95 in water/ft of packing.

REFERENCES

Bolles, W. L., and J. R. Fair, *Int. Chem. Eng. Symp. Ser. 56, 2,* 3.3/35 (1979).

Colburn, A. P., *Trans. AIChE, 35,* 211 (1939).

Cornell, D., W. G. Knapp, and J. R. Fair, *Chem. Eng. Prog., 56,* No. 7, 68 (1960a).

Cornell, D., W. G. Knapp, H. J. Close, and J. R. Fair, *Chem. Eng. Prog., 56,* No. 8, 48 (1960b).

Eckert, J. S., *Chem. Eng. Prog., 66,* No. 3, 39 (1970).

Eckert, J. S., *Chem. Eng., 82,* No. 8, 70 (1975).

FELLINGER, L., "Adsorption of Ammonia by Water and Acid in Various Standard Packings," Sc.D. thesis, Massachusetts Institute of Technology, 1941.

LEVA, M., *Tower Packings and Packed Tower Design*, 2nd ed., The U.S. Stoneware Co., Akron, Ohio, 1953.

ONDA, K., E. SADA, and Y. MURASE, *AIChE J.*, *5*, 235 (1959).

SHERWOOD, T. K., and F. A. L. HOLLOWAY, *Trans. AIChE*, *36*, 39 (1940).

VAN KREVELEN, D. W., and P. J. HOFTIJZER, *Chem. Eng. Prog.*, *44*, 532 (1948).

NOTATIONS

a = area per unit volume of packing, L^2/L^3
C = flooding correction factor from Figure 12-11
d_c = column diameter, L
D = flow rate of distillate, mol/t
D_L = liquid-phase diffusivity, L^2/t
F = packing factor, L^{-1}
H_L = height of a liquid-phase transfer unit, L
H_{OL} = height of a transfer unit based on the overall liquid-phase mass transfer coefficient, L
H_{OV} = height of a transfer unit based on the overall vapor-phase mass transfer coefficient, L
H_V = height of a gas-phase transfer unit, L
HETP = height of packing equivalent to a theoretical plate, L
k'_x = local mass transfer coefficient for the liquid phase, mol/tL^2 (mole fraction)
k'_y = local mass transfer coefficient for the vapor phase, mol/tL^2 (mole fraction)
K'_x = overall liquid-phase mass transfer coefficient, mol/tL^2 (mole fraction)
K'_y = overall vapor-phase mass transfer coefficient, mol/tL^2 (mole fraction)
L = flow rate of liquid stream, mol/tL^2
L_a, L_b = liquid flow rates at the top and bottom of the column, respectively, mol/tL^2
L'_m = liquid mass flow rate, M/tL^2
$\bar{L}$ = liquid flow rate written on a solute-free basis, mol/tL^2
m, m', m'' = slope of equilibrium curve
n = constant for different packings
N_A = molar flux of A relative to a fixed reference frame, mol/tL^2
N_L, N_V = number of transfer units
N_{OL}, N_{OV} = number of overall transfer units
S = cross-sectional area, L^2
Sc = Schmidt number
V = vapor flow rate, mol/tL^2
$\bar{V}$ = vapor flow rate written on a solute-free basis, mol/tL^2
V_a, V_b = vapor flow rates at the top and bottom of the column, respectively, mol/tL^2
x_A = mole fraction of A in the liquid
$x_{A,a}$, $x_{A,b}$ = mole fractions of A in the liquid at the top and bottom of the column

x_A^* = equilibrium mole fraction of A in the liquid
$\bar{X}_{A,a}, \bar{X}_{A,b}$ = solute-free concentrations of A in the liquid at the top and bottom of the tower
y_A = mole fraction of A in the vapor
$y_{A,a}, y_{A,b}$ = mole fractions of A in the vapor at the top and bottom of the column
y_A^* = equilibrium mole fraction of A in the vapor
$\bar{Y}_{A,a}, \bar{Y}_{A,b}$ = solute-free concentrations of A in the vapor at the top and bottom of the tower
Z = height of column packing, L

Greek Letters

β = bulk flow correction factor
μ_L = liquid viscosity, M/Lt
ρ_L = liquid density, M/L^3
ρ_V = vapor density, M/L^3
σ = surface tension, F/L
ϕ = parameter from Figures 12-7 through 12-10
ψ = parameter from Figures 12-12 through 12-15; ratio of density of water to density of liquid

PROBLEMS

12.1 A packed tower is to be used to absorb sulfur dioxide from air by contacting the gas mixture with pure water. The entering gas contains 10 mol% SO_2 and the existing gas is to contain no more than 0.8 mol% SO_2. The tower is to operate at 30°C and 2 atm absolute. Water is to enter the absorber at a flow rate equal to 1.2 times the minimum value. The gas mixture enters at a rate of 80 kgmol/m²·h. Calculate the height of the absorption tower if the gas and liquid mass transfer coefficients are $k'_y a$ = 50 kgmol/h·m³ and $k'_x a$= 1000 kgmol/h·m³, respectively.

12.2 Ammonia is to be absorbed from a mixture of air and ammonia by contacting the gas with water in a packed column. The gas mixture, which contains 5.0 mol% ammonia, enters the column at a flow rate of 100 kgmol/h·m² and the water flow rate is to be 1.5 times the minimum value. The gas and liquid mass transfer coefficients can be expressed as

$$k'_y a = 0.2(1 - y_A)V^{0.80}L^{0.20}$$

$$k'_x a = 0.35L^{0.80}$$

where V and L represent the molar flow rates of the gas and liquid, respectively, in units of kgmol/h·m². The tower operates at a constant temperature of 20°C and a pressure of 1 atm absolute. Calculate the height of the absorber if the concentration of ammonia in the exiting gas stream is to be 0.5 mol%.

12.3 A mixture that contains 40 mol% A and 60 mol% B is to be separated in a packed tower. The concentration of A in the overhead product is to be 90 mol%, and 95% of A that enters with the feed is to be recovered as overhead product. Also, 100 mol/h of feed enters the distillation column as a saturated liquid. A reflux ratio, L/D — 4.0, is to be used. The relative volatility of A to B is 2.0.

The mass transfer coefficients for this process are $k'_y a = 100$ kgmol/m³·s and $k'_x a = 200$ kgmol/m³·s. Calculate the height of the tower needed to make the separation. A total condenser and partial reboiler will be used. Assume equal molar overflow.

12.4 Sulfur dioxide is to be recovered from air by contacting the gas with pure water in a packed tower. The tower will be operated at 30°C and 1 atm absolute pressure. The entering gas contains 5% SO_2 and the exiting gas is to contain 0.5%. The gas flow rate entering the tower is 100 kgmol/h·m². If the overall mass transfer coefficient for the vapor phase is $K'_y a = 60$ kgmol/h·m³, determine **(a)** the minimum pure water rate (kgmol/h·m²) needed for this separation, and **(b)** the height of the tower. Water is to enter the absorber at a flow rate equal to 1.2 times the minimum value.

12.5 If the local mass transfer coefficients are constant over the column length and the equilibrium curve is nearly straight, an overall mass transfer coefficient can be evaluated and then used to determine the tower height. Assuming that this is the case for the separation process described in Problem 12.3, **(a)** determine the number of overall transfer units. **(b)** Determine the tower height if the overall and local mass transfer coefficients can be related by the expression

$$\frac{1}{K'_y a} = \frac{1}{k'_y a} + \frac{1.1}{k'_x a}$$

where the value 1.1 is the slope of the equilibrium curve in the dilute concentration range. **(c)** Compare your answer with that obtained in Problem 12.3.

12.6 A gas mixture containing 2% A and 98% B is to be contacted by a pure liquid in a packed absorption tower. The concentration of species A in the gas stream is to be reduced to 0.5%. The gas and liquid flow rates are 100 kgmol/h·m² and 2000 kgmol/h·m², respectively. The overall mass transfer coefficient for the gas phase is $K'_y a = 100$ kgmol/h·m³. Using the equilibrium data given below, determine the tower height.

y_A	0.003	0.023	0.043	0.0627	0.0826
x_A	0	0.002	0.004	0.006	0.008

Recall that if the equilibrium curve is straight and the solution is dilute, numerical integration is unnecessary.

12.7 Solute A is to be stripped from a liquid stream by contacting the liquid with a pure gas in a packed column. The liquid enters the column at an A-free rate of 150 kgmol/h and contains 7.0 mol% A. The gas enters the column at a rate of 500 kgmol/h. The equilibrium distribution of A in the gas and liquid is expressed as $y_A = 0.4x_A$. It is desired to reduce the concentration of A in the exiting liquid to 1.0 mol%. Determine the height of the column if the overall mass transfer coefficient for the liquid phase is 75 kgmol/h·m³.

12.8 A cocurrent absorption column is to be used to recover solute A from an inert gas. The concentration of A is to be reduced from $\bar{Y}_A = 0.08$ to $\bar{Y}_A = 0.01$ by contacting the gas with a pure solvent. The actual liquid rate is to be 1.2 times the minimum. The gas mixture enters the column at a rate of 100 kgmol/h. Determine **(a)** $(\bar{L}/\bar{V})_{min}$, **(b)** the actual $\bar{L}/\bar{V}$ ratio, **(c)** the concentration of A in the exiting solvent solution, and **(d)** the height of the column if the overall mass transfer coefficient for the vapor phase is 150 kgmol/h·m³. The equilibrium relationship can be expressed as $y_A = 4.0x_A$.

12.9 An absorber, which is packed with 1-in ceramic Raschig rings to a height of 10 ft, is to be used to absorb CO_2 from air by contacting the air–CO_2 mixture with water. The water is to enter the absorber at a rate of 2000 lb/h·ft². Assume that the gas mixture is very dilute and the absorption process is to be carried out at 20°C and 1 atm absolute pressure. **(a)** Determine the height of the liquid-phase transfer unit if the column operates at 70% of flooding, and **(b)** calculate the height of the liquid-phase transfer unit if 1-in Berl saddles are used as the packing.

12.10 Sulfur dioxide is to be removed from air by contacting the gas with water in a column packed with $\frac{3}{4}$-in ceramic Berl saddles. The entering gas contains 10 mol% SO_2. Water enters at a rate of 2.03×10^2 kgmol/h and the gas enters at a rate of 2.26×10^{-1} kgmol per kgmol of water. **(a)** Determine the diameter of the column at 70% of flooding, and **(b)** calculate the pressure drop through the column if the packing is dumped into the column and the height is 100 ft. The column is to operate at 30°C and a pressure of 2 atm absolute.

Design of Staged Columns 13

13.1 Introduction

The calculation methods developed and discussed in previous chapters have been made under the implicit assumption that a suitable apparatus is available to provide the intimate phase contact necessary to accomplish the interphase mass transfer. Some attention must be directed to the variables involved in estimating the number of such contacts required to accomplish the specified mass transfer and to determine the overall size of the apparatus (equipment) necessary for the specified flow rate. The most common types of interphase contacting equipment are those that contain an inert packing or those that contain trays. A preliminary design and size estimation for packed columns was discussed in Chapter 12. This chapter is concerned with the preliminary design and size estimation of towers that contain trays. The reader should keep in mind that the procedures discussed here are preliminary only, but should be of suitable quality for making budget estimates and preliminary cost analyses. The methods discussed here should not be considered final design techniques. For a final tray design, an expert with access to suitable computer programs and proprietary tray design methods should be consulted.

13.2 Plate Efficiency

A large portion of all mass transfer calculations is based on an "ideal," "theoretical," or "equilibrium" stage or tray. For an equilibrium tray, such as that shown in Figure 13-1 for a vapor–liquid system, the vapor and the liquid leaving

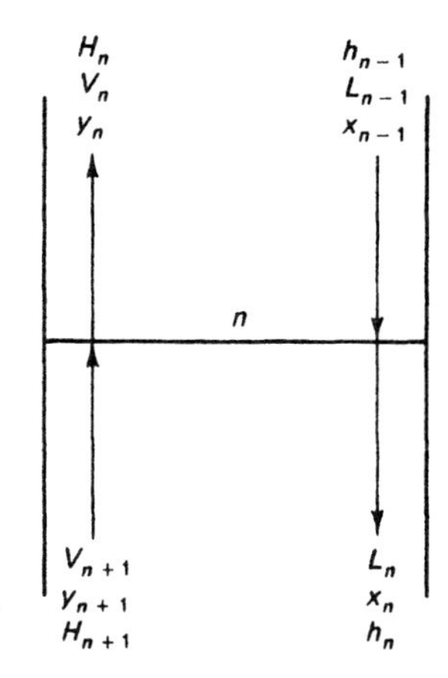

Figure 13-1 Theoretical tray for vapor–liquid contact

the tray are in thermodynamic equilibrium. Since they are at the same temperature and pressure, the compositions of the two streams that leave the tray may be related through the equilibrium distribution coefficient, K. Implicit in this statement is that the tray establishes not only mass transfer equilibrium but also heat transfer equilibrium. In general, however, the real tray will not perform as an equilibrium plate. Tray efficiencies can be related in several ways:

1. Through the mass transfer on the tray
2. Through the heat transfer on the tray
3. By overall column performance
4. By a symmetrical plate definition that simultaneously satisfies mass transfer, and material and enthalpy balance requirements, for the tray

Approaches 1 and 3 have been the most widely used but all four will be discussed briefly.

Mass transfer efficiencies

Murphree (1925) defined two efficiencies for each plate in the column. Assuming constant molar overflow, he developed a vapor efficiency given as

$$E_{MV} = \frac{y_n - y_{n+1}}{y_n^* - y_{n+1}} \times 100 \tag{13-1}$$

where E_{MV} = Murphree vapor efficiency for the tray,
y_n, y_{n+1} = vapor compositions leaving and entering a tray,
y_n^* = vapor phase composition that would be in equilibrium with the actual liquid leaving the tray.

The liquid-phase efficiency is similarly defined as

$$E_{ML} = \frac{x_n - x_{n-1}}{x_n^* - x_{n-1}} \times 100 \tag{13-2}$$

where E_{ML} = Murphree liquid efficiency for the tray,
x_n, x_{n-1} = liquid compositions defined as shown in Figure 13-1,
x_n^* = liquid composition that would be in equilibrium with the actual vapor leaving the tray.

Several problems arise when applying the Murphree efficiencies. The first, of course, is that they are not equal throughout the column; this is bothersome but not serious. Much more serious is the difficulty involved in calculating the composition of a stream that is supposed to be in equilibrium with a non-equilibrium stream.

Temperature efficiency

The temperature or thermal efficiencies for a tray have been attributed to Nord (1946) and Carey (1930). For the vapor phase

$$E_{TV} = \frac{T_n - T_{n+1}}{T_n^* - T_{n+1}} \times 100 \tag{13-3}$$

where E_{TV} = vapor temperature efficiency of the tray,
T_n, T_{n+1} = temperatures of the vapor streams denoted by the appropriate subscripts,
T_n^* = vapor temperature in equilibrium with that of the liquid leaving the actual plate.

The liquid thermal efficiency for the tray is

$$E_{tL} = \frac{t_n - t_{n-1}}{t_n^* - t_{n-1}} \times 100 \tag{13-4}$$

where E_{tL} = liquid thermal efficiency of the tray,
t_n, t_{n-1} = temperatures of the liquid streams denoted by the appropriate subscripts,
t_n^* = temperature of the liquid stream in equilibrium with that of the vapor leaving the actual plate.

The thermal or heat transfer efficiency has not been widely used and will not be discussed further in this chapter.

Overall efficiency

The overall tray efficiency of a column is simply defined as

$$E_o = \frac{\text{number of theoretical trays}}{\text{number of actual trays}} \times 100 \tag{13-5}$$

The overall tray efficiency, although widely used, is not representative of column operation because the different compositions on the various trays result in different tray efficiencies. In addition, each tray frequently will have a different tray efficiency for each component.

Generalized plate efficiency

The concept of a generalized tray efficiency that simultaneously satisfies the material balance, component balance, enthalpy balance, and equilibrium requirements was proposed by Standart (1965). The material balance is written as

$$V_{n+1} + L_{n-1} = V_n + L_n = V_n^* + L_n^* \tag{13-6}$$

The component balance around tray n is

$$V_{n+1}y_{n+1} + L_{n-1}x_{n-1} = V_n y_n + L_n x_n = V_n^* y_n^* + L_n^* x_n^* \tag{13-7}$$

The enthalpy balance for the tray is

$$V_{n+1}H_{n+1} + L_{n-1}h_{n-1} - Q_n = V_n H_n + L_n h_n = V_n^* H_n^* + L_n^* h_n^* \tag{13-8}$$

where V, L = total molar flow rate of vapor or liquid stream,
y, x = composition of vapor or liquid stream,
H, h = enthalpy of vapor or liquid stream,
$n, n+1, n-1$ = tray number,
$*$ = pseudoequilibrium tray defined by Eqs. (13-6) through (13-8),
Q = heat gained or lost by the tray.

Using these equations we can define tray efficiencies in terms of the overall material balance or for an individual component. Although satisfying from the standpoint that the equations above meet all of the requirements for tray efficiencies, efficiencies based on Eqs. (13-7) and (13-8) have not been widely used.

Estimating tray efficiencies

Over the years many procedures have been proposed for estimating tray efficiencies. Practically all of these are limited either to binary systems or to the efficiency of separation between the key components in a multicomponent system. MacFarland et al. (1972) improved the work of English and Van Winkle (1963) and presented an equation for predicting the Murphree tray efficiency. The equation of MacFarland et al. is expressed in terms of the liquid and vapor properties as

$$E_{MV} = 7.0(D_g)^{0.14}(\text{Sc})^{0.25}(\text{Re})^{0.08} \tag{13-9}$$

where E_{MV} = percent efficiency,
$D_g = \sigma_L/\mu_L U_V$,
$\text{Sc} = \mu_L/\rho_L D_{LK}$,
$\text{Re} = h_w U_V \rho_V/[(\mu_L)(\text{FA})]$,
and σ_L = surface tension, lb/h^2,
μ_L = liquid viscosity, lb/ft·h,
U_V = superficial vapor velocity, ft/h,

D_{LK} = diffusivity of light key component, ft^2/h,
h_W = height of weir, ft,
ρ_L = liquid density, lb/ft^3,
ρ_V = vapor density, lb/ft^3,
FA = fractional free area available for vapor flow.

To evaluate the terms in Eq. (13-9), MacFarland et al. (1972) suggested the following:

1. Calculate vapor density from the ideal gas equation.
2. Calculate the density of the liquid mixture as the mole fraction average molar volume.
3. Calculate liquid mixture viscosity by using the expression $\mu_{L,\text{mix}} = [\sum x_i(\mu_i)^{1/3}]^3$.
4. Calculate liquid surface tension using the Sugden Parachor method as shown in Perry et al. (1963).
5. Calculate the diffusivity of the liquid light key component by the dilute solution equation of Wilke and Chang (1955) or other suitable equations.

The equation of MacFarland et al. (1972) was derived from reported data on both bubble-cap and perforated-plate efficiencies and is applicable to both. However, their model is based on binary systems only.

The overall efficiency may be obtained from either a correlation or predictive procedure, or from operating data on a laboratory or pilot plant column. If experimental data are used, some intuition or judgment must be

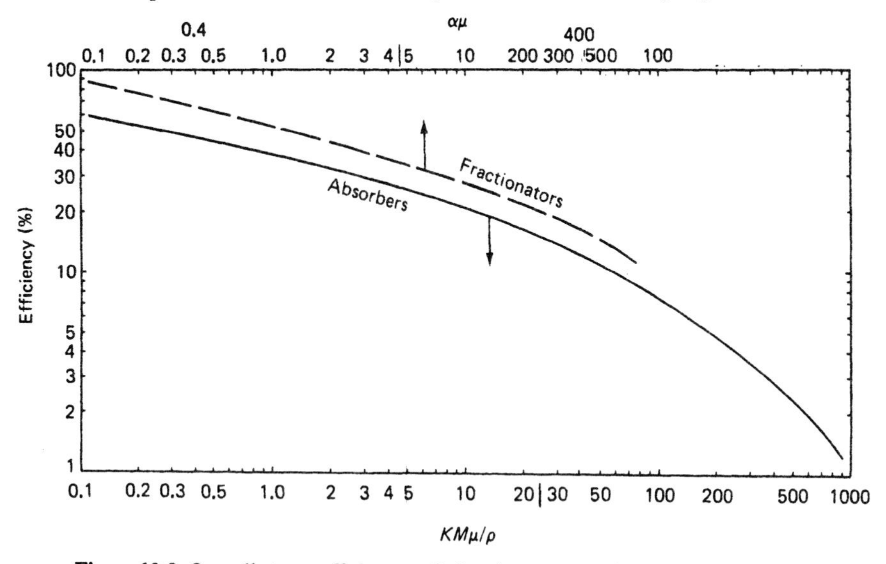

Figure 13-2 Overall tray efficiency of fractionators and absorbers (from Edmister, 1949)

applied to the experimentally determined overall efficiency in order to scale it from a small column to a large one. Several efficiencies have been used, but perhaps the best known is that proposed by O'Connell (1946). O'Connell's model has been modified graphically by Edmister (1949) and is shown in Figure 13-2. For fractionators, overall efficiencies obtained from Figure 13-2 will tend to be slightly conservative. When applying Figure 13-2 to fractionators, the X axis is the product of the feed viscosity and the relative volatility of the key components, both taken at the feed conditions. For absorbers, the X axis is $(KM\mu/\rho)$. The equilibrium constant is for the solute or key component being absorbed. The remaining terms are for the solvent. The viscosity μ has units of centipose, the density ρ has units of lb/ft^3, and M is the molecular weight.

13.3 Characteristics of Trayed Columns

Equilibrium trays are placed in vapor–liquid contacting columns to provide for intimate mixing of the vapor and liquid phases, and then provide for separation of the phases so that the vapor can flow upward and the liquid downward through the column. The mixing of the phases is provided by the flow of the liquid and vapor through holes or perforations in the tray. By spacing the trays from several inches to several feet apart in the column, the liquid and vapor are permitted to separate. The general flow pattern in the column is for the liquid to flow across the tray and to the tray below by passing through the downcomer. The vapor bubbles flow upward through the tray and to the tray above. In general, the tray may contain bubble caps, perforations, or a proprietary device usually referred to as a *valve*. Valve trays are used because the vapor flow area in the tray may be adjusted as the vapor flow rate through the tray changes.

Estimating areas for vapor flow through bubble-cap trays

For preliminary estimates of column diameter, the required column area can be broken into two parts: that required for vapor flow and that required for liquid flow. Figure 13-3 shows the correlation presented by Fair and Matthews (1958) for determining the vapor flow area on bubble-cap trays. In Figure 13-3 the liquid mass rate, L'_m, and gas mass rate, G, must be on the same time basis and the vapor and liquid densities, ρ_V and ρ_L, must be in consistent units. The superficial vapor velocity, U_0, is based on the empty tower cross section minus the area of one downcomer. The vapor velocity determined from Figure 13-3 is the maximum allowable vapor velocity for the column. This number, frequently referred to as the *flooding velocity*, with units of ft/s, will normally be derated to some extent in column design. If no other information is available, the maximum allowable velocity should be multiplied by about 0.85 to determine the design flow area for the vapor.

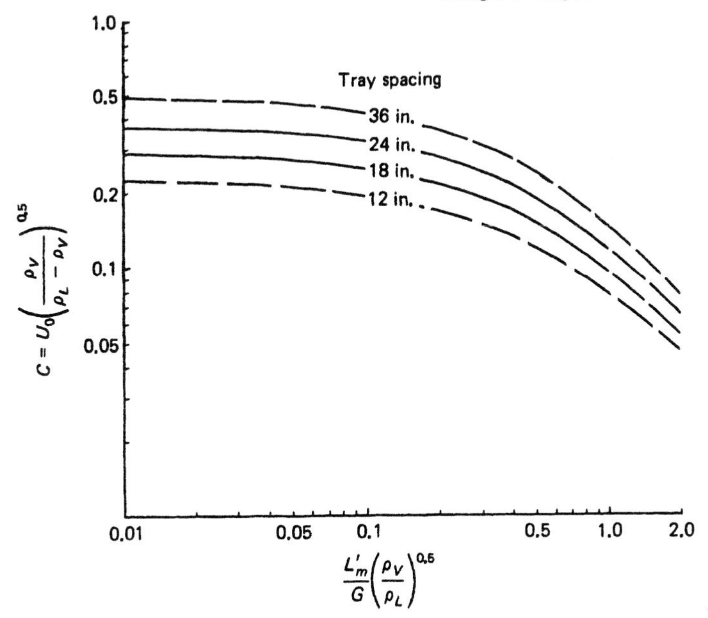

Figure 13-3 Correlation for estimating required vapor flow area for bubble-cap plates (from Fair and Matthews, 1958)

Vapor flow areas for sieve trays

Fair (1961) presented a correlation for estimating the required vapor flow area for sieve plates. His correlation, which is shown in Figure 13-4, is very similar in shape to that shown in Figure 13-3 for bubble-cap plates. The definition of terms given for Figure 13-4 is exactly the same as those defined for Figure 13-3. Figures 13-3 and 13-4 are limited to systems that exhibit little or no foaming and with interfacial tensions of 20 dyn/cm. For sieve trays the height of the weir must be less than 15% of the tray spacing and the bubbling area must occupy most of the area between the weirs. For cases in which the surface tension is not 20 dyn/cm, Fair (1961) proposed that the capacity factor obtained from Figure 13-4 be modified as

$$C = C_{\text{chart}}\left(\frac{\sigma}{20}\right)^{0.2} \tag{13-10}$$

Proprietary trays

A wide variety of proprietary trays are used in fractionators and absorbers. Each individual tray vendor generally will provide prospective customers with techniques for estimating the performance of a particular tray. Bolles (1976) and Thorngren (1978) have presented generalized procedures for estimating the performance of proprietary valve-type trays. The procedure proposed by Thorngren is more involved and has the disadvantage that the column must have

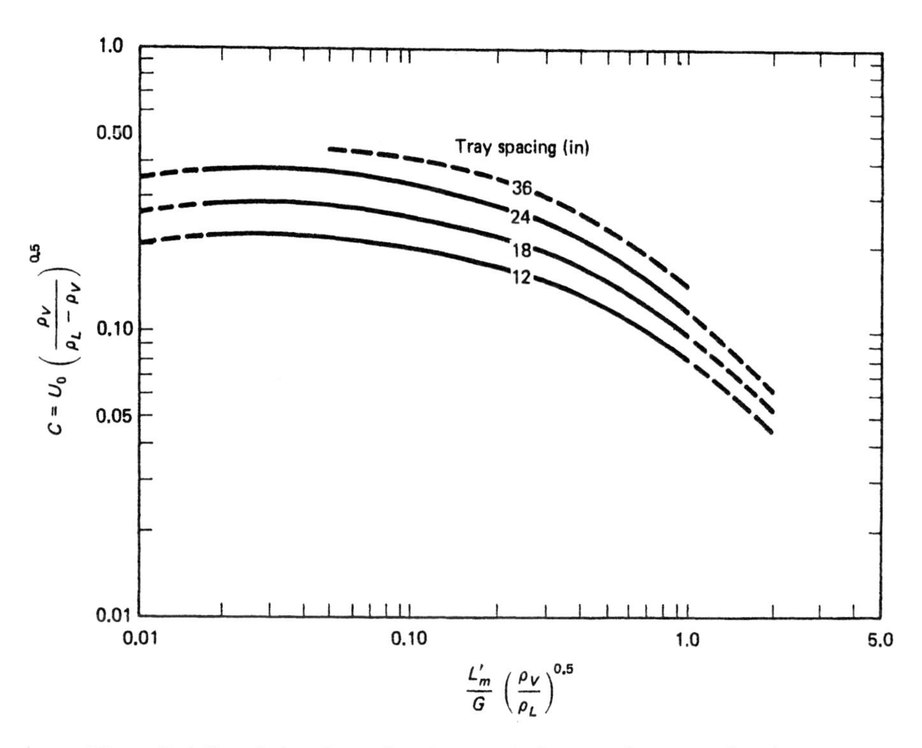

Figure 13-4 Correlation for estimating required vapor flow area for sieve plates (from Fair, 1961)

undergone preliminary sizing, with such things as the length of the weir and the length of the liquid flow path having been previously determined. The procedure suggested by Bolles is relatively simple and utilizes a correction for the vapor flooding velocity as determined from Figure 13-4 for sieve trays. The equation given by Bolles is

$$U_{Vnf} = CF_{sa}F_{st}\sqrt{\frac{\rho_L - \rho_V}{\rho_V}} \tag{13-11}$$

where U_{Vnf} = velocity of the vapor based on the net area at the flood point, ft/s,

C = value read from Figure 13-4,

F_{sa} = correction factor for slot area,

F_{st} = correction factor for surface tension.

The correction factors used in Eq. (13-11) are

$$F_{sa} = 5\left(\frac{A_{so}}{A_a}\right) + 0.5 \quad \text{for } 0.8 \le F_{sa} \le 1.0 \tag{13-12}$$

and

$$F_{st} = \left(\frac{\sigma}{20}\right)^{0.2} \quad \text{for } F_{st} \le 1.0 \tag{13-13}$$

where A_{so} = slot area with valves open, ft^2,
A_a = active area (column less two downcomers), ft^2,
σ = surface tension, dyn/cm.

Example 13.1

The liquid on a distillation tray consists of a binary mixture of 60% component A and 40% component B. The liquid mixture density is 0.75 g/cm³, the viscosity is 0.25 cP, and the interfacial tension is 20 dyn/cm. The diffusivity of the light key component is 2.3×10^{-4} ft²/h and the vapor density is 0.16 lb/ft³. The vapor velocity is 24,000 ft/h, the weir height is 2.5 in, and the fraction of free area on the tray is 0.1. Obtain an estimate of the tray efficiency.

Solution:

$$D_g = \frac{\sigma_L}{\mu_L U_V}$$

$$= \frac{20 \times 2.86 \times 10^4}{(0.25 \times 2.42)24{,}000}$$

$$= 39.4$$

$$\text{Sc} = \frac{\mu_L}{\rho_L D_{LK}}$$

$$= \frac{0.25 \times 2.42}{(0.75 \times 62.4)(2.3 \times 10^{-4})}$$

$$= 56.2$$

$$\text{Re} = \frac{h_W U_V \rho_V}{(\mu_L)(\text{FA})}$$

$$= \frac{(2.5/12)(24{,}000 \times 0.16)}{(0.25 \times 2.42)(0.1)}$$

$$= 1.3 \times 10^4$$

$$E_{MV} = 7(39.4)^{0.14}(56.2)^{0.25}(1.3 \times 10^4)^{0.08}$$

$$= 68.4\%$$

Example 13.2

In Example 13.1, the relative volatility of component A with respect to B is 1.6. Calculate the estimated overall tray efficiency.

Solution:

$$\alpha\mu = 1.6 \times 0.25 = 0.4$$

From Figure 13-2, the overall tray efficiency for a fractionator is 64%.

Example 13.3

The liquid in a distillation column flows at a rate of 35,011 lb/h and has a density of 38 lb/ft³. The vapor rate of flow is 25,007 lb/h and its density is 3.5 lb/ft³. Determine the required vapor flow area in the column if bubble-cap plates with 24-in spacings are used.

Solution:

$$\frac{L}{G}\left(\frac{\rho_V}{\rho_L}\right)^{0.5} = \frac{35{,}011}{25{,}007}\left(\frac{3.5}{38}\right)^{0.5} = 0.425$$

From Figure 13-3, $C = 0.204$ for a 24-in tray spacing. Thus

$$C = 0.204 = U_0\left(\frac{3.5}{38.0 - 3.5}\right)^{0.5}$$

$$U_0 = 0.641 \text{ ft/s}$$

Using an 85% approach to flooding, the allowable vapor velocity is

$$U_V = 0.641 \times 0.85 = 0.545 \text{ ft/s}$$

The required cross-sectional area is

$$A = \frac{25{,}007}{3.5 \times 3{,}600 \times 0.545} = 3.64 \text{ ft}^2$$

Since the area above is the required value for vapor flow, the area of one downcomer would have to be added to the above to obtain the total tower cross-sectional area.

REFERENCES

BOLLES, W. L., *Chem. Eng. Prog.*, *72*, No. 9, 43 (1976).

CAREY, S., thesis, Massachusetts Institute of Technology, 1930.

EDMISTER, W. C., *Petrol. Eng.*, *C45* (Jan. 1949).

ENGLISH, G. E., and M. VAN WINKLE, *Chem. Eng.*, *70*, No. 11, 241 (1963).

FAIR, J. R., *Petro/Chem. Eng.*, *33*, 45 (Sept. 1961).

FAIR, J. R., and R. L. MATTHEWS, *Petrol. Refiner*, *37*, No. 4, 153 (1958).

MACFARLAND, S. A., P. M. SIGMUND, and M. VAN WINKLE, *Hydrocarbon Process.*, *51*, No. 7, 111 (1972).

MURPHREE, E. V., *Ind. Eng. Chem.*, *17*, 747 (1925).

NORD, M., *Ind. Eng. Chem.*, *38*, 657 (1946).

O'CONNELL, H. E., *Trans. AIChE*, *42*, 741 (1946).

PERRY, R. H., C. H. CHILTON, and S. D. KIRKPATRICK, Eds., *Chemical Engineers' Handbook*, 4th ed., McGraw-Hill, New York, 1963.

STANDART, G., *Chem. Eng. Sci.*, *20*, 611 (1965).

THORNGREN, J. T., *Hydrocarbon Process.*, *57*, No. 8, 111 (1978).

WILKE, C. R., and P. CHANG, *AIChE J.*, *1*, 264 (1955).

NOTATIONS

A = cross-sectional area of the column, L^2
A_a = active area (column less two downcomers), L^2
A_{so} = slot area with valves open, L^2
C = constant from Figures 13-3 and 13-4
D_g = surface tension number defined by Eq. (13-9)
D_{LK} = diffusivity of light key component, L^2/t
E_{ML} = Murphree liquid-phase efficiency
E_{MV} = Murphree vapor-phase efficiency
E_o = overall tray efficiency of the column
E_{tL} = liquid thermal efficiency of the tray
E_{TV} = vapor thermal efficiency of the tray
F_{sa} = correction factor for slot area
F_{st} = correction factor for surface tension
FA = fractional free area available for vapor flow
G = vapor mass flow rate, M/t
h_n = enthalpy of the liquid stream leaving tray n, E/mol
h_n^* = enthalpy of the liquid stream leaving the pseudoequilibrium tray n, E/mol
h_W = height of weir, L
H_n = enthalpy of the vapor leaving tray n, E/mol
H_n^* = enthalpy of the vapor stream leaving the pseudoequilibrium stage n, E/mol
K = equilibrium constant for light key component
L_n = flow rate of liquid leaving tray n, mol/t
L'_m = liquid mass flow rate, M/t
L_n^* = flow rate of liquid leaving the pseudoequilibrium stage n, mol/t
M = molecular weight of the solvent, M/mol
n = tray number
Q_n = heat gain or loss by tray n, E/t
Re = Reynolds number defined by Eq. (13-9)
Sc = Schmidt number defined by Eq. (13-9)
t_n = temperature of the liquid stream leaving tray n
t_n^* = liquid temperature in equilibrium with that of the vapor leaving the actual tray n
T_n = temperature of the vapor stream leaving tray n
T_n^* = vapor temperature in equilibrium with that of the liquid leaving the actual tray n
U_0 = flooding velocity, L/t
U_V = superficial vapor velocity, L/t
U_{Vnf} = velocity of vapor based on net area at flood point, L/t
V_n = flow rate of vapor leaving tray n, mol/t
V_n^* = flow rate of vapor leaving the pseudoequilibrium stage n, mol/t
x_n = mole fraction of any component in the liquid leaving tray n
x_n^* = liquid-phase mole fraction that would be in equilibrium with the actual vapor leaving tray n
y_n = mole fraction of any component in the vapor leaving tray n
y_n^* = vapor-phase mole fraction that would be in equilibrium with the actual liquid leaving tray n

Greek Letters

α = relative volatility
μ_L = liquid viscosity, M/Lt
μ_V = vapor viscosity, M/Lt
ρ_L = liquid density, M/L^3
ρ_V = vapor density, M/L^3
σ_L = surface tension, F/L

PROBLEMS

13.1 An oil absorber utilizes a lean oil with the properties of octane to absorb propane from a natural gas stream. Estimate the overall tray efficiency if the absorber operates at 1000 psia and an average temperature of 0°F.

13.2 Calculate the required vapor flow area in Example 13.3 if valve trays are used in place of the bubble-cap trays. Assume that the slot area is 10% of the active column area.

13.3 An absorption column, which contains sieve plates with 24-in spacings, is to be used to remove ammonia from air at 20°C and 1 atm by contacting the air with water. Air enters the column at a rate of 25,000 lb/h and pure water enters at a flow rate of 40,000 lb/h. The entering gas contains 5 mol% ammonia. Determine the required gas flow area if the gas velocity is to be 70% of flooding.

13.4 A distillation column equipped with bubble-cap trays on 24-in spacings operates with a condenser pressure of 250 psia. The column has a total condenser, a partial reboiler, and 50 theoretical plates. The external reflux pump-back is 1.194 mol/mol feed, with the feed at the bubble point. At this reflux rate, the vapor boil-up in the reboiler is 1.81 mol/mol feed. The compositions of the feed, the distillate, and the bottom products are as follows:

Component	*Feed (mol%)*	*Distillate (mol%)*	*Bottoms (mol%)*
Propane	7.84	15.86	0.0
i-Butane	18.81	37.88	0.17
n-Butane	22.57	43.91	1.72
i-Pentane	27.59	2.15	52.45
n-Pentane	13.79	0.21	27.07
n-Hexane	9.40	0.0	18.59

(a) If the total feed rate to the column is 20,000 gal (60°F) per hour and the column operates at 85% of flooding, estimate the required vapor flow area for (i) the top tray in the column, and (ii) the bottom tray. **(b)** Estimate the overall tray efficiency for the column.

13.5 Estimate the Murphree vapor efficiency for the top and bottom trays for Problem 13.4. Assume that the fraction of free area on each tray is 0.12 and the weir height is 2.5 in.

Adsorption 14

14.1 Introduction

Adsorption involves the transfer of a material from one phase to a surface, where it is bound by intermolecular forces. Although adsorption is usually associated with the transfer from a gas or liquid to a solid surface, transfer from a gas to a liquid surface also occurs. The substance being concentrated on the surface is defined as the *adsorbate* and the material on which the adsorbate accumulates is defined as the *adsorbent.*

Adsorption is used in gas purification processes, such as in the removal of sulfur dioxide from a stack gas, and as a means of fractionating fluids which are difficult to separate by other separation methods. The oil and chemical industries make extensive use of adsorption in the cleanup and purification of wastewater streams, and for the dehydration of gases. The amount of adsorbate that can be collected on a unit of surface area is small. Thus a porous adsorbent with a large internal surface area is typically selected for industrial applications. Examples of adsorbents that meet this criteria are activated carbon, silica gel, molecular sieve, and activated alumina.

The design of adsorption equipment requires the selection of an adsorbent and information regarding the transfer of mass to the adsorbent surface. As part of the selection of an adsorbent, information describing the equilibrium capacity of the adsorbent is necessary. These data, which are obtained at constant temperature, are called *adsorption isotherms.* A description of the rate of mass transfer to the adsorbent can be obtained from breakthrough curves. These are obtained by passing a fluid that contains the adsorbate through a packed column filled with the selected adsorbent and monitoring the exit concentration. A discussion of equilibrium adsorption and rates of mass transfer to the adsorbent is presented in this chapter.

14.2 Equilibrium Considerations

The selection of an adsorbent includes a consideration of surface area as well as the type of solute and solvent involved in the adsorption process, since these relate to the types of bonds that are formed between the solid and fluid. Depending on the types of bonds that are formed between the adsorbate and the surface, adsorption is described as either physical adsorption or chemical adsorption. Physical adsorption results when the adsorbate adheres to the surface by van der Waals' forces (i.e., by dispersion and Coulombic forces). Although a displacement of electrons may exist, electrons are not shared between the adsorbent and adsorbate. During the adsorption process a quantity of heat, described as the *heat of adsorption*, is released. Since the quantity of heat released for physical adsorption is approximately equal to the heat of condensation, physical adsorption is often described as a condensation process. As expected, the quantity of material physically adsorbed increases as the adsorption temperature decreases. The nature of forces for physical adsorption is such that multiple layers of adsorbate will accumulate on the surface of the adsorbent.

The primary difference between physical adsorption and chemisorption is the nature of the bond that is formed between the adsorbed molecule and the adsorbent surface. Chemisorption is characterized by a sharing of electrons between the adsorbent and adsorbate which results in the liberation of a quantity of heat that is approximately equal to the heat of reaction. Because of the sharing of electrons with the surface, chemisorbed materials are restricted to the formation of a monolayer. Although chemical and physical adsorption are characterized by different thermal effects, a clear line between the two adsorption mechanisms does not exist. A large displacement of the electron cloud toward the adsorbent and the sharing of electrons often result in about the same heat effects. In some cases it has been observed that the quantity of adsorbate chemisorbed on a surface increases with increased temperature. Considering the nature of the bond, this effect might be expected.

Brunauer (1945) classified equilibrium isotherms for the adsorption of vapors into five principal forms, as shown in Figure 14-1. Type I is classified as the *Langmuir* type and is characterized by a monotonic approach to a limiting adsorption capacity that corresponds to the formation of a complete monolayer. This type is found for systems in which the adsorbate is chemisorbed. Type I isotherms have been observed for microporous adsorbents such as charcoal, silica gel, and molecular sieves in which the capillaries have a width of only a few molecular diameters. A type II isotherm is characteristic of the formation of multiple layers of adsorbate molecules on the solid surface. This type, which is

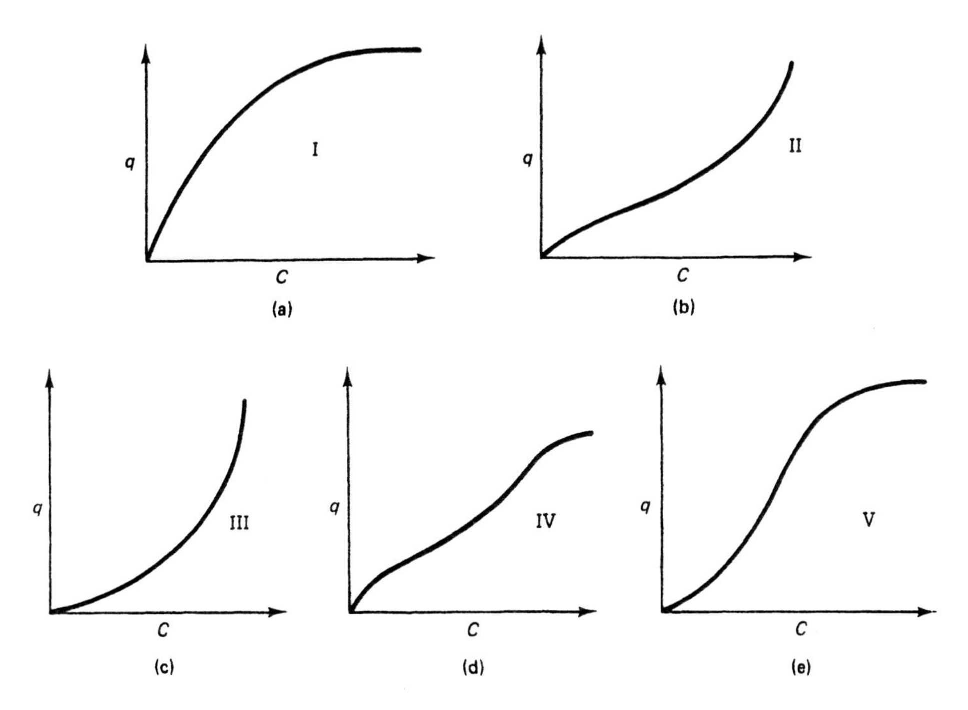

Figure 14-1 Brunauer's classification of adsorption isotherms, showing amount adsorbed versus final concentration in the fluid

known as the *BET* after Brunauer, Emmett, and Teller (1938), has been found to exist for nonporous solids. Type III isotherms, although similar to type II since they have been observed for nonporous solids, are relatively rare. The shape of type III isotherms also suggests the formation of multilayers. Types IV and V are considered to reflect capillary condensation since they level off as the saturation pressure of the adsorbate vapor is reached. Both types of isotherms exhibit a hysteresis loop when conducting desorption studies; porous adsorbents provide isotherms of this shape.

14.3 Adsorption Isotherm Models

Adsorption equilibrium studies, which correspond to no net mass transfer between phases, are used to determine the distribution of an adsorbate between the bulk fluid phase and the phase adsorbed on the surface of a solid adsorbent. The equilibrium distribution is generally measured at constant temperature and is referred to as an *equilibrium isotherm.* A number of mathematical models have been proposed to describe the adsorption process. In addition to monolayer and multilayer adsorption, models have been developed to describe situations in which the adsorbate occurs either locally on specific sites or is mobile over the surface of the adsorbent. Consideration has also been given to cases in which adsorbed molecules interact not only with the surface but also with each other. Some of the more widely used isotherm equations will be discussed in this section. For a comprehensive discussion of the various isotherms and the mechanisms of adsorption, the monographs of Brunauer (1945), Young and Crowell (1962), and Ross and Olivier (1964) are recommended.

Langmuir isotherm

Langmuir (1918) proposed a model that quantitatively described the volume of gas adsorbed onto an open surface such as mica. His model, which may be classified as applying to localized adsorption of monolayer coverage, includes the following assumptions:

1. All the sites of the solid have the same activity for adsorption.
2. There is no interaction between adsorbed molecules.
3. All of the adsorption occurs by the same mechanism, and each adsorbent complex has the same structure.
4. The extent of adsorption is no more than one monomolecular layer on the surface.

The derivation of his equation can be carried out by writing separately the rates of adsorption and desorption of the adsorbate on the surface. The surface is assumed to consist of the fraction covered by the adsorbed molecules, Θ, and the fraction of surface that is bare, $(1 - \Theta)$. Since adsorption is a rate process, the rate at which the surface is covered may be written as

$$r_a = k_a P(1 - \Theta)a_1 \tag{14-1}$$

where a_1 is a condensation factor which is set equal to 1, P is the pressure, and k_a is the rate constant for adsorption. The rate constant is a function of temperature and is related to the rate at which molecules condense on the surface by the expression

$$k_a = \frac{N_0}{(2\pi MRT)^{1/2}} \tag{14-2}$$

where M is the molecular weight, T is the absolute temperature, R is the gas constant, and N_0 is Avogadro's number.

The desorption rate is a function of the surface coverage and can be expressed as

$$r_d = k_d\Theta \tag{14-3}$$

The rate constant for desorption, k_d, is given in terms of a preexponential factor, k_1, and the energy of adsorption, E, by the equation

$$k_d = k_1 \exp\left(-\frac{E_1}{RT}\right) \tag{14-4}$$

The net rate of surface coverage is

$$\frac{d\Theta}{dt} = r_a - r_d = k_a P(1 - \Theta) - k_d \Theta \tag{14-5}$$

where we note that the fraction of surface coverage is a function of time. After integrating the equation above, we obtain

$$\Theta = \frac{(k_a/k_d)P}{1 + (k_a/k_d)P}\{1 - \exp[-(k_a + k_d)t]\} \tag{14-6}$$

When equilibrium exists (i.e., as time becomes very large), we obtain the Langmuir equation in terms of the fraction of surface coverage. Thus we have

$$\Theta = \frac{KP}{1 + KP} \tag{14-7}$$

where $K = k_a/k_d$ is by definition the adsorption equilibrium constant.

For equally energetic adsorption sites the energy of adsorption and K are independent of surface coverage. The fraction of surface covered can be replaced by V/V_m, where V_m denotes the volume of gas adsorbed in a monolayer per gram of solid. Since the assumption of monolayer coverage usually corresponds to chemisorption, V_m is the volume of adsorbate that occupies the active sites on the surface. We can write Eq. (14-7) in terms of volume as

$$V = \frac{V_m KP}{1 + KP} \tag{14-8}$$

The Langmuir equation can be reduced to a linear form at low pressure, since the term KP in the denominator will be much less than unity. Thus

$$V = V_m KP \tag{14-9}$$

At high pressure, however, V approaches the limiting value V_m.

In order to determine if equilibrium isotherm data can be modeled by the Langmuir equation, we write Eq. (14-8) in linear form as

$$\frac{P}{V} = \frac{1}{KV_m} + \frac{P}{V_m} \tag{14-10}$$

If the equation provides a valid fit to the data, a plot of P/V versus P should give a straight line and the two constants V_m and K may be evaluated from the slope and intercept, respectively. It is important to note that a good fit of the data by the model is a necessary but not sufficient condition to verify the accuracy of the assumptions imposed on the equation. Langmuir's equation might be expected to fit data that correspond to the shape of a type I isotherm. The Langmuir equation is frequently expressed in terms of the weight of adsorbate on the surface and concentrations other than pressure. Another form is

$$q_A = \frac{QK'C_A}{1 + K'C_A} \tag{14-11}$$

where q_A = equilibrium uptake by the adsorbent, g solute/g solid,
Q = weight of adsorbate for complete monolayer coverage, g solute/g solid,
C_A = concentration of solute in the fluid phase in equilibrium with the adsorbate concentration on the surface, mol/cm³,
K' = constant, cm³/mol.

Example 14.1

The adsorption of ethane on Linde molecular sieve, type 5A, was studied by Glessner and Myers (1969) at 35°C. (a) Using the data presented below, determine if the Langmuir equation can be used to model the data. (b) Calculate the total surface area of the solid.

P (mmHg)	*Uptake [cm³(STP)/g]*
0.17	0.059
0.95	0.318
5.57	1.638
12.09	3.613
111.32	24.236
220.87	34.278
300.05	38.340
401.25	41.779
500.18	44.037
602.74	45.693

Solution:

If the Langmuir equation can be used, a plot of P versus P/V should give a straight line.

P (mmHg)	*P/V (g·mmHg/cm³)*
0.17	2.88
0.95	2.99
5.57	3.40
12.09	3.35
111.32	4.59
220.87	6.44
300.05	7.83
401.25	9.60
500.18	11.36
602.74	13.19

From Figure 14-2, the intercept is

$$I = \frac{1}{KV_m} = 2.6$$

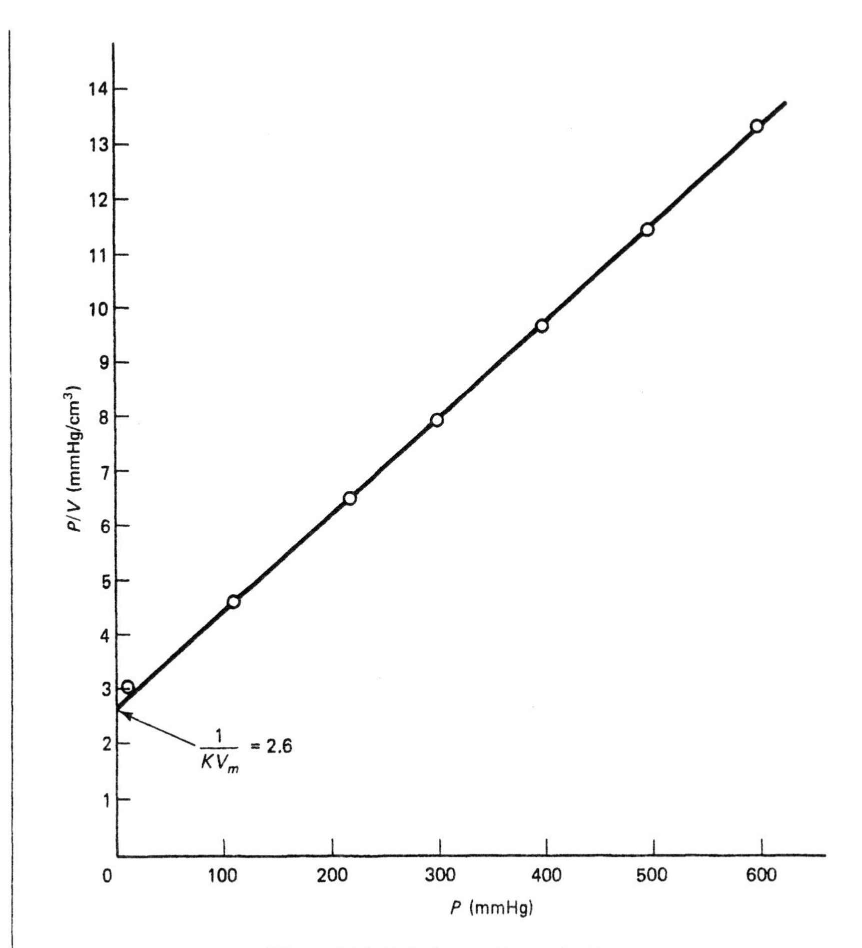

Figure 14-2 Solution to Example 14.1

and the slope is

$$\frac{1}{V_m} = 0.0175$$

Thus $V_m = 57.14 \text{ cm}^3$ gas/g solid

$$\frac{1}{KV_m} = 2.6$$

$$K = \frac{1}{2.6V_m} = 6.73 \times 10^{-3}(\text{mmHg})^{-1}$$

The projected area of a molecule on the surface can be calculated from the expression

$$\alpha = 1.09\left(\frac{M}{N_0\rho}\right)^{2/3}$$

where M is the molecular weight, ρ is the density of the adsorbed molecules, and N_0 is Avogadro's number. $\rho_{ethane} = 0.3549\ g/cm^3$ (liquefied C_2H_6), $M = 30.05$,

$$\alpha = 1.09\left[\frac{30.05}{(6.02 \times 10^{23})(0.3549)}\right]^{2/3} = 2.95 \times 10^{-15}$$

$$S_g = \frac{V_m N_0}{V}\alpha$$

$$= \frac{(57.14)(6.02 \times 10^{23})}{22{,}400}(2.95 \times 10^{-15})$$

$$= 4.53 \times 10^6\ cm^2/g\ solid$$

Freundlich isotherm

Another isotherm model that we will consider is a semiempirical equation attributed to Freundlich. This isotherm, frequently described as the *classical equation*, is widely used, particularly in the low to intermediate concentration range. It is expressed as

$$q = K(C)^{1/n} \tag{14-12}$$

where q is the uptake of adsorbate per unit weight of adsorbent, C is the equilibrium adsorbate concentration corresponding to q, K is a constant for the adsorbate–adsorbent system, and n is another constant that is restricted to values greater than unity. Although a fit to the Freundlich equation cannot be related to the adsorption mechanism, we can model equilibrium adsorption data by writing Eq. (14-12) in logarithmic form as

$$\log q = \log K + \frac{1}{n}\log C \tag{14-13}$$

Since Eq. (14-13) is linear, a plot of log q versus log C will yield a straight line with a slope of $1/n$ and an intercept equal to log K, provided that the data obey the equation.

Models based on equations of state

The adsorption of a liquid or vapor on a surface exerts an expanding pressure which opposes the surface tension of the clean surface. We can express this as

$$\pi = \gamma_0 - \gamma \tag{14-14}$$

where γ_0 is the surface tension of the clean surface, γ is the surface tension of the partially filled surface, and π is defined as the surface or spreading pressure. The amount of adsorbate that is contained in the surface phase can be related to surface tension by the Gibbs (1931) adsorption equation

$$\Gamma = -\frac{a}{RT}\frac{d\gamma}{da} \tag{14-15}$$

where a is the activity of the solute on the surface and Γ is defined as the surface excess concentration. It is simply written for our purposes as the amount of solute adsorbed per unit area,

$$\Gamma = \frac{q}{S} \tag{14-16}$$

where S is the specific surface area of the adsorbent (m^2/g) and q is defined as the moles adsorbed on the surface per gram of adsorbent. For dilute solutions we can write Eq. (14-15) in terms of concentration as

$$\Gamma = -\frac{C}{RT}\frac{d\gamma}{dC} = \frac{C}{RT}\frac{d\pi}{dC} = \frac{1}{RT}\frac{d\pi}{d\ln C} \tag{14-17}$$

We can express the Gibbs adsorption equation in terms of the spreading pressure for the vapor phase as

$$d\pi = RT\Gamma d\ln P = N_0 k_B T \Gamma d\ln P \tag{14-18}$$

where N_0 is Avogadro's number and k_B is the Boltzmann constant.

For limiting conditions it has been suggested that the adsorbed phase behaves on the surface as a compressed gas. For monolayer coverage or less, where the motion of the adsorbed layer is limited to two dimensions, equations of state, such as the ideal gas equation and van der Waals' equation, have been modified to describe the adsorbed phase. The two-dimensional analogue to the ideal gas equation, which may be called a *surface equation of state*, is

$$\pi A = RT \tag{14-19}$$

where π = spreading pressure, dyn/cm,
A = area per mole,
T = absolute temperature, K,
R = gas constant.

Considering the basic assumption embodied in the ideal gas equation, of no forces of interaction between molecules, we might assume that the equation above would apply for low surface coverage; this is indeed the case. By combining the ideal gas equation with the Gibbs isotherm, we can derive an isotherm equation as follows.

By differentiating the ideal gas surface equation at constant temperature, we obtain

$$A d\pi = -\pi dA \tag{14-20}$$

Upon substituting for π from Eq. (14-19), we get

$$A d\pi = -\frac{RT}{A} dA = -RT\, d\ln A \tag{14-21}$$

Introducing the Gibbs equation for the spreading pressure thus gives us

$$d\ln P = -d\ln A \tag{14-22}$$

Integration of Eq. (14-22) gives

$$\ln P = -\ln A + \ln K$$

or

$$P = \frac{K}{A} \tag{14-23}$$

where K is a constant of integration. Equation (14-23) relates the quantity of gas adsorbed per unit area to the pressure. We can relate the fraction of surface covered to pressure by dividing each side of the equation by the number of moles adsorbed when the surface is completely covered, n_m. Thus we have

$$P = \frac{Kn}{S_t} = \frac{Kn_m}{S_t}\Theta \tag{14-24}$$

or

$$\Theta = K'P \tag{14-25}$$

where Θ is the fraction of the surface covered by the adsorbate at pressure P and S_t is the total surface area. The equation above is often called the *linear* or *Henry's isotherm equation* and is valid only for very low surface coverage. For adsorption at low gas pressures it may be readily shown that the Langmuir equation reduces to the linear equation given above.

At higher gas pressures, experimental data suggest that the adsorbed phase behaves as a nonideal gas. In an attempt to account partially for this effect, Volmer (1925) modified the ideal gas equation by including a co-area correction factor as

$$\pi(A - A_0) = RT \tag{14-26}$$

where A_0 is the co-area correction factor and is associated with the excluded area per mole of adsorbate. The equilibrium isotherm equation derived from the equation of state above is

$$PK' = \frac{\Theta}{1-\Theta}\exp\left(\frac{\Theta}{1-\Theta}\right) \tag{14-27}$$

where K' is a constant. Kemball and Rideal (1946) have shown that the isotherm given by Eq. (14-27) could be used to describe the adsorption of various organic vapors on mercury.

In addition to including the effect of adding a co-area term as previously shown, the influence of attractive forces between adsorbed molecules can be described by employing a two-dimensional analogy of van der Waals' equation of state. Thus we have

$$\left(\pi + \frac{\alpha}{A^2}\right)(A - A_0) = RT \tag{14-28}$$

where α is an attraction constant. By combining Eq. (14-28) with Gibbs adsorption equation we can write the Hill–de Boer equilibrium isotherm as

$$PK' = \frac{\Theta}{1-\Theta}\exp\left(\frac{\Theta}{1-\Theta} - \frac{2\alpha\Theta}{RT\sigma_0}\right) \tag{14-29}$$

where $\sigma_0 = A_0/N_0$.

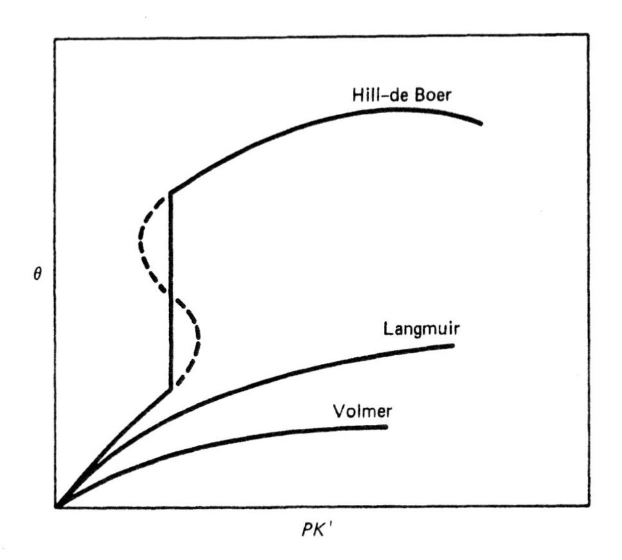

Figure 14-3 Comparison of adsorption models

A comparison of these models is given in Figure 14-3. Although the Hill–de Boer equation predicts the existence of an S-shaped isotherm at low pressures, the actual form of the isotherm is indicated by the solid line. This discontinuity corresponds to the transition of the adsorbed layer from a gas to a liquid. According to the Hill–de Boer model, this occurs at a surface coverage of $\Theta = \frac{1}{3}$. Volmer's isotherm, which takes into account the actual size of the molecule at maximum compression, resembles very closely Langmuir's equation.

Values of α and σ_0 have been related to the energy constant a and the volume term b in the three-dimensional van der Waals' equation by Hill (1946).

$$\alpha = a\left(\frac{9\pi}{256b}\right)^{1/3} \tag{14-30}$$

and

$$\sigma_0 = 2b\left(\frac{9\pi}{256b}\right)^{1/3} \tag{14-31}$$

The relationships above give relatively good agreement with experimental values when the adsorbate molecules are neither polarized nor oriented by the adsorbent. Parameters of the two- and three-dimensional van der Waals equations are given in Table 14-1 for several gases.

Multilayer adsorption models

In early studies experimental evidence suggested that multilayer adsorption occurred for many adsorbate–adsorbent systems. Multilayer formation is characteristic of physical adsorption but is more difficult to describe theoretically than monomolecular formation. Only a few theoretical models have contributed

TABLE 14-1. PARAMETERS FOR THE VAN DER WAALS EQUATION

	a (*ergs cm*^{3} *mol*$^{-2}$ × 10^{-12})	*b* (*cm*^{3} *mol*$^{-1}$)	α^{id} (*ergs cm*^{2} *molecule*$^{-2}$ × 10^{30})	σ_0^{id} (*cm*^{2} *molecule*$^{-1}$ × 10^{16})
Helium	0.034	23.7	1.32	11.1
Hydrogen	0.245	26.8	9.15	12.1
Argon	1.35	32.2	47.4	13.6
Krypton	2.33	39.8	76.0	15.7
Nitrogen	1.39	39.2	45.7	15.5
Oxygen	1.36	31.8	48.0	13.5
Chlorine	6.53	56.2	191	19.7
Carbon monoxide	1.49	39.8	48.6	15.7
Carbon dioxide	3.61	42.7	115	16.4
Sulfur dioxide	6.75	56.4	196	19.8
Hydrogen sulfide	4.45	42.9	142	16.4
Ammonia	4.2	37.0	141	15.0
Methane	2.26	42.8	75.0	16.4
Ethane	5.49	63.8	153	21.7
Ethylene	4.48	57.1	132	20.0
Propane	8.70	84.4	222	25.8
Cyclohexane	22.8	142.4	488	36.6
Benzene	18.1	115.4	427	32.7
Chloroform	15.2	102.2	368	29.4
Carbon tetrachloride	20.4	138.3	441	35.9
Chlorobenzene	25.5	145.3	542	37.1
Diethyl ether	17.5	134.4	380	34.9
Acetone	14.0	99.4	337	28.9
Water	5.46	30.5	196	13.1

significantly to a better understanding of the formation of multilayers. Because a discussion of all of these models is beyond the scope of this text, only the potential theory as proposed by Polanyi (1920) and the model developed by Brunauer et al. (1938) will be discussed. Although these models are conceptually different, they are of practical importance in determining the pore volume and surface area of an adsorbent.

Potential Theory: The potential theory is somewhat different from models discussed previously because of its thermodynamic approach. This theory was initially applied to multilayer gas adsorption and assumes that a potential field exists at the surface of a solid which exerts a strong attractive force on the surrounding gas. A cross section of the adsorbed phase and equipotential surfaces, as proposed by the potential theory, is shown in Figure 14-4. The potential on the surface of the adsorbent is ϵ_0, with potential surfaces farther away from the surface denoted by $\epsilon_1, \epsilon_2, \ldots, \epsilon_\infty$. The adsorption potential is defined as the work done by the adsorption forces in bringing a molecule from the gas phase to some equipotential surface in the adsorbed phase. Since the

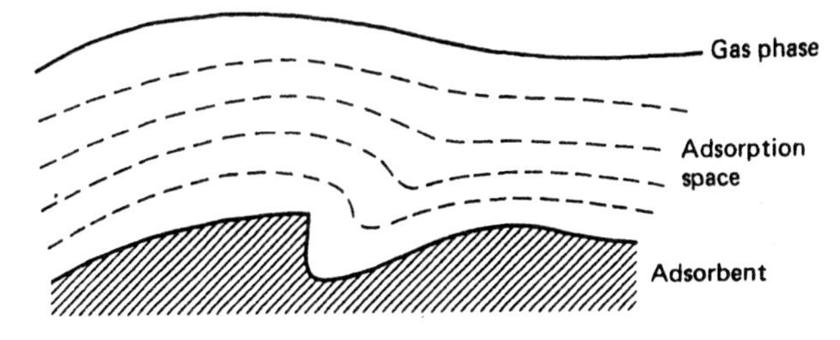

Figure 14-4 Adsorption potential model

molecules in the adsorbed phase are compressed by successive layers of molecules, the adsorption potential is equated to the difference in free energy between the adsorbed species and the saturated liquid. This is given as a function of position in the potential field as

$$\epsilon(x) = \int_{P_x}^{P^\circ} V\,dP \tag{14-32}$$

where P° is the vapor pressure of the liquid and P_x is the pressure of the gas phase. By assuming that the gas phase is ideal and the fluid in the adsorbed phase is incompressible, the adsorption potential becomes the isothermal work of compression and is written as

$$\epsilon(x) = \int_{P_x}^{P^\circ} RT\frac{dP}{P} = RT\ln\frac{P^\circ}{P_x} \tag{14-33}$$

The space between each equipotential surface and the solid surface corresponds to the volume adsorbed $\phi(x)$. The maximum potential occurs at the adsorbent surface with a corresponding volume adsorbed of $\phi(x) = 0$. At the outermost potential surface, the potential is zero and the volume adsorbed is a maximum. The volume adsorbed at any point is

$$\phi(x) = \frac{q}{\rho} \tag{14-34}$$

where q is the weight of the adsorbate at the system pressure and ρ is the density of the fluid adsorbate at the adsorption temperature. Following the assumption of Polanyi that the adsorption potential is independent of temperature, pairs of q and P_x obtained from isotherm data can be used to construct a characteristic curve that relates $\epsilon(x)$ to either $\phi(x)$ or q for a given solid. Provided the adsorbate–adsorbent system gives a single characteristic curve over a sufficiently wide pressure range, we can use the characteristic curve to develop isotherms at other temperatures. A characteristic curve showing the adsorption of n-butane on silica gel is given in Figure 14-5.

Dubinin and co-workers (Dubinin, 1960, 1967; Bering et al., 1966, 1972) extended the potential theory to include the adsorption of different gases on the same solid. They postulated that for similar compounds the adsorption potentials would have a constant ratio for equal adsorption volumes. Thus

$$\frac{\epsilon_1}{\epsilon_2} = \gamma \tag{14-35}$$

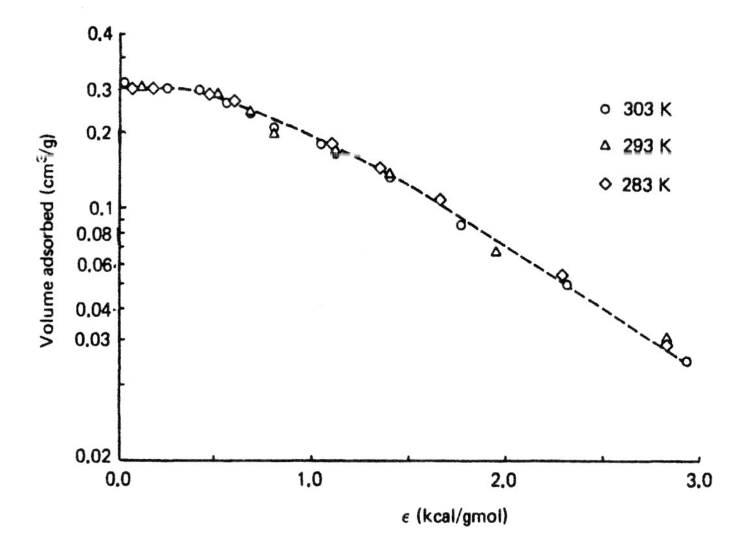

Figure 14-5 Characteristic curve of *n*-butane (from Al-Sahhaf et al., 1981)

where γ is defined as the affinity coefficient, and ϵ_1 and ϵ_2 are the adsorption potentials of similar vapors on the same adsorbent at equal adsorbed volumes. Since the affinity coefficient can be estimated as the ratio of the molar volumes of the adsorbed phase, we obtain

$$\frac{\epsilon_1}{\epsilon_2} = \frac{[RT\ln(P^\circ/P_x)]_1}{[RT\ln(P^\circ/P_x)]_2} = \frac{\bar{V}_1}{\bar{V}_2} \tag{14-36}$$

By rearranging Eq. (14-36) we get

$$\left(\frac{RT}{\bar{V}}\ln\frac{P^\circ}{P_x}\right)_1 = \left(\frac{RT}{\bar{V}}\ln\frac{P^\circ}{P_x}\right)_2 \tag{14-37}$$

Affinity coefficients for several adsorbates are compared to the ratio of the molar volumes in Table 14-2. As seen, γ is very close to the ratio of the liquid molar

TABLE 14-2. AFFINITY COEFFICIENTS RELATIVE TO BENZENE, $\gamma = 1$ (YOUNG AND CROWELL, 1962)

Vapor	γ *(Expt.)*	$\bar{V}/\bar{V}$ *(Benzene)*
C_6H_{12}	1.04	1.21
C_7H_{16}	1.50	1.65
CH_3Cl	0.56	0.59
CH_2Cl_2	0.66	0.71
$CHCl_3$	0.88	0.90
CCl_4	1.07	1.09
C_2H_5Cl	0.78	0.80
CH_3OH	0.40	0.46
C_2H_5OH	0.61	0.65
HCOOH	0.60	0.63

volumes. Another approximate relationship for γ is given in terms of the attraction constant α of van der Waals' equation.

$$\gamma = \frac{\alpha_1}{\alpha_2} \tag{14-38}$$

At temperatures above the critical and for pressures at which the ideal gas equation does not adequately describe the vapor phase, the Polanyi equation was modified by Lewis et al. (1950) to give

$$\left(\frac{RT}{\bar{V}} \ln \frac{f^\circ}{f}\right)_1 = \left(\frac{RT}{\bar{V}} \ln \frac{f^\circ}{f}\right)_2 \tag{14-39}$$

where f° is the fugacity of the vapor at its vapor pressure and f is the fugacity of the vapor at the adsorption pressure. Both are evaluated at the system temperature T. In the equation above, the molar volume of the liquid is evaluated at a temperature such that the vapor pressure of the pure liquid is equal to the adsorption pressure. Grant et al. (1962) studied the adsorption of normal alkanes on activated carbon above the critical temperature and demonstrated that by extrapolating subcritical vapor pressure data to the supercritical region a good fit to Eq. (14-39) could be obtained.

BET Isotherm: Because experimental evidence suggested that multilayer adsorption occurred for many adsorbate–adsorbent systems, Brunauer et al. (1938) extended Langmuir's approach to account for this phenomenon. Their model, which has come to be known as the *BET equation*, includes the basic assumptions of the Langmuir equation with the exceptions that multilayer adsorption will occur and the heat of adsorption for the first layer will be different from the value for all succeeding layers. The heat of adsorption for these layers is equal to the latent heat of condensation of the liquid adsorbate. Although this model does not permit lateral interaction between the molecules on the surface, vertical interaction between the adsorbed layer and the gas phase is necessary to achieve multilayer coverage.

Assuming that evaporation can occur only from an exposed surface, a consideration of adsorption equilibrium with the bare surface and with the adsorbed monolayer can be used to obtain a general multilayer equation. A physical picture of the BET model is shown in Figure 14-6. The BET equation is considered to be a general method for practical surface area determination by using nitrogen at 77K as the adsorbate. The surface area can be determined by writing the BET equation as

$$\frac{P}{V(P^\circ - P)} = \frac{1}{V_m C} + \frac{(C-1)P}{C V_m P^\circ} \tag{14-40}$$

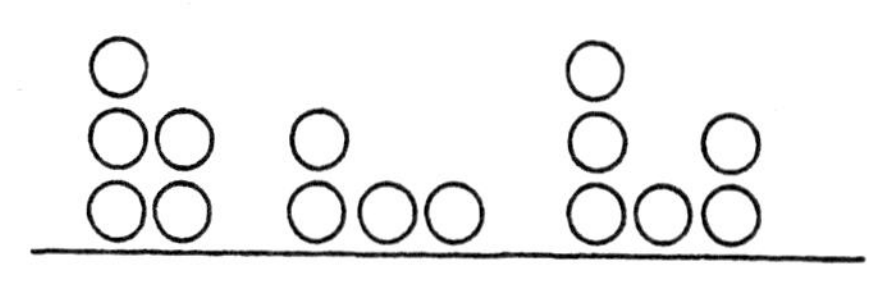

Figure 14-6 BET adsorption model

A plot of $P/V(P^\circ - P)$ versus P/P° should give a straight line with

$$\text{slope} = \frac{C-1}{V_m C} \tag{14-41}$$

and an intercept

$$I = \frac{1}{V_m C} \tag{14-42}$$

The volume of adsorbed gas that corresponds to a monomolecular layer can be obtained by solving Eqs. (14-41) and (14-42). By combining these two equations, we obtain

$$V_m = \frac{1}{\text{slope} + I} \tag{14-43}$$

Heat of adsorption

When a gas or liquid is adsorbed on a surface, a quantity of heat defined as the heat of adsorption is released. The heat of adsorption depends not only on the adsorbate–adsorbent system but also on the conditions under which the determination was made. For our purposes, the discussion here will be limited to the isosteric heat of adsorption. As shown by Ross and Olivier (1962), the isosteric heat of adsorption may be expressed as

$$\Delta H_{st} = RT^2\left(\frac{\partial \ln P}{\partial T}\right)_\Theta = -R\left[\frac{\partial \ln P}{\partial(1/T)}\right]_\Theta \tag{14-44}$$

The equation above permits us to determine the isosteric heat of adsorbtion at any surface coverage from adsorption isotherm data. If the adsorbate is condensed on the surface, it can be readily shown that for the case of low surface coverage, for which the ideal gas law is obeyed, the isosteric heat of adsorption is equal to the latent heat of vaporization.

Example 14.2

Using the data given below, (extracted from Glessner, 1969), for the adsorption of carbon dioxide on type 5A molecular sieve (a) show that the data for 35°C and 55°C can be described by a single characteristic curve as required by the potential theory. (b) Calculate the isosteric heat of adsorption.

$T = 35°C$		$T = 45°C$		$T = 55°C$	
P (*mmHg*)	*Uptake* [$cm^3(STP)/g$]	*P* (*mmHg*)	*Uptake* [$cm^3(STP)/g$]	*P* (*mmHg*)	*Uptake* [$cm^3(STP)/g$]
0.028	0.755	0.085	0.935	0.21	1.26
0.15	2.17	0.26	2.06	0.44	2.09
0.35	3.73	0.48	3.16	0.78	3.11
0.62	5.46	0.70	4.16	0.99	3.66
0.92	7.07	0.75	4.28	1.12	3.81
0.94	7.14	—	—	1.18	4.01

Solution:

(a) According to the potential theory a plot of uptake versus $T \ln(P°/P_x)$ is independent of temperature. Using the vapor pressures of 77.6 atm at 35°C and 111.6 atm at 55°C, the following table was prepared.

$T = 35°C$		$T = 55°C$	
Uptake (cm^3/g)	$T \ln(P°/P_x)$ (K)	*Uptake* (cm^3/g)	$T \ln(P°/P_x)$ (K)
0.775	4485	1.26	4234
2.17	3968	2.09	3991
3.73	3706	3.11	3804
5.46	3540	3.66	3726
7.07	3409	3.81	3685
7.14	3402	4.01	3668

The resulting potential curve is shown in Figure 14-7.

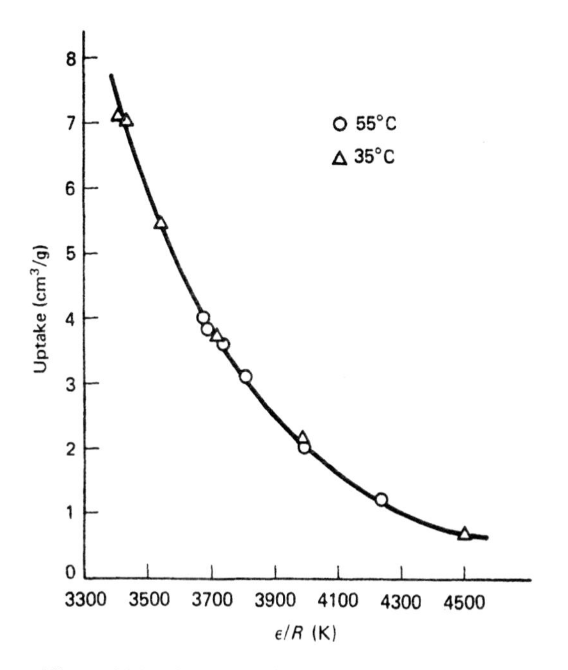

Figure 14-7 Characteristic curve for Example 14.2

(b) The heat of adsorption can be calculated by using Eq. (14-44) and Figure 14-8. The heat of adsorption is obtained from the slope of the lines.

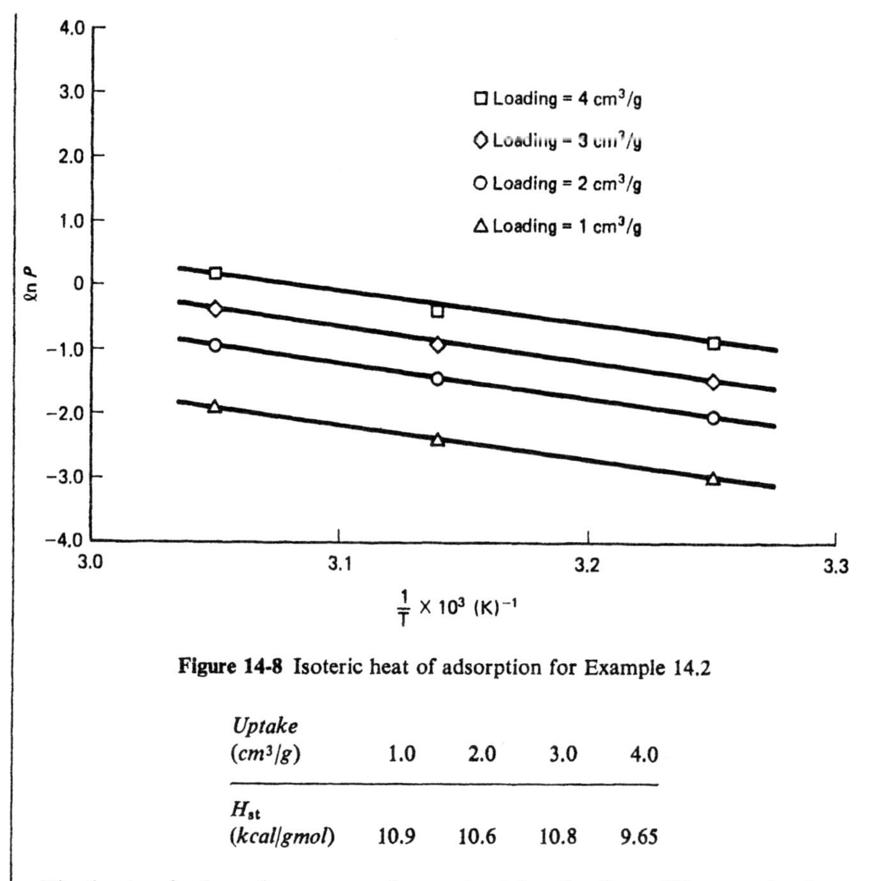

Figure 14-8 Isoteric heat of adsorption for Example 14.2

Uptake (cm^3/g)	1.0	2.0	3.0	4.0
H_{st} *(kcal/gmol)*	10.9	10.6	10.8	9.65

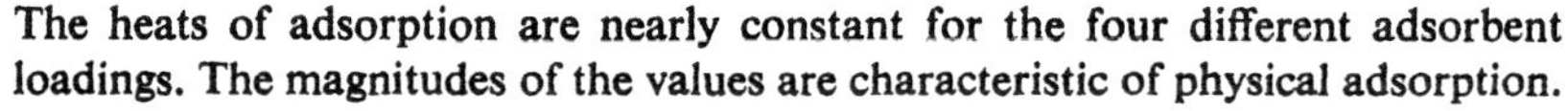

The heats of adsorption are nearly constant for the four different adsorbent loadings. The magnitudes of the values are characteristic of physical adsorption.

14.4 Fundamentals of Dynamic Adsorption

The analysis of packed bed adsorbers is based on the development of effluent concentration–time curves, which are a function of adsorber geometry and operating conditions, and equilibrium adsorption data. The effluent concentration–time curve is usually referred to as the *breakthrough curve* and is obtained by flowing a fluid that contains an adsorbable solute with an initial concentration C_0 through a bed packed with clean or regenerated adsorbent. As the flow of the fluid continues, the bed becomes saturated at a given position and a concentration distribution is established within the bed as shown in Figure 14-9.

At time t_t the solute first appears in the effluent stream. Time t_b is defined as the time required to reach the breakpoint concentration, indicated as C_b. This corresponds to the maximum allowable concentration in the effluent. Time

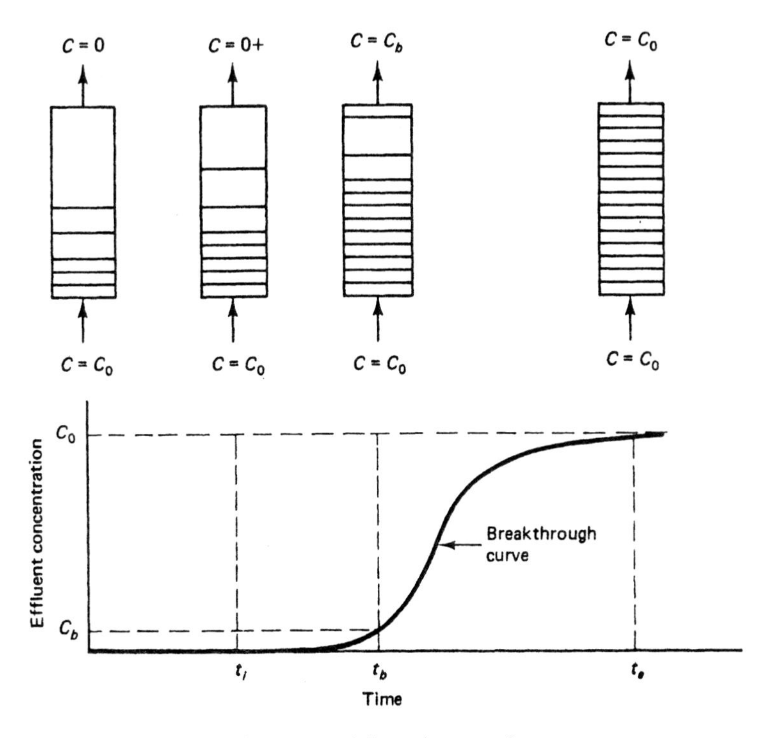

Figure 14-9 Adsorption wavefront

t_e is the time at which the bed becomes saturated with adsorbate. At this time the bed is exhausted and must be regenerated. The time period from t_l to t_e corresponds to the thickness of the adsorption or mass transfer zone in the bed and is related to the mechanism of the adsorption process. It is readily seen that the area behind the breakthrough curve represents the quantity of adsorbate retained in the column. This corresponds to a point on the equilibrium isotherm. If the isotherm can be represented by the Langmuir equation, this point is expressed as

$$q^{\infty} = \frac{QK'C_0}{1 + K'C_0} \quad \text{or} \quad C_s^{\infty} = \frac{Q'K'C_0}{1 + K'C_0} \tag{14-45}$$

where C_0 is the concentration of solute in the influent solution, and q^{∞} and C_s^{∞} are the saturation capacities of the adsorbate in the bed. Consistent units must be used in the equations above.

For the purpose of discussing dynamic adsorption, we will classify the equilibrium isotherms as either (a) favorable, (b) linear, or (c) unfavorable. These are shown in Figure 14-10. If the isotherm is concave in the direction of the fluid concentration, as shown by curve (a), layers of high concentration in the bed move faster than areas of low concentration. This results in the adsorption zone becoming thinner as the wavefront moves through the bed, and gives a breakthrough curve that is self-sharpening. For a favorable isotherm, the

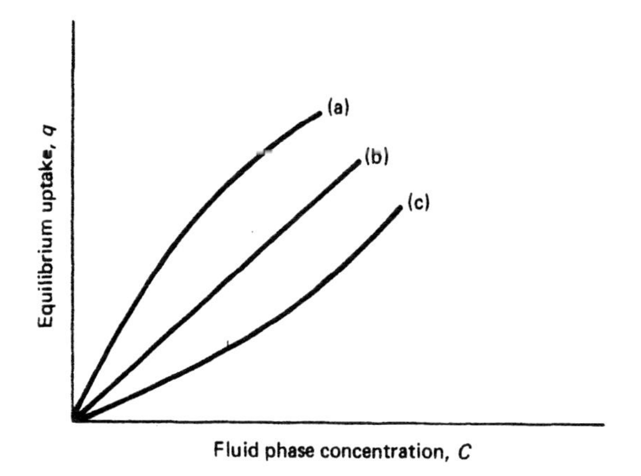

Figure 14-10 Shape of equilibrium isotherms

breakthrough curve develops and moves through the packed column in a constant pattern. The unfavorable isotherm results in a breakthrough curve that becomes more diffuse as it traverses the bed length. For nonequilibrium adsorption, a favorable isotherm will yield a constant pattern breakthrough curve after a period of time.

Material balance for packed adsorbers

Adsorption in packed beds can be modeled by using the shell balance method to derive an equation that describes the transport of mass from the flowing fluid phase to the fixed adsorbent particles. For the differential column section ΔZ shown in Figure 14-11, the general mass balance for solute A is

$$(\text{rate of } A \text{ in}) - (\text{rate of } A \text{ out}) = (\text{rate of } A \text{ accumulation})$$

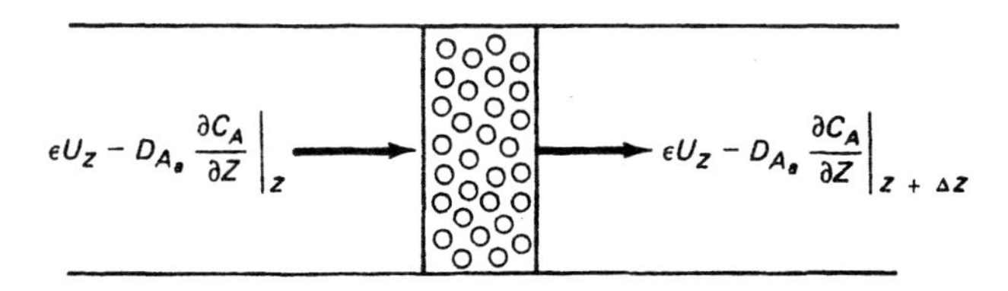

Figure 14-11 Fixed-bed adsorber

Since the solute is transported through the differential element by both diffusion and bulk flow, and is accumulated in the adsorbent particles and interstices, the mass balance becomes

$$\epsilon S\left[-D_{A_e}\frac{\partial C_A}{\partial Z} + C_A U_Z\right]_{Z,t} - \epsilon S\left[-D_{A_e}\frac{\partial C_A}{\partial Z} + C_A U_Z\right]_{Z+\Delta Z,t}$$
$$= \left[S\,\Delta Z\epsilon\,\frac{\partial C_A}{\partial t}\right]_Z + \left[S\,\Delta Z(1-\epsilon)\frac{\partial C_{As}}{\partial t}\right]_Z \quad (14\text{-}46)$$

where ϵ = void fraction in the bed,
S = cross-sectional area of the column,
D_{A_a} = effective axial diffusion coefficient of A,
C_A = concentration of A in the fluid phase, mol/cm^3,
C_{As} = average concentration of A in the solid phase, mol/cm^3,
U_Z = interstitial velocity.

Rearranging the equation above and dividing both sides by $\epsilon S\,\Delta Z$ gives

$$\frac{D_{A_a}\dfrac{\partial C_A}{\partial Z}\bigg|_{Z+\Delta Z,t} - D_{A_a}\dfrac{\partial C_A}{\partial Z}\bigg|_{Z,t}}{\Delta Z} - \frac{C_A U_Z\Big|_{Z+\Delta Z,t} - C_A U_Z\Big|_{Z,t}}{\Delta Z} = \frac{\partial C_A}{\partial t}\bigg|_Z + \frac{1-\epsilon}{\epsilon}\frac{\partial C_{As}}{\partial t}\bigg|_Z \tag{14-47}$$

By applying the limiting process for ΔZ and assuming that the axial diffusion coefficient and velocity are constant, we obtain

$$\left(D_{A_a}\frac{\partial^2 C_A}{\partial Z^2}\right)_t - \left(U_Z\frac{\partial C_A}{\partial Z}\right)_t = \left(\frac{\partial C_A}{\partial t}\right)_Z + \left(\frac{1-\epsilon}{\epsilon}\frac{\partial C_{As}}{\partial t}\right)_Z \tag{14-48}$$

To simplify the solution to Eq. (14-48), we can assume dilute adsorbate concentration in the fluid. This results in a nearly isothermal operation and eliminates the need for solving the accompanying energy balance for the column. Further, if we assume a small pressure drop through the bed and a plug flow velocity profile, the equations of continuity and motion can also be eliminated. In developing the mass balance for the column, we have assumed that radial concentration gradients do not exit. Even with these simplifying assumptions, solutions to Eq. (14-48) are difficult to obtain. In most industrial adsorbers, equilibrium between the adsorbate and adsorbent is not reached. At low flow rates, for which equilibrium conditions are approached, the axial dispersion term is significant and must be considered along with the bulk flow term for both gas and liquid phase adsorption. For higher flow rates, axial dispersion usually becomes unimportant but equilibrium is not reached in the bed.

A solution to Eq. (14-48) requires information regarding the transfer of the solute to the adsorbent. It has been postulated that mass transfer to the adsorbent is controlled by one or a combination of the following mechanisms:

1. External mass transfer
2. Adsorption onto the surface of the adsorbent
3. Internal mass transfer through the fluid phase which occupies the pores of the adsorbent
4. Internal mass transfer along the solid surfaces of the pores of the adsorbent

Two approaches have been used in solving the mass transfer equation for a packed adsorber. One is to solve the differential equation that describes the transfer of the adsorbate into the column, coupled with the equation that describes diffusion into a single particle. This method is difficult, and numerical techniques are frequently used to obtain a solution. The second approach is mechanistic and entails the assumption of a model to represent the rate of mass transfer of the adsorbate from the fluid to the adsorbent. Using a model to represent one or more of the mechanisms above and introducing equilibrium data for the adsorbate–adsorbent system, we can integrate the differential mass balance to produce the unsteady-state concentration profile in the packed bed. From the breakthrough curve, it is then possible to predict the uptake of adsorbate.

Mass transfer coefficients for the various mechanisms are presented in a manner similar to that shown by Lightfoot et al. (1962). These are

$$\frac{\partial C_{As}}{\partial t} = \frac{k_f a}{1-\epsilon}(C_A - C_{Ai}) \tag{14-49a}$$

or

$$\frac{\partial q_A}{\partial t} = \frac{k_f a}{\rho_b}(C_A - C_{Ai}) \tag{14-49b}$$

$$\frac{\partial C_{As}}{\partial t} = k_s a(C_{Asi} - C_{As}) \tag{14-50a}$$

or

$$\frac{\partial q_A}{\partial t} = k_s a(q_{Ai} - q_A) \tag{14-50b}$$

$$\frac{\partial C_{As}}{\partial t} = \frac{K_f a}{1-\epsilon}(C_A - C_A^*) \tag{14-51a}$$

or

$$\frac{\partial q_A}{\partial t} = \frac{K_f a}{\rho_b}(C_A - C_A^*) \tag{14-51b}$$

where k_f is the fluid-phase mass transfer coefficient, k_s is the solid-phase mass transfer coefficient, K_f is the overall mass transfer coefficient, $\rho_s = \rho_b/(1-\epsilon)$, and $q_A = C_{As}/\rho_s$. The concentrations used in Eqs. (14-49) through (14-51) are as follows: C_{Ai} is the interfacial concentration in the fluid phase, q_{Ai} is the interfacial concentration in the solid phase, and C_A^* is the concentration of the fluid phase that is in equilibrium with the solid. The relationships between the fluid and solid phase concentrations are shown in Figure 14-12. The overall mass transfer coefficient is related to the local mass transfer coefficients, as shown previously for gas–liquid systems, by the expression

$$\frac{1}{K_f a} = \frac{1}{k_f a} + \frac{1}{m' k_s a(1-\epsilon)} \tag{14-52}$$

where m' is obtained from the equilibrium isotherm as follows:

$$m' = \frac{C_{Asi} - C_{As}}{C_{Ai} - C_A^*} \tag{14-53}$$

Local mass transfer coefficients for the fluid phase can be predicted for packed beds by using the methods presented in Chapter 6. According to Helf-

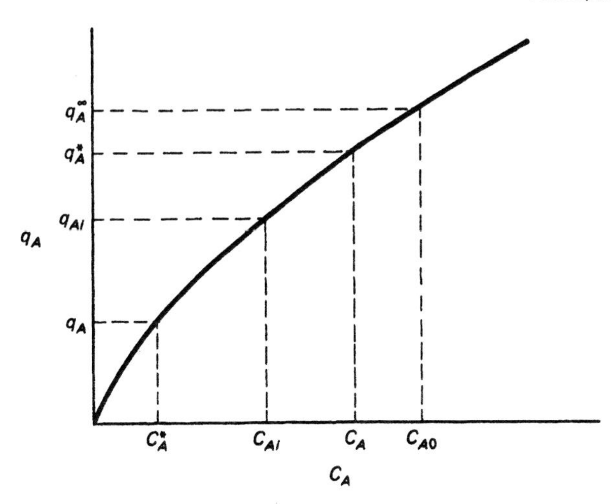

Figure 14-12 Concentration differences

ferich (1962), the local mass transfer coefficient in the solid phase can be predicted for spherical particles by using the apparent diffusivity of the solute through the solid particle with the expression

$$k_s = \frac{10D_A}{d_p(1-\epsilon)} = \frac{k_s'}{1-\epsilon} \tag{14-54}$$

where d_p is the diameter of the particle. Coefficients for nonspherical particles can be predicted by replacing d_p with $6/S_0$, where S_0 is the surface area of the particle. Lightfoot et al. (1962) use the mass transfer coefficient $k_s'/(1-\epsilon)$ instead of k_s for the solid phase.

14.5 Prediction of Breakthrough Curves Using Linear Adsorption Isotherms

Control of mass transfer by a single resistance

One of the earliest mechanistic models used to predict breakthrough curves was proposed by Hougen and Marshall (1947). Their model was developed for cases in which mass transfer to the solid adsorbent was controlled by a fluid-phase mass transfer coefficient. In addition, their model was based on the assumption that equilibrium between the fluid and solid phases could be described by a linear isotherm equation.

Beginning with Eq. (14-48) and neglecting axial diffusion, we obtain

$$-\frac{\epsilon U_Z}{1-\epsilon}\left(\frac{\partial C_A}{\partial Z}\right)_t - \frac{\epsilon}{1-\epsilon}\left(\frac{\partial C_A}{\partial t}\right)_Z = \left(\frac{\partial C_{As}}{\partial t}\right)_Z \tag{14-55}$$

Introducing $\rho_s = \rho_b/(1-\epsilon)$ and $q_A = C_{As}/\rho_s$ into Eq. (14-55) gives

$$-\frac{\epsilon U_Z}{\rho_b}\left(\frac{\partial C_A}{\partial Z}\right)_t - \frac{\epsilon}{\rho_b}\left(\frac{\partial C_A}{\partial t}\right)_Z = \left(\frac{\partial q_A}{\partial t}\right)_Z \tag{14-56}$$

If the fluid content of the bed is small compared to the total volume of fluid throughput, the second term in Eq. (14-56) may be neglected, as indicated by Hougen and Marshall (1947). Thus the mass balance for the packed bed is

$$-\frac{\epsilon U_Z}{\rho_b}\left(\frac{\partial C_A}{\partial Z}\right)_t = \left(\frac{\partial q_A}{\partial t}\right)_Z \tag{14-57}$$

The change in adsorbate content of the adsorbent and of the fluid may be expressed in terms of the rate of adsorption. The rate of uptake for this case is given as

$$\left(\frac{\partial q_A}{\partial t}\right)_Z = \frac{K_f a}{\rho_b}(C_A - C_A^*) \tag{14-58}$$

For a linear isotherm, the equilibrium fluid-phase concentration is related to the concentration in the solid by the expressions

$$q_A = K_D C_A^* \qquad q_A^\infty = K_D C_{A0} \qquad q_{At} = K_D C_{At} \tag{14-59}$$

where K_D is the distribution coefficient and q_A^∞ is the saturation capacity of the bed when the concentration of the solute in the influent is C_{A0}. Equation (14-58) and the mass balance equation can be simplified by rewriting them in terms of the following dimensionless variables.

$$\zeta = \frac{ZK_f a}{\epsilon U_Z} \quad \text{(bed-length parameter)} \tag{14-60}$$

$$\tau = \frac{K_f a}{K_D \rho_b}\left(t - \frac{Z}{U_Z}\right) \quad \text{(time parameter)} \tag{14-61}$$

$$\bar{X} = \frac{C_A}{C_{A0}} \tag{14-62}$$

$$\bar{Q} = \frac{q_A}{q_A^\infty} = \frac{q_A}{K_D C_{A0}} \tag{14-63}$$

We can express Eqs. (14-57) and (14-58) in terms of the new variables as

$$\frac{\partial \bar{Q}}{\partial \tau} = \bar{X} - \bar{Q} \tag{14-64}$$

and

$$\frac{\partial \bar{X}}{\partial \zeta} = -\bar{X} + \bar{Q} \tag{14-65}$$

For the case in which the influent concentration is constant at the entrance and the initial solute concentration in the bed is zero, the boundary conditions are

$$C_A = C_{A0} \quad \text{at } Z = 0 \qquad \bar{X} = 1 \quad \text{at } \zeta = 0 \quad \text{for all } \tau \tag{14-66}$$

$$q_A = 0 \quad \text{at } t - \frac{Z}{U_Z} = 0 \qquad \bar{Q} = 0 \quad \text{at } \tau = 0 \quad \text{for all } \zeta \tag{14-67}$$

A solution may be obtained by using Laplace transforms. The solution is

$$\bar{X} = 1 - \int_0^{\zeta} e^{-(\tau+\zeta)} J_0(i\sqrt{4\zeta\tau})\, d\zeta = \bar{J}(\tau, \zeta) \tag{14-68}$$

or

$$\bar{Q} = \int_0^{\tau} e^{-(\tau+\zeta)} J_0(i\sqrt{4\zeta\tau})\, d\tau \tag{14-69}$$

where J_0 is a zero-order Bessel function of the first kind. A graphical solution to the equation above has been presented by Hougen and Marshall (1947). An expanded version of the Hougen and Marshall graphical solution is shown in Figure 14-13. A solution to the equation above was first obtained by Anzelius (1926) for the analogous heat transfer problem.

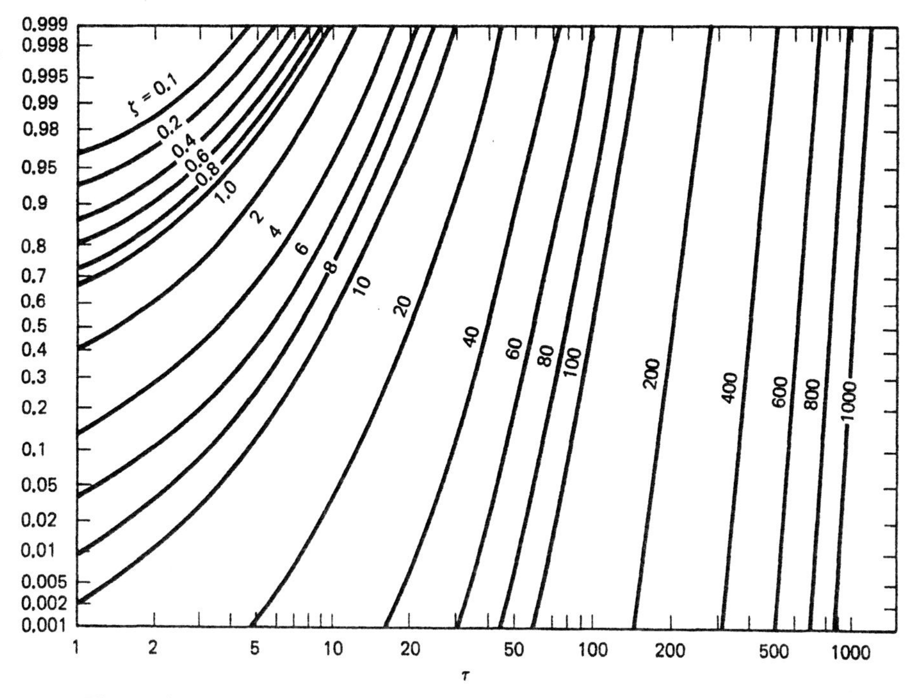

Figure 14-13 Effluent concentrations in packed adsorbers (from Vermeulen et al., 1973)

The approach of Hougen and Marshall (1947) can be extended to include other rate-controlling mechanisms. For the case in which diffusion within the solid particle is rate controlling, the rate of adsorbate uptake is expressed by

$$\left(\frac{\partial q_A}{\partial t}\right)_z = k_s a(q_{Ai} - q_A) \tag{14-70}$$

For a linear isotherm as expressed by Eq. (14-59), the mass balance and Eq. (14-70) can be written in dimensionless form by redefining the bed-length and time parameters as

$$\zeta = \frac{Zk_s a K_D \rho_b}{\epsilon U_Z} \quad \text{(bed-length parameter)} \tag{14-71}$$

and

$$\tau = k_s a\left(t - \frac{Z}{U_Z}\right) \quad \text{(time parameter)} \tag{14-72}$$

After introducing the nondimensional variables above into Eqs. (14-57) and (14-70), Eqs. (14-64) and (14-65) are again obtained. These equations can then be solved with the boundary conditions given by Eqs. (14-66) and (14-67) to obtain the solution proposed by Hougen and Marshall.

For the case in which first-order reversible surface kinetics is used to describe the change in solute concentration of the solid phase, the rate of uptake is

$$\left(\frac{\partial q_A}{\partial t}\right)_Z = k_1 C_A - k_2 q_A = k_1\left(C_A - \frac{q_A}{K_D}\right) \tag{14-73}$$

The mass balance and Eq. (14-73) can be written in dimensionless form by introducing the following variables:

$$\zeta = \frac{k_1 \rho_b Z}{\epsilon U_Z} \quad \text{(bed-length parameter)} \tag{14-74}$$

and

$$\tau = \frac{k_1}{K_D}\left(t - \frac{Z}{U_Z}\right) \quad \text{(time parameter)} \tag{14-75}$$

Equations (14-57) and (14-73) thus reduce to Eqs. (14-64) and (14-65) with the same boundary conditions. The Hougen and Marshall solution is again obtained but with different expressions for the bed-length and time parameters.

For the case in which a second-order reversible surface reaction can be used to describe the rate of solute uptake by the solid, the rate expression is

$$\left(\frac{\partial q_A}{\partial t}\right)_Z = k_1' C_A(Q_m - q_A) - k_2 q_A \tag{14-76}$$

$$= k_1'\left[C_A(Q_m - q_A) - \frac{q_A}{K'}\right] \tag{14-77}$$

This model, which was originally proposed by Thomas (1944) for the Langmuir isotherm equation, can also be used to obtain the Hougen and Marshall type of solution. If the solute concentration, C_{A0}, in the influent is small, the Langmuir isotherm equation becomes $q_A^* = K' Q_m C_{A0} = K_D C_{A0}$, where Q_m is the moles in a monomolecular layer per gram of solid. The dimensionless variables for this model are

$$\zeta = \frac{k_1' Q_m \rho_b Z}{\epsilon U_Z} \quad \text{(bed-length parameter)} \tag{14-78}$$

and

$$\tau = \frac{k_1' Q_m C_{A0}}{q_A^\infty}\left(t - \frac{Z}{U_Z}\right) \quad \text{(time parameter)} \tag{14-79}$$

Equations of the form solved by Hougen and Marshall can again be obtained, and the solution presented in Figure 14-13 can be used.

An approximation of the $\bar{J}(\zeta, \tau)$ function for large values of τ and ζ was reported by Thomas (1944). The approximation may be written in terms of the error function as

$$\bar{J}(\zeta, \tau) = \frac{1}{2}\left[1 - \text{erf}(\sqrt{\zeta} - \sqrt{\tau}) + \frac{e^{-(\sqrt{\zeta}-\sqrt{\tau})^2}}{\sqrt{\pi}(\sqrt{\tau} + \sqrt[4]{\zeta\tau})}\right] \quad (14\text{-}80)$$

As indicated by Vermeulen et al. (1973), the equation above is accurate to within 1% when $\zeta\tau \geq 36$. When $\zeta\tau \geq 3600$, which is frequently the case in industrial applications, the last term may be neglected, thus giving the approximation suggested by Klinkenberg (1948).

$$\bar{J}(\zeta, \tau) = \tfrac{1}{2}\,\text{erfc}(\sqrt{\zeta} - \sqrt{\tau}) \quad \text{for } \tau < \zeta \quad (14\text{-}81)$$

and

$$\bar{J}(\zeta, \tau) = \tfrac{1}{2}[1 + \text{erf}(\sqrt{\zeta} - \sqrt{\tau})] \quad \text{for } \tau > \zeta \quad (14\text{-}82)$$

Example 14.3

In an example proposed by Lee and Cummings (1967), compressed air is to be dried with silica gel in a 5.8-ft-high packed bed. The air enters the bed at 35°F and 80 lb/in^2 with a humidity of 93%. The density of the influent is 0.518 lb/ft^3 and the superficial velocity through the bed is 88 ft/min. Other properties of the influent are:

$$\mu_f = 0.041 \text{ lb/ft·h} \quad \text{and} \quad D_{w-a} = 0.134 \text{ ft}^2\text{/h}$$

The equilibrium isotherm is linear and the distribution coefficient at the existing conditions is

$$K_D\rho_s = 10.6 \times 10^4 \frac{\text{lb } H_2O/\text{ft}^3 \text{ solid}}{\text{lb } H_2O/\text{ft}^3 \text{ air}}$$

Characteristics of the packing are as follows:

Average particle diameter, $d_p = 0.0127$ ft

Packed-bed density, $\rho_b = 45$ lb/ft^3

Void fraction, $\epsilon = 0.35$

External area, $a = 231.0$ ft^2/ft^3

Shape factor, $\psi = 0.91$

Using the information provided, predict the shape of the breakthrough curve if the operation is isothermal. The gas-phase mass transfer coefficient is $K_f a = 1.43 \times 10^4 (h^{-1})(\text{ft}^3 \text{ air/ft}^3 \text{ bed})$.

Solution:

$$\zeta = \frac{ZK_fa}{\epsilon U_Z} = \frac{(5.8)(1.43 \times 10^4)}{88(60)} = 15.7$$

From Figure 14-13:

$\bar{X} = \frac{C_A}{C_{A0}}$	τ	$t = \frac{\tau(K_D\rho_s)(1-\epsilon)}{K_fa} + \frac{Z}{U_Z}$
0.01	4.5	21.74
0.05	7.5	36.2
0.10	8.0	38.64
0.2	10.0	48.3
0.6	15.0	72.45
0.8	19.0	91.77
0.95	23.0	111.09
0.97	27.0	130.41

The break time, which is selected as $\bar{X} = 0.05$, is seen to be 36.2 h and the bed exhaustion time is 111.1 h. This compares to calculated values for the break time and exhaustion time of 59.8 h and 92.8 h, respectively, as found by Lee and Cummings for the same process. The discrepancy in the predicted values is the result of using different correlations for predicting mass transfer coefficients. The coefficient used in the calculation above is based on the experimental data of Eagleton and Bliss (1953), whereas the coefficient used by Lee and Cummings was obtained from Eq. (6-154).

In Example 14.3 the breakthrough curve was predicted on the basis of isothermal adsorption. However, the heat of adsorption that is liberated when a solute is adsorbed on a surface will cause the temperature of the packed bed to increase. The amount of heat liberated is directly proportional to the amount of solute in the influent stream. The increase in bed temperature that might be expected due to the heat liberated from the adsorption process is partially offset by the nonadsorbable carrier gas that flows through the bed. Because of the higher bed temperature, the equilibrium adsorption capacity of the adsorbent will be reduced and will result in a shorter breakthrough time. Lee and Cummings (1967) developed an equation that will correct the break time for non-isothermal effects for the adsorption of water on silica gel.

The slope of the breakthrough curve increases with increasing bed-length modulus, ζ, as seen in Figure 14-13. Thus when the fluid-phase mass transfer coefficient is very large and offers little resistance to the transfer of mass to the solid, the value for the bed-length modulus will be large and the breakthrough curve slope should be expected to be very steep. This is usually the case for gases. Liquids, however, frequently exhibit much flatter breakthrough curves due to the presence of a stagnant liquid film on the solid surface.

Combined internal and external resistances

Frequently, the rate-controlling step for adsorption in packed beds is intraparticle diffusion instead of the resistance created by the fluid film, as shown in the preceding example. For this case, a concentration gradient exists inside the adsorbent and must be considered in conjunction with the film resistance if the breakthrough curve is to be predicted accurately. Rosen (1952, 1954) modeled the breakthrough curve for the case in which both film resistance and intraparticle diffusion are rate controlling, without assuming a mechanism a priori. Although the Rosen model is restricted to linear equilibrium isotherms and spherical adsorbent particles, it has been used successfully for many experimental systems by introducing an equivalent spherical radius and a linear approximation to the isotherm over the concentration region of interest.

If axial diffusion is negligible and the fluid velocity is uniform over the cross section of the column, the mass balance is given by Eq. (14-56).

$$-\frac{\epsilon U_Z}{\rho_b}\left(\frac{\partial C_A}{\partial Z}\right)_t - \frac{\epsilon}{\rho_b}\left(\frac{\partial C_A}{\partial t}\right)_Z = \left(\frac{\partial q_A}{\partial t}\right)_Z \tag{14-56}$$

We can simplify the equation above by introducing a change of variable

$$\Theta = t - \frac{Z}{U_Z} \tag{14-83}$$

and

$$\chi = \frac{Z\rho_b}{\epsilon U_Z} \tag{14-84}$$

The resulting equation thus becomes

$$\frac{\partial C_A}{\partial \chi} + \frac{\partial q_A}{\partial \Theta} = 0 \tag{14-85}$$

with boundary conditions

$$\text{BC1:} \quad C_A = C_{A0} \quad \text{at } \chi = 0 \quad \text{for all } \Theta \tag{14-86}$$

$$\text{BC2:} \quad q_A = 0 \quad \text{at } \Theta = 0 \quad \text{for all } \chi \tag{14-87}$$

Equation (14-85) describes the rate of adsorption as a function of the concentration change along the bed length. External mass transfer through the stagnant film surrounding the particle is described by Eq. (14-49).

The balance for uptake by diffusion into the sphere is given by the unsteady-state equation

$$\frac{\partial q_i}{\partial \Theta} = \frac{D_A}{r^2}\frac{\partial}{\partial r}\left(r^2 \frac{\partial q_i}{\partial r}\right) \tag{14-88}$$

where the boundary conditions are

$$\text{BC1:} \quad q_i = 0 \quad \text{at } \Theta = 0 \quad \text{all } \chi, r \tag{14-89}$$

$$\text{BC2:} \quad \frac{\partial q_i}{\partial r} = 0 \quad \text{at } r = 0 \quad \text{all } \chi, \Theta \tag{14-90}$$

$$\text{BC3:} \quad q_i = q_s = K_D C_i(\chi, \Theta) \quad \text{at } r = R \tag{14-91}$$

Equation (14-91) describes the linear equilibrium between the interfacial gas concentration and the concentration of solute on the solid surface. To complete the formulation of the problem, we relate the average concentration in the adsorbent to the internal concentration at radius r by

$$q_A = \frac{\int_0^R q_i r^2 \, dr}{\int_0^R r^2 \, dr} = \frac{3}{R^3} \int_0^R q_i r^2 \, dr \tag{14-92}$$

The system of equations above has been solved for a packed bed that is initially free of adsorbate, and gives effluent concentration curves as a function of time and bed length following a step increase in the influent concentration from zero to C_{A0} at $t = 0$. The solution is

$$\frac{C_A}{C_{A0}} = \frac{1}{2} + \psi(\nu, X, Y) \tag{14-93}$$

where

$$\nu = \frac{D_A K_D \rho_s}{RK_f} \quad \text{(film-resistance parameter)} \tag{14-94}$$

$$X = \frac{3D_A K_D \rho_s Z}{mU_Z R^2} \quad \text{(bed-length parameter)} \tag{14-95}$$

$$Y = \frac{2D_A}{R^2}\left(t - \frac{Z}{U_Z}\right) \quad \text{(contact-time parameter)} \tag{14-96}$$

and $m = \epsilon/(1 - \epsilon)$. Tabulated and graphical solutions are given by Rosen. An asymptotic expression for the solution which is valid for large values of X (greater than 40) is

$$\frac{C_A}{C_{A0}} = \frac{1}{2}\left\{1 + \text{erf}\left[\frac{(3Y/2X) - 1}{2\sqrt{\nu/X}}\right]\right\} \tag{14-97}$$

Example 14.4

To demonstrate the use of this model, let us reconsider Example 14.3. Assuming that $D_A = 2 \times 10^{-8}$ ft^2/s, density of influent $\rho_f = 0.518$ lb/ft^3, and viscosity of influent $\mu = 0.041$ lb/ft·h, calculate the breakthrough and bed depletion times.

Solution:

If we divide Eq. (14-96) by Eq. (14-95), we obtain

$$\begin{aligned} t &= \frac{3K_D \rho_s Z}{2mU_Z} \frac{Y}{X} + \frac{Z}{U_Z} \\ &= \frac{3(10.6 \times 10^4)(5.8)}{2[0.35/(1 - 0.35)](88/0.35)(60)} \frac{Y}{X} + \frac{5.8}{(88/0.35)60} \\ &= 113.5 \frac{Y}{X} + 0.0004 \text{ h} \end{aligned}$$

Dividing the film-resistance parameter by the bed-length parameter gives

$$\frac{\nu}{X} = \frac{mU_Z R}{3ZK_f} = \frac{\left(\frac{0.35}{1-0.35}\right)\left(\frac{88}{0.35}\right)(60)\left(\frac{0.0127}{2}\right)}{3(5.8)(1.43 \times 10^4/231)} = 4.79 \times 10^{-2}$$

By rearranging Eq. (14-97), we have

$$\text{erf } E = 2\left(\frac{C_A}{C_{A0}}\right) - 1$$

where

$$E = \frac{\frac{3}{2}(Y/X) - 1}{2\sqrt{\nu/X}}$$

Thus

$$\frac{Y}{X} = \frac{2}{3}\left(2E\sqrt{\frac{\nu}{X}} + 1\right)$$

$$= \frac{2}{3}(2E\sqrt{4.79 \times 10^{-2}} + 1)$$

$$= 0.292E + 0.667$$

Using values of the error function from Chapter 4, we can calculate values for Y/X and t for various values of C_A/C_{A0}. At the break time, $C_A/C_{A0} = 0.05$,

$$\text{erf } E = 2(0.05) - 1 = -0.9$$

$$E = -1.16$$

and

$$\frac{Y}{X} = 0.292(-1.16) + 0.667 = 0.328$$

Therefore,

$$t_b = 113.5(0.328) + 0.0004$$

$$= 37.2 \text{ h}$$

At bed exhaustion, $C_A/C_{A0} = 0.95$. Thus

$$\text{erf } E = 2(0.95) - 1 = 0.9$$

$$E = 1.16$$

and

$$\frac{Y}{X} = 0.292(1.16) + 0.667 = 1.006$$

Therefore,

$$t_e = 113.5(1.006) + 0.0004$$

$$= 114.2 \text{ h}$$

We see that the breakthrough and bed exhaustion times predicted by the Rosen model and Hougen and Marshall model are nearly identical.

14.6 Prediction of Breakthrough Curves Using the Langmuir Equation and Kinetic Model

In adsorption, both the external film resistance and internal diffusion within the particle usually contribute to the rate of solute uptake. However, the general approach taken by Thomas (1944), in which he assumed surface kinetics was rate limiting, has proven to be the most versatile method for designing adsorption beds. The mechanistic model of Thomas was originally derived for ion exchange, but was shown by Hiester and Vermeulen (1952) to be useful for cases in which rate controlling steps other than surface kinetics applied. For the Thomas model, equilibrium between the fluid and solid is described by the Langmuir isotherm. Several kinetic models will be considered in this section.

Second-order kinetics controlling

The kinetic derivation for adsorption described by the Langmuir model can be derived in a manner similar to that shown in Section 14.3, where the net rate of uptake is equal to the rate of adsorption minus the rate of desorption. Thus for the adsorption of solute A, we have

$$\frac{\partial C_s}{\partial t} = k_a C(Q' - C_s) - k_d C_s \tag{14-98}$$

or

$$\frac{\partial C_s}{\partial t} = k_a\left[C(Q' - C_s) - \frac{1}{K'}C_s\right] \tag{14-99}$$

where Q' is the maximum capacity of the adsorbent and K' is the adsorption equilibrium constant. The subscript A that has been used to indicate the adsorbate has been omitted in the equations above. From Eq. (14-45),

$$Q' = \frac{1 + K'C_0}{K'C_0} C_s^\infty$$

Upon substituting Q' from the equation above into Eq. (14-99), we obtain

$$\frac{\partial(C_s/C_s^\infty)}{\partial t} = k_a \frac{1 + K'C_0}{K'}\left[\frac{C}{C_0}\left(1 - \frac{C_s}{C_s^\infty}\right) - \frac{1}{1 + K'C_0}\frac{C_s}{C_s^\infty}\left(1 - \frac{C}{C_0}\right)\right] \tag{14-100}$$

For Langmuir kinetics we thus have

$$\frac{\partial(C_s/C_s^\infty)}{\partial t} = \Delta_a\left[\frac{C}{C_0}\left(1 - \frac{C_s}{C_s^\infty}\right) - r^*\frac{C_s}{C_s^\infty}\left(1 - \frac{C}{C_0}\right)\right] \tag{14-101}$$

where

$$\Delta_a = k_a \frac{1 + K'C_0}{K'} \tag{14-102}$$

and

$$r^* = \frac{1}{1 + K'C_0} \tag{14-103}$$

The equation above can be written in dimensionless form by defining bed-length and time parameters. The dimensionless groups are

$$\zeta = \frac{Z(1 - \epsilon)C_s^\infty \Delta_a}{\epsilon U_Z C_0} \quad \text{(bed-length parameter)} \tag{14-104}$$

$$\tau = \Delta_a \Theta = \Delta_a\left(t - \frac{Z}{U_Z}\right) \quad \text{(time parameter)} \tag{14-105}$$

$$\bar{Q} = \frac{C_s}{C_s^\infty} \tag{14-106}$$

$$\bar{X} = \frac{C}{C_0} \tag{14-107}$$

Upon introducing the dimensionless groups above into Eq. (14-101), we obtain

$$\frac{\partial \bar{Q}}{\partial \tau} = \bar{X}(1 - \bar{Q}) - r^*\bar{Q}(1 - \bar{X}) \tag{14-108}$$

If the general material balance equation, Eq. (14-55), is written in dimensionless form by using the dimensionless variables above, it becomes

$$-\left(\frac{\partial \bar{Q}}{\partial \tau}\right)_\zeta = \left(\frac{\partial \bar{X}}{\partial \zeta}\right)_\tau \tag{14-109}$$

Upon combining Eqs. (14-108) and (14-109), we obtain

$$\frac{\partial \bar{Q}}{\partial \tau} = \bar{X}(1 - \bar{Q}) - r^*\bar{Q}(1 - \bar{X}) \tag{14-110}$$

and

$$\frac{\partial \bar{X}}{\partial \zeta} = -\bar{X}(1 - \bar{Q}) + r^*\bar{Q}(1 - \bar{X}) \tag{14-111}$$

As in the Hougen and Marshall model, when the bed is initially free of adsorbate the boundary conditions are

$$\bar{X} = 1 \quad \text{at } \zeta = 0 \quad \text{for all } \tau \tag{14-112}$$

$$\bar{Q} = 0 \quad \text{at } \tau = 0 \quad \text{for all } \zeta \tag{14-113}$$

Thomas obtained a solution to the set of equations above for the concentrations of the fluid and solid phases. The resulting solutions are

$$\frac{C}{C_0} = \frac{\bar{J}(r^*\zeta, \tau)}{\bar{J}(r^*\zeta, \tau) + [1 - \bar{J}(\zeta, r^*\tau)] \exp[(r^* - 1)(\tau - \zeta)]} \tag{14-114}$$

and

$$\frac{C_s}{C_s^\infty} = \frac{1 - \bar{J}(\tau, r^*\zeta)}{\bar{J}(r^*\zeta, \tau) + [1 - \bar{J}(\zeta, r^*\tau)] \exp[(r^* - 1)(\tau - \zeta)]} \tag{14-115}$$

where $\bar{J}$ is the function presented for the Hougen and Marshall model. It can be readily shown that for the case in which $r^* = 1$, which corresponds to a linear isotherm, the Thomas and Hougen and Marshall solutions are the same.

External film resistance controlling

If the approach taken by Thomas is to be of general use, then other rate-controlling mechanisms must be related to the solution above. When an external film resistance is controlling,

$$\frac{\partial C_s}{\partial t} = \frac{k_f a}{1 - \epsilon}(C - C_i) \tag{14-49a}$$

where C_i is the concentration of the solute in the fluid on the exterior of the surface. Since the adsorbate concentration throughout the porous solid is uniform, C_i is related to the solid concentration by the equilibrium isotherm. According to the equation above, when the rate of adsorption is equal to zero, $C = C_i$. By substituting into Eq. (14-101) for $\partial(C_s/C_s^\infty)/\partial t = 0$, we obtain

$$C_t = \frac{C_s}{C_s^\infty} \frac{r^* C_0}{1 + (C_s/C_s^\infty)(r^* - 1)} \tag{14-116}$$

Thus, if we introduce the definition of C_t given by Eq. (14-116) into Eq. (14-49a), we get

$$\frac{\partial(C_s/C_s^\infty)}{\partial t} = \frac{k_f a C_0}{C_s^\infty(1-\epsilon)[1 + (C_s/C_s^\infty)(r^* - 1)]}\left[\frac{C}{C_0}\left(1 - \frac{C_s}{C_s^\infty}\right) - r^* \frac{C_s}{C_s^\infty}\left(1 - \frac{C}{C_0}\right)\right] \tag{14-117}$$

$$= \Delta_E\left[\frac{C}{C_0}\left(1 - \frac{C_s}{C_s^\infty}\right) - r^* \frac{C_s}{C_s^\infty}\left(1 - \frac{C}{C_0}\right)\right] \tag{14-118}$$

The equation above can be written in dimensionless form by using the variables

$$\zeta = \frac{Z(1-\epsilon)C_s^\infty \Delta_E}{\epsilon U_Z C_0} \tag{14-119}$$

$$\tau = \Delta_E \Theta = \frac{k_f a C_0}{C_s^\infty(1-\epsilon)[1 + (C_s/C_s^\infty)(r^* - 1)]}\left(t - \frac{Z}{U_Z}\right) \tag{14-120}$$

$$\bar{Q} = \frac{C_s}{C_s^\infty} \tag{14-121}$$

$$\bar{X} = \frac{C}{C_0} \tag{14-122}$$

After writing Eq. (14-118) and the general mass balance in nondimensional form, we obtain Eqs. (14-110) and (14-111) and the boundary conditions given by (14-112) and (14-113). Thus the solutions are given by Eqs. (14-114) and (14-115), but with the dimensionless variables defined as above. Although Δ_E contains the variable C_s/C_s^∞, Hiester and Vermeulen (1952) suggested that an average value be used. For $r^* < 1$, $C_s/C_s^\infty = 0.5$, and for $r^* > 1$, $C_s/C_s^\infty = 1/(r^* + 1)$.

Internal particle resistance

The preceding approach can be extended to the case in which diffusion within the adsorbent particle is rate controlling. The rate of adsorbate uptake is expressed by Eq. (14-50a) as

$$\frac{\partial C_s}{\partial t} = k_s a(C_{st} - C_s) \tag{14-50a}$$

where C_{st} is the interfacial concentration of the adsorbate on the surface of the solid and is assumed to be in equilibrium with the interfacial concentration in the fluid phase. To obtain the differential equations describing this case, we substitute C_{st} for C_s in Eq. (14-101) and obtain

$$C_{st} = \frac{C_s^\infty}{(1 - r^*) + r^* C_0/C} \tag{14-123}$$

Substitution of this relationship into Eq. (14-50a) gives

$$\frac{\partial(C_s/C_s^\infty)}{\partial t} = \frac{k_s a}{[r^* + (C/C_0)(1 - r^*)]}\left[\frac{C}{C_0}\left(1 - \frac{C_s}{C_s^\infty}\right) - r^*\frac{C_s}{C_s^\infty}\left(1 - \frac{C}{C_0}\right)\right] \tag{14-124}$$

To reduce the expression above to the form given by Eqs. (14-110) and (14-111), the dimensionless time parameter is defined by

$$\tau = \Delta_s \Theta = \frac{k_s a}{r^* + (C/C_0)(1 - r^*)}\left(t - \frac{Z}{U_Z}\right) \tag{14-125}$$

Again the solution is represented by Eqs. (14-114) and (14-115). Hiester and Vermeulen (1952) suggested that an average value be used for the variable C/C_0 in the expression above. If the entire breakthrough curve is to be determined, a value of $C/C_0 = 0.5$ should be used.

Combined internal and external resistances

We can combine the internal and external resistances by defining a general rate constant as

$$\kappa = \Delta K'_{\text{lim}} \tag{14-126}$$

where κ is a general rate constant and K'_{lim} is a limiting distribution coefficient between the fluid and solid.

$$K'_{\text{lim}} = \frac{C_s^\infty \rho_b}{C_0 \epsilon} \tag{14-127}$$

For resistance due to the presence of an external film,

$$\kappa_E = \frac{k_f a C_0}{C_s^\infty(1 - \epsilon)[1 + (C_s/C_s^\infty)(r^* - 1)]}\frac{C_s^\infty \rho_b}{C_0 \epsilon} \tag{14-128}$$

If the resistance inside the particle is rate controlling,

$$\kappa_S = \frac{k_s a}{r^* + (C/C_0)(1 - r^*)}\frac{C_s^\infty \rho_b}{C_0 \epsilon} \tag{14-129}$$

The overall rate coefficient for the case in which both the external film resistance and intraparticle resistance are important is thus given as

$$\frac{1}{\kappa} = \frac{1}{\kappa_E} + \frac{1}{\kappa_S} \tag{14-130}$$

As previously noted, the kinetic model proposed by Thomas (1944) and extended by Hiester and Vermeulen (1952) is probably the most versatile and most accurate method for describing adsorption in a packed bed. However, it is considerably more difficult to use than the methods proposed by either Rosen (1952, 1954) or Hougen and Marshall (1947).

14.7 Design of Packed-Bed Adsorbers by the LUB/Equilibrium Section Method

One procedure that is widely used in adsorber column design is described as the LUB/equilibrium concept method. In this method the packed-bed adsorber is viewed as consisting of two sections, the equilibrium section and the LUB (length of unused bed) section. The size of the equilibrium section is found from equilibrium adsorption data at the bed design temperature. The length of the equilibrium section represents the shortest bed length possible and can be described as the stoichiometric length since the adsorbent in the equilibrium section of the bed is assumed to be in equilibrium with the adsorbate in the fluid. The stoichiometric wavefront moves through the bed as a step function.

The equilibrium concept does not provide an accurate estimate of the bed length since the length of the mass transfer zone is not known. Because of the presence of the mass transfer zone, all of the adsorbent behind the actual wavefront will not be at its maximum capacity. Therefore, we must add an additional quantity of adsorbent to the bed to compensate for the presence of the mass transfer zone. This equivalent quantity of adsorbent is described as the LUB. The stoichiometric wavefront relative to the actual stable wavefront is shown in Figure 14-14. In this figure, t_b is defined as the time when the leading edge of the breakthrough curve leaves the bed, t_e is the time when the trailing edge of the wavefront leaves the bed, and t_s is the time when the stoichiometric wavefront would leave the bed. The stoichiometric front is found by equating the unused bed capacity behind the front to the used capacity ahead of the front. The stoichiometric time is found from Figure 14-14 by adjusting t_s until the areas

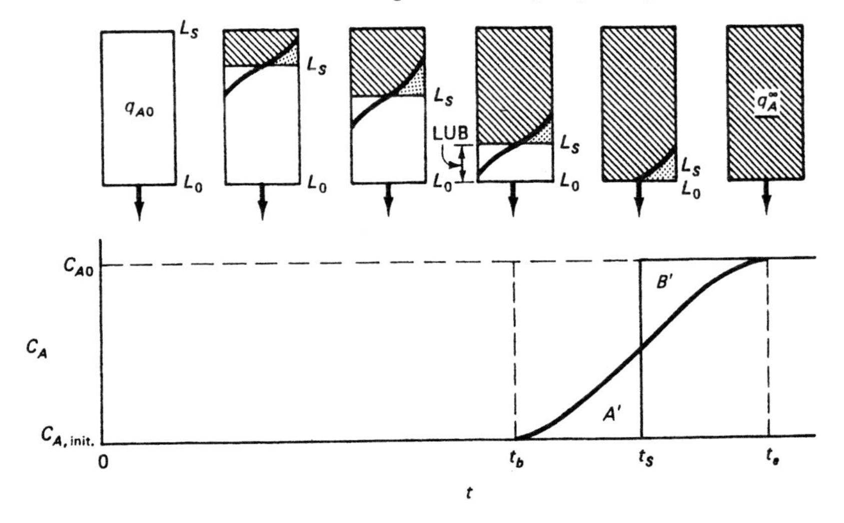

Figure 14-14 Stoichiometric front relative to the mass transfer zone

indicated by A' and B' are equal. At breakthrough, the leading edge of the actual wavefront relative to the stoichiometric front is

$$\text{LUB} = L_0 - L_s \tag{14-131}$$

where L_0 is the total bed length and L_s is the distance the stoichiometric front has moved through the bed.

The important assumptions on which the LUB/equilibrium section concept is based have been summarized by Collins (1967) as follows.

1. The holdup of the adsorbable component in the adsorbent voids is small compared to the equilibrium adsorbate loading.
2. The flow rate, temperature, and concentration of the feed are constant.
3. The temperature, composition, and velocity do not vary in the radial direction.
4. The bed temperature and initial adsorbate loading are uniform.
5. The initial bed and feed temperatures are equal.
6. Chemical reactions do not occur.
7. The bed operation is isothermal.
8. The mass transfer zone is stable.

The LUB method is developed by considering the velocities of the stoichiometric wavefront and the actual breakthrough curve. For any time, the length of the equilibrium or stoichiometric section is

$$L_s = u't \tag{14-132}$$

where u' is the velocity of the mass transfer wavefront. For the breakthrough time, t_b, we have

$$L_s = u't_b \tag{14-133}$$

At t_b the stoichiometric front has not yet moved through the entire length of the bed. At t_s, the length of the stoichiometric wavefront will exit the bed and $L_s = L_0$. Thus

$$L_s = u't_s = L_0 \tag{14-134}$$

From Eq. (14-131),

$$\text{LUB} = L_0 - L_s = u'(t_s - t_b) \tag{14-135}$$

Since $L_s = L_0 = u't_s$, we obtain

$$u' = \frac{L_0}{t_s}$$

Therefore, the length of the unused bed is

$$\text{LUB} = L_0\left(\frac{t_s - t_b}{t_s}\right) \tag{14-136}$$

To complete the model we must make a material balance across the equilibrium section of the bed. As shown by Collins (1967), the length of the stoichiometric wavefront can be expressed as

$$L_s = \frac{\epsilon U_Z(C_{A0} - C^*_{A,\text{init}})t_b M_A}{\rho_b(q_A^\infty - q_{A0})} \tag{14-137}$$

where q_A^∞ = saturation capacity of the adsorbent, g solute/g solid,
q_{A0} = initial concentration of solute on the adsorbent, g solute/g solid,
C_{A0} = concentration of solute in the influent, gmol/cm³,
$C^*_{A,\text{init}}$ = concentration of solute in equilibrium with q_{A0}, gmol/cm³,
M_A = molecular weight of the adsorbate.

From the equation above the velocity of the stoichiometric wavefront is

$$u' = \frac{L_s}{t_b} = \frac{\epsilon U_Z(C_{A0} - C^*_{A,\text{init}})M_A}{\rho_b(q_A^\infty - q_{A0})} \tag{14-138}$$

Examination of the material balance used by Collins (1967) shows that the maximum driving force for mass transfer was used for both phases and that the driving force is constant. The approximate nature of the material balance above should be recognized since the concentration differences vary with time and position. The LUB/equilibrium method does partially correct for the thickness of the mass transfer zone and is widely used. However, if this method is to be used successfully, an effluent breakthrough trace should be developed.

REFERENCES

AL-SAHHAF, T. A., E. D. SLOAN, and A. L. HINES, *Ind. Eng. Chem. Process Des. Dev.*, *20*, 658 (1981).

ANZELIUS, A., *Z. Angew. Math. Mech.*, *6*, 291 (1926).

BERING, B. P., M. M. DUBININ, and V. V. SPERPINSKY, *J. Colloid Interface Sci.*, *21*, 378 (1966).

BERING, B. P., M. M. DUBININ, and V. V. SPERPINSKY, *J. Colloid Interface Sci.*, *38*, 186 (1972).

BLAKLY, R. L., and B. N. TAYLOR, *Chem. Eng. Prog. Sym. Ser. 96*, *65*, 93 (1969).

BRUNAUER, S., *The Adsorption of Gases and Vapors*, Princeton University Press, Princeton, N.J., 1945.

BRUNAUER, S., P. H. EMMETT, and E. TELLER, *J. Am. Chem. Soc.*, *60*, 309 (1938).

COLLINS, J. J., *Chem. Eng. Prog. Sym. Ser. 74*, *63*, 31 (1967).

DUBININ, M. M., *Chem. Rev.*, *60*, 235 (1960).

DUBININ, M. M., *J. Colloid Interface Sci.*, *23*, 487 (1967).

EAGLETON, L. C., and H. BLISS, *Chem. Eng. Prog.*, *49*, 543 (1953).

EMMETT, P. H., *Catalysis*, Vol. 1, Reinhold, New York, 1954.

FARRIER, D. S., A. L. HINES, and S. W. WANG, *J. Colloid Interface Sci.*, *69*, 233 (1979).

GIBBS, J. W., *The Collected Works of J. W. Gibbs*, Vol. 1, Longman, Green, New York, 1931.

GLESSNER, A. J., "Sorption of Pure Gases and Binary Mixtures in Molecular Sieves," Ph.D. thesis, University of Pennsylvania, 1969.

GLESSNER, A. J., and A. L. MYERS, *Chem. Eng. Prog. Sym. Ser. 96*, *65*, 73 (1969).

GRANT, R. J., M. MANES, and S. B. SMITH, *AIChE J.*, *8*, 403 (1962).

HELFFERICH, F., *Ion Exchange*, McGraw-Hill, New York, 1962.

HIESTER, N. K., and T. VERMEULEN, *Chem. Eng. Prog.*, *48*, 505 (1952).

HILL, T. L., *J. Chem. Phys.*, *14*, 441 (1946).

HOUGEN, O. A., and W. R. MARSHALL, *Chem. Eng. Prog.*, *43*, 197 (1947).

KEMBALL, C., and E. K. RIDEAL, *Proc. R. Soc. (Lond.)*, *A187*, 53 (1946).

KLINKENBERG, A., *Ind. Eng. Chem.*, *40*, 1970 (1948).

LANGMUIR, I., *J. Am. Chem. Soc.*, *40*, 1361 (1918).

LEE, H., and W. P. CUMMINGS, *Chem. Eng. Prog. Sym. Ser. 74*, *63*, 42 (1967).

LEWIS, W. K., E. R. GILLILAND, B. CHERTOW, and W. P. CADOGAN, *Ind. Eng. Chem.* *42*, 1326 (1950).

LIGHTFOOT, E. N., R. J. SANCHEZ-PALMA, and D. O. EDWARDS, in *New Chemical Engineering Separation Techniques*, p. 99, Wiley-Interscience, New York, 1962.

POLANYI, M., *Z. Phys.*, *2*, 111 (1920).

ROSEN, J. B., *J. Chem. Phys.*, *20*, 387 (1952).

ROSEN, J. B., *Ind. Eng. Chem.*, *46*, 1590 (1954).

ROSS, S., and J. P. OLIVIER, *On Physical Adsorption*, Wiley-Interscience, New York, 1964.

ROUX, A., A. A. HUANG, Y. H. MA, and I. ZWIEBEL, *AIChE Sym. Ser. 134*, *69*, 53 (1973).

THOMAS, H. C., *J. Am. Chem. Soc.*, *66*, 1664 (1944).

VERMEULEN, T., G. KLEIN, and N. K. HIESTER, in J. H. PERRY, Ed., *Chemical Engineers' Handbook*, McGraw-Hill, New York, 1973.

VOLMER., M., *Z. Phys. Chem.*, *115*, 253 (1925).

YOUNG, D. M., and A. D. CROWELL, *Physical Adsorption of Gases*, Butterworth, London, 1962.

NOTATIONS

a = activity of the solute on the surface, dimensionless; constant in Van der Waals' equation, erg L^3/mol^2; external surface area of solid, L^2 solid/L^3 bed

a_1 = condensation factor, dimensionless

A = area per mole defined in Eq. (14-19), L^2 gas/mol solute

A_0 = co-area correction factor, L^2 gas/mol solute

b = constant in the van der Waals equation, L^3/mol

C = concentration of solute in fluid, mol solute/L^3 fluid; constant in BET equation [Eq. (14-40)], dimensionless

C_A = concentration of A in the fluid phase in equilibrium with the adsorbate concentration on the surface, mol A/L^3 fluid

C_{Ai} = interfacial concentration of A in the fluid phase, mol A/L^3 fluid

C_{As} = average concentration of A in the solid phase, mol A/L^3 solid

C_{Asi} = interfacial concentration of A in the solid phase, mol A/L^3 solid

C_{A0} = concentration of A in the influent, mol A/L^3 fluid

C_A^* = concentration of A in the fluid phase that is in equilibrium with the solid, mol A/L^3 fluid

$C_{A,\text{init}}^*$ = concentration of A in equilibrium with q_{A0}, mol A/L^3 fluid

C_s^∞ = concentration of solute on the surface of solid at saturation, mol solute/L^3 solid [Eq. (14-45)]; g solute/g solid [Eq. (14-137)]

d_p = diameter of solid particle, L

D_{A_a} = effective axial diffusion coefficient of A, L^2/t

E = dimensionless parameter defined in Eq. (14-97), dimensionless

f = fugacity of the gas at the adsorption pressure, F/L^2

f° = fugacity of the gas at its vapor pressure, F/L^2

$\bar{J}$ = concentration parameter defined in Eq. (14-68), dimensionless

k_a = rate constant for adsorption, L^3 fluid/mol solute t

k_B = Boltzmann constant, 1.38×10^{-16} erg/molecule K

k_d = rate constant for desorption, $1/t$

k_fa = fluid-phase mass transfer coefficient, L^3 fluid/L^3 bed t

k_sa = solid-phase mass transfer coefficient, $1/t$

k_1 = preexponential factor, $1/t$; rate constant, L^3 fluid/M solid t

k_2 = rate constant, $1/t$

k_1' = rate constant, L^3 fluid/mol solute t

K = adsorption equilibrium constant, L^2/F; integration constant, dimensionless

K' = constant in Eq. (14-27), L^2/F; equilibrium constant, L^3 fluid/mol solute

K_D = distribution coefficient, L^3 fluid/M solid

K_fa = overall mass transfer coefficient, L^3 fluid/L^3 bed t

K_{lim}' = limiting distribution coefficient between the fluid and solid, M solid/L^3 solid

L_0 = total bed length, L

L_s = distance the stoichiometric front has moved through the bed, L

LUB = length of unused bed, L

m = $\epsilon/(1-\epsilon)$, L^3 void/L^3 solid

m' = constant defined in Eq. (14-53), L^3 fluid/M solid

M_A = molecular weight of A, M solute/mol solute

n = constant in Eq. (14-12), dimensionless; actual number of moles adsorbed on the surface, mol

n_m = number of moles of solute adsorbed when the solid surface is completely covered, mol

N_0 = Avogadro's number, 6.023×10^{23} molecules/gmol

P = pressure, F/L^2

P° = pressure of the liquid solute, F/L^2
P_x = pressure of the gas phase, F/L^2
q_A = equilibrium uptake, $M\ A/M$ solid [Eq. (14-34)]; moles adsorbed on the surface of the adsorbent, mol A/M solid
q_{A0} = initial concentration of solute on the adsorbent, $M\ A/M$ solid
q^∞ = saturation capacity of the adsorbate in the bed, mol solute/L^3 solid; M solute/M solid [Eq. (14-137)]
q_{Ai} = interfacial concentration in solid phase, mol A/M solid
Q = weight of adsorbate for complete monolayer coverage, M solute/M solid
Q_m = moles adsorbed in a monomolecular layer, mol solute/M solid
$\bar{Q}$ = concentration parameter, dimensionless
Q' = maximum capacity of the adsorbent, mol solute/L^3 solid
r_a = adsorption rate, molecules/tL^2 solid
r_d = desorption rate, molecules/tL^2 solid
r^* = parameter defined in Eq. (14-103), dimensionless
R = gas constant, appropriate units; radius of the solid particle, L
S = specific surface area of the adsorbent, L^2/M solid; cross-sectional area of a column, L^2
S_g = total surface area of the solid, L^2/M solid
S_t = total surface area, L^2
S_0 = surface area of a particle, L^2 solid/M solid
t_b = time required for the solute in the effluent concentration to reach the breakpoint concentration, t
t_e = time at which the bed becomes saturated with adsorbate, t
t_i = time at which the solute first appears in the effluent stream, t
t_s = time when the stoichiometric wavefront would leave the bed, t
T = absolute temperature, K
u' = velocity of the mass transfer wavefront, L bed/t
U_Z = interstitial velocity, L bed/t
V = volume of gas adsorbed, L^3 gas/M solid
$\bar{V}_i$ = molar volume of the liquid adsorbate at state i, L^3 gas/mol solute
V_m = volume of gas adsorbed in a monolayer, L^3 gas/M solid
X = bed-length parameter, dimensionless
$\bar{X}$ = concentration parameter, dimensionless
Y = contact-time parameter, dimensionless
Z = bed length, L

Greek Letters

α = projected area of a molecule on the surface, L^2/molecule; attraction constant, erg L^2/molecule^2
γ = affinity coefficient, dimensionless; surface tension of the partially filled surface, F/L
γ_0 = surface tension of a clean surface, F/L
Γ = surface excess concentration, mol solute/L^2 solid
Δ_a = parameter defined in Eq. (14-102), $1/t$
Δ_E = parameter defined in Eq. (14-120), $1/t$

Δ_s = parameter defined in Eq. (14-125), $1/t$
ΔH_{st} = isosteric heat of adsorption, E/mol
ϵ = void fraction in the adsorption bed, L^3 void/L^3 bed
$1 - \epsilon$ = volume fraction of solid in the bed, L^3 solid/L^3 bed
$\epsilon(X)$ = adsorption potential, E/mol
ζ = bed-length parameter, dimensionless
Θ = fraction of the surface covered by adsorbed molecules, dimensionless; time variable, t
κ = general rate constant, M solid/L^3 solid t
κ_E = resistance due to the presence of external film, M solid/L^3 solid t
κ_S = resistance inside the particle, M solid/L^3 solid t
ν = film resistance parameter, dimensionless
π = spreading pressure due to adsorption, F/L
ρ = density of liquid adsorbate, M solute/L^3 solution
ρ_b = density of the adsorption bed, M solid/L^3 bed
ρ_s = particle density, M solid/L^3 solid
σ_0 = area per molecule defined as $\sigma_0 = A_0/N_0$, L^2/molecule
τ = time variable, dimensionless
$\phi(x)$ = volume of solute adsorbed at any point r, L^3 solute/M solid
χ = parameter defined in Eq. (14-84), M solid t/L^3 fluid

PROBLEMS

14.1 The adsorption of sulfur dioxide on mordenites was studied by Roux et al. (1973) at 0°C. **(a)** Using the data below, determine the Langmuir constants, and **(b)** calculate the total surface area of the solid. The density of liquid SO_2 at 0°C in the adsorbed phase is 1.43 g/cm³.

P_{SO_2} *(mmHg)*	*Uptake (mmol/g)*
5	1.75
10	2.20
15	2.40
20	2.62
30	2.75
40	2.85
50	3.00
60	3.05
70	3.12

14.2 The equilibrium adsorption isotherm for nitrogen on silica gel was obtained by Emmett (1954) at −195.8°C. Using the data below, determine the change of spreading pressure as a function of the pressure of nitrogen in the vapor phase. The surface area of the adsorbent (silica gel) is 550 m²/g.

P_{N_2} (mmHg)	Uptake [cm³(STP)/g]
27.69	60.61
146.15	94.55
200.00	101.82
261.54	115.15
292.31	120.00
361.54	134.55
433.85	157.58

14.3 Ethylene is to be separated from butadiene by flowing the gas mixture through a packed bed of molecular sieve at a temperature of 15°C and a pressure of 1.0 atm. From an analysis of the data it was found that only 20% of the adsorbent surface was covered. Assuming that the Hill–de Boer model can be used to describe the adsorption system, estimate the fraction of surface coverage at a pressure of 10 atm and 15°C.

14.4 Emmett (1954) studied the adsorption of argon on 0.606 g of silica gel at −183°C. Using the BET equation and the data below, calculate the surface area of the adsorbent.

P (mmHg)	Volume Adsorbed [cm³(STP)]
78.46	55.03
176.92	72.73
224.62	80.00
378.46	106.67
432.31	117.58
515.38	138.18
584.62	166.06

The surface area of the adsorbent can be related to the volume of gas adsorbed in the monolayer by the expression

$$S_g = 4.35 \times 10^4 V_m$$

where S_g is the surface area per unit weight of the adsorbent (cm²/g solid) and V_m is the volume of gas adsorbed in the monolayer (cm³/g).

14.5 The adsorption of phenol from an aqueous solution onto XAD-8 resin was studied by Farrier et al. (1979). The isosteric heat of adsorption can be expressed by

$$\frac{\Delta H_{st}}{R} = -\left[\frac{\partial(\ln C_f)}{\partial\left(\frac{1}{T}\right)}\right]_q$$

where C_f is the equilibrium concentration and q is the uptake. From the data below, calculate the heat of adsorption at $q = 1.0$ mg of phenol per gram of solid.

Temperature (°C)	*q (mg/g solid)*	C_f *(mg/ml solution)*
0	0.5	0.00082
	1.0	0.002
	2.0	0.005
25	0.5	0.0015
	1.0	0.005
	2.1	0.008
75	0.5	0.005
	1.0	0.010
	2.0	0.020

14.6 The adsorption of carbon dioxide on molecular sieve, type 5A, was studied by Blakly and Taylor (1969) at three different temperatures. **(a)** Using the data below, determine if the Polanyi potential theory can be used to model the data, and **(b)** calculate the isosteric heat of adsorption.

	Equilibrium Uptake (lb CO_2/lb solid)		
P_{CO_2} *(mmHg)*	$T = 32°F$	$T = 50.9°F$	$T = 64.4°F$
1.0	0.043	0.032	0.024
2.0	0.066	0.048	0.038
3.0	0.078	0.062	0.047
4.0	0.086	0.072	0.055
5.0	0.092	0.080	0.063
6.0	0.098	0.086	0.070
7.0	0.102	0.090	0.076
8.0	0.105	0.093	0.080
9.0	0.108	0.098	0.083

14.7 Starting with Eqs. (14-64) and (14-65) and the boundary conditions given by Eqs. (14-66) and (14-67), derive the dimensionless concentration profile for adsorption in a packed bed.

14.8 An adsorption tower packed with activated charcoal is to be used to remove trace quantities of radioactive CO_2 from an air stream at 25°C and 1 atm. The air enters the tower at a superficial velocity of 4 cm/s. Using the Rosen model, calculate the height of packing needed for this separation if the breakthrough time is to be 2 h. Assume that breakthrough occurs when the effluent air stream contains 5% of the entering CO_2. The properties are as follows:

$$D_{CO_2\text{-air}} = 0.161 \text{ cm}^2/\text{s}$$

void fraction = 0.43

particle diameter = 0.3 cm

$$\text{distribution coefficient, } K_D\rho_s = 155 \frac{\text{g CO}_2/\text{cm}^3 \text{ solid}}{\text{g CO}_2/\text{cm}^3 \text{ gas}}$$

14.9 A 16.8-cm-long adsorption bed packed with silica gel is to be used to remove ammonia from a helium gas at a pressure of 2.4 atm absolute and 25°C. The initial concentration of NH_3 in the influent gas is 9.442×10^{-3} kg/m³, and the flow rate through the bed is 1.642×10^4 cm³/h. Using the adsorption data below, predict the breakthrough curve if the operation is isothermal. Assume that the rate of adsorption is controlled by second-order kinetics and that the Thomas model can be applied to this system. Compare your results to the experimental data given here.

C/C_0	t (min/g solid)
0.002	26.50
0.151	30.52
0.234	32.25
0.350	33.56
0.465	34.34
0.583	35.52
0.695	36.50
0.818	37.93
0.930	39.24
0.990	40.42

The system properties are:

mass transfer coefficient $k_a = 0.44\left(\frac{1}{\text{h}}\right)\left(\frac{1}{\text{mg}}\right)$

equilibrium adsorption constant $K' = 0.68596$ liter/mg

bed porosity = 0.418

mass of adsorbent = 5.35 g

maximum adsorption capacity $C_s^\infty = 91.72$ mg/g

bulk density = 426.5 g/liter

bed length = 16.8 cm

bed diameter = 0.975 cm

14.10 Starting with Eq. (14-55) and the dimensionless groups given in Eqs. (14-104) through (14-107), derive Eq. (14-109). Recall that $C = f(\zeta, \tau)$.

Viscosity of Gases and Liquids

VISCOSITIES OF GASES, COORDINATES FOR FIGURE A-1

Gas	*X*	*Y*	*Gas*	*X*	*Y*
Acetic acid	7.7	14.3	Freon-113	11.3	14.0
Acetone	8.9	13.0	Helium	10.9	20.5
Acetylene	9.8	14.9	Hexane	8.6	11.8
Air	11.0	20.0	Hydrogen	11.2	12.4
Ammonia	8.4	16.0	$3H_2 + 1N_2$	11.2	17.2
Argon	10.5	22.4	Hydrogen bromide	8.8	20.9
Benzene	8.5	13.2	Hydrogen chloride	8.8	18.7
Bromine	8.9	19.2	Hydrogen cyanide	9.8	14.9
Butene	9.2	13.7	Hydrogen iodide	9.0	21.3
Butylene	8.9	13.0	Hydrogen sulfide	8.6	18.0
Carbon dioxide	9.5	18.7	Iodine	9.0	18.4
Carbon disulfide	8.0	16.0	Mercury	5.3	22.9
Carbon monoxide	11.0	20.0	Methane	9.9	15.5
Chlorine	9.0	18.4	Methyl alcohol	8.5	15.6
Chloroform	8.9	15.7	Nitric oxide	10.9	20.5
Cyanogen	9.2	15.2	Nitrogen	10.6	20.0
Cyclohexane	9.2	12.0	Nitrosyl chloride	8.0	17.6
Ethane	9.1	14.5	Nitrous oxide	8.8	19.0
Ethyl acetate	8.5	13.2	Oxygen	11.0	21.3
Ethyl alcohol	9.2	14.2	Pentane	7.0	12.8
Ethyl chloride	8.5	15.6	Propane	9.7	12.9
Ethyl ether	8.9	13.0	Propyl alcohol	8.4	13.4
Ethylene	9.5	15.1	Propylene	9.0	13.8
Fluorine	7.3	23.8	Sulfur dioxide	9.6	17.0
Freon-11	10.6	15.1	Toluene	8.6	12.4
Freon-12	11.1	16.0	2,3,3-Trimethylbutane	9.5	10.5
Freon-21	10.8	15.3	Water	8.0	16.0
Freon-22	10.1	17.0	Xenon	9.3	23.0

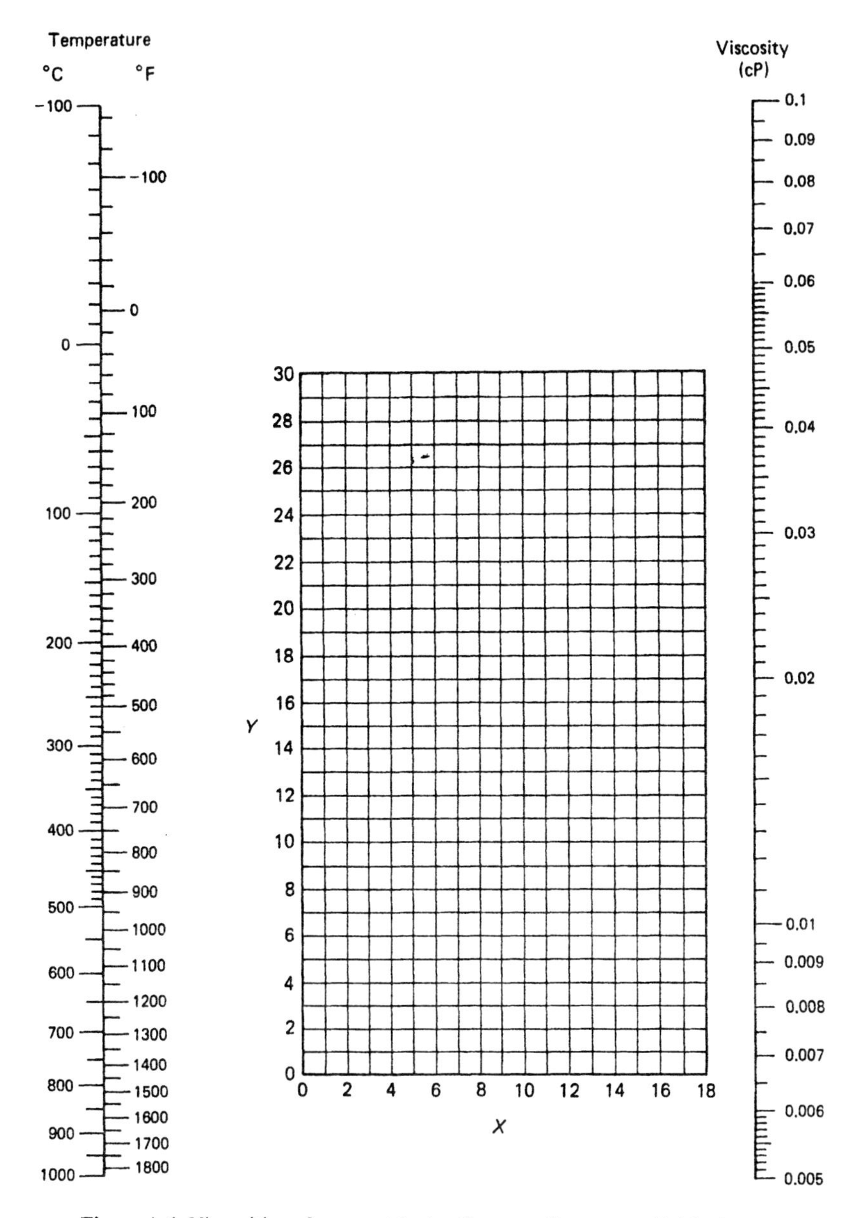

Figure A-1 Viscosities of gases at 1 atm (for coordinates, see Table A-1.)

VISCOSITIES OF LIQUIDS, COORDINATES FOR FIGURE A-2[a]

Liquid	*X*	*Y*	*Liquid*	*X*	*Y*
Acetaldehyde	15.2	4.8	Ethyl formate	14.2	8.4
Acetic acid, 100%	12.1	14.2	Ethyl iodide	14.7	10.3
Acetic acid, 70%	9.5	17.0	Ethylene glycol	6.0	23.6
Acetic anhydride	12.7	12.8	Formic acid	10.7	15.8
Acetone, 100%	14.5	7.2	Freon-11	14.4	9.0
Acetone, 35%	7.9	15.0	Freon-12	16.8	5.6
Allyl alcohol	10.2	14.3	Freon-21	15.7	7.5
Ammonia, 100%	12.6	2.0	Freon-22	17.2	4.7
Ammonia, 26%	10.1	13.9	Freon-113	12.5	11.4
Amyl acetate	11.8	12.5	Glycerol, 100%	2.0	30.0
Amyl alcohol	7.5	18.4	Glycerol, 50%	6.9	19.6
Aniline	8.1	18.7	Heptane	14.1	8.4
Anisole	12.3	13.5	Hexane	14.7	7.0
Arsenic trichloride	13.9	14.5	Hydrochloric acid, 31.5%	13.0	16.6
Benzene	12.5	10.9	Isobutyl alcohol	7.1	18.0
Bimethyl oxalate	12.3	15.8	Isobutyric acid	12.2	14.4
Biphenyl	12.0	18.3	Isopropyl alcohol	8.2	16.0
Brine, $CaCl_2$, 25%	6.6	15.9	Kerosene	10.2	16.9
Brine, NaCl, 25%	10.2	16.6	Linseed oil, raw	7.5	27.2
Bromine	14.2	13.2	Mercury	18.4	16.4
Bromotoluene	20.0	15.9	Methanol, 100%	12.4	10.5
Butyl acetate	12.3	11.0	Methanol, 90%	12.3	11.8
Butyl alcohol	8.6	17.2	Methanol, 40%	7.8	15.5
Butyric acid	12.1	15.3	Methyl acetate	14.2	8.2
Carbon dioxide	11.6	0.3	Methyl chloride	15.0	3.8
Carbon disulfide	16.1	7.5	Methyl ethyl ketone	13.9	8.6
Carbon tetrachloride	12.7	13.1	Naphthalene	7.9	18.1
Chlorobenzene	12.3	12.4	Nitric acid, 95%	12.8	13.8
Chloroform	14.4	10.2	Nitric acid, 60%	10.8	17.0
Chlorosulfonic acid	11.2	18.1	Nitrobenzene	10.6	16.2
o-Chlorotoluene	13.0	13.3	Nitrotoluene	11.0	17.0
m-Chlorotoluene	13.3	12.5	Octane	13.7	10.0
p-Chlorotoluene	13.3	12.5	Octyl alcohol	6.6	21.1
m-Cresol	2.5	20.8	Pentachloroethane	10.9	17.3
Cyclohexanol	2.9	24.3	Pentane	14.9	5.2
Dibromoethane	12.7	15.8	Phenol	6.9	20.8
Dichloroethane	13.2	12.2	Phosphorus tribromide	13.8	16.7
Dichloromethane	14.6	8.9	Phosphorus trichloride	16.2	10.9
Diethyl oxalate	11.0	16.4	Propionic acid	12.8	13.8
Dipropyl oxalate	10.3	17.7	Propyl alcohol	9.1	16.5
Ethyl acetate	13.7	9.1	Propyl bromide	14.5	9.6
Ethyl alcohol, 100%	10.5	13.8	Propyl chloride	14.4	7.5
Ethyl alcohol, 95%	9.8	14.3	Propyl iodide	14.1	11.6
Ethyl alcohol, 40%	6.5	16.6	Sodium	16.4	13.9
Ethylbenzene	13.2	11.5	Sodium hydroxide, 50%	3.2	25.8
Ethyl bromide	14.5	8.1	Stannic chloride	13.5	12.8
Ethyl chloride	14.8	6.0	Sulfur dioxide	15.2	7.1
Ethyl ether	14.5	5.3	Sulfuric acid, 110%	7.2	27.4

VISCOSITIES OF LIQUIDS, COORDINATES FOR FIGURE A-2 (CONTINUED)

Liquid	*X*	*Y*	*Liquid*	*X*	*Y*
Sulfuric acid, 98%	7.0	24.8	Trichloroethylene	14.8	10.5
Sulfuric acid, 60%	10.2	21.3	Turpentine	11.5	14.9
Sulfuryl chloride	15.2	12.4	Vinyl acetate	14.0	8.8
Tetrachloroethane	11.9	15.7	Water	10.2	13.0
Tetrachloroethylene	14.2	12.7	*o*-Xylene	13.5	12.1
Titanium tetrachloride	14.4	12.3	*m*-Xylene	13.9	10.6
Toluene	13.7	10.4	*p*-Xylene	13.9	10.9

[a]From R. H. Perry and C. H. Chilton, Eds., *Chemical Engineers' Handbook*, 5th ed., McGraw-Hill, New York, 1973.

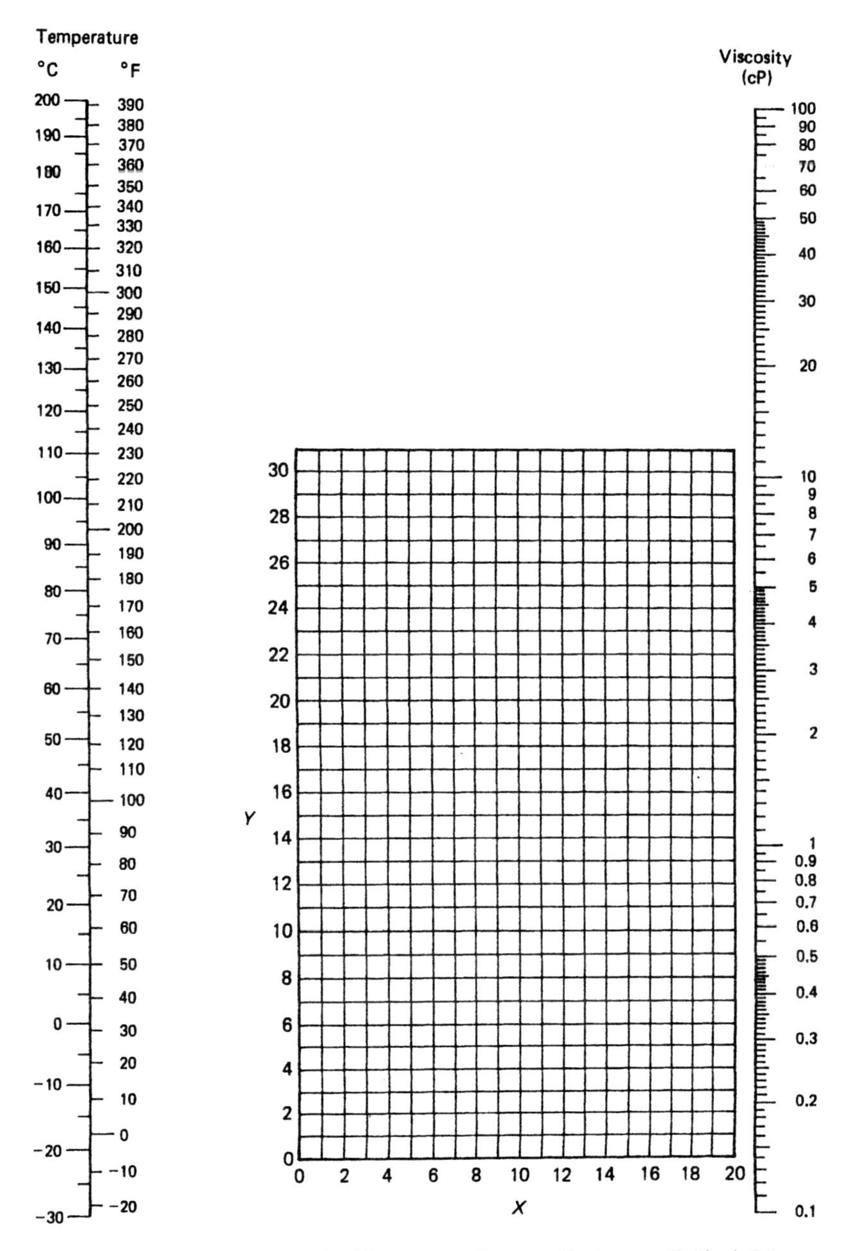

Figure A-2 Viscosities of liquids at 1 atm (for coordinates, see Table A-2.)

Equilibrium Data

APPENDIX

TABLE B-1. EQUILIBRIUM DATA FOR NH_3–WATER[a]

Mass NH_3 *per 100 Masses* H_2O	*Partial Pressure of* NH_3 *in Vapor (mmHg)*			
	20°*C*	30°*C*	40°*C*	50°*C*
100				
90				
80				
70				
60	945			
50	686			
40	470	719		
30	298	454	692	
25	227	352	534	825
20	166	260	395	596
15	114	179	273	405
10	69.6	110	167	247
7.5	50.0	79.7	120	179
5	31.7	51.0	76.5	115
4	24.9	40.1	60.8	91.1
3	18.2	29.6	45.0	67.1
2	12.0	19.3	30.0	44.5
1			15.4	22.2

[a]From T. K. Sherwood, *Ind. Eng. Chem.*, *17*, 745 (1925).

TABLE B-2. EQUILIBRIUM DATA FOR SO_2–WATER[a]

Mass SO_2 *per 100 Masses* H_2O	*Partial Pressure of* SO_2 *in Vapor (mmHg)*			
	20°C	30°C	40°C	50°C
20				
15				
10	698			
7.5	517	688		
5.0	336	452	665	
2.5	161	216	322	458
1.5	92	125	186	266
1.0	59	79	121	172
0.7	39.0	52	87	116
0.5	26.0	36	57	82
0.3	14.1	19.7		
0.1	3.2	4.7	7.5	12.0
0.05	1.2	1.7	2.8	4.7
0.02	0.5	0.6	0.8	1.3

[a]From T. K. Sherwood, *Ind. Eng. Chem.*, *17*, 745 (1925).

TABLE B-3. EQUILIBRIUM DATA FOR ACETONE–ETHANOL AT 1 ATM[a]

	Mole Percent Acetone			*Mole Percent Acetone*	
Temp. (°C)	x_A	y_A	*Temp.* (°C)	x_A	y_A
78.3	0.0	0	63.6	40.0	60.5
75.4	5.0	15.5	61.8	50.0	67.4
73.0	10.0	26.2	60.4	60.0	73.9
71.0	15.0	34.8	59.1	70.0	80.2
69.0	20.0	41.7	58.0	80.0	86.5
67.3	25.0	47.8	57.0	90.0	92.9
65.9	30.0	52.4	56.1	100.0	100.0
64.7	35.0	56.6			

[a]Data from R. H. Perry, C. H. Chilton, and S. D. Kirkpatrick, Eds., *Chemical Engineers' Handbook*, 4th ed., McGraw-Hill, New York, 1963.

TABLE B-4. EQUILIBRIUM DATA FOR METHANOL–WATER AT 1 ATM[a]

Temp. (°C)	Mole Percent Methanol x_A	Mole Percent Methanol y_A	Temp. (°C)	Mole Percent Methanol x_A	Mole Percent Methanol y_A
100.0	0.0	0.0	75.3	40.0	72.9
96.4	2.0	13.4	73.1	50.0	77.9
93.5	4.0	23.0	71.2	60.0	82.5
91.2	6.0	30.4	69.3	70.0	87.0
89.3	8.0	36.5	67.6	80.0	91.5
87.7	10.0	41.8	66.0	90.0	95.8
84.4	15.0	51.7	65.0	95.0	97.9
81.7	20.0	57.9	64.5	100.0	100.0
78.0	30.0	66.5			

[a]Data from R. H. Perry, C. H. Chilton, and S. D. Kirkpatrick, Eds., *Chemical Engineers' Handbook*, 4th ed., McGraw-Hill, New York, 1963.

TABLE B-5. EQUILIBRIUM DATA FOR ETHANOL–WATER AT 1 ATM[a]

Temp. (°C)	Mole Percent Ethanol x_A	Mole Percent Ethanol y_A	Temp. (°C)	Mole Percent Ethanol x_A	Mole Percent Ethanol y_A
100.0	0.0	0.0	81.5	32.73	58.26
95.5	1.90	17.00	80.7	39.65	61.22
89.0	7.21	38.91	79.8	50.79	65.64
86.7	9.66	43.75	79.7	51.98	65.99
85.3	12.38	47.04	79.3	57.32	68.41
84.1	16.61	50.89	78.74	67.63	73.85
82.7	23.37	54.45	78.41	74.72	78.15
82.3	26.08	55.80	78.15	89.43	89.43

[a]Data from R. H. Perry, C. H. Chilton, and S. D. Kirkpatrick, Eds., *Chemical Engineers' Handbook*, 4th ed., McGraw-Hill, New York, 1963.

TABLE B-6. HENRY'S LAW CONSTANTS IN AQUEOUS SOLUTIONS[a, b]

	$H_A \times 10^{-4}$ (*atm/mol fraction*)									
Temp. (°C)	CO_2	CO	C_2H_6	C_2H_4	He	H_2	H_2S	CH_4	N_2	O_2
0	0.0728	3.52	1.26	0.552	12.9	5.79	2.68	2.24	5.29	2.55
10	0.104	4.42	1.89	0.768	12.6	6.36	3.67	2.97	6.68	3.27
20	0.142	5.36	2.63	1.02	12.5	6.83	4.83	3.76	8.04	4.01
30	0.186	6.20	3.42	1.27	12.4	7.29	6.09	4.49	9.24	4.75
40	0.233	6.96	4.23	—	12.1	7.51	7.45	5.20	10.4	5.35
50	0.283	7.61	5.00	—	11.5	7.65	8.84	5.77	11.3	5.88

[a]Data from R. H. Perry, C. H. Chilton, and S. D. Kirkpatrick, Eds., *Chemical Engineers' Handbook*, 4th ed., McGraw-Hill, New York, 1963.
[b]$\bar{P}_A = H_A x_A$, where $\bar{P}_A$ = partial pressure of A in the gas phase (atm), x_A = mole fraction, and H_A = Henry's constant.

TABLE B-7. TERNARY EQUILIBRIUM DATA FOR WATER–ACETIC ACID–METHYL ISOBUTYL KETONE AT 25°C[a]

Water Layer (wt %)			*Methyl Isobutyl Ketone Layer (wt %)*		
Water	*Acetic Acid*	*Methyl Isobutyl Ketone*	*Water*	*Acetic Acid*	*Methyl Isobutyl Ketone*
98.45	0.0	1.55	2.12	0.0	97.88
95.45	2.85	1.70	2.80	1.87	95.33
85.8	11.7	2.5	5.4	8.9	85.7
75.7	20.5	3.8	9.2	17.3	73.5
67.8	26.2	6.0	14.5	24.6	60.9
55.0	32.8	12.2	22.0	30.8	47.2
42.9	34.6	22.5	31.0	33.6	35.4

[a]From T. K. Sherwood, J. E. Evans, and J. V. A. Longcor, *Ind. Eng. Chem.*, *31*, 1144 (1939).

TABLE B-8. TERNARY EQUILIBRIUM DATA FOR ETHYLBENZENE–STYRENE–DIETHYLENE GLYCOL AT 25°C[a]

Ethylbenzene Layer (wt %)			*Ethylene Glycol Layer (wt %)*		
Ethylbenzene	*Styrene*	*Diethylene Glycol*	*Ethylbenzene*	*Styrene*	*Diethylene Glycol*
90.56	8.63	0.81	9.85	1.64	88.51
80.40	18.67	0.93	9.31	3.49	87.20
70.49	28.51	1.00	8.72	5.48	85.80
60.93	37.98	1.09	8.07	7.45	84.48
52.96	45.84	1.20	7.31	9.49	83.20
53.55	45.25	1.20	7.35	9.25	83.40
41.51	57.09	1.40	6.06	12.54	81.40
43.29	55.32	1.39	6.30	12.00	81.70
21.60	76.60	1.8	3.73	18.62	77.65

[a]From M. G. Boobar et al., *Ind. Eng. Chem.*, *43*, 2922 (1951).

Equilibrium K Values

APPENDIX

C

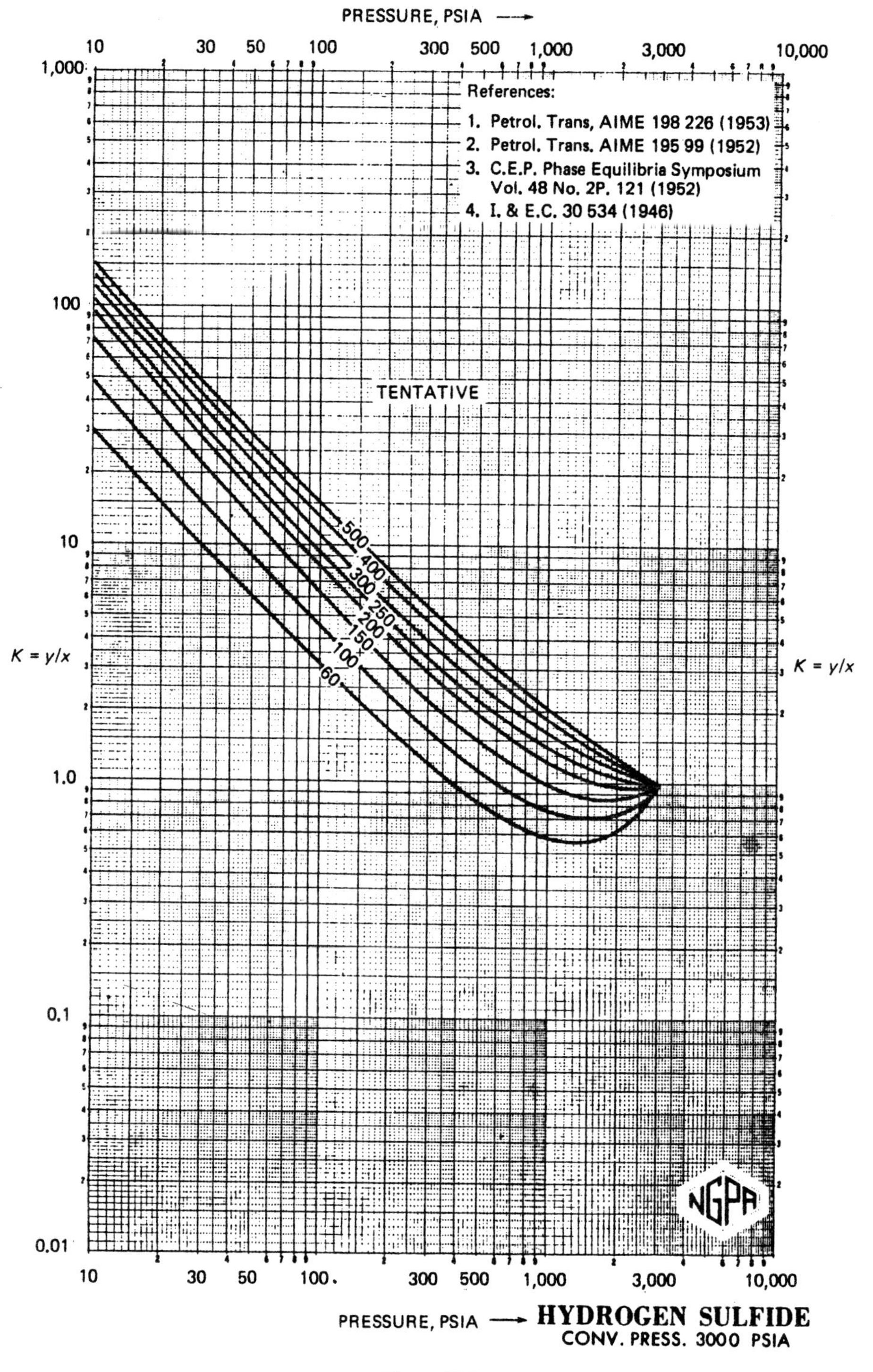
PRESSURE, PSIA
10
30
50
100
300
500
1,000
3,000
10,000
1,000
100
10
1.0
0.1
0.01
K = y/x
References:
1. Petrol. Trans, AIME 198 226 (1953)
2. Petrol. Trans. AIME 195 99 (1952)
3. C.E.P. Phase Equilibria Symposium Vol. 48 No. 2P. 121 (1952)
4. I. & E.C. 30 534 (1946)
TENTATIVE
500
400
300
250
200
150
100
60
NGPA
HYDROGEN SULFIDE
CONV. PRESS. 3000 PSIA

Figure C-1

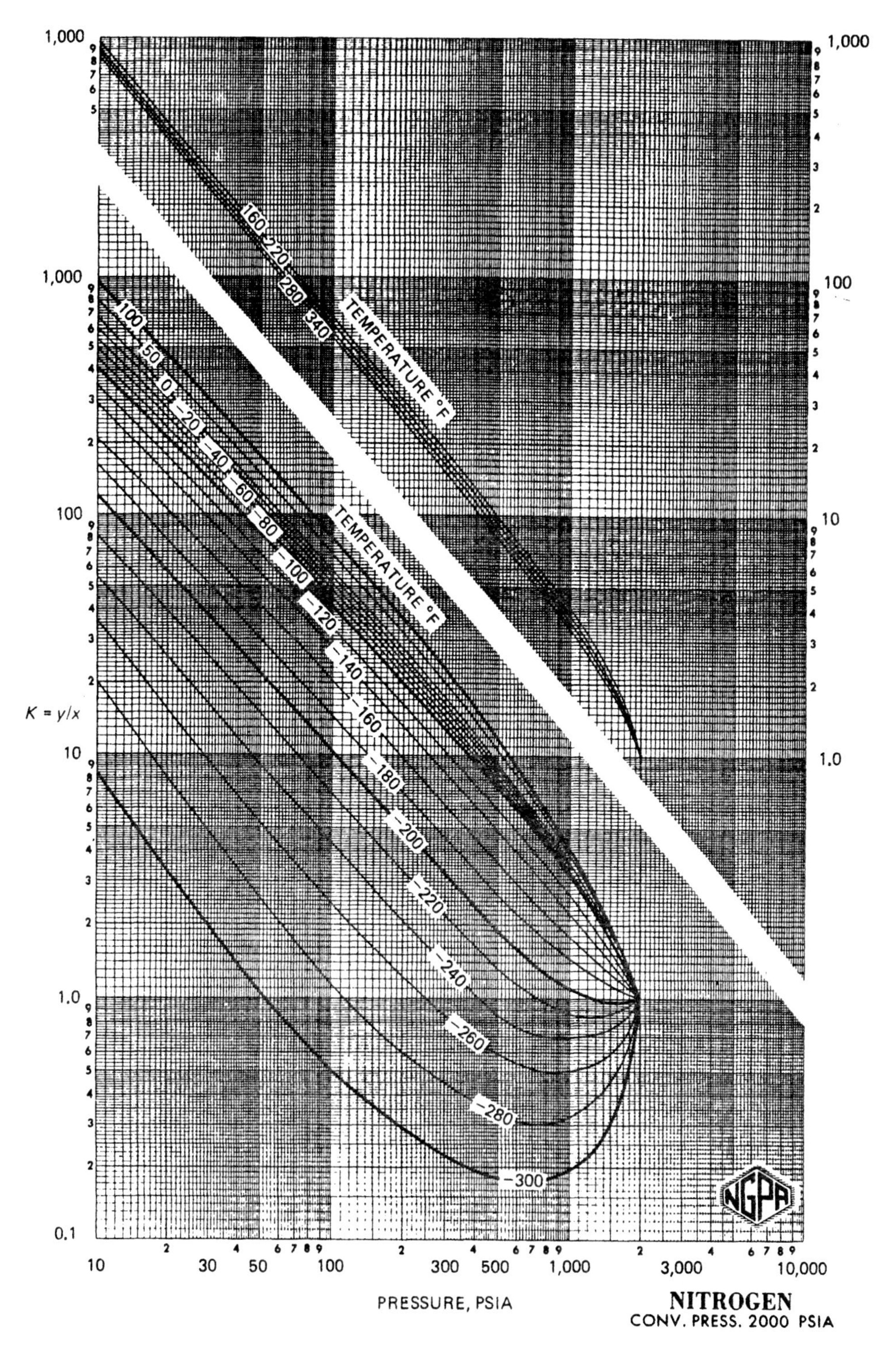

K = y/x
PRESSURE, PSIA
TEMPERATURE °F
TEMPERATURE °F
160
220
280
340
100
50
0
−20
−40
−60
−80
−100
−120
−140
−160
−180
−200
−220
−240
−260
−280
−300
NITROGEN
CONV. PRESS. 2000 PSIA
NGPA

Figure C-2

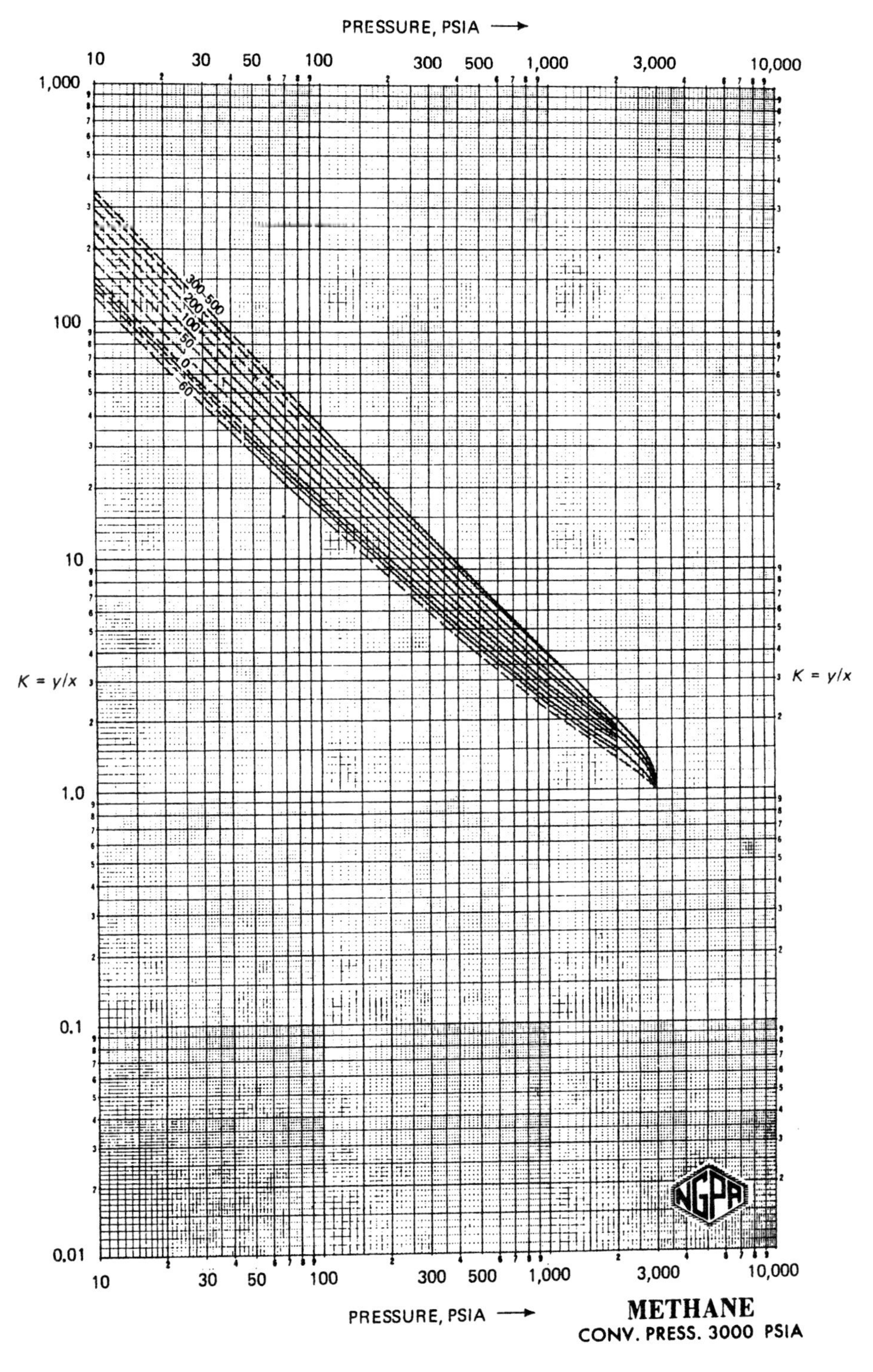
PRESSURE, PSIA
10
30
50
100
300
500
1,000
3,000
10,000
1,000
100
10
1.0
0.1
0.01
K = y/x
K = y/x
300-500
200
100
50
0
-60
PRESSURE, PSIA
NGPA
METHANE
CONV. PRESS. 3000 PSIA

Figure C-3

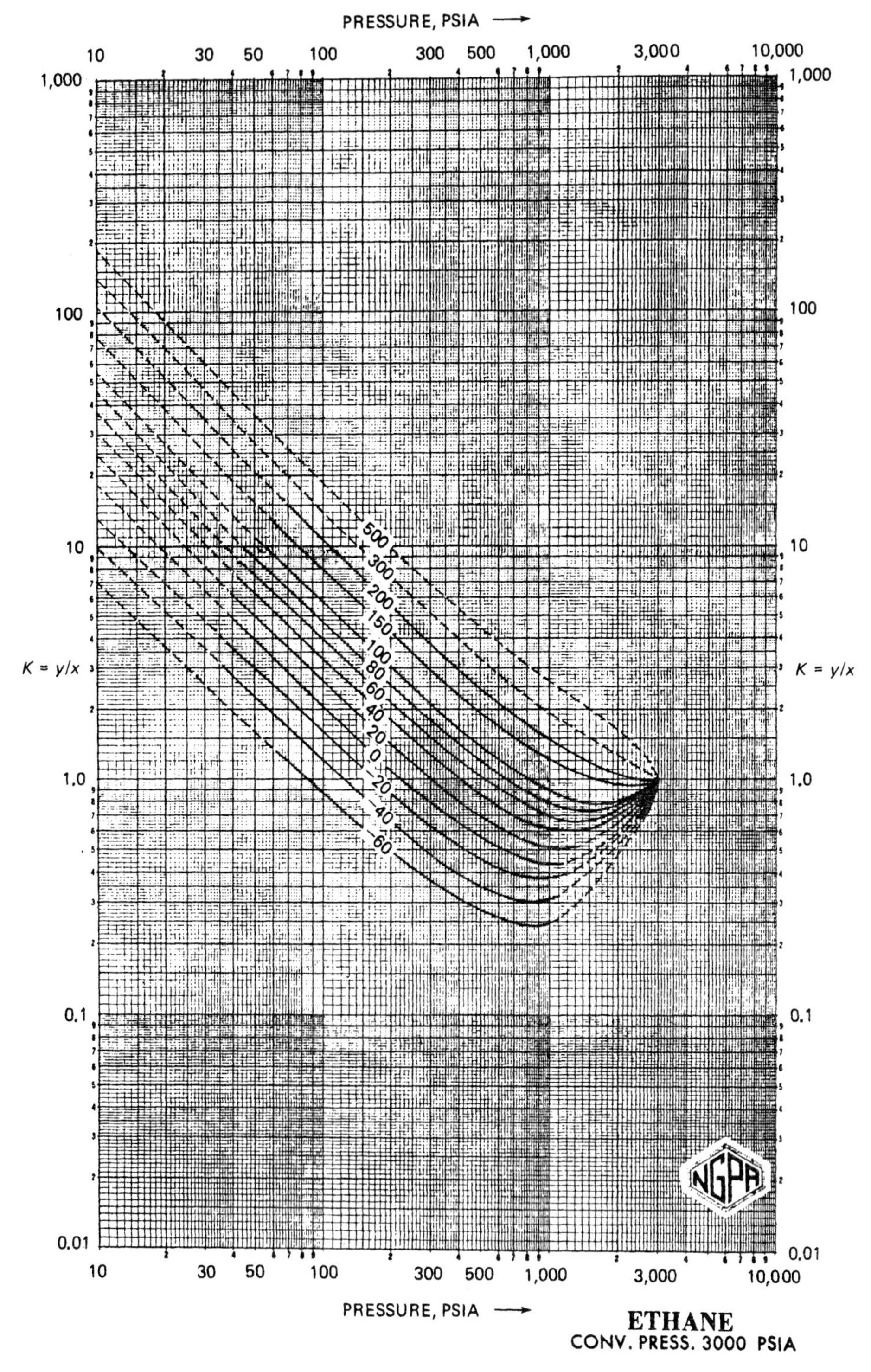
PRESSURE, PSIA
10
30
50
100
300
500
1,000
3,000
10,000
1,000
100
10
1.0
0.1
0.01
K = y/x
500
300
200
150
100
80
60
40
20
0
-20
-40
-60
NGPA
ETHANE
CONV. PRESS. 3000 PSIA

Figure C-4

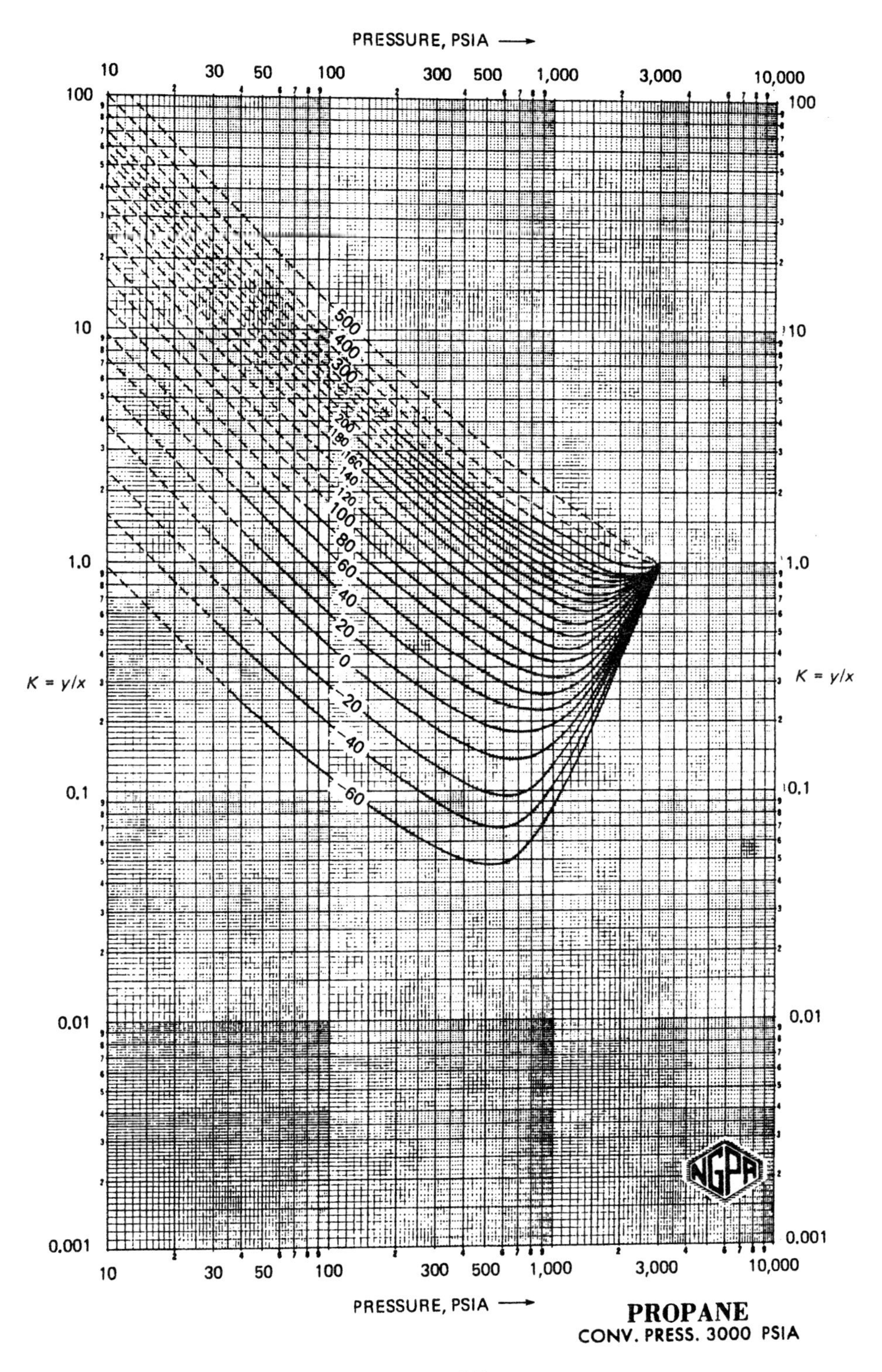
PRESSURE, PSIA
10
30
50
100
300
500
1,000
3,000
10,000
100
10
1.0
0.1
0.01
0.001
K = y/x
500
400
300
200
180
160
140
120
100
80
60
40
20
0
-20
-40
-60
NGPA
PROPANE
CONV. PRESS. 3000 PSIA

Figure C-5

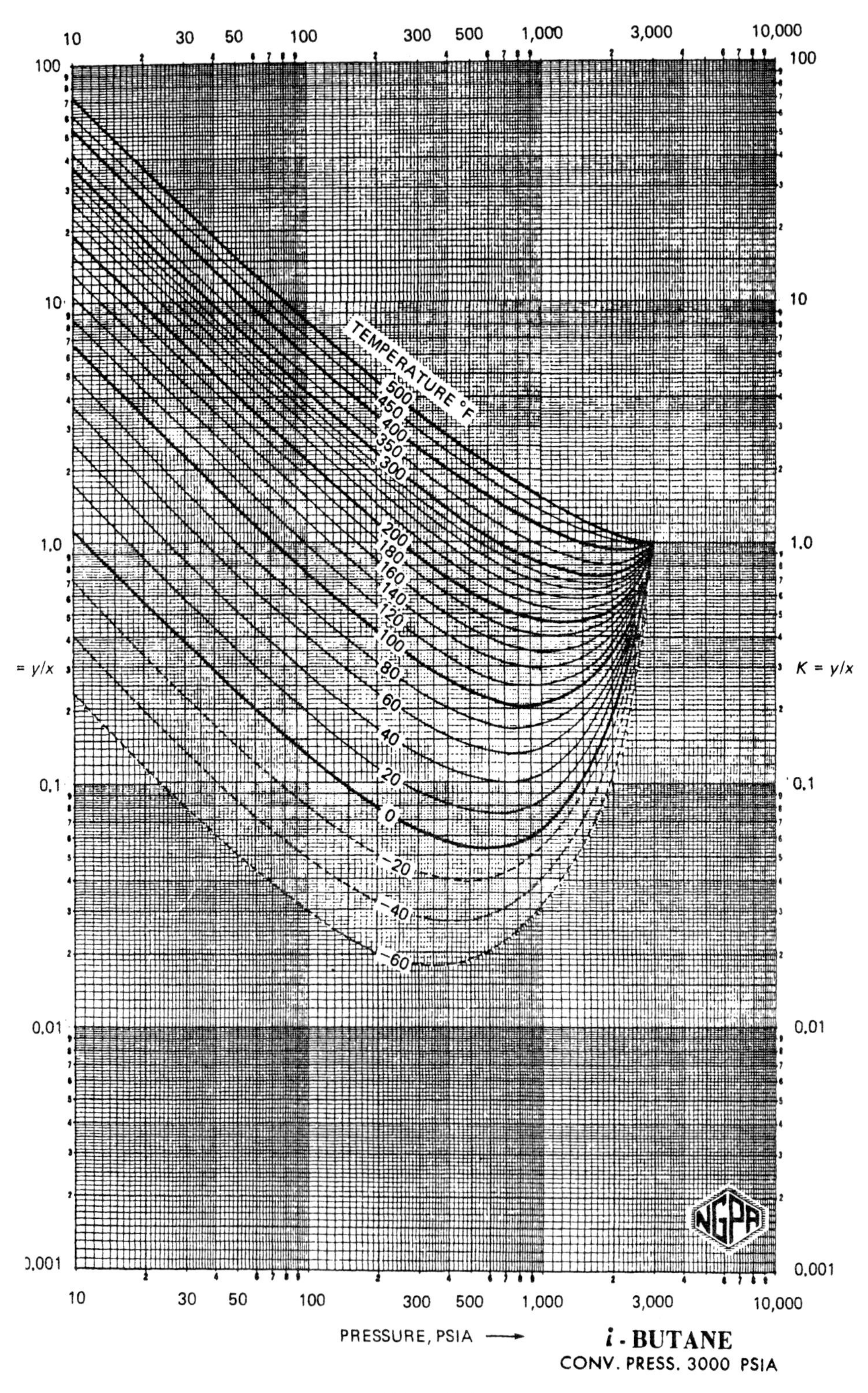
10
30
50
100
300
500
1,000
3,000
10,000
100
10
1.0
0.1
0.01
0.001
= y/x
K = y/x
TEMPERATURE °F
500
450
400
350
300
200
180
160
140
120
100
80
60
40
20
0
−20
−40
−60
PRESSURE, PSIA
i - BUTANE
CONV. PRESS. 3000 PSIA
NGPA

Figure C-6

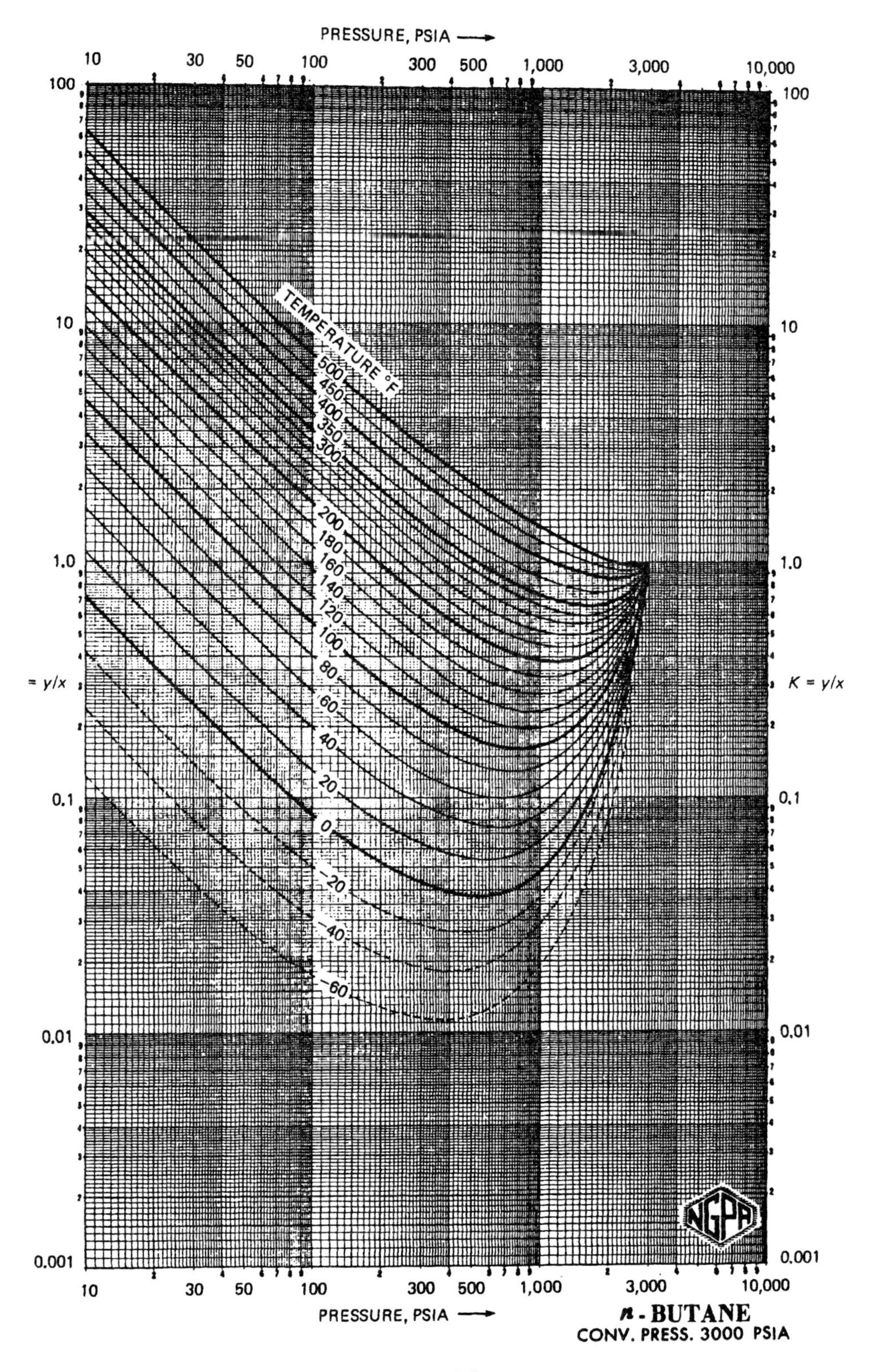

PRESSURE, PSIA
10
30
50
100
300
500
1,000
3,000
10,000
100
10
1.0
0.1
0.01
0.001
K = y/x
TEMPERATURE °F
500
450
400
350
300
200
180
160
140
120
100
80
60
40
20
0
−20
−40
−60
NGPA
n - BUTANE
CONV. PRESS. 3000 PSIA

Figure C-7

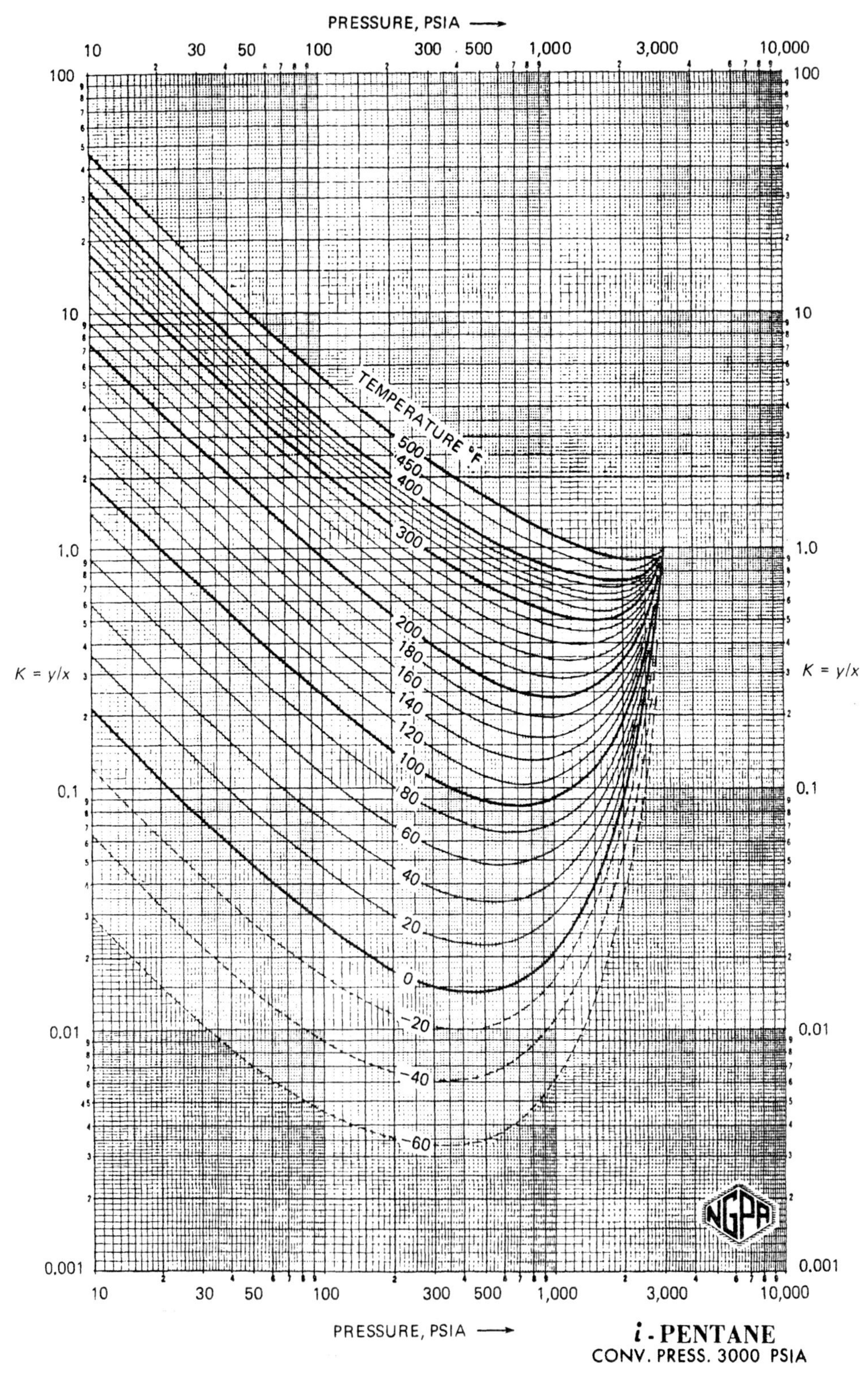

PRESSURE, PSIA
10
30
50
100
300
500
1,000
3,000
10,000
100
10
1.0
0.1
0.01
0.001
K = y/x
TEMPERATURE °F
500
450
400
300
200
180
160
140
120
100
80
60
40
20
0
-20
-40
-60
NGPA
i-PENTANE
CONV. PRESS. 3000 PSIA

Figure C-8

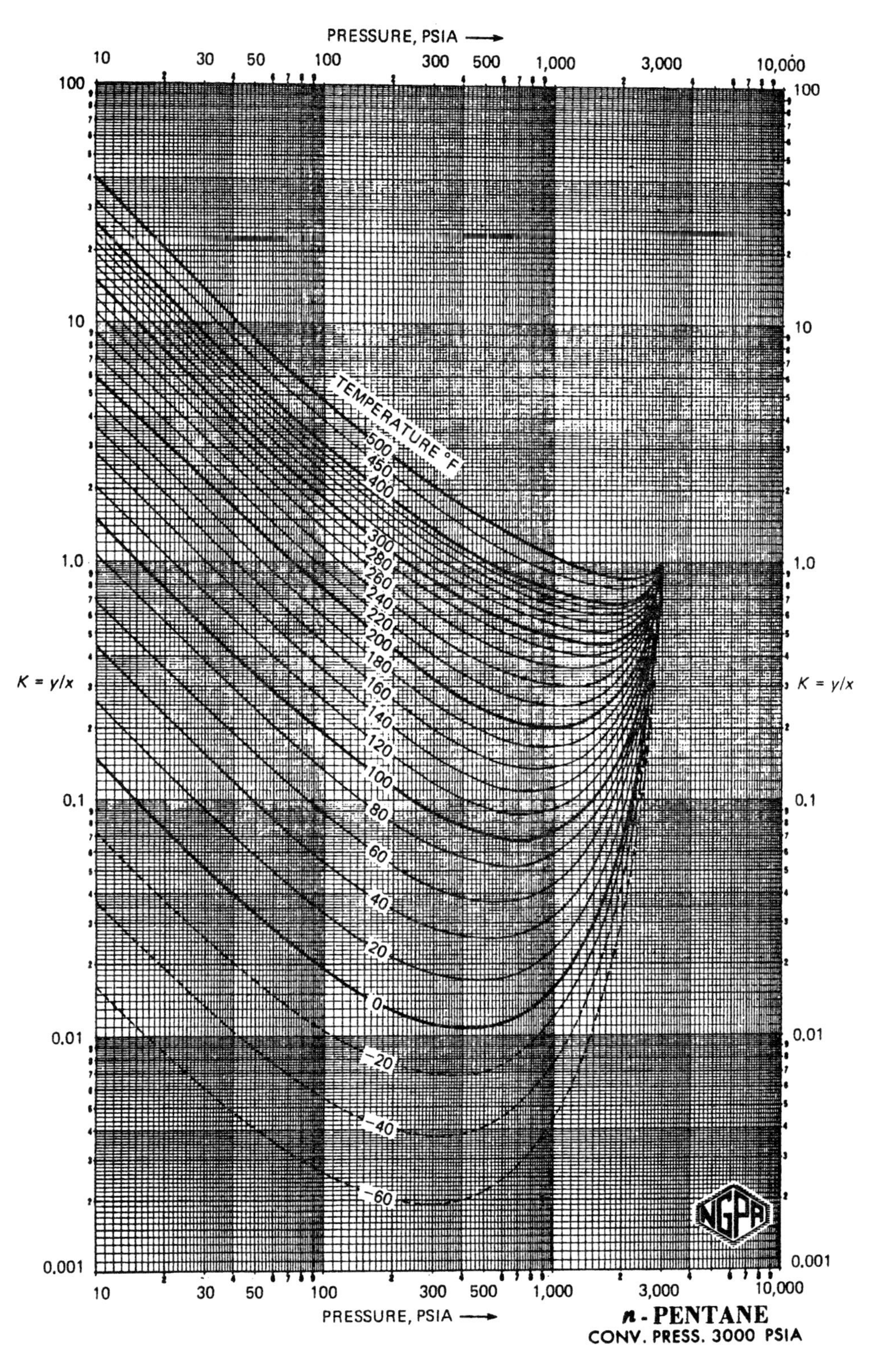
PRESSURE, PSIA
10
30
50
100
300
500
1,000
3,000
10,000
100
10
1.0
0.1
0.01
0.001
K = y/x
TEMPERATURE °F
500
450
400
300
280
260
240
220
200
180
160
140
120
100
80
60
40
20
0
-20
-40
-60
NGPA
n-PENTANE
CONV. PRESS. 3000 PSIA

Figure C-9

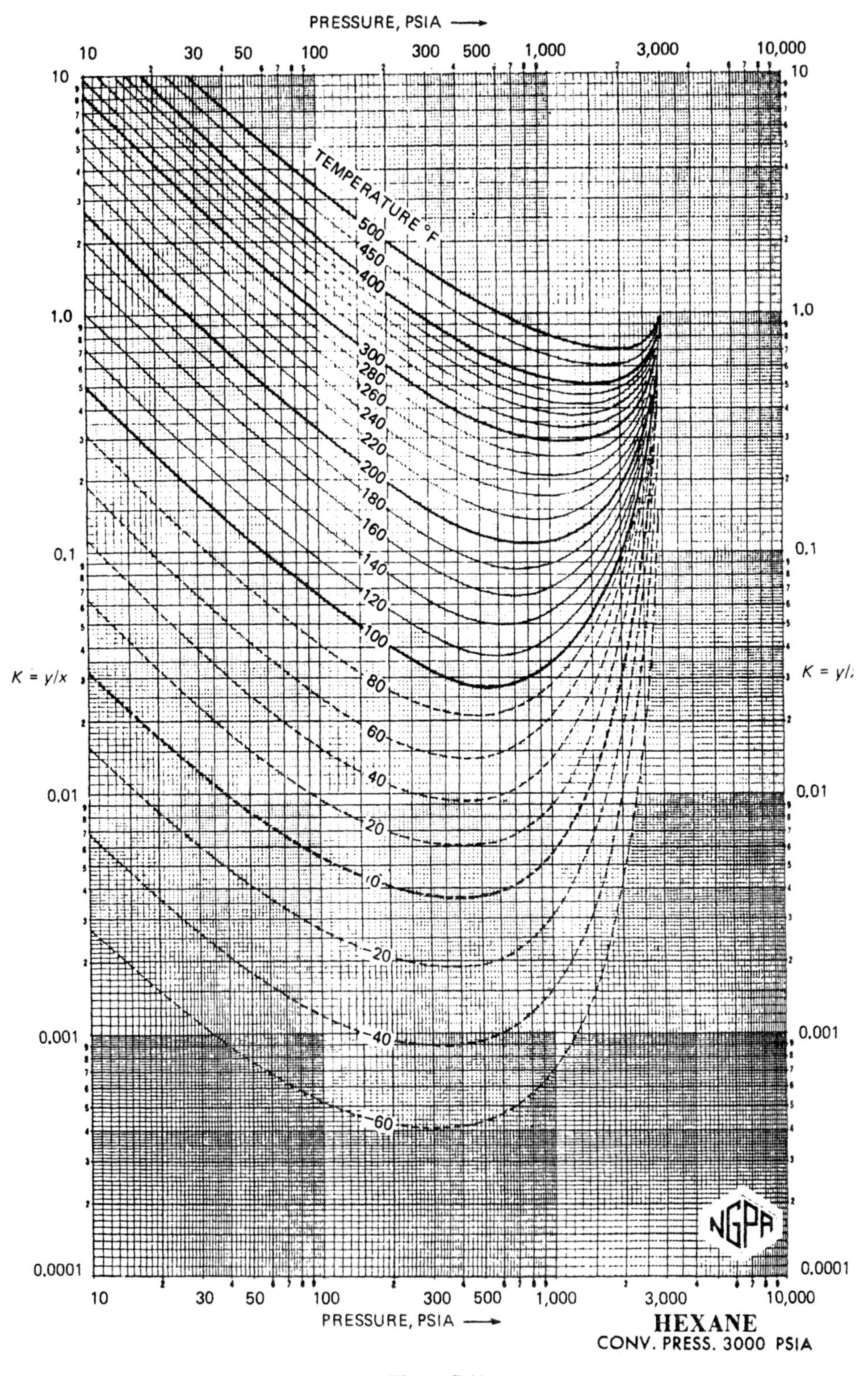
PRESSURE, PSIA
10
30
50
100
300
500
1,000
3,000
10,000
TEMPERATURE °F
500
450
400
300
280
260
240
220
200
180
160
140
120
100
80
60
40
20
10
-20
-40
-60
10
1.0
0.1
0.01
0.001
0.0001
K = y/x
NGPA
HEXANE
CONV. PRESS. 3000 PSIA

Figure C-10

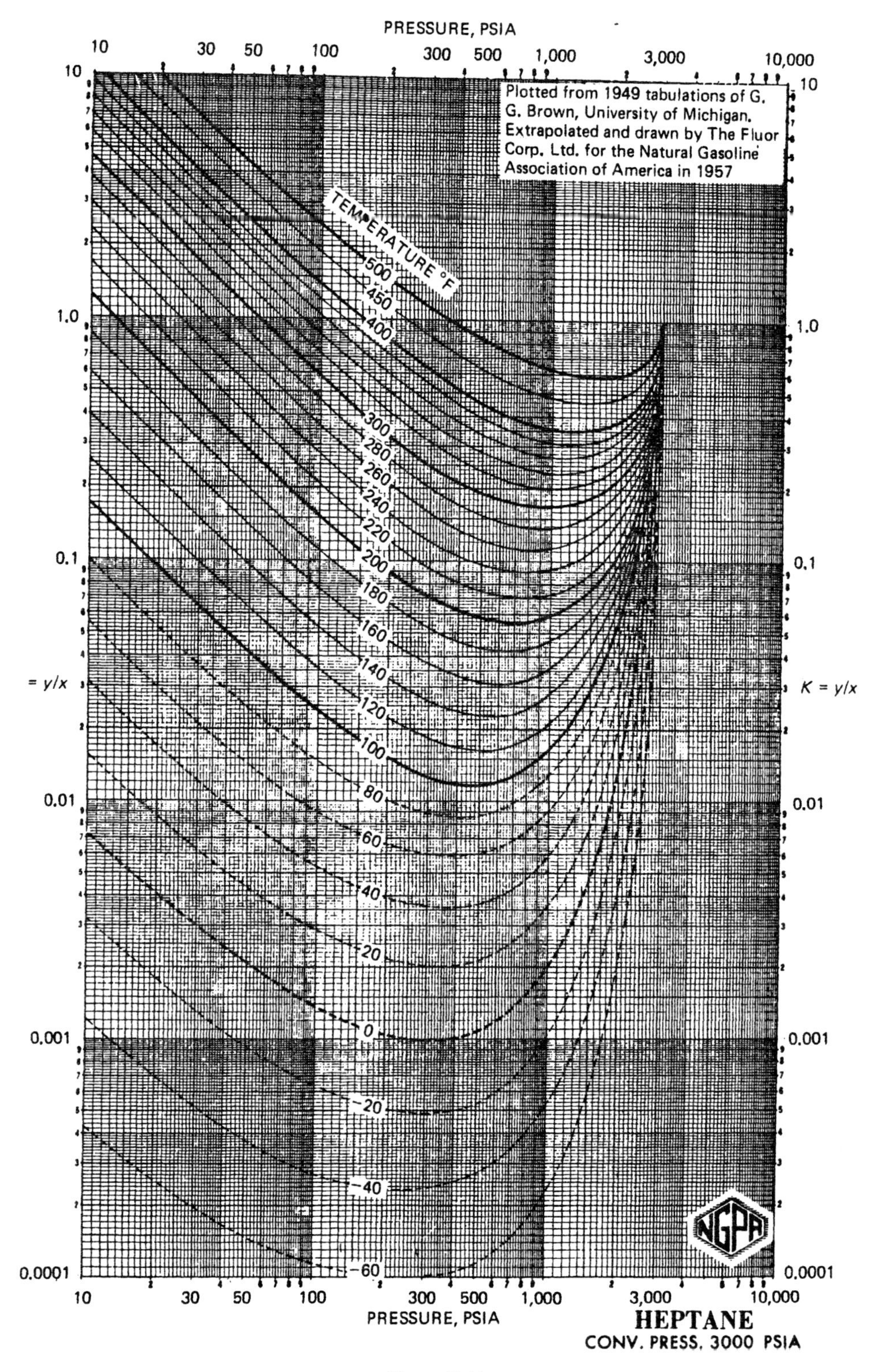

PRESSURE, PSIA
10
30
50
100
300
500
1,000
3,000
10,000
Plotted from 1949 tabulations of G. G. Brown, University of Michigan. Extrapolated and drawn by The Fluor Corp. Ltd. for the Natural Gasoline Association of America in 1957
TEMPERATURE °F
500
450
400
300
280
260
240
220
200
180
160
140
120
100
80
60
40
20
0
−20
−40
−60
1.0
0.1
0.01
0.001
0.0001
= y/x
K = y/x
NGPA
HEPTANE
CONV. PRESS. 3000 PSIA

Figure C-11

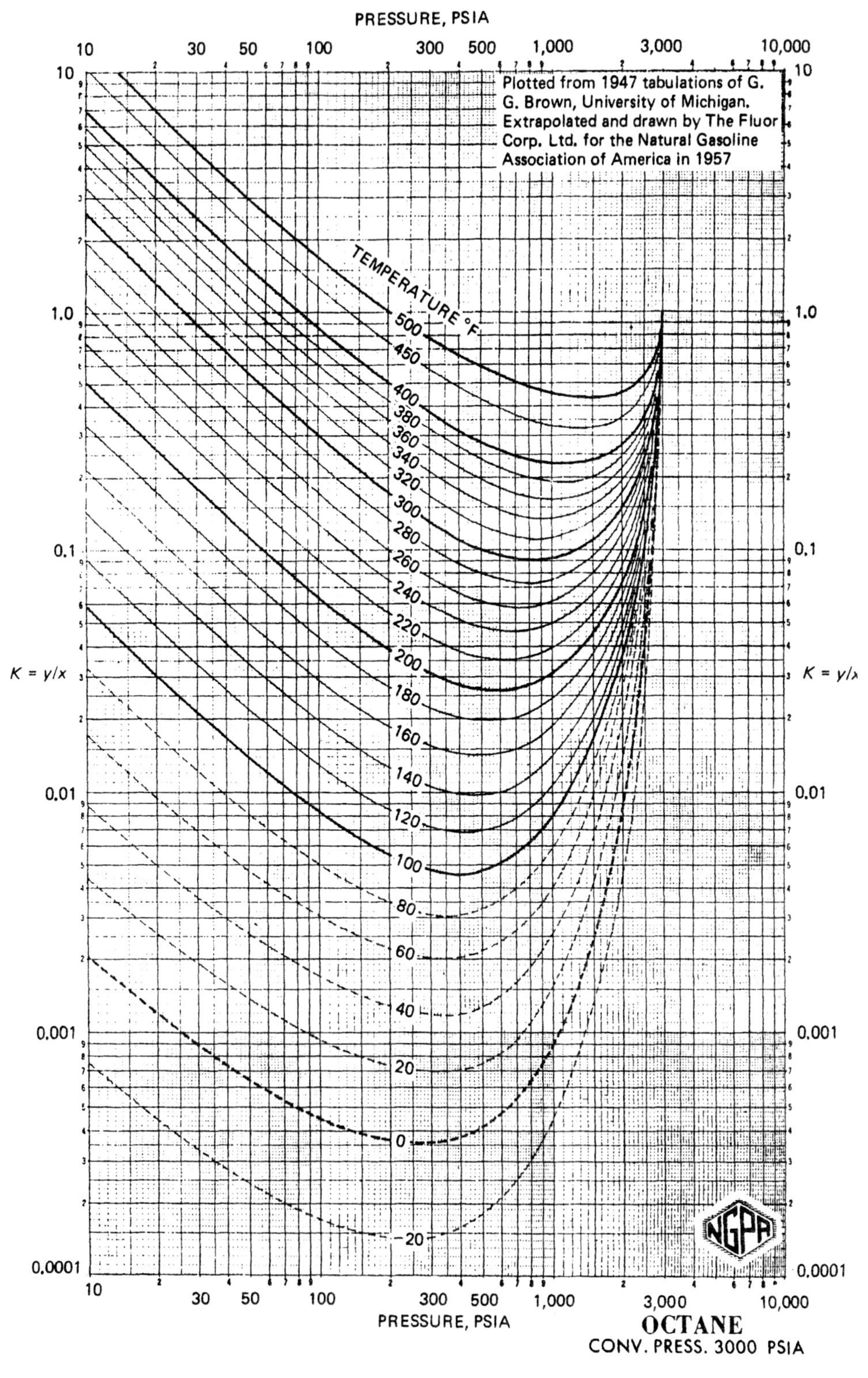
PRESSURE, PSIA
10
30
50
100
300
500
1,000
3,000
10,000
Plotted from 1947 tabulations of G. G. Brown, University of Michigan. Extrapolated and drawn by The Fluor Corp. Ltd. for the Natural Gasoline Association of America in 1957
TEMPERATURE °F
500
450
400
380
360
340
320
300
280
260
240
220
200
180
160
140
120
100
80
60
40
20
0
-20
K = y/x
10
1.0
0.1
0.01
0.001
0.0001
NGPA
OCTANE
CONV. PRESS. 3000 PSIA

Figure C-12

Enthalpy Data

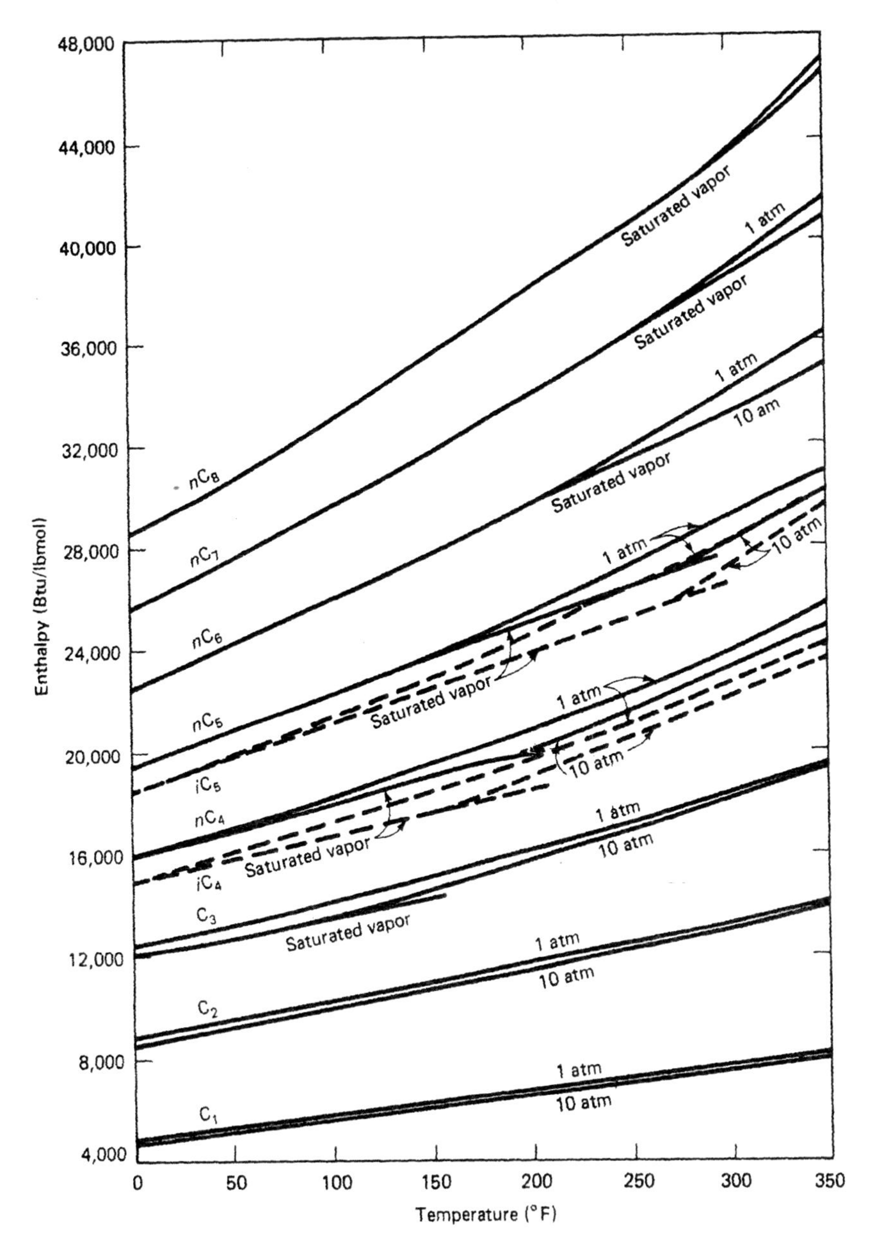

Figure D-1 Enthalpies of hydrocarbon gases (from J. B. Maxwell, *Data Book on Hydrocarbons*, Van Nostrand, Princeton, N.J., 1950)

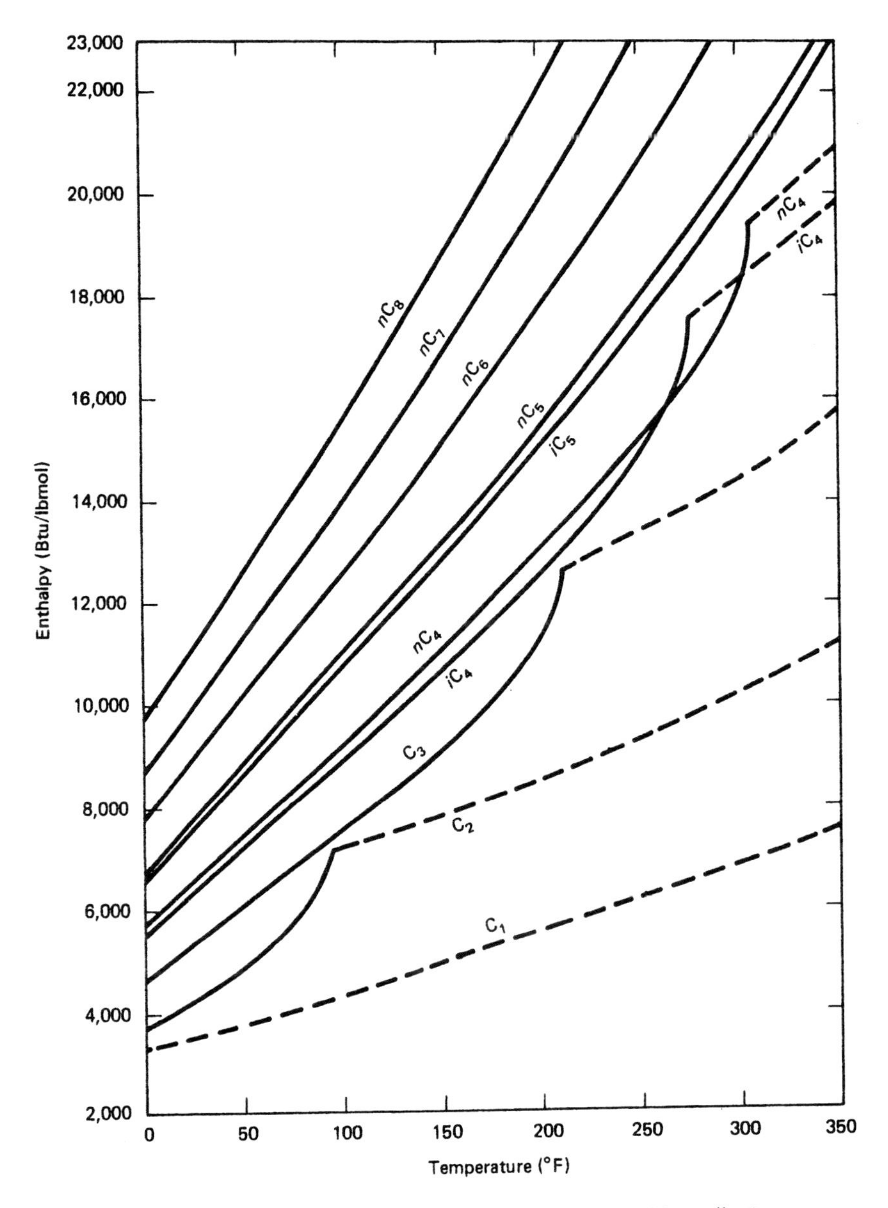

Figure D-2 Enthalpies of hydrocarbon liquids (from J. B. Maxwell, *Data Book on Hydrocarbons*, Van Nostrand, Princeton, N.J., 1950)

Unit Conversion Factors and Constants

TABLE E-1. CONVERSION FACTORS

To Convert from:	*To:*	*Multiply by:*
Length		
Å (angstrom)	m (meter)	1.000×10^{-10}
cm	m	1.000×10^{-2}
ft	m	3.048×10^{-1}
in	m	2.540×10^{-2}
μm (micrometer)	m	1.000×10^{-6}
Area		
cm^2	m^2	1.000×10^{-4}
ft^2	m^2	9.290×10^{-2}
in^2	m^2	6.452×10^{-4}
Volume		
bbl (42 gal)	m^3	1.590×10^{-1}
cm^3	m^3	1.000×10^{-6}
ft^3	m^3	2.832×10^{-2}
gal (U.S.)	m^3	3.785×10^{-3}
Mass		
lb_m	kg (kilogram)	4.536×10^{-1}
ton (long, 2240 lb_m)	kg	1.016×10^{3}
ton (short, 2000 lb_m)	kg	9.072×10^{2}
Density		
g/cm^3	kg/m^3	1.000×10^{3}
lb_m/ft^3	kg/m^3	1.602×10^{1}

TABLE E-1. (CONTINUED)

To Convert from:	*To:*	*Multiply by:*
Force		
lb_f	N (newton)	4.448
kg_f	N	9.807
dyn	N	1.000×10^{-5}
Pressure		
in Hg (60°F)	N/m^2(Pa, pascal)	3.377×10^3
in H_2O (60°F)	N/m^2(Pa)	2.488×10^2
lb_f/ft^2	N/m^2(Pa)	4.788×10^1
lb_f/in^2 (psi)	N/m^2(Pa)	6.895×10^3
torr (mmHg, 0°C)	N/m^2(Pa)	1.333×10^2
Viscosity		
P (poise, g/cm·s)	Pa·s	1.000×10^{-1}
cP (centipoise)	Pa·s	1.000×10^{-3}
St (stoke)	m^2/s	1.000×10^{-4}
cSt (centistoke)	m^2/s	1.000×10^{-6}
Energy terms		
Btu	J (joule)	1.055×10^3
$Btu/lb_m \cdot °F$	J/kg·K	4.187×10^3
Btu/h	W (watt)	2.931×10^{-1}
Btu/s	W	1.055×10^3
$Btu/h \cdot ft^2 \cdot °F$	$J/s \cdot m^2 \cdot K$	5.678
$Btu/h \cdot ft^2$	$J/s \cdot m^2$	3.155
Btu/h·ft·°F	J/s·m·K	1.731
cal	J	4.187
cal/g·°C	J/kg·K	4.187×10^3
erg	J	1.000×10^{-7}
$ft \cdot lb_f$	J	1.356
Temperature		
°F	K (kelvin)	$T_K = (T_F + 459.67)/1.8$
°R	K	$T_K = T_R/1.8$

TABLE E-2. CONSTANTS

Gas constants	$R = 82.06$ atm·cm³/gmol·K
	$= 0.7302$ atm·ft³/lbmol·°R
	$= 1.987$ Btu/lbmol·°R
	$= 1.987$ cal/gmol·K
	$= 1545$ $ft \cdot lb_f$/lbmol·°R
	$= 8314$ N·m/kgmol·K
Avogadro's number	$N_0 = 6.02 \times 10^{23}$ molecules/gmol
Boltzmann's constant	$K = R/N_0 = 1.38 \times 10^{-16}$ erg/molecule·K
Planck's constant	$h = 6.62 \times 10^{-27}$ erg·s

Author Index

A

B

C

D

E

F

G

H

P

R

S

T

U

V

W

Y

Z

Subject Index

A

D

E

F

G

H

I

J

L

M

N

O

P

R

S

T

U

V

W